Lecture Notes in Computer Science　12125

More information about this series at http://www.springer.com/series/7407

Daniel Bienstock · Giacomo Zambelli (Eds.)

Integer Programming and Combinatorial Optimization

21st International Conference, IPCO 2020
London, UK, June 8–10, 2020
Proceedings

 Springer

Editors
Daniel Bienstock
Department of IEOR
Columbia University
New York, NY, USA

Giacomo Zambelli
Department of Management
London School of Economics
and Political Science
London, UK

ISSN 0302-9743   ISSN 1611-3349   (electronic)
Lecture Notes in Computer Science
ISBN 978-3-030-45770-9   ISBN 978-3-030-45771-6   (eBook)
https://doi.org/10.1007/978-3-030-45771-6

LNCS Sublibrary: SL1 – Theoretical Computer Science and General Issues

This Springer imprint is published by the registered company Springer Nature Switzerland AG
The registered company address is: Gewerbestrasse 11, 6330 Cham, Switzerland

Preface

This volume collects the 33 extended abstracts presented at IPCO 2020, the 21st Conference on Integer Programming and Combinatorial Optimization, held during June 8–10, 2020, in London, UK. IPCO is under the auspices of the Mathematical Optimization Society, and it is an important forum for presenting the latest results of theory and practice of the various aspects of discrete optimization. The first IPCO conference took place at the University of Waterloo in May 1990, and the London School of Economics and Political Science is hosting the 21st such event – in online form, due to unavoidable circumstances.

In response to the call for papers, we received 126 submissions. The 16-person Program Committee met in Aussois, France, in January 2020. Each submission was reviewed by at least three Program Committee members. There were many high-quality submissions, of which the committee selected 33 to appear in the conference proceedings. We expect the full versions of the extended abstracts appearing in this Springer *Lecture Notes in Computer Science* (LNCS) series to be submitted for publication in refereed journals, and a special issue of *Mathematical Programming Series B* is on the way. For the first time, IPCO featured a Best Paper Award, which was awarded to Kim-Manuel Klein for the paper "About the Complexity of Two-Stage Stochastic IPs."

This year, IPCO was preceded (again, in online format) by a Summer School during May 6–7, 2020, with lectures by Bertrand Guenin, Santanu Dey, and Laura Sanita. We thank them warmly for their contributions. We would also like to thank:

- The authors who submitted their research to IPCO
- The members of the Program Committee, who spent much time and energy reviewing the submissions
- The expert additional reviewers whose opinions were crucial in the paper selection
- The members of the Local Organizing Committee, who made this conference possible
- The Mathematical Optimization Society and in particular the members of its IPCO Steering Committee: Jens Vygen, Oktay Günlük, and Jochen Koenemann, for their help and advice
- EasyChair for making paper management simple and effective
- EasyChair and Springer for their efficient cooperation in producing this volume

We would also like to thank the following sponsors for their financial support: Amazon, FICO, MOSEK, Springer, The Optimization Firm, and the Department of Mathematics (London School of Economics and Political Science).

March 2020

Daniel Bienstock
Giacomo Zambelli

Organization

Program Committee

Tobias Achterberg	Gurobi and ZIB, Germany
Alper Atamtürk	UC Berkeley, USA
Amitabh Basu	Johns Hopkins University, USA
Daniel Bienstock (PC Chair)	Columbia University, USA
Claudia D'Ambrosio	LIX, CNRS, and École Polytechnique, France
Daniel Dadush	CWI, The Netherlands
Alberto del Pia	University of Wisconsin-Madison, USA
Santanu S. Dey	Georgia Tech, USA
Yuri Faenza	Columbia University, USA
Jon Lee	University of Michigan, USA
Jeff Linderoth	University of Wisconsin-Madison, USA
Michele Monaci	University of Bologna, Italy
Sebastian Pokutta	Georgia Tech, USA
Eduardo Uchoa	Universidade Federal Fluminense, Brazil
László Végh	LSE, UK
Giacomo Zambelli	LSE, UK

Local Committee

Ahmad Abdi	LSE, UK
Neil Olver	LSE, UK
László Végh	LSE, UK
Giacomo Zambelli	LSE, UK

Contents

Idealness of k-wise Intersecting Families

Ahmad Abdi[1]([✉]), Gérard Cornuéjols[2], Tony Huynh[3], and Dabeen Lee[4]

[1] Department of Mathematics, LSE, London, UK
a.abdi1@lse.ac.uk
[2] Tepper School of Business, Carnegie Mellon University, Pittsburgh, USA
gc0v@andrew.cmu.edu
[3] School of Mathematics, Monash University, Melbourne, Australia
tony.bourbaki@gmail.com
[4] Institute for Basic Science, Daejeon, South Korea
dabeenl@ibs.re.kr

Abstract. A clutter is *k-wise intersecting* if every k members have a common element, yet no element belongs to all members. We conjecture that every 4-wise intersecting clutter is non-ideal. As evidence for our conjecture, we prove it in the binary case. Two key ingredients for our proof are Jaeger's 8-flow theorem for graphs, and Seymour's characterization of the binary matroids with the sums of circuits property. As further evidence for our conjecture, we also note that it follows from an unpublished conjecture of Seymour from 1975.

1 Introduction

Let V be a finite set of *elements*, and $\mathcal{C}$ be a family of subsets of V called *members*. The family $\mathcal{C}$ is a *clutter* over *ground set* V, if no member contains another one [11]. A *cover of* $\mathcal{C}$ is a subset $B \subseteq V$ such that $B \cap C \neq \varnothing$ for all $C \in \mathcal{C}$. A cover is *minimal* if it does not contain another cover. The family of minimal covers forms another clutter over the ground set V, called the *blocker of* $\mathcal{C}$ and denoted $b(\mathcal{C})$. It is well-known that $b(b(\mathcal{C})) = \mathcal{C}$ [11,15]. Consider for $w \in \mathbb{Z}_+^V$ the dual pair of linear programs

$$
\begin{array}{llll}
\min & w^\top x & \max & \mathbf{1}^\top y \\
\text{s.t.} & \sum (x_u : u \in C) \geq 1 \quad \forall C \in \mathcal{C} & \text{s.t.} & \sum (y_C : u \in C \in \mathcal{C}) \leq w_u \quad \forall u \in V \\
& x \geq 0 & & y \geq 0
\end{array}
$$

where the left and right LPs are denoted (P) and (D), respectively. If the dual (D) has an integral optimal solution for every right-hand-side vector $w \in \mathbb{Z}_+^V$, then $\mathcal{C}$ is said to have the *max-flow min-cut (MFMC) property* [7]. By the theory of *totally dual integral* linear systems, for every MFMC clutter, the primal (P) also admits an integral optimal solution for every cost vector $w \in \mathbb{Z}_+^V$ [12]. This is why the class of MFMC clutters is a natural host to many beautiful *min-max theorems* in Combinatorial Optimization [8]. Let us elaborate.

The *packing number of* $\mathcal{C}$, denoted $\nu(\mathcal{C})$, is the maximum number of pairwise disjoint members. Note that $\nu(\mathcal{C})$ is equal to the maximum value of an integral

© Springer Nature Switzerland AG 2020
D. Bienstock and G. Zambelli (Eds.): IPCO 2020, LNCS 12125, pp. 1–12, 2020.
https://doi.org/10.1007/978-3-030-45771-6_1

feasible solution to (D) for $w = 1$. Furthermore, the covers correspond precisely to the $0-1$ feasible solutions to (P). The *covering number of* $\mathcal{C}$, denoted $\tau(\mathcal{C})$, is the minimum cardinality of a cover. Notice that $\tau(\mathcal{C})$ is equal to the minimum value of an integral feasible solution to (P) for $w = 1$. Also, by Weak LP Duality, $\tau(\mathcal{C}) \geq \nu(\mathcal{C})$. The clutter $\mathcal{C}$ *packs* if $\tau(\mathcal{C}) = \nu(\mathcal{C})$ [9]. Observe that if a clutter is MFMC, then it packs.

If the primal (P) has an integral optimal solution for every cost vector $w \in \mathbb{Z}_+^V$, then $\mathcal{C}$ is said to be *ideal* [10]. Ideal clutters form a rich class of clutters, one that contains the class of MFMC clutters, as discussed above. This containment is strict, and in fact, the richest examples of ideal clutters are those that are not MFMC [14,19]. Furthermore, unlike MFMC clutters, if a clutter is ideal, then so is its blocker [8,13,18].

A clutter is *intersecting* if every two members intersect yet no element belongs to every member [2]. That is, a clutter $\mathcal{C}$ is intersecting if $\tau(\mathcal{C}) \geq 2$ and $\nu(\mathcal{C}) = 1$. In particular, an intersecting clutter does not pack, and therefore is not MFMC. Intersecting clutters, however, may be ideal. For instance, the clutter

$$Q_6 := \{\{1,3,5\}, \{1,4,6\}, \{2,3,6\}, \{2,4,5\}\},$$

whose elements are the edges and whose members are the triangles of K_4, is an intersecting clutter that is ideal [23]. In fact, Q_6 is the smallest intersecting clutter which is ideal ([1], Proposition 1.2). It is worth pointing out that if a clutter *and* its blocker are both intersecting, then the clutter must be non-ideal [3].

In this paper, we propose a sufficient condition for non-idealness that is purely combinatorial. We say that $\mathcal{C}$ is k-wise *intersecting* if every k members have a common element, yet no element belongs to all members. Note that for $k = 2$, this notion coincides with the notion of intersecting clutters. Furthermore, for $k \geq 3$, a k-wise intersecting clutter is also $(k-1)$-wise intersecting. The following is our main conjecture.

Conjecture 1. There exists an integer $k \geq 4$ such that every k-wise intersecting clutter is non-ideal.

A clutter is *binary* if the symmetric difference of any odd number of members contains a member [17]. Equivalently, a clutter is binary if every member and every minimal cover have an odd number of elements in common [17]. In particular, if a clutter is binary, then so is the blocker. Many rich classes of clutters are in fact binary [8]. For example, given a graph $G = (V, E)$ and distinct vertices s and t, the clutter of st-paths over ground set E is binary. The clutter Q_6 is also binary. As evidence for Conjecture 1, our main result is that it holds for all binary clutters.

Theorem 2. *Every 4-wise intersecting binary clutter is non-ideal.*

We also show that 4 cannot be replaced by 3 in Conjecture 1, even for binary clutters.

Proposition 3. *There exists an ideal 3-wise intersecting binary clutter.*

The example from Proposition 3 comes from the Petersen graph, and also coincides with the clutter T_{30} from [21], §79.3e. It has 30 elements and is the smallest such example that we are aware of.

Finally, as further evidence for Conjecture 1, we also show that it follows from an unpublished conjecture by Seymour from 1975 that was documented in [21], §79.3e.

1.1 Paper Outline

In Sect. 2, we show that a special class of clutters, called *cuboids* [1,5], sit at the heart of Conjecture 1. Cuboids allow us to reformulate Conjecture 1 in terms of set systems.

In Sect. 3, we prove Proposition 3 and Theorem 2. Two key ingredients of our proof of Theorem 2 are Jaeger's 8-*flow Theorem* [16] for graphs, and Seymour's characterization of the binary matroids with the *sums of circuits property* [25].

In Sect. 4, we propose a line of attack for tackling Conjecture 1, inspired by the recent work of [4]. We also discuss two applications of Theorem 2 to ideal binary clutters. Each application goes hand-in-hand with two strengthenings of Conjecture 1. One strengthening is proposed by us, which we believe is the right strategy for tackling Conjecture 1. The other conjecture is the unpublished conjecture by Seymour.

2 Cuboids

Let $n \in \mathbb{N}$ and $S \subseteq \{0,1\}^n$. The *cuboid of S*, denoted cuboid(S), is the clutter over ground set $[2n] := \{1, \ldots, 2n\}$ whose members have incidence vectors $(p_1, 1 - p_1, \ldots, p_n, 1 - p_n)$ over all $(p_1, \ldots, p_n) \in S$. We say that a clutter is a *cuboid* if it is isomorphic to cuboid(S), for some S.

Observe that for each $C \in$ cuboid(S), $|C \cap \{2i - 1, 2i\}| = 1$ for all $i \in [n]$. In particular, every member of cuboid(S) has size n (hence cuboid(S) is a clutter) and $\tau($cuboid$(S)) \leq 2$. Cuboids were introduced in [5] and further studied in [1].

We now describe what it means for cuboid(S) to be k-wise intersecting. We say that the points in S *agree on a coordinate* if $S \subseteq \{x : x_i = a\}$ for some coordinate $i \in [n]$ and some $a \in \{0,1\}$.

Remark 4. Let $S \subseteq \{0,1\}^n$. Then cuboid(S) is a k-wise intersecting clutter if, and only if, the points in S do not agree on a coordinate yet every k points do.

Next, we describe what it means for cuboid(S) to be ideal. Let conv(S) denote the convex hull of S. An inequality of the form $\sum_{i \in I} x_i + \sum_{j \in J} (1 - x_j) \geq 1$, for some disjoint $I, J \subseteq [n]$, is called a *generalized set covering inequality* [8]. The set S is *cube-ideal* if every facet of conv(S) is defined by $x_i \geq 0$, $x_i \leq 1$, or a generalized set covering inequality [1].

Theorem 5 ([1], **Theorem 1.6**). *Let $S \subseteq \{0,1\}^n$. Then* cuboid(S) *is an ideal clutter if, and only if, S is a cube-ideal set.*

As a result, Conjecture 1 for cuboids reduces to the following conjecture:

Conjecture 6. There is a constant $k \geq 4$ such that for every cube-ideal set, either all the points agree on a coordinate, or there is a subset of at most k points that do not agree on a coordinate.

Surprisingly, we now show that Conjecture 6 is equivalent to Conjecture 1! Let $\mathcal{C}$ be a clutter over ground set V. To *duplicate an element u of $\mathcal{C}$* is to introduce a new element $\bar{u}$, and replace $\mathcal{C}$ by the clutter over ground set $V \cup \{\bar{u}\}$, whose members are $\{C : C \in \mathcal{C}, u \notin C\} \cup \{C \cup \{\bar{u}\} : C \in \mathcal{C}, u \in C\}$. A *duplication of $\mathcal{C}$* is a clutter obtained from $\mathcal{C}$ by repeatedly duplicating elements. It is easily checked that a clutter is ideal if and only if some duplication of it is ideal. Moreover, a clutter is k-wise intersecting if and only if some duplication of it is k-wise intersecting.

Let I and J be disjoint subsets of V. The *minor $\mathcal{C} \setminus I/J$* obtained after *deleting* I and *contracting* J is the clutter over ground set $V - (I \cup J)$ whose members are the minimal sets in $\{C - J : C \in \mathcal{C}, C \cap I = \varnothing\}$. It is well-known that $b(\mathcal{C} \setminus I/J) = b(\mathcal{C})/I \setminus J$ [22]. If $J = \varnothing$, then $\mathcal{C} \setminus I/J = \mathcal{C} \setminus I$ is called a *deletion minor*.

If every $k \geq 2$ members of a clutter have a common element, then so do every k members of a deletion minor. Furthermore, for every element $v \in V$, $\tau(\mathcal{C}) \geq \tau(\mathcal{C} \setminus v) \geq \tau(\mathcal{C}) - 1$, where $\tau(\mathcal{C} \setminus v) = \tau(\mathcal{C}) - 1$ if and only if v belongs to some minimum cover of $\mathcal{C}$. Motivated by these observations, we say that a clutter $\mathcal{C}$ is *tangled* if $\tau(\mathcal{C}) = 2$ and every element belongs to a minimum cover.

We require the following facts about tangled clutters.

Proposition 7. *Let $\mathcal{C}$ be a binary tangled clutter. Then $\mathcal{C}$ is a duplication of a cuboid.*

Proof. If $\{e, f\}$ is a minimum cover, then for each $C \in \mathcal{C}$, $|C \cap \{e, f\}|$ must be odd and therefore 1, since $\mathcal{C}$ is a binary clutter. As a result, if $\{e, f\}, \{e, g\}$ are both minimum covers, then f, g must be duplicated elements. Moreover, if every element is contained in exactly one minimum cover, then $\mathcal{C}$ must be a cuboid. These two observations, along with the fact that $\mathcal{C}$ is a tangled clutter, imply that $\mathcal{C}$ is a duplication of a cuboid. □

Remark 8. Let $\mathcal{C}$ be a k-wise intersecting clutter. Let $\mathcal{C}'$ be a deletion minor of $\mathcal{C}$ that is minimal subject to $\tau(\mathcal{C}') \geq 2$. Then $\mathcal{C}'$ is a tangled k-wise intersecting clutter.

Moreover, if a clutter is ideal, then so is every minor of it [23]. Thus, it suffices to prove Conjecture 1 for tangled clutters.

Theorem 9 ([4], **Theorem 5.5**). *Let $\mathcal{C}$ be an ideal tangled clutter. Then*

$$\mathcal{C}' := \{C \in \mathcal{C} : |C \cap \{u, v\}| = 1 \ \forall \ \{u, v\} \in b(\mathcal{C})\}$$

is also an ideal tangled clutter. Moreover, $\mathcal{C}'$ is a duplication of some cuboid.

We are now ready to prove that Conjectures 6 and 1 are equivalent.

Theorem 10. *Conjecture 6 is true for k if, and only if, Conjecture 1 is true for k.*

Proof. We already showed ($\Leftarrow$). It remains to prove ($\Rightarrow$). Suppose Conjecture 1 is false for some $k \geq 4$. That is, there is an ideal k-wise intersecting clutter $\mathcal{C}$. Let $\mathcal{C}'$ be a deletion minor of $\mathcal{C}$ that is minimal subject to $\tau(\mathcal{C}') \geq 2$. By Remark 8, $\mathcal{C}'$ is an ideal tangled k-wise intersecting clutter. Let

$$\mathcal{C}'' := \{C \in \mathcal{C}' : |C \cap \{u,v\}| = 1 \ \forall \ \{u,v\} \in b(\mathcal{C}')\}.$$

By Theorem 9, $\mathcal{C}''$ is an ideal tangled clutter that is a duplication of some cuboid, say cuboid(S). As every k members of $\mathcal{C}'$ have a common element, so do every k members of $\mathcal{C}''$, so the latter is k-wise intersecting. As a result, cuboid(S) is an ideal k-wise intersecting clutter, so by Remark 4 and Theorem 5, S is a cube-ideal set whose points do not agree on a coordinate yet every k points do. Therefore, S refutes Conjecture 6 for k, as required. $\square$

3   Proof of Theorem 2

Let $S \subseteq \{0,1\}^n$. For $x,y \in \{0,1\}^n$, $x \vartriangle y$ denotes the coordinate-wise sum of x,y modulo 2. We say that S is a *vector space over $GF(2)$*, or simply a *binary space*, if $a \vartriangle b \in S$ for all $a,b \in S$. Notice that a nonempty binary space necessarily contains $\mathbf{0}$.

Remark 11 ([4], Remark 7.5). Let $\mathbf{0} \in S \subseteq \{0,1\}^n$. If cuboid($S$) is a binary clutter, then S is a binary space.

3.1   The 8-flow Theorem

Let $G = (V,E)$ be a graph where loops and parallel edges are allowed, where every loop is treated as an edge not incident to any vertex. A *cycle* is a subset $C \subseteq E$ such that every vertex is incident with an even number of edges in C. A *bridge of G* is an edge e that does not belong to any cycle. The *cycle space of G* is the set

$$\text{cycle}(G) := \{\chi_C : C \subseteq E \text{ is a cycle}\} \subseteq \{0,1\}^E$$

where χ_C denotes the incidence vector of C. As $\varnothing$ is a cycle, and the symmetric difference of any two cycles is also a cycle, it follows that cycle(G) is a binary space. We require the following two results on cycle spaces of graphs:

Remark 12. Let $G = (V,E)$ be a graph. Then the points in cycle(G) agree on a coordinate if, and only if, G has a bridge. Moreover, for all $k \in \mathbb{N}$, cycle(G) has a subset of at most $k + 1$ points that do not agree on a coordinate if, and only if, G has at most k cycles the union of which is E.

Theorem 13 ([1], **Corollary 2.6 and Theorem 2.8**). *The cycle space of every graph is a cube-ideal set.*

We need the following version of the celebrated 8-*Flow Theorem* of Jaeger [16].

Theorem 14 ([16]). *Every bridgeless graph $G=(V,E)$ contains at most 3 cycles the union of which is E. That is, given the set* $\mathrm{cycle}(G) \subseteq \{0,1\}^E$, *either all the points agree on a coordinate, or there is a subset of at most 4 points that do not agree on a coordinate.*

One may wonder whether the $3,4$ in Theorem 14 may be replaced by $2,3$? The answer is no, due to the Petersen graph (see Fig. 1a):

Remark 15 (see [26]). The edge set of the Petersen graph is not the union of 2 cycles.

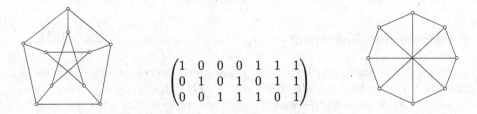

$$\begin{pmatrix} 1 & 0 & 0 & 0 & 1 & 1 & 1 \\ 0 & 1 & 0 & 1 & 0 & 1 & 1 \\ 0 & 0 & 1 & 1 & 1 & 0 & 1 \end{pmatrix}$$

Fig. 1. (a) Petersen. (b) Fano. (c) Wagner.

As a consequence, an ideal 3-wise intersecting clutter does exist:

Proof of Proposition 3. Let S be the cycle space of the Petersen graph, and let $\mathcal{C} := \mathrm{cuboid}(S)$. By Remark 15, the Petersen is a bridgeless graph that does not have 2 cycles the union of which is the edge set, so by Remark 12, the points in S do not agree on a coordinate, but every subset of $2+1=3$ points do. Moreover, S is a cube-ideal set by Theorem 13. Therefore, by Remark 4 and Theorem 5, $\mathcal{C}$ is an ideal 3-wise intersecting clutter, as required.   □

The cuboid of the cycle space of the Petersen graph has already shown up in the literature, and is denoted T_{30} by Schrijver [21], §79.3e. Consider the graph obtained from Petersen by subdividing every edge once, and let T be any vertex subset of even cardinality containing all the new vertices. Then the clutter of minimal T-joins of this graft is precisely T_{30}. This construction is due to Seymour ([24], p. 440).

3.2 Sums of Circuits Property

For background and notation regarding binary matroids, we refer the reader to Appendix A.

A binary matroid has the *sums of circuits property* if its cycle space is a cube-ideal set. This notion is due to Seymour [24].[1] By Theorem 13, graphic matroids have the sums of circuits property [24]. Seymour also proved a decomposition theorem [25] for binary matroids with the sums of circuits property. It turns out they can all be produced from graphic matroids and two other matroids, which we now describe. The *Fano matroid* F_7 is the binary matroid represented by the matrix in Fig. 1b. The second matroid is $M(V_8)^*$, where V_8 is the graph in Fig. 1c. Seymour [25] showed that F_7 and $M(V_8)^*$ both have the sums of circuits property.

To generate all binary matroids with the sums of circuits property, we require three composition rules. Let M_1, M_2 be binary matroids over ground sets E_1, E_2, respectively. We denote by $M_1 \triangle M_2$ the binary matroid over ground set $E_1 \triangle E_2$ whose cycles are all subsets of $E_1 \triangle E_2$ of the form $C_1 \triangle C_2$, where C_i is a cycle of M_i for $i \in [2]$. Then $M_1 \triangle M_2$ is a *1-sum* if $E_1 \cap E_2 = \varnothing$; $M_1 \triangle M_2$ is a *2-sum* if $E_1 \cap E_2 = \{e\}$, where e is neither a loop nor a coloop of M_1 or M_2; and $M_1 \triangle M_2$ is a *Y-sum* if $E_1 \cap E_2$ is a cocircuit of cardinality 3 in both M_1 and M_2 and contains no circuit in M_1 or M_2.

Theorem 16 ([25], (6.4), (6.7), (6.10) and (16.4)). *Let M be a binary matroid with the sums of circuits property. Then M is obtained recursively by means of 1-sums, 2-sums and Y-sums starting from copies of $F_7, M(V_8)^*$ and graphic matroids.*

We are ready to prove Conjecture 6 for cube-ideal binary spaces.

Theorem 17. *Every binary matroid without a coloop and with the sums of circuits property has at most 3 cycles the union of which is the ground set.*

Proof. A *3-cycle cover* of a binary matroid is three (not necessarily distinct) cycles whose union is the ground set.

Claim. Both F_7 and $M(V_8)^*$ have 3-cycle covers.

Subproof. Given the matrix representation of F_7 in Fig. 1a, label the columns $1, \ldots, 7$ from left to right. Then $\varnothing, \{1, 2, 3, 7\}, \{4, 5, 6\}$ a 3-cycle cover of F_7. Next, label the vertices of V_8 so that the outer 8-cycle is labelled $1, \ldots, 8$. Then $M(V_8)^*$ has a 3-cycle cover given by the following cuts of V_8: $\delta(\{1, 6, 7, 8\})$, $\delta(\{1, 7\}), \delta(\{2, 4\})$, where $\delta(X)$ is the set of edges with exactly one end in X.◊

Claim. Let M, M_1, M_2 be binary matroids such that $M = M_1 \triangle M_2$ and $M_i, i \in [2]$ has a 3-cycle cover. Then the following statements hold:

(i) If M is a 1-sum of M_1, M_2, then M has a 3-cycle cover.
(ii) If M is a 2-sum of M_1, M_2, then M has a 3-cycle cover.
(iii) If M is a Y-sum of M_1, M_2, then M has a 3-cycle cover.

---

[1] Seymour's definition appears different from ours, but they are equivalent by [1], Corollary 2.6.

Subproof. For $i \in [2]$, let E_i be the ground set of M_i and C_1^i, C_2^i, C_3^i be a 3-cycle cover of M_i. Clearly, **(i)** holds. For **(ii)**, let $E_1 \cap E_2 = \{e\}$. We may assume $e \in C_1^i$ for all $i \in [2]$. By replacing C_2^i by $C_1^i \bigtriangleup C_2^i$ if necessary, we may assume $e \notin C_2^i$ for all $i \in [2]$. Similarly, we may assume $e \notin C_3^i$ for $i \in [2]$. But now $\{C_j^1 \bigtriangleup C_j^2 : j \in [3]\}$ is a 3-cycle cover of M. For **(iii)**, suppose $E_1 \cap E_2 = \{e, f, g\}$. Since $\{e, f, g\}$ is a cocircuit of both M_1, M_2, and since cocircuits and circuits of a binary matroid have an even number of elements in common, $|C_j^i \cap \{e, f, g\}| \in \{0, 2\}$ for all i, j. Therefore, after possibly relabeling e, f, g simultaneously in M_1 and M_2, and after possibly relabeling C_1^i, C_2^i, C_3^i for all i, we may assume that

- $C_1^i \cap \{e, f, g\} = \{e, f\}$ for all $i \in [2]$, and
- $C_2^i \cap \{e, f, g\} = \{e, g\}$ or $\{f, g\}$ for all $i = [2]$.

For $i \in [2]$, after possibly replacing C_2^i with $C_2^i \bigtriangleup C_1^i$, we may assume $C_2^i \cap \{e, f, g\} = \{e, g\}$. For $i \in [2]$, after possibly replacing C_3^i with $C_3^i \bigtriangleup C_1^i, C_3^i \bigtriangleup C_2^i$ or $C_3^i \bigtriangleup C_1^i \bigtriangleup C_2^i$, we may assume $C_3^i \cap \{e, f, g\} = \varnothing$. But now $\{C_j^1 \bigtriangleup C_j^2 : j \in [3]\}$ is a 3-cycle cover of M, as required.◇

We leave the proof of the following claim as an easy exercise for the reader.

Claim. Let M, M_1, M_2 be binary matroids such that $M = M_1 \bigtriangleup M_2$, where $\bigtriangleup$ is either a 1-, 2- or Y-sum. If M has no coloop, then neither do M_1, M_2.◇

The proof is complete by combining the above claims with Theorems 14 and 16. □

3.3   Proof of Theorem 2

Proof of Theorem 2. We prove the contrapositive statement. Let $\mathcal{C}$ be an ideal binary clutter such that $\tau(\mathcal{C}) \geq 2$. We need to exhibit ≤ 4 members without a common element. Let $\mathcal{C}'$ be a deletion minor of $\mathcal{C}$ that is minimal subject to $\tau(\mathcal{C}') \geq 2$. It suffices to exhibit ≤ 4 members of $\mathcal{C}'$ without a common element. Notice that $\mathcal{C}'$ is ideal, and as a minor of a binary clutter, it is also binary [22]. Moreover, by our minimality assumption, $\mathcal{C}'$ is a tangled clutter. Thus, by Proposition 7, $\mathcal{C}'$ is a duplication of a cuboid, say cuboid(S) where we may choose S so that $\mathbf{0} \in S$. It suffices to exhibit ≤ 4 members of cuboid(S) without a common element.

Note that cuboid(S) is an ideal binary cuboid with $\tau(\text{cuboid}(S)) \geq 2$. So, by Theorem 5 and Remark 11, S is a cube-ideal binary space whose points do not agree on a coordinate. By Theorem 17, S has ≤ 4 points that do not agree on a coordinate, thereby yielding ≤ 4 members of cuboid(S) without a common element, as required. □

4   Applications and Two More Conjectures

4.1   Embedding Projective Geometries

In this section, we propose a strengthening of Conjecture 1. We begin by motivating our strengthening. Conjecture 1 predicts that for some $k \geq 4$, every ideal clutter with covering number at least two has k members without a common element. By moving to a deletion minor, if necessary, we may assume that our ideal clutter is tangled. Our stronger conjecture predicts that the clutter must actually have 2^{k-1} members that "correspond to a projective geometry", and of these members, k many will not have a common element.

Let A be the $(k-1) \times (2^{k-1}-1)$ matrix whose columns are all the nonzero vectors in $\{0,1\}^{k-1}$. The binary matroid represented by A is called a *projective geometry* over $GF(2)$, and is denoted $PG(k-2,2)$. Let $r := 2^{k-1}-1$. Recall that $\operatorname{cocycle}(PG(k-2,2)) \subseteq \{0,1\}^r$ is the row space of A generated over $GF(2)$. Note that the $k-1$ points in $\operatorname{cocycle}(PG(k-2,2))$ corresponding to the rows of A agree on precisely one coordinate, which is set to 1. These $k-1$ points, together with the zero point $\mathbf{0}$, yield k points that do not agree on a coordinate. This yields the following remark.

Remark 18. There are k points of $\operatorname{cocycle}(PG(k-2,2))$ that do not agree on a coordinate. In particular, $\operatorname{cuboid}(\operatorname{cocycle}(PG(k-2,2)))$ has k members without a common element.

Let $\mathcal{C}$ be an ideal tangled clutter. We say that $\mathcal{C}$ *embeds the projective geometry* $PG(k-2,2)$ if a subset of $\mathcal{C}$ is a *duplication* of $\operatorname{cuboid}(\operatorname{cocycle}(PG(k-2,2)))$. This notion is due to [4]. We propose the following conjecture.

Conjecture 19. There exists an integer $k \geq 4$ such that every ideal tangled clutter embeds one of $PG(0,2), \ldots, PG(k-2,2)$.

In Appendix B we show that Conjecture 19 is indeed a strengthening of Conjecture 1.

Proposition 20. *If Conjecture 19 holds for k, then Conjecture 1 holds for k.*

Proposition 21 ([4], **Proposition 7.4**). *Let S be a binary space of $GF(2)$-rank r whose points do not agree on a coordinate. Then $\operatorname{cuboid}(S)$ embeds one of $PG(0,2), \ldots, PG(r-1,2)$.*

As an application of Theorem 2, we now prove Conjecture 19 for $k=3$ for the class of binary clutters.

Theorem 22. *Every ideal binary tangled clutter embeds $PG(0,2)$, $PG(1,2)$, or $PG(2,2)$.*

Proof. Let $\mathcal{C}$ be a binary tangled clutter. By Proposition 7, $\mathcal{C}$ is a duplication of a cuboid, say $\operatorname{cuboid}(S)$ for some S containing $\mathbf{0}$. It suffices to show that $\operatorname{cuboid}(S)$ embeds one of the three projective geometries. Note that $\operatorname{cuboid}(S)$ is an ideal binary cuboid with $\tau(\operatorname{cuboid}(S)) \geq 2$. By Theorem 5 and Remark 11, S is a cube-ideal binary space whose points do not agree on a coordinate. By Theorem 17, S has a subset of at most 3 points that do not agree on a coordinate. Let S' be the binary space generated by these points. Note that $S' \subseteq S$, the points in S' do not agree on a coordinate, and S' has $GF(2)$-rank at most 3. By Proposition 21, $\operatorname{cuboid}(S')$, and therefore $\operatorname{cuboid}(S)$, embeds one of $PG(0,2), PG(1,2), PG(2,2)$, as desired. $\square$

We now give an application of Theorem 22. Let G be a bridgeless graph. By applying Theorem 22 to cuboid(cycle(G)), G has 8 cycles where every edge is used in exactly 4 of the cycles. Since one of the 8 cycles may be assumed to be $\varnothing$, G has 7 cycles such that each edge is in exactly 4 of the cycles. This is Proposition 6 of [6].

4.2 Dyadic Fractional Packings

We finish by deriving another consequence of Theorem 2. Let $\mathcal{C}$ be a clutter over ground set V. A *fractional packing of* $\mathcal{C}$ is a vector $y \in \mathbb{R}_+^{\mathcal{C}}$ such that $\sum (y_C : C \in \mathcal{C}, v \in C) \leq 1$ for all $v \in V$. The *value of* y is $\mathbf{1}^\top y$. For $n \in \mathbb{N}$, the vector y is $\frac{1}{n}$-*integral* if every entry is $\frac{1}{n}$-integral.

Proposition 23 ([4], **follows from Theorem 1.16**). *For every* $k \in \mathbb{Z}_{\geq 0}$, cuboid(cocycle($PG(k,2)$)) *has a* $\frac{1}{2^k}$-*integral packing of value* 2.

This, combined with Theorem 22, implies the following:

Theorem 24. *Every ideal binary clutter* $\mathcal{C}$ *with* $\tau(\mathcal{C}) \geq 2$ *has a* $\frac{1}{4}$-*integral packing of value* 2.

Proof. Let $\mathcal{C}'$ be a deletion minor of $\mathcal{C}$ that is minimal subject to $\tau(\mathcal{C}') \geq 2$. Then $\mathcal{C}'$ is an ideal binary tangled clutter, so by Theorem 22, $\mathcal{C}'$ embeds one of $PG(0,2), PG(1,2), PG(2,2)$. By Proposition 23, $\mathcal{C}'$, and therefore $\mathcal{C}$, has a $\frac{1}{4}$-integral packing of value 2, as required.   □

In fact, Seymour conjectures a far-reaching generalization of this theorem:

Conjecture 25 Seymour 1975, see [21], *§79.3e).* Every ideal clutter $\mathcal{C}$ has a $\frac{1}{4}$-integral packing of value $\tau(\mathcal{C})$.

This conjecture is open even for binary clutters, and in particular, for the clutter of minimal T-joins of a graft ([8], Conjecture 2.15).

Proposition 26. *If Conjecture 25 is true, then so is Conjecture 1 for* $k = 5$.

Proof. Assume Conjecture 25 is true. Let $\mathcal{C}$ be an ideal clutter with $\tau(\mathcal{C}) \geq 2$. Let $\mathcal{C}'$ be a deletion minor of $\mathcal{C}$ with $\tau(\mathcal{C}') = 2$. Since $\mathcal{C}'$ is also ideal, it has a $\frac{1}{4}$-integral packing $y \in \mathbb{R}_+^{\mathcal{C}'}$ of value 2. Notice that $y_C \in \left\{0, \frac{1}{4}, \frac{2}{4}, \frac{3}{4}, 1\right\}$ for each $C \in \mathcal{C}'$. In particular, $|\{C : y_C > 0\}| \leq 8$. Pick a minimal subset $\mathcal{C}'' \subseteq \{C : y_C > 0\}$ such that $\sum_{C \in \mathcal{C}''} y_C > 1$. Then $|\mathcal{C}''| \leq 5$, and it is easily checked that the members of $\mathcal{C}''$ cannot have a common element. As a result, $\mathcal{C}'$, and therefore $\mathcal{C}$, has a subset of at most 5 members without a common element, as required.   □

A   Binary Matroids

For basics on matroids, we refer the reader to Oxley [20]. Let E be a finite set, $S \subseteq \{0,1\}^E$ a binary space, and $S^\perp$ the orthogonal complement of S, that is,

$S^\perp = \{y \in \{0,1\}^E : y^\top x \equiv 0 \pmod 2 \ \forall\, x \in S\}$. Notice that $S^\perp$ is another binary space, and that $(S^\perp)^\perp = S$. Therefore, there exists a $0-1$ matrix A whose columns are labeled by E such that $S = \{x \in \{0,1\}^E : Ax \equiv \mathbf{0} \pmod 2\}$, and $S^\perp$ is the row space of A generated over $GF(2)$.

Let $\mathcal{S} := \{C \subseteq E : \chi_C \in S\}$. The pair $M := (E, \mathcal{S})$ is a *binary matroid*, and the matrix A is a *representation of M*. We call E the *ground set of M*. The sets in $\mathcal{S}$ are the *cycles of M*, and $\mathcal{S}$ is the *cycle space of M*, denoted by $\mathrm{cycle}(M)$. The minimal nonempty sets in $\mathcal{S}$ are the *circuits of M*, and the circuits of cardinality one are *loops*.

Let $\mathcal{S}^\perp := \{D \subseteq E : \chi_D \in S^\perp\}$. The binary matroid $M^* := (E, \mathcal{S}^\perp)$ is the *dual of M*. Notice that $(M^*)^* = M$. The sets in $\mathcal{S}^\perp$ are the *cocycles of M*, and $\mathcal{S}^\perp$ is the *cocycle space of M*, denoted by $\mathrm{cocycle}(M)$. The minimal nonempty sets in $\mathcal{S}^\perp$ are the *cocircuits of M*, and the cocircuits of cardinality one are *coloops of M*.

Remark 27. Let M be a binary matroid. Then the points in $\mathrm{cycle}(M)$ agree on a coordinate if, and only if, M has a coloop. Moreover, for every integer $k \geq 1$, $\mathrm{cycle}(M)$ has a subset of at most $k + 1$ points that do not agree on a coordinate if, and only if, M has at most k cycles the union of which is E.

Let $G = (V, E)$ be a graph. The binary matroid whose cycle space is $\mathrm{cycle}(G)$ is a *graphic matroid*, and is denoted $M(G)$. Notice the one-to-one correspondence between the cycles of $M(G)$ and the cycles of G, between the loops of $M(G)$ and the loops of G, between the cocycles of $M(G)$ and the cuts of G, and between the coloops of $M(G)$ and the bridges of G. Therefore, Remark 27 is an extension of Remark 12.

B   Proof of Proposition 20

Proof of Proposition 20. Assume Conjecture 19 holds for k. Let $\mathcal{C}$ be an ideal clutter with $\tau(\mathcal{C}) \geq 2$. Let $\mathcal{C}'$ be a deletion minor of $\mathcal{C}$ that is minimal subject to $\tau(\mathcal{C}') \geq 2$. Then $\mathcal{C}'$ is an ideal tangled clutter. Thus, $\mathcal{C}'$ embeds $PG(n-2, 2)$ for some $n \in \{2, \ldots, k\}$. That is, a duplication of $\mathrm{cuboid}(PG(n-2, 2))$ is a subset of $\mathcal{C}'$. By Remark 18, $\mathrm{cuboid}(PG(n-2, 2))$ has n members without a common element, so the duplication, and therefore $\mathcal{C}'$, must have n members without a common element. Thus, $\mathcal{C}$ has $n \leq k$ members without a common element, as required.   □

References

1. Abdi, A., Cornuéjols, G., Guričanová, N., Lee, D.: Cuboids, a class of clutters. J. Comb. Theor. Ser. B **142**, 144–209 (2019)
2. Abdi, A., Cornuéjols, G., Lee, D.: Intersecting restrictions in clutters. Combinatorica (to appear)
3. Abdi, A., Cornuéjols, G., Lee, D.: Identically self-blocking clutters. In: Lodi, A., Nagarajan, V. (eds.) IPCO 2019. LNCS, vol. 11480, pp. 1–12. Springer, Cham (2019). https://doi.org/10.1007/978-3-030-17953-3_1

4. Abdi, A., Cornuéjols, G., Superdock, M.: Projective geometries, simplices and clutters (to be submitted)
5. Abdi, A., Pashkovich, K., Cornuéjols, G.: Ideal clutters that do not pack. Math. Oper. Res. **43**(2), 533–553 (2017)
6. Bermond, J.C., Jackson, B., Jaeger, F.: Shortest coverings of graphs with cycles. J. Combin. Theor. Ser. B **35**(3), 297–308 (1983)
7. Conforti, M., Corneujois, G.: Clutters that pack and the max flow min cut property: a conjecture. Technical report, The Fourth Bellairs Workshop on Combinatorial Optimization (1993)
8. Cornuéjols, G.: Combinatorial Optimization: Packing and Covering, vol. 74. SIAM (2001)
9. Cornuéjols, G., Guenin, B., Margot, F.: The packing property. Math. Program. **89**(1, Ser. A), 113–126 (2000)
10. Cornuéjols, G., Novick, B.: Ideal 0, 1 matrices. J. Combin. Theor. Ser. B **60**(1), 145–157 (1994)
11. Edmonds, J., Fulkerson, D.R.: Bottleneck extrema. J. Comb. Theor. **8**, 299–306 (1970)
12. Edmonds, J., Giles, R.: A min-max relation for submodular functions on graphs. In: Studies in Integer Programming, Annals of Discrete Mathematics, vol. 1, pp. 185–204 (1977). (proceedings of Workshop, Bonn, 1975)
13. Fulkerson, D.R.: Blocking polyhedra. In: Graph Theory and its Applications, pp. 93–112. Academic Press, New York (1970). (Proc. Advanced Sem., Math. Research Center, Univ. of Wisconsin, Madison, Wis., 1969)
14. Guenin, B.: A characterization of weakly bipartite graphs. J. Combin. Theor. Ser. B **83**(1), 112–168 (2001)
15. Isbell, J.R.: A class of simple games. Duke Math. J. **25**, 423–439 (1958)
16. Jaeger, F.: Flows and generalized coloring theorems in graphs. J. Combin. Theor. Ser. B **26**(2), 205–216 (1979)
17. Lehman, A.: A solution of the Shannon switching game. J. Soc. Ind. Appl. Math. **12**, 687–725 (1964)
18. Lehman, A.: On the width-length inequality. Math. Program. **16**(2), 245–259 (1979)
19. Lucchesi, C.L., Younger, D.H.: A minimax theorem for directed graphs. J. London Math. Soc. **17**(3), 369–374 (1978)
20. Oxley, J.: Matroid Theory. Oxford Graduate Texts in Mathematics, 2nd edn., vol. 21. Oxford University Press, Oxford (2011)
21. Schrijver, A.: Combinatorial Optimization. Polyhedra and Efficiency, Algorithms and Combinatorics, vol. 24. Springer, Heidelberg (2003)
22. Seymour, P.D.: The forbidden minors of binary clutters. J. London Math. Soc. **12**(3), 356–360 (1975/76)
23. Seymour, P.D.: The matroids with the max-flow min-cut property. J. Combin. Theor. Ser. B **23**(2–3), 189–222 (1977)
24. Seymour, P.D.: Sums of circuits. In: Graph Theory and Related Topics, pp. 341–355. Academic Press, New York-London (1979). (Proc. Conf., Univ. Waterloo, Waterloo, Ont., 1977)
25. Seymour, P.D.: Matroids and multicommodity flows. Eur. J. Combin. **2**(3), 257–290 (1981)
26. Tutte, W.T.: On the algebraic theory of graph colorings. J. Comb. Theor. **1**, 15–50 (1966)

Flexible Graph Connectivity
Approximating Network Design Problems Between 1- and 2-Connectivity

David Adjiashvili[1], Felix Hommelsheim[2(✉)], and Moritz Mühlenthaler[3]

[1] Department of Mathematics, ETH Zürich, Zürich, Switzerland
[2] Fakultät für Mathematik, TU Dortmund University, Dortmund, Germany
`felix.hommelsheim@math.tu-dortmund.de`
[3] Laboratoire G-SCOP, Grenoble INP, Univ. Grenoble-Alpes, Grenoble, France

Abstract. Graph connectivity and network design problems are among the most fundamental problems in combinatorial optimization. The minimum spanning tree problem, the two edge-connected spanning subgraph problem (2-ECSS) and the tree augmentation problem (WTAP) are all examples of fundamental well-studied network design tasks that postulate different initial states of the network and different assumptions on the reliability of network components. In this paper we motivate and study *Flexible Graph Connectivity* (FGC), a problem that mixes together both the modeling power and the complexities of all aforementioned problems and more. In a nutshell, FGC asks to design a connected network, while allowing to specify different reliability levels for individual edges.

In this paper we develop a general algorithmic approach for approximating FGC that yields approximation algorithms with ratios that are close to the known best bounds for many special cases, such as 2-ECSS and WTAP. Our algorithm and analysis combine various techniques including a weight-scaling algorithm, a charging argument that uses a variant of exchange bijections between spanning trees and a factor revealing min-max-min optimization problem.

Keywords: Connectivity augmentation · Approximation algorithms · Network design

1 Introduction

Many real-world design and engineering problems can be modeled as either graph connectivity or network design problems. Routing, city planning, communication infrastructure are just a few examples where network design problems are omnipresent. To realistically model a real-life problem as a network design problem, it is imperative to consider issues of *reliability*, namely the capacity of the systems' resources to withstand disturbances, failures, or even adversarial attacks. This aspect motivates the study of many classical network design problems, as well as robust counterparts of many connectivity problem.

© Springer Nature Switzerland AG 2020
D. Bienstock and G. Zambelli (Eds.): IPCO 2020, LNCS 12125, pp. 13–26, 2020.
https://doi.org/10.1007/978-3-030-45771-6_2

The problem studied in this paper, which we call *Flexible Graph Connectivity* (FGC), lies in the intersection of classical network design and robust optimization. It encapsulates several well studied problems that have received significant attention from the research community, such as the minimum spanning tree problem, the 2-edge-connected spanning subgraph problem (2-ECSS) [9,13,15,20], the weighted tree augmentation problem (WTAP) [1,8,9,14,17–19], and the matching augmentation problem [6]. As such, FGC is APX-hard and it encompasses all of the technical challenges associated with approximating these problems simultaneously. In a sense, minimum spanning tree and 2-ECSS represent two far ends of a spectrum of possible network design tasks that can be modeled with FGC and the other mentioned problems lie in between. We argue that by translating attributes of a real-life network design problem, one is much more likely to encounter a problem from the aforementioned spectrum, as opposed to one of its more famous extreme cases.

The problem FGC is formally defined as follows. The input is given by an undirected connected graph $G = (V, E)$, non-negative edge weights $w \in \mathbb{Q}_{\geq 0}^E$ and a set of edges $\overline{F} \subseteq E$ called *safe* edges. Let $F := E \setminus \overline{F}$ be the *unsafe* edges. The task is to compute a minimum-weight edge set $S \subseteq E$ with the property that $(V, S - f)$ is a connected graph for every $f \in F$.

We briefly illustrate why minimum spanning tree, 2-ECSS, and WTAP are all special cases of FGC. Clearly, if all edges of the input graph are safe, then an optimal solution is a minimum-weight spanning tree. If, on the other hand, all edges of the input graph are unsafe, then an optimal solution is a minimum-weight 2-edge-connected spanning subgraph. Finally, if the unsafe edges form a weight-zero spanning tree T of the input graph, then an optimal solution is a minimum-cost tree augmentation of T. The goal of this paper is to provide approximation algorithms for FGC and our main result is the following theorem.

Theorem 1. FGC *admits a polynomial-time 2.523-approximation algorithm.*

In the spirit of recent advancements for WTAP [1,8,14,19], our results also extend to bounded-weight versions of the problem. A bounded-weight FGC instance is one whose weights all range between 1 and some fixed constant $M \in \mathbb{N}$. For bounded-weight FGC we obtain the following result.

Theorem 2. *Bounded-weight* FGC *admits a polynomial-time 2.404-approximation algorithm.*

We elaborate on the techniques used to prove these theorems as well as their connection to results on 2-ECSS and WTAP later on. We start by giving our motivation for studying this problem.

1.1 Importance of Non-uniform Models for Network Reliability

The vast majority of network design problems are motivated by reliability requirements imposed on real-life networks. For example, the k-edge connected spanning subgraph problem [10,11] asks to construct a connected network that can withstand a failure of at most $k - 1$ edges. In the *k-edge connectivity augmentation problem* we are given a k-edge connected graph and the goal is to add a minimum weight-set of edges such that the resulting graph is $(k + 1)$-edge connected. It is shown in [7] that k-edge connectivity augmentation can be reduced to WTAP for odd k and to cactus augmentation for even k [5,12]. The more general *survivable network design problem* asks to construct a network that admits a prescribed number of edge-disjoint paths between every pair of nodes [13,15].

While all of the above problems are important for modeling reliability in the design of real-world networks, they also neglect the inherent inhomogeneity of resources (such as nodes, links etc.) in such environments. This inhomogeneity stems from various factors, such as geographical region, available construction material, proximity to hazards, the ability to defend the asset and more. To incorporate this aspect of real-world problems it is imperative to differentiate between different resources, not only in terms of their cost, but also in terms of their reliability. In the simplest setup, we would like to distinguish between *safe* and *unsafe* resources (edges, nodes etc.) and postulate that failures can only occur among the unsafe resources. With FGC we adopt this model and study the basic graph connectivity problem.

In the robust optimization literature, several non-uniform failure models have been proposed. Adjiashvili, Stiller and Zenklusen [4] proposed the bulk-robust model, in which a solution has to be chosen that withstands the failure of any input prescribed set of scenarios, each comprising a subset of the resources. Since subsets can be specified arbitrarily, bulk-robustness can be used to model high correlations between failures of individual elements, as well as highly non-uniform scenarios. FGC falls into the category of bulk-robust network design problems, since we can model the reliability criterion by creating failure scenarios, one per unsafe edge in the bulk-robust setup and require the graph to be connected. In fact, the existing results [4] on bulk-robust optimization imply the existence of $\log n$ ratio for the problem. In this paper we improve this significantly.

For further related work and connections to robust optimization we refer the reader to the full version of the paper [3].

1.2   Complexity of FGC and Its Relationship to 2-ECSS and WTAP

As was pointed out before, some classical and well studied network design problems are special cases of FGC, including 2-ECSS, and WTAP. Thus, approximating FGC, is at least as challenging as approximating the latter two problems. In this section we present some evidence that FGC might actually be significantly harder. For the general case of both 2-ECSS and WTAP the iterative rounding algorithm of Jain [15] provides an approximation factor of 2, which is best known. It is important to note that 2-ECSS subsumes WTAP in the general (weighted) case. Nevertheless, it is both instructive and useful to relate FGC, to both problems. In particular, this enables us to improve the approximation ratio for the bounded-weight version of FGC.

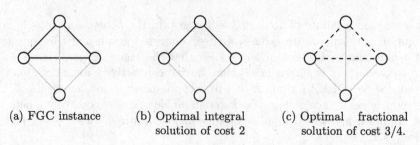

(a) FGC instance   (b) Optimal integral   (c) Optimal   fractional
                       solution of cost 2   solution of cost 3/4.

Fig. 1. FGC instance for which (1) has integrality gap 8/3. Gray edges are unsafe and have cost 0. Black edges are safe and have cost 1. The dashed (safe) edges are fractional with value 1/4.

In contrast, for the unweighted versions of both problems (and in the case of WTAP more generally for bounded weights), a long line of results has generated numerous improvements beyond ratio two, leading to the currently best known bounds of 4/3 for unweighted 2-ECSS [20], 1.46 for unweighted tree augmentation [14] and 1.5 for WTAP with bounded weights [8,14]. The case of unit (or bounded) weights is where techniques for approximating 2-ECSS and WTAP start to differ significantly. In both cases, the known best bounds are achieved by combining LP-based techniques with clever combinatorial tools. Nevertheless, there seems to be very little intersection in both the nature of used LPs and the overall approaches, as techniques suitable for one problem do no seem to provide competitive ratios for the other.

Consequently, there are several implications for approximating FGC. Firstly, achieving an approximation factor better than two is an ambitious task, as it would simultaneously improve the long-standing best known bounds for both 2-ECSS and WTAP. At the same time, for achieving a factor two, it may be possible to use classical tools for survivable network design [15]. We show that, at least with the natural LPs, this is impossible, as the integrality gap of such LPs can be significantly larger than 2. Consider the following natural generalization of the cut-based formulation for survivable network design to FGC.

$$\text{minimize}\ \ w^T x$$
$$\text{subject to}\ \ \sum_{f \in \delta(S) \cap F} x_f + \sum_{e \in \delta(S) \cap \overline{F}} 2x_e \geq 2 \ \ \text{for all}\ \emptyset \subsetneq S \subsetneq V \tag{1}$$
$$x_e \in \{0,1\} \ \text{for all}\ e \in E.$$

In essence, the IP formulation (1) states that each cut in the graph needs to contain at least one safe edge or at least two edges, which indeed is the feasibility condition for FGC. One can show that many important properties that are central in Jain's [15] analysis still hold, e.g., the possibility to perform uncrossing for tight constraints at a vertex LP solution. These properties might become useful for devising a pure LP-based algorithm for FGC. However, for the instance shown in Fig. 1, the integrality gap of formulation (1) is at least 8/3.

Table 1. Approximation ratios for FGC and some special cases.

Problem	General weights	Bounded weights	Unweighted
FGC	2.523 (Theorem 1)	2.404 (Theorem 2)	3/2 ([3, Theorem 3])
2-ECSS	2 [15]	2 [15]	4/3 [20]
WTAP	2 [9]	3/2 [14]	1.46 [14]

Finally, the recent advances achieved for unweighted and bounded-weight versions of 2-ECSS and WTAP seem to be unsuitable to directly tackle FGC, as they were not found to provide good ratios for both 2-ECSS and WTAP simultaneously. It is natural to conclude that a good ratio for FGC can only be achieved by a combination of techniques suitable for both 2-ECSS and WTAP.

To summarize, it seems that while the only known techniques for simultaneously approximating both 2-ECSS and WTAP within a factor two rely on rounding natural linear programs relaxations of a more general network design problem (such as the survivable network design problem [13,15]), the integrality gaps of such natural LPs for FGC are significantly larger than two. At the same time, due to its strong motivation, it is desirable to achieve a factor close to two, which is the state of the art for both 2-ECSS and WTAP.

In this paper we show that this goal can be achieved by properly combining algorithms for 2-ECSS, and WTAP. Our algorithms are simple and black-box to such an extent that results for restricted versions of WTAP (e.g., bounded cost) can be directly applied to FGC with the same restrictions, thus leading to improved bounds for these restricted versions of FGC as well. At the same time, the analysis is complex and requires careful charging arguments, generalizations of the notion of exchange bijections of spanning trees and factor revealing optimization problems, as we elaborate next.

1.3   Main Techniques and an Overview of the Algorithm

We present here a high-level overview of some of the technical ingredients that go into our algorithm and analysis used to prove Theorem 1. The algorithm carefully combines the following three rather simple algorithms for FGC, each having an approximation ratio significantly worse than the one exhibited in Theorem 1. Each of the algorithms establishes 2-edge-connectivity in a modified graph, where a subset of safe edges is contracted. In order to be able to establish 2-edge connectivity, we need to add a parallel unsafe edge e' for each safe edge e of the same cost. It can easily be observed that optimal solutions for the new instance are also optimal solution for the old instance and vice versa.

Algorithm A Compute a 2-edge connected spanning subgraph.

Algorithm B Compute a minimum spanning tree and make it 2-connected by solving the corresponding WTAP instance.

Algorithm C Compute a minimum spanning tree, contract its safe edges, then compute a 2-edge connected spanning subgraph. Return the union of this solution and the safe edges of the spanning tree.

It is not hard to show that Algorithms A, B and C are polynomial-time approximation algorithms for FGC, with approximation ratios of $4, 3$ and 5, respectively, given that 2-ECSS and WTAP admit polynomial-time 2-approximation algorithms. We defer the details to the later sections and instead present a road map for proving the main result.

Our approximate solution is obtained from returning the best of many solutions, each computed by one of the above three algorithms on an instance that is computed from the original instance by appropriately scaling the costs of the safe edges. The motivation for making safe edges cheaper is that buying a similarly priced unsafe edge instead likely incurs extra costs, since one safe edge or at least two edges have to cross each cut. The technical challenge is to determine the most useful scaling factors.

The main idea in the analysis is to relate the costs of edges in an optimal solution to the costs of edges in the computed solutions based on a generalization of *exchange bijections* between spanning trees. Exchange bijections are bijections between two bases of a matroid (e.g., spanning trees in a connected graph). We introduce our generalized notion of α-*monotone exchange bijections*, where α is a scaling factor used in the algorithm, and prove that they always exist between spanning trees of the optimal and computed solutions. We then combine the properties of such bijections with additional technical ideas to derive an upper bound on the cost of the computed solutions of algorithms A, B and C. The bound is given in terms of several parameters that represent proportions of costs associated with parts of the computed and an unknown optimal solution, defined through the exchange bijections.

The final step is to combine all obtained upper bounds. Since we have the choice of selecting the scaling factors, but have no control over the remaining parameters appearing in the upper bounds, we can compute a conservative upper bound on the approximation ratio by solving a three-stage *factor-revealing min-max-min optimization problem*. The inner minimum is taken over the upper bounds on the values of the solutions computed by algorithms A, B, and C. The maximum is taken over the parameters that depend on an unknown optimal solution. Finally, the outer minimum is taken over the choice of scaling factors. One interesting aspect of our factor revealing optimization problem is that its solution gives not only a bound on the approximation ratio of the algorithm (as in, e.g., [2,16]), but it also suggests optimal instance-independent scaling factors to be used by the algorithm itself. One can show that by computing the scaling factors it is possible to constrain further the unknown parameters in the problem, thus obtaining better instance-specific bounds. While we can compute these factors in polynomial time, we will not elaborate on this approach in the paper.

Since the overall factor-revealing optimization problem is a three-stage min-max-min program, we can only provide analytic proof for its optimal value for very small sizes. However, we are still able to give an analytic bound of 2.523 in order to prove Theorem 1 by combining only algorithms A and C, but using the optimal choices of scaling factors for a given instance. To achieve the factor 2.404 for bounded weight instances we use all three algorithms A, B and C to bound the optimal solution of the min-max-min problem. Clearly, using all three algorithms yields better bounds, but we cannot give an analytic upper bound. Instead we give a computational upper bound on the min-max-min problem using the baron solver [21]. An overview of our approximation guarantees along with some related results can be found in Table 1.

1.4 Notation and Organization

Unless stated otherwise graphs are loopless but may have parallel edges. Let $G = (V, E)$ be a graph with vertex set V and edge set E. We denote by $E(G)$ the edge set of G. For an edge e we may write $G - e$ (resp., $G + e$) for the graph $(V, E \setminus \{e\})$ (resp., $(V, E \cup \{e\})$). For an edge set $E' \subseteq E$ we denote by G/E' the graph obtained from G by contracting the edges in E'. We denote by λ (resp., τ) the ratio of an approximation algorithm for 2-ECSS (resp., WTAP).

For the remainder of this paper we fix an instance $I = (G, w, \overline{F})$ of FGC and some optimal solution $Z^* \subseteq E(G)$ of I. To avoid technicalities, we add to each safe edge a parallel unsafe edge of the same cost. It is readily seen that this modification preserves optimal solutions. Observe that the solution Z^* has the following structure. For some $r \in \mathbb{N}$, the graph $(V(G), Z^*)$ consists of r 2-edge-connected components $C_1, C_2, \ldots, C_r$ that are joined together by safe edges $E' := \{\overline{f}_1, \overline{f}_2, \ldots, \overline{f}_{r-1}\} \subseteq \overline{F}$ in a tree-like fashion. That is, if we contract each component C_i to a single vertex, the remaining graph is a tree T^* with edge set E'. On the other hand, if we contract E', we obtain a 2-edge-connected spanning subgraph of the resulting graph. We let $\delta := w(E')/\mathrm{OPT}(I)$. That is, the value δ is the proportion of the cost of the safe cut edges E' relative to the total cost of the optimal solution Z^*.

2 The Algorithm

We start by formally defining algorithms A, B and C, and give bounds for their individual ratios (except for Algorithm C, which requires results from the full version [3, Section 4]). We then state our algorithm for FGC, which combines all three algorithms together with a careful weight scaling.

Algorithm A: A $(2 + 2\delta)$-Approximation Algorithm. Algorithm A computes (in polynomial time) a λ-approximate 2-edge connected spanning subgraph, e.g., by Jain's algorithm [15]. Algorithm A then removes all unsafe edges that are parallel to safe edges and returns the resulting edge set. Observe that the returned solution is feasible. We now argue that Algorithm A is a $(2 + 2\delta)$-approximation algorithm. By adding a copy of each edge in E' to the optimal solution Z^* we obtain a 2-edge-connected spanning subgraph H_1^* of G. The cost $w(H_1^*)$ of H_1^* is given by

$$w(H_1^*) = \sum_{e \in Z^* \setminus E'} w(e) + 2 \cdot \sum_{e \in E(Z^*) \cap E'} w(e) \ .$$

and since $w(H_1) \leq 2w(H_1^*)$, it follows that $w(H_1) \leq 2w(H_1^*) = 2(1 - \delta) \cdot \mathrm{OPT}(I) + 4\delta\mathrm{OPT}(I) = (2 + 2\delta) \cdot \mathrm{OPT}(I)$ as claimed. Observe that Algorithm A performs best if the weight of the safe edges E' is small.

Algorithm B: A 3-Approximation Algorithm. Algorithm B computes a minimum spanning tree T of G and then computes a τ-approximate solution to the WTAP instance (G, T', w), where the tree T' is obtained by contracting each safe edge of T. The solution of the WTAP instance together with the tree T is a feasible solution for I Since $w(T) \leq \mathrm{OPT}(I)$, the best available algorithms for WTAP (see Table 1) give a 3-approximation for FGC, and a 5/2-approximation for FGC on bounded-weight instances.

Algorithm C: Algorithm C first computes a minimum spanning tree T of G. Let G' be the graph obtained from G by contracting the safe edges of T, that is, $G' := G/(E(T) \cap \overline{F})$. Algorithm C then computes a λ-approximate 2-edge-connected spanning subgraph $H' \subseteq E(G')$ and returns the edge set $(E(T) \cap \overline{F}) \cup H'$. It is readily seen that Algorithm C computes a feasible solution and that the safe edges of T have cost at most $\mathrm{OPT}(I)$. Using exchange bijections from [3, Section 4], it can be shown that $w(H') \leq 4 \cdot \mathrm{OPT}(I)$, so Algorithm C is a 5-approximation algorithm for FGC.

Algorithm 1, our main approximation algorithm for FGC, proceeds as follows. It first computes suitable scaling factors $W \subseteq [0, 1]$ (called "threshold values") for the costs of the safe edges (see [3, Proposition 8]). Then, it runs Algorithm A to obtain solution Z^A. We say that we run Algorithm B (resp., C) *with scaling factor* α if the minimum spanning tree in Algorithm B (resp., C) is computed with respect to weights obtained from w by scaling the costs of the safe edges by α. Algorithm 1 runs Algorithms B and C with each scaling factor $\alpha \in W \cup \{0, 1\}$ and returns a solution of minimal weight among all the different solutions computed by Algorithms A, B, and C.

Clearly Algorithm 1 computes a feasible solution. It runs in polynomial time if there are polynomially many threshold values in the set W and if this set can be computed in polynomial time. Proofs are deferred to the full version [3, Section 4]. Using properties of the computed threshold values we show that the selection of the scaling factors in Algorithm 1 is best possible.

Lemma 1. *Let $W' \subseteq [0, 1]$ and let $\mathcal{A}'(I)$ be the weight of the solution returned by Algorithm 1 run with W set to W'. Then $\mathcal{A}(I) \leq \mathcal{A}'(I)$.*

An analysis of the approximation ratio of Algorithm 1 is given in Sect. A. Here, we provide a high-level overview. Our starting point is Lemma 1, which allows us to assume that Algorithm 1 tries *all* scaling factors in $[0, 1]$. Let us denote by $\mathcal{A}(I)$ the weight of the solution returned by Algorithm 1. We show that the approximation ratio of Algorithm 1 is bounded from above by the optimal

---

Algorithm 1: Improved Approximation Algorithm for FGC

---

input : Instance $I = (G, w, \overline{F})$ of FGC

compute threshold values $W := \{\alpha_e \mid e \in E(G)\} \cup \{0, 1\}$

run Algorithm A on I to obtain solution Z^A

for *threshold value* $\alpha \in W$ **do**

    run algorithms B and C with scaling factor α to obtain solutions Z_α^B
    and Z_α^C, respectively

return *solution with lowest cost among* Z^A *and* $\{Z_\alpha^B, Z_\alpha^C \mid \alpha \in W\}$

---

value of a min-max-min optimization problem. For an instance I of FGC and some $N \in \mathbb{N}$, the optimization problem has the following data.

- Scaling factors $\alpha_1, \alpha_2, \ldots, \alpha_N \in [0, 1]$. Due to the discussion above, in our analysis of Algorithm 1 we are free to choose these values.
- Parameters $\beta_1, \beta_2, \ldots, \beta_N, \gamma_1, \gamma_2, \ldots, \gamma_N, \delta \in [0, 1]$, which depend on the structure of an optimal solution and satisfy $\sum_{j=1}^{N} \beta_j + \sum_{j=1}^{N} \gamma_j = 1$.
- Functions f^A, as well as $f_1^B, f_2^B, \ldots, f_N^B$ and $f_1^C, f_2^C, \ldots, f_N^C$ that bound from above in terms of $\alpha_i, \beta_i, \gamma_i, 1 \leq i \leq N$, λ and τ, the cost of the solutions computed by algorithms A, B, and C, respectively.

Precise definitions of the parameters and the functions will be given later. We note that Algorithm B is only used in the proof of Theorem 2. Technically, we show that for a proper choice of functions f^A, f_i^B and f_i^C, the ratio of Algorithm 1 is bounded by the optimal value of the following optimization problem.

$$\min_{\substack{\alpha_i \in [0,1]\,:\,1 \leq i \leq N}} \quad \max_{\substack{\beta_i \in [0,1]\,:\,1 \leq i \leq N \\ \gamma_i \in [0,1]\,:\,1 \leq i \leq N}} \quad \min_{1 \leq i \leq N} \quad \{f^A(\cdot), f_i^B(\cdot), f_i^C(\cdot)\}$$

$$\text{subject to} \quad \sum_{j=1}^{N} \beta_j + \sum_{j=1}^{N} \gamma_j = 1 \tag{2}$$

Next, we present the main technical tools we use in order to derive the functions f^A, f_i^B and f_i^C that occur in (2). We first show that safe edges and unsafe edges exhibit a "threshold" behavior with respect to MSTs if the costs are scaled by some $\alpha \in [0, 1]$. Furthermore, we show that (i) the corresponding threshold values can be computed in polynomial time, which is essential to ensure that Algorithm 1 runs in polynomial time and (ii) they are the best choice of scaling factors for Algorithm 1, which allows us to assume in our analysis that we execute Algorithm 1 for *all* scaling factors $\alpha \in [0, 1]$. For $\alpha \in [0, 1]$, we denote by

$$w_\alpha(e) = \begin{cases} \alpha \cdot w(e) & \text{if } e \in \overline{F}, \text{and} \\ w(e) & \text{otherwise} \end{cases}$$

the weight function obtained from w by scaling the costs of the safe edges by α. A spanning tree T is called α-*minimum spanning tree* (α-MST) if $E(T)$ has minimal weight with respect to w_α.

Consider changing the scaling factor α smoothly from 0 to 1. We observe that for any safe edge e, if there is an α-MST containing e, then there is also an α'-MST containing e for any $\alpha' \leq \alpha$. On the other hand, if there is an α-MST containing an unsafe edge f then there is also an α'-MST containing f for any $\alpha \leq \alpha' \leq 1$. Next, we formally define the notion of thresholds and state that they always exist.

Definition 1. *Let $e \in E$ and $\alpha_e \in [0,1]$. We say that α_e is a lower threshold for e if for any $\alpha \in [0,1]$ there is an α-MST containing e if and only if $\alpha \geq \alpha_e$. If e is in no α-MST for $0 \leq \alpha \leq 1$, we let the lower threshold value of e be ∞. Similarly, α_e is an upper threshold for e if for $\alpha \in [0,1]$ there is an α-MST containing e if and only if $\alpha \leq \alpha_e$. The threshold values of an instance $I = (G, w, \overline{F})$ is defined as $\{\alpha_e \mid e \in E(G)\}$.*

Lemma 2. *For each unsafe edge $f \in F$ there is a lower threshold $\alpha_f \in [0,1] \cup \{\infty\}$. For each safe edge $e \in \overline{F}$ there is an upper threshold $\alpha_e \in [0,1]$.*

It is easily seen that there are $O(|V(G)|^2)$ threshold values. In fact, one can show that there are at most $|V(G)| - 1$ different threshold values [3, Proposition 1]. This implies in particular that Algorithm 1 runs in polynomial time. To prove that threshold values are optimal scaling factors (Lemma 1), consider a scaling factor $\alpha \in [0,1]$ and an α-MST T_α computed by Algorithm B or C. Choose a smallest interval $[\alpha_L, \alpha_R]$ containing α, such that α_L and α_R are threshold values and observe that T_α is either a α_L-MST or a α_R-MST. Hence, the solution returned by Algorithm 1 is at least as good as one for scaling factor α.

In our analysis of Algorithm 1, we use a charging argument based on the notion of monotone exchange bijections, which we now introduce. Let G be a connected graph and let T and T' be spanning trees of G. A bijection $\varphi : E(T') \to E(T)$ is called *exchange bijection* if for each $e \in E(T')$, the graph $T' - e + \varphi(e)$ is a spanning tree of G. An exchange bijection φ is *monotone* if for each edge $e \in E(T')$ we have $w(e) \leq w(\varphi(e))$. For any two spanning trees T and T' a canonical exchange bijection exists: Note that the edge sets of spanning trees of G are the bases of the graphic matroid $M(G)$ of G. By the strong basis exchange property of matroids there is a bijection between $E(T) \setminus E(T')$ and $E(T') \setminus E(T)$ with the required properties, which can be extended to an exchange bijection by mapping each item in $E(T) \cap E(T')$ to itself. Furthermore, if T' is an MST then for any spanning tree T', a canonical exchange bijection is monotone.

We generalize monotone exchange bijections as follows.

Definition 2. *Let $\alpha \in [0,1]$ and let T, T' be spanning trees of G. An exchange bijection $\varphi : E(T') \to E(T)$ is α-monotone if for each edge $e \in E(T')$ we have*

1. $w(e) \leq \frac{1}{\alpha} w(\varphi(e))$ *if $e \in \overline{F}$ and $\varphi(e) \in F$, and*
2. $w(e) \leq w(\varphi(e))$ *if either $e, \varphi(e) \in \overline{F}$ or $e, \varphi(e) \in F$, and*
3. $w(e) \leq \alpha w(\varphi(e))$ *if $e \in F$ and $\varphi(e) \in \overline{F}$.*

We show that for any spanning tree T of G, there is an α-monotone exchange bijection from an α-MST to T.

Lemma 3. *Let $\alpha \in [0,1]$, let T_α be an α-MST of G and let T be any spanning tree of G. Then there is an α-monotone exchange bijection $\varphi : E(T_\alpha) \to E(T)$.*

A   Proof Sketch of Theorem 1

In this section we give a sketch of an analytic upper bound of 2.523 on the approximation ratio of Algorithm 1. For this purpose it suffices to only consider Algorithms A and C. That is, using α-monotone exchange bijections from Sect. 2, we determine functions $f^A(\cdot)$ and $f^C(\cdot)$ for the optimization problem (2), where $f^C(\cdot)$ depends on a selection of scaling factors and some other parameters to be introduced shortly. Recall that according to Lemma 1, the selection of scaling factors in Algorithm 1 is optimal. Surprisingly, a worst-case instance for our bounds $f^A(\cdot)$ and $f^C(\cdot)$ in fact has a single threshold value which is $1/\lambda$. However, to obtain the approximation ratio of 2.523 it is crucial to execute Algorithm 1 with all threshold values of the given instance.

Let $\mathcal{I}(N)$ be a class of instances of FGC with at most N threshold values (see Definition 1). In the following, suppose that $I \in \mathcal{I}(N)$ and recall that an optimal solution $Z^* \subseteq E(G)$ of I consists of r 2-edge-connected components $C_1, C_2, \ldots, C_r$ that are joined together by safe edges $E' := \{\overline{f}_1, \overline{f}_2, \ldots, \overline{f}_{r-1}\} \subseteq \overline{F}$ in a tree-like fashion. Moreover, for any spanning tree $T \subseteq Z^*$ we have $E' \subseteq T$.

Observe that since there is an unsafe edge for each safe edge of same weight in G, we have that each threshold value is in $[0,1]$. Let $0 \leq \alpha_1 \leq \alpha_2 \leq \ldots \leq \alpha_N \leq 1$ be the N threshold values of I in non-decreasing order. In order to prepare our analysis, we consider for $i \in \{1, 2, \ldots, N\}$ an α_i-MST T_i, an α_i-monotone exchange bijection $\varphi_i : T_i \to T$ (which exist due to Lemma 3) and a weight function $w_i := w_{\alpha_i}$. For $2 \leq i \leq N$ we choose φ_i such that for each $e \in E(T_{i-1}) \cap E(T_i)$ we have $\varphi_{i-1}(e) = \varphi_i(e)$. This can be done due to Corollary 12 in [3]. In order to define the parameters of the optimization problem (2), for $1 \leq i \leq N$, we partition the edge set of the α_i-MST T_i into four parts D_i, O_i, F_i, and S_i as follows.

- $D_i := \{e \in E(T_i) \cap F \mid \varphi_i(e) \in E'\}$
- $O_i := \{e \in E(T_i) \cap \overline{F} \mid \varphi_i(e) \in E'\}$
- $F_i := \{e \in E(T_i) \cap F \mid \varphi_i(e) \in E(T) \setminus E'\}$
- $S_i := \{e \in E(T_i) \cap \overline{F} \mid \varphi_i(e) \in E(T) \setminus E'\}$

The parameters of (2) are given as follows. For $1 \leq i \leq N$ we let $E_i^{\overline{F}}$ (resp., E_i^F) be the set of edges in E' (resp., $E(T) - E'$) that have threshold value α_i. That is, $E_i^{\overline{F}} := \{e \in E' \mid \alpha_e = \alpha_i\}$ and $E_i^F := \{e \in E(T) - E' \mid \alpha_e = \alpha_i\}$. For $1 \leq i \leq N$ we let $\beta_i = w(E_i^{\overline{F}})/\mathrm{OPT}(I)$ and $\gamma_i = w(E_i^F)/\mathrm{OPT}(I)$ be the fraction of the weight of the optimal solution that is contributed by the edges in $E_i^{\overline{F}}$ (resp., E_i^F). Finally, let $\xi \in [0,1]$ be the the fraction of the weight of the optimal solution that does not correspond to the tree T; i.e., $\xi := \frac{w(Z^*) - w(T)}{\mathrm{OPT}(I)}$. The following properties of β_i, γ_i, $1 \leq i \leq N$, are readily verified:

1. $\beta_1, \beta_2, \ldots \beta_N, \gamma_1, \gamma_2, \ldots \gamma_N, \xi \in [0,1]$,
2. $\sum_{j=1}^{N} \beta_j = \frac{w(E')}{\text{OPT}(I)}$,
3. $\sum_{j=1}^{N} \gamma_j = \frac{w(T-E')}{\text{OPT}(I)}$ and
4. $\xi = 1 - \sum_{j=1}^{N} \beta_j - \sum_{j=1}^{N} \gamma_j$.

We now bound the cost of the solutions Z_i^C and Z^A returned by Algorithm C (resp., Algorithm A) in terms of the parameters.

Lemma 4. *Suppose we run Algorithm 1 with the optimal threshold values $W = \{\alpha_i\}_{1 \leq i \leq N}$. Let Z_i^C be the solution to the instance $(G, w_i, \overline{F})$ of FGC computed by Algorithm C in Algorithm 1. Then*

$$w(Z_i^C) \leq \left(1 + \sum_{j=1}^{i-1} (\lambda - 1 + \lambda\alpha_j)\beta_j + (\lambda - 1) \cdot \sum_{j=1}^{N} \gamma_j + \sum_{j=i}^{N} \frac{\gamma_j}{\alpha_j}\right) \cdot \text{OPT}(I).$$

Lemma 5. *Suppose we run Algorithm 1 with the optimal threshold values $W = \{\alpha_i\}_{1 \leq i \leq N}$. Let Z^A be the solution to the instance $(G, w, \overline{F})$ of FGC computed by Algorithm A in Algorithm 1. Then*

$$w(Z^A) \leq \left(\lambda + \lambda \cdot \sum_{j=1}^{N} \alpha_j\beta_j\right) \cdot \text{OPT}(I).$$

With the bounds from Lemmas 4 and 5 and by applying standard techniques we can simplify problem (2) to

$$\max \lambda \cdot \left(1 + \sum_{j=1}^{N} \alpha_j\hat{\beta}_j\right)$$

$$\text{subject to } \sum_{j=1}^{N} \hat{\beta}_j \cdot (1 + \alpha_j(\lambda - 1 + \lambda\alpha_j)) = 1, \tag{3}$$

$$0 \leq \alpha_1 \leq \alpha_2 \leq \ldots \leq \alpha_N \leq 1,$$

$$\hat{\beta}_j \in [0,1] \text{ for all } j \in \{1, \ldots, N\}.$$

Theorem 3. *The approximation guarantee of Algorithm 1 for instances with at most N threshold values is upper bounded by the optimal value of optimization problem (3).*

We solve Problem (3) analytically and observe that the optimal value does not depend on N. Hence we obtain the claimed approximation ratio of 2.523 for $\lambda = 2$.

Theorem 4. *Algorithm 1 has an approximation guarantee of $\frac{\lambda \cdot (\lambda + 2\sqrt{\lambda})}{2\sqrt{\lambda} + \lambda - 1}$.*

References

1. Adjiashvili, D.: Beating approximation factor two for weighted tree augmentation with bounded costs. ACM Trans. Alg. (TALG) **15**(2), 19 (2018)
2. Adjiashvili, D., Bosio, S., Weismantel, R., Zenklusen, R.: Time-expanded packings. In: Esparza, J., Fraigniaud, P., Husfeldt, T., Koutsoupias, E. (eds.) ICALP 2014. LNCS, vol. 8572, pp. 64–76. Springer, Heidelberg (2014). https://doi.org/10.1007/978-3-662-43948-7_6
3. Adjiashvili, D., Hommelsheim, F., Mühlenthaler, M.: Flexible graph connectivity (2019). https://arxiv.org/abs/1910.13297
4. Adjiashvili, D., Stiller, S., Zenklusen, R.: Bulk-robust combinatorial optimization. Math. Program. **149**(1–2), 361–390 (2015). https://doi.org/10.1007/s10107-014-0760-6
5. Byrka, J., Grandoni, F., Ameli, A.J.: Breaching the 2-approximation barrier for connectivity augmentation: a reduction to Steiner tree (2019). https://arxiv.org/abs/1911.02259
6. Cheriyan, J., Dippel, J., Grandoni, F., Khan, A., Narayan, V.: The matching augmentation problem: a $\frac{7}{4}$-approximation algorithm. Math. Program. (2019). https://doi.org/10.1007/s10107-019-01394-z
7. Dinits, E., Karzanov, A., Lomonosov, M.: On the structure of a family of minimal weighted cuts in a graph. Studies in Discrete Optimization, pp. 290–306 (1976)
8. Fiorini, S., Groß, M., Könemann, J., Sanità, L.: Approximating weighted tree augmentation via Chvátal-gomory cuts. In: Proceedings of the Twenty-Ninth Annual ACM-SIAM Symposium on Discrete Algorithms. SODA 2018, pp. 817–831. Society for Industrial and Applied Mathematics (2018)
9. Frederickson, G.N., JáJá, J.: Approximation algorithms for several graph augmentation problems. SIAM J. Comput. **10**(2), 270–283 (1981). https://doi.org/10.1137/0210019
10. Gabow, H.N., Gallagher, S.R.: Iterated rounding algorithms for the smallest k-edge connected spanning subgraph. SIAM J. Comput. **41**(1), 61–103 (2012)
11. Gabow, H.N., Goemans, M.X., Tardos, É., Williamson, D.P.: Approximating the smallest k-edge connected spanning subgraph by LP-rounding. Networks **53**(4), 345–357 (2009). https://doi.org/10.1002/net.20289
12. Gálvez, W., Grandoni, F., Ameli, A.J., Sornat, K.: On the cycle augmentation problem: hardness and approximation algorithms. In: Bampis, E., Megow, N. (eds.) WAOA 2019. LNCS, vol. 11926, pp. 138–153. Springer, Cham (2020). https://doi.org/10.1007/978-3-030-39479-0_10
13. Goemans, M.X., Goldberg, A.V., Plotkin, S.A., Shmoys, D.B., Tardos, É., Williamson, D.P.: Improved approximation algorithms for network design problems. In: Proceedings of the Fifth Annual ACM-SIAM Symposium on Discrete Algorithms, pp. 223–232 (1994)
14. Grandoni, F., Kalaitzis, C., Zenklusen, R.: Improved approximation for tree augmentation: saving by rewiring. In: Proceedings of the 50th Annual ACM SIGACT Symposium on Theory of Computing. STOC 2018, pp. 632–645. ACM, New York (2018). https://doi.org/10.1145/3188745.3188898
15. Jain, K.: A factor 2 approximation algorithm for the generalized Steiner network problem. Combinatorica **21**(1), 39–60 (2001). https://doi.org/10.1007/s004930170004
16. Jain, K., Mahdian, M., Markakis, E., Saberi, A., Vazirani, V.V.: Greedy facility location algorithms analyzed using dual fitting with factor-revealing LP. J. ACM **50**(6), 795–824 (2003). https://doi.org/10.1145/950620.950621

17. Kortsarz, G., Nutov, Z.: A simplified 1.5-approximation algorithm for augmenting edge-connectivity of a graph from 1 to 2. ACM Trans. Algorithms **12**(2), 23 (2016). https://doi.org/10.1145/2786981
18. Kortsarz, G., Nutov, Z.: LP-relaxations for tree augmentation. Discrete Appl. Math. **239**, 94–105 (2018). https://doi.org/10.1016/j.dam.2017.12.033
19. Nutov, Z.: On the tree augmentation problem. In: 25th Annual European Symposium on Algorithms (ESA 2017), vol. 87, p. 61. Schloss Dagstuhl-Leibniz-Zentrum für Informatik (2017)
20. Sebő, A., Vygen, J.: Shorter tours by nicer ears: 7/5-approximation for the graph-TSP, 3/2 for the path version, and 4/3 for two-edge-connected subgraphs. Combinatorica **34**(5), 597–629 (2014). https://doi.org/10.1007/s00493-014-2960-3
21. The Optimization Firm: Baron (2019). https://minlp.com/baron

Faster Algorithms for Next Breakpoint and Max Value for Parametric Global Minimum Cuts

Hassene Aissi[1]([✉]), S. Thomas McCormick[2], and Maurice Queyranne[2]

[1] Paris Dauphine University, Paris, France
aissi@lamsade.dauphine.fr
[2] Sauder School of Business at the University of British Columbia,
Vancouver, Canada
{tom.mccormick,maurice.queyranne}@sauder.ubc.ca

Abstract. The parametric global minimum cut problem concerns a graph $G = (V, E)$ where the cost of each edge is an affine function of a parameter $\mu \in \mathbb{R}^d$ for some fixed dimension d. We consider the problems of finding the next breakpoint in a given direction, and finding a parameter value with maximum minimum cut value. We develop strongly polynomial algorithms for these problems that are faster than a naive application of Megiddo's parametric search technique. Our results indicate that the next breakpoint problem is easier than the max value problem.

Keywords: Parametric optimization · Global minimum cut

1 Introduction

Connectivity is a central subject in graph theory and has many practical applications in, e.g., communication and electrical networks. We consider the parametric global minimum cut problem in graphs. A *cut* X in an undirected graph $G = (V, E)$ is a non-trivial vertex subset, i.e., $\emptyset \neq X \subset V$. It cuts the set $\delta(X) = \{e \in E : e \cap X \neq \emptyset \neq e \setminus X\}$ of edges.

In the parametric global minimum cut problem, we are given an undirected graph $G = (V, E)$ where the cost $c_\mu(e)$ of each edge $e \in E$ is an affine function of a d-dimensional parameter $\mu \in \mathbb{R}^d$, i.e., $c_\mu(e) = c^0(e) + \sum_{i=1}^d \mu_i c^i(e)$, where $c^0, \ldots, c^d : E \to \mathbb{Z}$ are $d + 1$ cost functions defined on the set of edges. By not imposing a sign condition on these functions, we may handle, as in [28, Section 3.5], situations where some characteristics, measured by functions c^i, improve with μ while other deteriorate. We assume that the dimension d of the parameter space is a fixed constant. The *cost* of cut C for the edge costs c_μ is $c_\mu(C) \equiv c_\mu(\delta(C)) = \sum_{e \in \delta(C)} c_\mu(e)$. Define $M_0 = \{\mu \in \mathbb{R}^d \mid c_\mu(e) \geq 0 \text{ for all}$

H. Aissi—This research benefited from the support of the FMJH Program PGMO and from the support of EDF, Thales, Orange et Criteo.

© Springer Nature Switzerland AG 2020
D. Bienstock and G. Zambelli (Eds.): IPCO 2020, LNCS 12125, pp. 27–39, 2020.
https://doi.org/10.1007/978-3-030-45771-6_3

$e \in E\}$ a closed and convex subset of the parameter space where the parametric costs of all the edges are non-negative. Throughout the paper we consider only μ belonging to a nonempty simplex $M \subset M_0$. As usual we denote $|E|$ by m and $|V|$ by n.

For any $\mu \in M$, let C_μ^* denote a cut with a minimum cost $Z(\mu) \equiv c_\mu(C_\mu^*)$ for edge costs c_μ. Function $Z := Z(\mu)$ is a piecewise linear concave function [27]. Its graph is composed by a number of facets (linear pieces) and breakpoints (vertices). In order to avoid dealing with a trivial problem, Z is assumed to have at least one breakpoint. The maximum number of facets of the graph of Z is called the *combinatorial facet complexity* of Z. Mulmuley [23, Theorem 3.10] considers the case $d = 1$ and gives a large strongly polynomial bound on the combinatorial facet complexity of the global minimum cut problem. In [4, Theorem 4], the authors improved and extended this result to a constant dimension d and give a strongly polynomial bound $O\left(m^d n^2 \log^{d-1} n\right)$. By combining this result with several existing computational geometry algorithms, the authors give a slow algorithm for constructing function Z for general d, and a $O(mn^4 \log n + n^5 \log^2 n)$ algorithm when $d = 1$. In the particular case where cost functions $c^0, \ldots, c^d$ are nonnegative, Karger [16] gives a significantly tighter bound $O\left(n^{d+2}\right)$ on the combinatorial facet complexity and shows that function Z can be computed using a randomized algorithm in $O\left(n^{2d+2} \log n\right)$ time. These results are summarized in rows 5 and 6 of Table 1. Note that in contrast with the parametric global minimum cut problem, Carstensen [6] shows that the combinatorial facet complexity of the minimum $s - t$-cut problem (namely, to find a minimum cost cut X that separates two given vertices s and t, in the sense that $|X \cap \{s, t\}| = 1$) is exponential, even for a single parameter.

In this paper, we consider the following parametric problems:

$P_{\mathrm{NB}}(M)$ Given a simplex $M \subset \mathbb{R}^d$, a value $\mu^0 \in M \subset \mathbb{R}^d$, and a direction $\nu \in \mathbb{Z}^d$.
      Find the next breakpoint $\mu^{\mathrm{NB}} \in M$ of Z after μ^0 in direction ν, if any.
$P_{\max}(M)$ Given a simplex $M \subset \mathbb{R}^d$, find a value $\mu^* \in M$ such that $Z(\mu^*) = \max_{\mu \in M} Z(\mu)$.

In contrast to $P_{\max}(M)$, $P_{\mathrm{NB}}(M)$ is a one-dimensional parametric optimization problem as it considers the restriction of function Z to some direction $\nu \in \mathbb{Z}^d$. This problem corresponds to the *ray shooting problem* which is a standard topic in sensitivity analysis [13, Section 30.3] to identify ranges of optimality and related quantities. Given $\lambda \geqslant 0$, the cost $c_{\mu^0 + \lambda \nu}$ of each edge $e \in E$ in the direction ν and with step λ is defined by $c_{\mu^0 + \lambda \nu}(e) = c^0(e) + \sum_{i=1}^d (\mu_i^0 + \lambda \nu_i) c^i(e)$. Let $\bar{c}^0(e) = c^0(e) + \sum_{i=1}^d \mu_i^0 c^i(e)$ and $\bar{c}^1(e) = \sum_{i=1}^d \nu_i c^i(e)$. The edge costs can be rewritten as $c_{\mu^0 + \lambda \nu}(e) = \bar{c}^0(e) + \lambda \bar{c}^1(e)$. For any cut $\emptyset \neq C \subset V$, its cost for the edge costs $c_{\mu^0 + \lambda \nu}$ is a function $c_{\mu^0 + \lambda \nu}(C) = \bar{c}^0(C) + \lambda \bar{c}^1(C)$ of variable λ. For any $\mu \in M$, let $Z'(\mu, \nu)$ denote the right derivative of Z in direction ν at μ.

$P_{\max}(M)$ arises in the context of network reinforcement problem. Consider the following 2-player game of reinforcing a graph against an attacker. Given a graph $G = (V, E)$ where each edge $e \in E$ has a capacity $c^0(e)$, the Graph player wants to reinforce the capacities of the edges in E by buying $d+1$ resources subject to a budget B. The Graph player can spend $\$\mu_i \geqslant 0$ on each resource i to increase the capacities of all edges to $c_\mu(e) = c^0(e) + \sum_{i=1}^{d+1} \mu_i c^i(e)$, where all functions c^i are assumed to be non-negative. The Attacker wants to remove some edges of E in order to cut the graph into two pieces at a minimum cost. Therefore, these edges correspond to an optimal cut $\delta(C_\mu^*)$ and their removal cost is $Z(\mu)$. The Graph player wants to make it as expensive as possible for the Attacker to cut the graph, and so he wants to solve $P_{\max}(M)$. It is optimal for the Graph player to spend all the budget, and thus to spend $\mu_{d+1} = B - \sum_{i=1}^{d} \mu_i$ on resource $d+1$. Therefore, the cost of removing edge e as a function of the amounts spent on the first d resources is $c_\mu(e) = c^0(e) + \sum_{i=1}^{d} \mu_i c^i(e) + \left(B - \sum_{i=1}^{d} \mu_i\right) c^{d+1}(e) = \left(c^0(e) + B c^{d+1}(e)\right) + \sum_{i=1}^{d} \mu_i \left(c^i(e) - c^{d+1}(e)\right)$. Note that $c^i(e) - c^{d+1}(e)$ may be negative. This application illustrates how negative parametric edge costs may arise even when all original data are non-negative.

Clearly problems $P_{\max}(M)$ and $P_{\mathrm{NB}}(M)$ can be solved by constructing function Z. However, the goal of this paper is to give much faster strongly polynomial algorithms for these problems without explicitly constructing the whole function Z. This extended abstract omits most proofs; see the full version [3] for details.

1.1 Related Works

The results mentioned in this section are summarized in the first four rows of Table 1. We concentrate on *strongly* polynomial bounds here.

Table 1. New results in this paper are in red. Compare these to the non-parametric lower bounds in green, and the various upper bounds in blue.

Problem	Deterministic	Randomized
Non-param Global MC	[25, 31] $O(mn + n^2 \log n)$	[16] $O(m \log^3 n)$ ([18] $O(n^2 \log^3 n)$)
All α-approx for $\alpha < \frac{4}{3}$	[26] $O(n^4)$	[16] $O(n^2 \log n)$
Megiddo P_{NB} ($\sim d = 1$)	[31] $O(n^5 \log n)$	[32, 18] $O(n^2 \log^5 n)$
Megiddo $P_{\max}$ ($\sim$ gen'l d)	[31] $O(n^{2d+3} \log^d n)$	[32, 18] $O(n^2 \log^{4d+1} n)$
All of $Z(\mu)$ for $d = 1$	[4] $O(mn^4 \log n + n^5 \log^2 n)$	[17] $O(n^4 \log n)$
All of $Z(\mu)$ for gen'l d	[4] [big]	[17] $O(n^{2d+2} \log n)$
This paper P_{NB} ($\sim d = 1$)	[25, 31] $O(mn + n^2 \log n)$	[15] $O(n^2 \log^3 n)$
This paper $P_{\max}$ ($\sim$ gen'l d)	$O(n^4 \log^{d-1} n)$	Open

The standard (non-parametric) global minimum cut is a special case of the parametric global minimum cut, i.e., for some fixed value $\mu \in M$. Nagamochi and Ibaraki [24] and Stoer and Wagner [29] give a deterministic algorithm for this problem that runs in $O(mn + n^2 \log n)$ time. Karger and Stein [17] give a faster randomized algorithm that runs in $\tilde{O}(n^2)$ time. Karger [15] improves the running time and gives an $\tilde{O}(m)$ time algorithm.

Given $\alpha > 1$, a cut is called α-*approximate* if its cost is at most at factor of α larger than the optimal value. A remarkable property of the global minimum cut problem is that there exists a strongly polynomial number of near-optimal cuts. Karger [15] showed that the number of α-approximate cuts is $O(n^{\lfloor 2\alpha \rfloor})$. Nagamochi et al. [26] give a deterministic $O(m^2 n + mn^{2\alpha})$ time algorithm for enumerating them. For the particular case $1 < \alpha < \frac{4}{3}$, they improved this running time to $O(m^2 n + mn^2 \log n)$. Nagamochi and Ibaraki [25, Corollary 4.14] further reduced the running time to $O(n^4)$. The fastest randomized algorithm to enumerate all the near-optimal cuts, which is an $\tilde{O}(n^{\lfloor 2\alpha \rfloor})$ time algorithm by Karger and Stein [17], is faster than the best deterministic algorithm.

Megiddo's parametric searching method [19,20] is a powerful technique to solve parametric optimization problems. Megiddo's approach was originally designed to handle one-dimensional parametric problems. Cohen and Megiddo [9] extend it to fixed dimension $d > 1$, see also [2]. The crucial requirement is that the underlying non-parametric problem must have an *affine* algorithm, that is all numbers manipulated are affine functions of parameter μ. This condition is not restrictive, as many combinatorial optimization algorithms have this property; e.g., minimum spanning tree [12], matroid and polymatroid optimization [30], maximum flow [10]. The technique can be summarized as follows in the special case $d = 1$. Megiddo's approach simulates the execution of an affine algorithm $\mathcal{A}$ on an unknown target value $\bar{\mu}$ ($= \mu^{\mathrm{NB}}$ or μ^*) by considering it as a symbolic constant. During the course of execution of $\mathcal{A}$, if we need to determine the sign of some function f at $\bar{\mu}$, we compute the root r of f. The key point is that by testing if $\bar{\mu} = r$, $\bar{\mu} < r$, or $\bar{\mu} > r$, we can determine the sign of $f(\bar{\mu})$. This operation is called a *parametric test* and requires calling algorithm $\mathcal{A}$ with parameter value fixed at r.

Tokuyama [30] considers the analogue of problem $P_{\max}(M)$ for several geometric and graph problems, called the *minimax (or maximin) parametric optimization*, and gives efficient algorithms for them based on Megiddo's approach. He observes that the randomized algorithm of Karger [14] is affine. In order to improve the running time, Tokuyama implemented Megiddo's technique using the parallel algorithm of Karger and Stein [17] which solves the minimum cut problem in $\tilde{O}(\log^3 n)$ randomized parallel time using $O(\frac{n^2}{\log^2 n})$ processors. The resulting randomized algorithm for $P_{\max}(M)$ has a $O(n^2 \log^{4d+1} n)$ running time. The result was stated only for $P_{\max}(M)$ but it is easy to see that the same running time can be obtained for $P_{\mathrm{NB}}(M)$. We show in [3] that Stoer and Wagner's algorithm [29] is affine and can be combined with Megiddo's approach in order to solve $P_{\mathrm{NB}}(M)$ and $P_{\max}(M)$. This gives deterministic algorithms that run in $O(n^{2d+3} \log^d n)$ and $O(n^5 \log n)$ time for $P_{\max}(M)$ and $P_{\mathrm{NB}}(M)$ respectively.

1.2   Our Results

Our new results are summarized in rows 7 and 8 of Table 1.

The algorithms based on Megiddo's approach typically introduce a slowdown with respect to the non-parametric algorithm. For $d = 1$, these algorithms perform similar parametric tests and solve problems $P_{\mathrm{NB}}(M)$ and $P_{\max}(M)$ with the same running time. This gives the impression that these problems have the same complexity in this special case. The main contribution of the paper is to extend the techniques of Nagamochi and Ibaraki [24] and Stoer and Wagner [29] and Karger [14] to handle parametric edge costs. We give faster deterministic and randomized algorithms for problems $P_{\mathrm{NB}}(M)$ and $P_{\max}(M)$ which are not based on Megiddo's approach. We show that problem $P_{\mathrm{NB}}(M)$ can be solved with the same running time as the non-parametric global minimum cut (Theorems 1 and 2). We give for problem $P_{\max}(M)$ a much faster deterministic algorithm exploiting the key property that all near-optimal cuts can be enumerated in strongly polynomial time (Theorem 3). The algorithm builds upon a scaling technique given in [4]. The differences in how we tackle problems $P_{\mathrm{NB}}(M)$ and $P_{\max}(M)$ illustrate that $P_{\mathrm{NB}}(M)$ might be significantly easier than $P_{\max}(M)$.

Notice that our new algorithms for $P_{\mathrm{NB}}(M)$ in row 7 of Table 1 are *optimal*, in the sense that their running times match the best-known running times of the non-parametric versions of the problem (up to log factors). That is, the times quoted in row 7 of Table 1 are (nearly) the same as those in row 1 (with the exception that we do not match the Karger's speedup from $\tilde{O}(n^2)$ to $\tilde{O}(m)$ for the non-parametric randomized case).

2   Problem $P_{\mathrm{NB}}(M)$

We discuss in Sects. 2.1 and 2.2 efficient deterministic and randomized algorithms for solving problem $P_{\mathrm{NB}}(M)$ respectively. These algorithms are based on edge contractions. A preliminary step is to compute an upper bound $\bar{\lambda} > 0$ such that the next breakpoint μ^{NB} satisfies $\mu^{\mathrm{NB}} = \mu^0 + \lambda^{\mathrm{NB}}\nu$ for some $\lambda^{\mathrm{NB}} \in [0, \bar{\lambda}]$. We show that $\bar{\lambda} := \sum_{e \in E} |\bar{c}^0(e)|$ suffices, and we show also how to compute the right derivative $Z'(\mu^0, \nu)$ of Z in direction ν at μ^0.

2.1   A Deterministic Contraction Algorithm

We describe in this section a deterministic algorithm for $P_{\mathrm{NB}}(M)$ based on the concept of *pendant pair*. We call an ordered pair (u, v) of vertices in G a pendant pair for edge costs $c_\mu(e)$ for some $\mu \in M$ if $\min\{c_\mu(X) : \emptyset \subset X \subset V$ separating u *and* v$\} = c_\mu(\delta(v))$.

The algorithm proceeds in $n - 1$ phases and computes iteratively the next breakpoint μ^{NB}, if any, or claims that it does not exist. In the former case, the algorithm refines, at each iteration r, an upper bound $\bar{\lambda}^{\mathrm{NB}}$ of λ^{NB} by choosing some $\lambda^r \in [0, \bar{\lambda}]$ and merging a pendant pair (u^r, v^r) in G^r for edge costs $c_{\mu^0 + \lambda^r \nu}(e)$. The process continues until the residual graph contains only one node. All the details are summarized in Algorithm 1.

Since cuts $\delta(v)$ for all $v \in V^r$ are also cuts in G, it follows that $Z(\mu) \leqslant Z^r(\mu)$ for any $\mu \in M$. In particular, $L(\lambda) = Z(\mu^0 + \lambda\nu) \leqslant Z^r(\mu^0 + \lambda\nu)$ for any $\lambda \in [0, \lambda^{\mathrm{NB}}]$. By the definition of λ^r, this implies that

$$\lambda^{\mathrm{NB}} \leqslant \lambda^{\mathrm{r}} \text{ and } \mathrm{L}(\lambda^{\mathrm{r}}) \leqslant \mathrm{Z}^{\mathrm{r}}(\mu^0 + \lambda^{\mathrm{r}}\nu). \tag{1}$$

Computing the lower envelope of $O(n)$ linear functions and getting function Z^r takes $O(n \log n)$ time [5]. Therefore, the running time of an iteration r of Algorithm 1 is dominated by the time of computing a pendant pair in $O(m + n \log n)$ time [29]. The added running time of the $n - 1$ iterations of the while loop takes $O(mn + n^2 \log n)$. Note that this corresponds to the same running time of computing a non-parametric minimum cut [29]. Since the test performed in Step 11 requires the computation of a minimum cut in graph G with edge costs $c_{\mu^0 + \bar{\lambda}\nu}(e)$, it follows that the overall running time of Algorithm 1 is $O(mn + n^2 \log n)$. The following result summarizes the running time of our contraction algorithm.

Theorem 1. *Algorithm 1 solves $P_{NB}(M)$ in $O(mn + n^2 \log n)$ time.*

Algorithm 1. Deterministic Parametric Edge Contraction for $P_{NB}(M)$

Require: graph $G = (V, E)$, edge costs $c^0, \dots, c^d$, a direction ν, an upper bound $\bar{\lambda}$, the optimal value $Z(\mu^0)$, and the slope $Z'(\mu^0, \nu)$
Ensure: next breakpoint μ^{NB} if any
 1: let $E^0 \leftarrow E$, $V^0 \leftarrow V$, $G^0 \leftarrow G$, $r \leftarrow 0$, $\bar{\lambda}^{NB} \leftarrow \bar{\lambda}$
 2: **while** $|V^r| > 1$ **do**
 3:     define functions $L(\lambda) := Z(\mu^0) + \lambda Z'(\mu^0, \nu)$, $Z^r(\mu) := \min_{v \in V^r} c_\mu(\delta(v))$, compute, if any, $\hat{\lambda}^r := \min\{\lambda > 0 : Z^r(\mu^0 + \lambda \nu) \leqslant L(\lambda)\}$, and let $\lambda^r := \begin{cases} \min\{\bar{\lambda}, \hat{\lambda}^r\} & \text{if } \hat{\lambda}^r \text{ exists} \\ \bar{\lambda} & \text{otherwise} \end{cases}$
 4:     **if** $\lambda^r < \bar{\lambda}^{NB}$ **then**
 5:         set $\bar{\lambda}^{NB} \leftarrow \lambda^r$
 6:     **end if**
 7:     compute a pendant pair (u^r, v^r) in G^r for edge costs $c_{\mu^0 + \lambda^r \nu}(e)$ using the algorithm given in [29]
 8:     merge nodes u^r and v^r and remove self-loops
 9:     set $r \leftarrow r + 1$ and let $G^r = (V^r, E^r)$ denote the resulting graph
10: **end while**
11: **if** $L(\bar{\lambda}) > \min_C \{c_{\mu^0 + \bar{\lambda}\nu}(C) : \emptyset \neq C \subset V\}$ **then**
12:     return $\mu^{NB} = \mu^0 + \bar{\lambda}^{NB} \nu$
13: **else**
14:     the next breakpoint does not exist
15: **end if**

2.2 A Randomized Contraction Algorithm

The algorithm performs a number of random edge contractions and iteratively solves the next breakpoint problem. At each iteration r, the algorithm chooses some $\tilde{\mu}^r \in M$ and randomly selects an edge $e \in E^r$ with probability $\frac{c_{\tilde{\mu}^r}(e)}{c_{\tilde{\mu}^r}(E_r)}$ to be contracted. The point $\tilde{\mu}^r$ is defined as the intersection of functions $L(\lambda) := Z(\mu^0) + \lambda Z'(\mu^0, \nu)$ and $UB^r(\lambda) := \frac{1}{|V_r|} c_{\mu^0 + \lambda \nu}(E_r)$ and may vary from one iteration to the next. The choice of the appropriate value of $\tilde{\mu}^r$ is crucial to ensure the high success probability of solving the problem, and is the main contribution of this algorithm. The random edges contraction sequence continues until obtaining a graph G' with two nodes. If the next breakpoint μ^{NB} exists, then the algorithm returns it after computing an optimal cut $C^*_{\mu^{\mathrm{NB}}}$ for edge costs $c_{\mu^{\mathrm{NB}}}(e)$ defining the right derivative $Z'(\mu^{\mathrm{NB}}, \nu)$ of Z in direction ν at μ^{NB}. Otherwise, the algorithm claims that it does not exist. All the details are summarized in Algorithm 2.

Algorithm 2. Randomized Parametric Edge Contraction for $P_{\mathrm{NB}}(M)$

Require: graph $G = (V, E)$, edge costs $c^0, \ldots, c^d$, a direction ν, an upper bound $\bar{\lambda}$, the optimal value $Z(\mu^0)$, and the slope $Z'(\mu^0, \nu)$
Ensure: next breakpoint μ^{NB} if any
1: let $E^0 \leftarrow E$, $V^0 \leftarrow V$, $G^0 \leftarrow G$, $r \leftarrow 0$
2: **while** $|V^r| > 2$ **do**
3:     compute the intersection point λ^r of functions $L(\lambda) := Z(\mu^0) + \lambda Z'(\mu^0, \nu)$ and $UB^r(\lambda) := \frac{1}{|V^r|} \sum_{v \in V^r} c_{\mu^0 + \lambda \nu}(\delta(\{v\}))$
4:     **if** $\lambda^r \in [0, \bar{\lambda}]$ **then**
5:         set $\tilde{\mu}^r = \mu^0 + \lambda^r \nu$
6:     **else**
7:         set $\tilde{\mu}^r = \mu^0 + \bar{\lambda} \nu$
8:     **end if**
9:     choose an arbitrary edge $e \in E_r$ with probability $\frac{c_{\tilde{\mu}^r}(e)}{c_{\tilde{\mu}^r}(E^r)}$
10:    $r \leftarrow r + 1$
11:    contract e by merging all its vertices and removing self-loops
12:    let $G^r = (V^r, E^r)$ denote the resulting graph
13: **end while**
14: let C denote the unique cut in the final graph G' and define $\mu^0 + \bar{\lambda}^{\mathrm{NB}} \nu$ as the intersection value of functions $L(\lambda)$ and $c_{\mu^0 + \lambda \nu}(C)$
15: **if** $\bar{\lambda}^{\mathrm{NB}} > 0$ **then**
16:    return $\mu^{\mathrm{NB}} = \mu^0 + \bar{\lambda}^{\mathrm{NB}} \nu$
17: **else**
18:    the next breakpoint does not exist
19: **end if**

We say that an edge e in G_r *survives* at the current contraction if it is not chosen to be contracted. An edge $e \in G$ survives at the end of iteration r if it survives all the r edge contractions. A cut C *survives* at the end of iteration r if every edge $e \in \delta(C)$ has survived. We show that a fixed optimal cut $C^*_{\mu^{\mathrm{NB}}}$ is returned by Algorithm 2 with probability at least $\binom{n}{2}^{-1}$. This error probability is the same as for the original (non-parametric) contraction algorithm [14,17].

Theorem 2. *Algorithm 2 solves $P_{\mathrm{NB}}(M)$ with high probability in $O(n^2 \log^3 n)$ time.*

3   Problem $P_{\max}(M)$

Our $P_{\max}(M)$ algorithm uses the following geometric tools [22]: an *arrangement* $A(\mathcal{H})$, formed by a set $\mathcal{H}$ of hyperplanes in $\mathbb{R}^d$, corresponds to a division of $\mathbb{R}^d$ into $O(|\mathcal{H}|^d)$ d-dimensional convex regions called *cells*. Given a simplex P in $\mathbb{R}^d$, let $A(\mathcal{H}) \cap P$ denote the restriction of the arrangement $A(\mathcal{H})$ to P. We use a version of *point location in arrangements* (PLA) in our algorithm [9, 30].

$P_{\mathrm{reg}}(\mathcal{H}, P, \bar{\mu})$ Given a simplex P, a set $\mathcal{H}$ of hyperplanes in $\mathbb{R}^d$, and a target value $\bar{\mu} \in \mathbb{R}^d$, locate a d-dimensional simplex $R \subseteq A(\mathcal{H}) \cap P$ containing $\bar{\mu}$.

Fix a constant $1 < \varepsilon < \sqrt{\frac{4}{3}}$ and let $\beta = \frac{\varepsilon^2 - 1}{m} > 0$. Compute $p = 1 + \lceil \log \frac{m^2}{\varepsilon^2 - 1} / \log \varepsilon^2 \rceil$ so that $\beta \varepsilon^{2(p-1)} > m$, and observe that $p = O(\log n)$. For a given edge $\bar{e} \in E$, define the $p+2$ affine functions $g_i : \mathbb{R}^d \to \mathbb{R}$ by $g_0(\bar{e}, \mu) = 0$, $g_i(\bar{e}, \mu) = \beta \varepsilon^{2(i-1)} c_\mu(\bar{e})$ for $i = 1, \ldots, p$, and $g_{p+1}(\bar{e}, \mu) = +\infty$.

Algorithm 3. Deterministic algorithm for $P_{\max}(M)$

Require: graph $G = (V, E)$, edge costs $c^0, \ldots, c^d$, and $1 < \varepsilon < \sqrt{\frac{4}{3}}$
Ensure: the optimal value μ^*
1: let $A(\mathcal{H}_1)$ denote the arrangement formed by the set $\mathcal{H}_1$ of hyperplanes $H_{e,e'} = \{\mu \in \mathbb{R}^d : c_\mu(e) = c_\mu(e')\}$ for any pair of edges $e, e' \in E$
2: solve $P_{\mathrm{reg}}(\mathcal{H}_1, M, \mu^*)$ and compute a d-dimensional simplex $R_1 \subseteq A(\mathcal{H}_1) \cap M$ containing μ^*
3: choose arbitrarily μ^1 in the interior of R_1, compute a maximum spanning tree T of G for edge costs $c_{\mu^1}(e)$, and let $\bar{e}$ be an edge in T such that $c_{\mu^1}(\bar{e}) = \arg\min_{e \in T} c_{\mu^1}(e)$
4: let $\pi(e)$ denote the rank of edge $e \in E$ according to the increasing edges costs order in R_1 (ties are broken arbitrary)
5: **if** $\min_{\mu \in R_1} c_\mu(\bar{e}) = 0$ **then**
6:    let $\tilde{e}$ be an edge such that $\pi(\tilde{e}) \in \arg\min_{e \in E}\{\pi(e) : c_e(\mu) > 0$ for all $\mu \in R_1\}$ and $R_1' = \{\mu \in R_1 : g_p(\bar{e}, \mu) \geqslant c_\mu(\tilde{e})\}$
7:    set $R_1 \longleftarrow R_1'$
8: **end if**
9: let $A(\mathcal{H}_2)$ denote the arrangement formed by the set $\mathcal{H}_2$ of hyperplanes $H_i(e) = \{\mu \in \mathbb{R}^d : c_\mu(e) = g_i(\bar{e}, \mu)\}$ for any edges $e \in E$ and for $i = 1, \ldots, p$
10: solve $P_{\mathrm{reg}}(\mathcal{H}_2, R_1, \mu^*)$ and compute a d dimensional simplex $R_2 \subseteq A(\mathcal{H}_2) \cap R_1$ containing μ^*
11: choose arbitrarily $\mu^2 \in R_2$ and compute the set $\mathcal{C}$ of all the ε-approximate cuts for edge costs $c_{\mu^2}(e)$
12: let $A(\mathcal{H}_3)$ denote the arrangement formed by the set $\mathcal{H}_3$ of hyperplanes $H_{C,C'} = \{\mu \in \mathbb{R}^d : c_\mu(C) = c_\mu(C')\}$ for any pair of cuts $C, C' \in \mathcal{C}$
13: solve $P_{\mathrm{reg}}(\mathcal{H}_3, R_2, \mu^*)$ and return μ^*

In order to overcome the difficulty that edges costs $c_\mu(e)$ may have different total orders for different values of μ, Algorithm 3 restricts the parametric search to a d-dimensional simplex R_1 containing μ^* where $c_\mu(e)$ have the same total order for any $\mu \in R_1$. Let $\pi(e)$ denote the rank of edge $e \in E$ according to the increasing edges costs order in R_1. The algorithm needs to divide R_1 into smaller regions using as in Mulmuley [23] the relationship between cuts and spanning trees. However, the proof of Mulmuley's result is complicated and yields a large number of regions. Consider an arbitrary μ^1 in the interior of R_1 and compute a maximum spanning tree T for costs $c_{\mu^1}(e)$. Let $\bar{e}$ denote an edge in T such that $c_{\mu^1}(\bar{e}) = \arg\min_{e \in T} c_{\mu^1}(e)$. Since functions $c_\mu(e)$ may intersect only at the boundaries of R_1, for any edge $e \in T \setminus \{\bar{e}\}$ exactly one of the following cases occurs: i) $c_{\mu^1}(e) = c_{\mu^1}(\bar{e})$, and therefore $c_\mu(e) = c_\mu(\bar{e})$ for all $\mu \in R_1$, or ii) $c_{\mu^1}(e) > c_{\mu^1}(\bar{e})$, and therefore $c_\mu(e) \geqslant c_\mu(\bar{e})$ for all $\mu \in R_1$. In either cases, edge $\bar{e}$ satisfies

$$c_\mu(\bar{e}) = \min_{e \in T} c_\mu(e) \text{ for all } \mu \in R_1. \tag{2}$$

Since every cut in G intersects T in at least one edge, by (2) we have the following lower bound on the minimum cut value.

$$c_\mu(\bar{e}) \leqslant Z(\mu) \text{ for all } \mu \in R_1. \tag{3}$$

Let $\bar{C}$ denote the cut formed by deleting $\bar{e}$ from T. By the cut optimality condition, we obtain the following upper bound on the minimum cut value.

$$Z(\mu) \leqslant c_\mu(\bar{C}) = \sum_{e \in \delta(\bar{C})} c_\mu(e) \leqslant m c_\mu(\bar{e}) < g_p(\bar{e}, \mu), \tag{4}$$

where the last inequality follows from the definition of function $g_p(\bar{e}, \mu)$. Let $A(\mathcal{H}_2)$ denote the arrangement formed by the set $\mathcal{H}_2$ of hyperplanes $H_i(e) = \{\mu \in \mathbb{R}^d : c_\mu(e) = g_i(\bar{e}, \mu)\}$ for any edges $e \in E$ and for $i = 1, \ldots, p$. Suppose first that $c_\mu(\bar{e}) > 0$ for all $\mu \in R_1$, then by (3) we have $Z(\mu) > 0$ for all $\mu \in R_1$. In this case, we may apply the technique given in [4, Theorem 4] to compute all the optimal cuts for the unknown edge costs $c_{\mu^*}(e)$. Consider a d-dimensional simplex $R_2 \subseteq A(\mathcal{H}_2) \cap R_1$ containing μ^* and any optimal cut $C_{\mu^*}^*$ for edge costs $c_{\mu^*}(e)$. Since functions $g_p(\bar{e}, \mu)$ and $c_\mu(e)$, for all $e \in E$, may intersect only at the boundaries of R_2, it follows by (4) that $c_\mu(e) \leqslant g_p(\bar{e}, \mu)$ for all $e \in \delta(C_{\mu^*}^*)$ and all $\mu \in R_2$. By construction of the arrangement $A(\mathcal{H}_2) \cap R_1$, for every edge e in $\delta(C_{\mu^*}^*)$ there exists some $q \in \{0, \ldots, p\}$ such that

$$g_q(\bar{e}, \mu) \leqslant c_\mu(e) \leqslant g_{q+1}(\bar{e}, \mu) \text{ for all } \mu \in R_2. \tag{5}$$

The following result shows that not all functions $c_\mu(e)$ of the edges in $\delta(C_\mu^*)$ are below function $g_1(\bar{e}, \mu)$ for all $\mu \in R_2$.

Lemma 1. *For any $\bar{\mu} \in R_2$ and any optimal cut $C_{\bar{\mu}}^*$ for edge costs $c_{\bar{\mu}}(e)$, there exists at least an edge $e \in \delta(C_{\bar{\mu}}^*)$ satisfying $c_\mu(e) \geqslant g_1(\bar{e}, \mu)$ for all $\mu \in R_2$.*

By (5) and Lemma 1, one can use the same arguments as in [4, Theorem 4] and get the following result.

Lemma 2. *If $Z(\mu) > 0$ for all $\mu \in R_1$, then any specific optimal cut $C_{\mu^*}^*$ for edge costs $c_{\mu^*}(e)$ is an ε-approximate cut for edge costs $c_\mu(e)$ for every $\mu \in R_2$.*

The optimal value μ^* is defined by the intersection of parametric functions $c_\mu(C)$ of at least $d + 1$ optimal cuts C for edge costs $c_{\mu^*}(e)$. If the condition of Lemma 2 holds, the enumeration of these solutions can be done by picking some μ^2 in a simplex $R_2 \subseteq A(\mathcal{H}_2) \cap R_1$ containing μ^* and computing the set $\mathcal{C}$ of all the ε-approximate cuts for edge costs $c_{\mu^2}(e)$. Note that this set is formed by $O(n^2)$ cuts [26]. Naturally, μ^* can be obtained by computing the lower envelope of the parametric functions $c_\mu(C)$ for all the cuts $C \in \mathcal{C}$. However, this will take an excessive $O(n^{2d}\alpha(n))$ running time [11], where $\alpha(n)$ is the inverse of Ackermann's function. Instead, observe that μ^* is a vertex of at least $d + 1$ cells of the arrangement $A(\mathcal{H}_3) \cap R_2$ formed by the set $\mathcal{H}_3$ of hyperplanes $H_{C,C'} = \{\mu \in \mathbb{R}^d : c_\mu(C) = c_\mu(C')\}$ for any pair of cuts $C, C' \in \mathcal{C}$. Therefore, μ^* is a vertex of any simplex containing it and included in a cell of the arrangement $A(\mathcal{H}_3) \cap R_2$. By solving PLA problem in $A(\mathcal{H}_3) \cap R_2$, μ^* can be computed more efficiently.

In order to complete the algorithm, we need to handle the case where $\min_{\mu \in R_1} c_\mu(\bar{e}) = 0$. It is sufficient to consider in this case a restriction $R_1' \subset R_1$ containing μ^* such that $\min_{\mu \in R_1'} c_\mu(\bar{e}) > 0$. The following results show how to construct such a restriction.

Lemma 3. *There exists at least an edge $\hat{e} \in E$ such that $c_\mu(\hat{e}) > 0$ for all $\mu \in R_1$.*

Let $\tilde{e}$ be an edge such that $\pi(\tilde{e}) \in \arg\min_{e \in E}\{\pi(e) : c_\mu(e) > 0 \text{ for all } \mu \in R_1\}$ and $R_1' = \{\mu \in R_1 : g_p(\bar{e}, \mu) \geqslant c_\mu(\tilde{e})\}$. By Lemma 3, edge $\tilde{e}$ exists and we have $g_p(\bar{e}, \mu) = \beta \varepsilon^{2(p-1)} c_\mu(\bar{e}) > 0$ for all $\mu \in R_1'$. This shows that $c_\mu(\bar{e}) > 0$ for all $\mu \in R_1'$ and thus, the condition of Lemma 2 holds in R_1'. It remains now to show that $\mu^* \notin R_1 \setminus R_1'$.

Lemma 4. *Function Z has no breakpoint in $R_1 \setminus R_1'$.*

Let $T(d)$ denote the running time of Algorithm 3 for solving $P_{\max}(M)$ with d parameters and $T(0) = O(n^2 \log n + nm)$ denote the running time of computing a minimum (non-parametric) global cut using the algorithm given in [29]. The input of a call to problem PLA requires $O(n^4)$ hyperplanes. Therefore, by Lemma 5 given in the appendix, the $\Theta(1)$ calls to problem PLA can be solved recursively in $O(\log(n)T(d-1) + n^4)$ time. The enumeration of all the $O(n^2)$ approximate cuts can be done in $O(n^4)$ time [25, Corollary 4.14]. Therefore, the running time of Algorithm 3 is given by the following recursive formula.

$$T(d) = O(\log(n)T(d-1) + n^4) = O(\log^d(n)T(0) + \log^{d-1}(n)n^4) = O(\log^{d-1}(n)n^4).$$

Theorem 3. *Algorithm 3 solves $P_{\max}(M)$ in $O(\log^{d-1}(n)n^4)$ time.*

4   Conclusion

As shown in Table 1, our improved algorithms are significantly faster than what one could otherwise get from just using Megiddo's algorithm [20,21]. As mentioned in Sect. 1.2, our results for $P_{\mathrm{NB}}(M)$ are close to being the best possible, as they are only log factors slower than the best known non-parametric algorithms. One exception is that we don't quite match Karger's [15] speedup to near-linear time for the randomized case, and we leave this as an open problem.

Our deterministic algorithm for $P_{\max}(M)$ is also close to best possible, though in a weaker sense. It uses the ability to compute all α-optimal cuts in $O(n^4)$ time, and otherwise is only log factors slower than $O(n^4)$. The conspicuous open problem here is to find a faster randomized algorithm for $P_{\max}(M)$ when $d > 1$.

5   Appendix

5.1   Geometric tools

A classical problem in computational geometry called *point location in arrangements* (PLA) is useful to our algorithm. PLA has been widely used in various contexts such as linear programming [1,18] or parametric optimization [9,30]. For more details, see [22, Chapter 5].

Given a simplex P, an arrangement $A(\mathcal{H})$ formed by a set $\mathcal{H}$ of hyperplanes in $\mathbb{R}^d$, let $A(\mathcal{H}) \cap P$ denote the restriction of the arrangement $A(\mathcal{H})$ to P. The goal of PLA is to construct a data structure in order to quickly locate a cell of $A(\mathcal{H}) \cap P$ containing an unknown target value $\bar{\mu}$. Solving PLA requires the explicit construction of the arrangement $A(\mathcal{H})$ which can be done in an excessive $O(|\mathcal{H}|^d)$ running time [22, Theorem 6.1.2]. For our purposes, it is sufficient to solve the following simpler form of PLA.

$P_{\mathrm{reg}}(\mathcal{H}, P, \bar{\mu})$ Given a simplex P, a set $\mathcal{H}$ of hyperplanes in $\mathbb{R}^d$, and a target value $\bar{\mu}$, locate a d-dimensional simplex $R \subseteq A(\mathcal{H}) \cap P$ containing a target and unknown value $\bar{\mu}$.

Cohen and Megiddo [9] consider the problem $Max(f)$ of maximizing a concave function $f : \mathbb{R}^d \to \mathbb{R}$ with fixed dimension d and give, under some conditions, a polynomial time algorithm. This algorithm also uses problem $P_{\mathrm{reg}}(\mathcal{H}, P, \bar{\mu})$ as a subroutine, where in this context the target value $\bar{\mu}$ is the optimal value of $Max(f)$. Let $T(d)$ denote the time required to solve $Max(f)$ with d parameters and $T(0)$ denote the running time of evaluating f at any value in $\mathbb{R}^d$. The authors solve $P_{\mathrm{reg}}(\mathcal{H}, P, \bar{\mu})$ recursively using multidimensional parametric search technique. See also [7,8,30].

Lemma 5. *Given a simplex P, a set $\mathcal{H}$ of hyperplanes in $\mathbb{R}^d$, and a target and unknown value $\bar{\mu}$, $P_{\mathrm{reg}}(\mathcal{H}, P, \bar{\mu})$ can be solved in $O(\log(|\mathcal{H}|)T(d-1) + |\mathcal{H}|)$ time.*

References

1. Agarwal, P.K., Sharir, M., Toledo, S.: An efficient multi-dimensional searching technique and its applications. Technical report CS-1993-20, Department of Computer Science, Duke University (1993)
2. Agarwal, P.K., Sharir, M.: Efficient algorithms for geometric optimization. ACM Comput. Surv. (CSUR) **30**(4), 412–458 (1998)
3. Aissi, H., McCormick, S.T., Queyranne, M.: Faster algorithms for next breakpoint and max value for parametric global minimum cuts. arXiv:1911.11847 (2019)
4. Aissi, H., Mahjoub, A.R., McCormick, S.T., Queyranne, M.: Strongly polynomial bounds for multiobjective and parametric global minimum cuts in graphs and hypergraphs. Math. Program. **154**(1–2), 3–28 (2015). https://doi.org/10.1007/s10107-015-0944-8
5. Boissonnat, J.D., Yvinec, M.: Algorithmic Geometry. Cambridge University Press, Cambridge (1998)
6. Carstensen, P.J.: Complexity of some parametric integer and network programming problems. Math. Program. **26**(1), 64–75 (1983). https://doi.org/10.1007/BF02591893
7. Chazelle, B., Friedman, J.: A deterministic view of random sampling and its use in geometry. Combinatorica **10**(3), 229–249 (1990). https://doi.org/10.1007/BF02122778
8. Clarkson, K.L.: New applications of random sampling in computational geometry. Discret. Comput. Geom. **2**(2), 195–222 (1987). https://doi.org/10.1007/BF02187879
9. Cohen, E., Megiddo, N.: Maximizing concave functions in fixed dimensions. In: Pardalos, P.M. (ed.) Complexity in Numerical Optimization, pp. 74–87. World Scientific Publishing, Singapore (1993)
10. Cohen, E., Megiddo, N.: Algorithms and complexity analysis for some flow problems. Algorithmica **11**(3), 320–340 (1994). https://doi.org/10.1007/BF01240739
11. Edelsbrunner, H., Herbert, H., Guibas, L.J., Sharir, M.: The upper envelope of piecewise linear functions: algorithms and applications. Discret. Comput. Geom. **4**(1), 311–336 (1989)
12. Fernández-Baca, D.: Multi-parameter minimum spanning trees. In: Gonnet, G.H., Viola, A. (eds.) LATIN 2000. LNCS, vol. 1776, pp. 217–226. Springer, Heidelberg (2000). https://doi.org/10.1007/10719839_22
13. Fernández-Baca, D., Venkatachalam, B.: Sensitivity analysis in combinatorial optimization. In: Gonzalez, T. (ed.) Handbook of Approximation Algorithms and Metaheuristics. Chapman and Hall/CRC Press, Boca Raton (2007)
14. Karger, D.R.: Global min-cuts in RNC, and other ramifications of a simple min-cut algorithm. In: Proceedings of the Fourth Annual ACM-SIAM Symposium on Discrete Algorithms, pp. 21–30 (1993)
15. Karger, D.R.: Minimum cuts in near-linear time. J. ACM **47**(1), 46–76 (2000)
16. Karger, D.R.: Enumerating parametric global minimum cuts by random interleaving. In: Proceedings of the Forty-Eight Annual ACM Symposium on Theory of Computing, pp. 542–555 (2016)
17. Karger, D.R., Stein, C.: A new approach to the minimum cut problem. J. ACM **43**(4), 601–640 (1996)
18. Matoušek, J., Schwarzkopf, O.: Linear optimization queries. In: Proceedings of the Eighth ACM Symposium on Computational Geometry, pp. 16–25 (1992)

19. Megiddo, N.: Combinatorial optimization with rational objective functions. Math. Oper. Res. **4**(4), 414–424 (1979)
20. Megiddo, N.: Applying parallel computation algorithms in the design of serial algorithms. J. ACM **30**, 852–865 (1983)
21. Megiddo, N.: Linear programming in linear time when the dimension is fixed. J. ACM **31**, 114–127 (1984)
22. Mulmuley, K.: Computational Geometry: An Introduction Through Randomized Algorithms. Prentice-Hall, Upper Saddle River (1994)
23. Mulmuley, K.: Lower bounds in a parallel model without bit operations. SIAM J. Comput. **28**(4), 1460–1509 (1999)
24. Nagamochi, H., Ibaraki, T.: Computing edge-connectivity in multigraphs and capacitated graphs. SIAM J. Discret. Math. **5**(1), 54–66 (1992)
25. Nagamochi, H., Ibaraki, T.: Algorithmic Aspects of Graph Connectivity. Cambridge University Press, Cambridge (2008)
26. Nagamochi, H., Nishimura, K., Ibaraki, T.: Computing all small cuts in undirected networks. SIAM J. Discret. Math. **10**, 469–481 (1997)
27. Nemhauser, G.L., Wolsey, L.A.: Integer and Combinatorial Optimization. Wiley, Hoboken (1999)
28. Radzik, T.: Parametric flows, weighted means of cuts, and fractional combinatorial optimization. In: Pardalos, P. (ed.) Complexity in Numerical Optimization, pp. 351–386. World Scientific Publishing, Singapore (1993)
29. Stoer, M., Wagner, F.: A simple min-cut algorithm. J. ACM **44**(4), 585–591 (1997)
30. Tokuyama, T.: Minimax parametric optimization problems and multi-dimensional parametric searching. In: Proceedings of the Thirty-Third Annual ACM Symposium on Theory of Computing, pp. 75–83 (2001)

Optimizing Sparsity over Lattices and Semigroups

Iskander Aliev[1], Gennadiy Averkov[2(✉)], Jesús A. De Loera[3], and Timm Oertel[1]

[1] Cardiff University, Cardiff, UK
[2] Brandenburg University of Technology Cottbus-Senftenberg, Senftenberg, Germany
averkov@b-tu.de
[3] University of California, Davis, USA

Abstract. Motivated by problems in optimization we study the *sparsity* of the solutions to systems of linear Diophantine equations and linear integer programs, i.e., the number of non-zero entries of a solution, which is often referred to as the ℓ_0-norm. Our main results are improved bounds on the ℓ_0-norm of sparse solutions to systems $A\boldsymbol{x} = \boldsymbol{b}$, where $A \in \mathbb{Z}^{m \times n}$, $\boldsymbol{b} \in \mathbb{Z}^m$ and $\boldsymbol{x}$ is either a general integer vector (lattice case) or a non-negative integer vector (semigroup case). In the lattice case and certain scenarios of the semigroup case, we give polynomial time algorithms for computing solutions with ℓ_0-norm satisfying the obtained bounds.

1 Introduction

This paper discusses the problem of finding sparse solutions to systems of linear Diophantine equations and integer linear programs. We investigate the ℓ_0-norm $\|\boldsymbol{x}\|_0 := |\{i : x_i \neq 0\}|$, a function widely used in the theory of *compressed sensing* [6,9], which measures the sparsity of a given vector $\boldsymbol{x} = (x_1, \ldots, x_n)^\top \in \mathbb{R}^n$ (it is clear that the ℓ_0-norm is actually not a norm).

Sparsity is a topic of interest in several areas of optimization. The ℓ_0-norm minimization problem over reals is central in the theory of the classical compressed sensing, where a linear programming relaxation provides a guaranteed approximation [8,9]. Support minimization for solutions to Diophantine equations is relevant for the theory of compressed sensing for discrete-valued signals [11,12,17]. There is still little understanding of discrete signals in the compressed sensing paradigm, despite the fact that there are many applications in which the signal is known to have discrete-valued entries, for instance, in wireless communication [22] and the theory of error-correcting codes [7]. Sparsity was also investigated in integer optimization [1,10,20], where many combinatorial optimization problems have useful interpretations as sparse semigroup problems. For example, the edge-coloring problem can be seen as a problem in the semigroup generated by matchings of the graph [18]. Our results provide natural out-of-the-box sparsity bounds for problems with linear constraints and integer variables in a general form.

© Springer Nature Switzerland AG 2020
D. Bienstock and G. Zambelli (Eds.): IPCO 2020, LNCS 12125, pp. 40–51, 2020.
https://doi.org/10.1007/978-3-030-45771-6_4

1.1   Lattices: Sparse Solutions of Linear Diophantine Systems

Each integer matrix $A \in \mathbb{Z}^{m \times n}$ determines the lattice $\mathcal{L}(A) := \{Ax : x \in \mathbb{Z}^n\}$ generated by the columns of A. By an easy reduction via row transformations, we may assume without loss of generality that the rank of A is m.

Let $[n] := \{1, \ldots, n\}$ and let $\binom{[n]}{m}$ be the set of all m-element subsets of $[n]$. For $\gamma \subseteq [n]$, consider the $m \times |\gamma|$ submatrix A_γ of A with columns indexed by γ. One can easily prove that the determinant of $\mathcal{L}(A)$ is equal to

$$\gcd(A) := \gcd \left\{ \det(A_\gamma) : \gamma \in \binom{[n]}{m} \right\}.$$

Since $\mathcal{L}(A_\gamma)$ is the lattice spanned by the columns of A indexed by γ, it is a sublattice of $\mathcal{L}(A)$. We first deal with a natural question: *Can the description of a given lattice $\mathcal{L}(A)$ in terms of A be made sparser by passing from A to A_γ with γ having a smaller cardinality than n and satisfying $\mathcal{L}(A) = \mathcal{L}(A_\gamma)$?* That is, we want to discard some of the columns of A and generate $\mathcal{L}(A)$ by $|\gamma|$ columns with $|\gamma|$ being possibly small.

For stating our results, we need several number-theoretic functions. Given $z \in \mathbb{Z}_{>0}$, consider the prime factorization $z = p_1^{s_1} \cdots p_k^{s_k}$ with pairwise distinct prime factors $p_1, \ldots, p_k$ and their multiplicities $s_1, \ldots, s_k \in \mathbb{Z}_{>0}$. Then the number of prime factors $\sum_{i=1}^{k} s_i$ counting the multiplicities is denoted by $\Omega(z)$. Furthermore, we introduce $\Omega_m(z) := \sum_{i=1}^{k} \min\{s_i, m\}$. That is, by introducing m we set a threshold to account for multiplicities. In the case $m = 1$ we thus have $\omega(z) := \Omega_1(z) = k$, which is the number of prime factors in z, not taking the multiplicities into account. The functions Ω and ω are called *prime Ω-function* and *prime ω-function*, respectively, in number theory [15]. We call Ω_m the *truncated prime Ω-function*.

Theorem 1. *Let $A \in \mathbb{Z}^{m \times n}$, with $m \leq n$, and let $\tau \in \binom{[n]}{m}$ be such that the matrix A_τ is non-singular. Then the equality $\mathcal{L}(A) = \mathcal{L}(A_\gamma)$ holds for some γ satisfying $\tau \subseteq \gamma \subseteq [n]$ and*

$$|\gamma| \leq m + \Omega_m \left(\frac{|\det(A_\tau)|}{\gcd(A)} \right). \tag{1}$$

Given A and τ, the set γ can be computed in polynomial time.

One can easily see that $\omega(z) \leq \Omega_m(z) \leq \Omega(z) \leq \log_2(z)$ for every $z \in \mathbb{Z}_{>0}$. The estimate using $\log_2(z)$ gives a first impression on the quality of the bound (1). It turns out, however, that $\Omega_m(z)$ is much smaller on the average. Results in number theory [15, §22.10] show that the average values $\frac{1}{z}(\omega(1) + \cdots + \omega(z))$ and $\frac{1}{z}(\Omega(1) + \cdots + \Omega(z))$ are of order $\log \log z$, as $z \to \infty$.

As an immediate consequence of Theorem 1 we obtain

Corollary 2. *Consider the linear Diophantine system*

$$Ax = b, \ x \in \mathbb{Z}^n \tag{2}$$

with $A \in \mathbb{Z}^{m \times n}$, $b \in \mathbb{Z}^m$ and $m \leq n$. Let $\tau \in \binom{[n]}{m}$ be such that the $m \times m$ matrix A_τ is non-singular. If (2) is feasible, then (2) has a solution x satisfying the sparsity bound

$$\|x\|_0 \leq m + \Omega_m \left(\frac{|\det(A_\tau)|}{\gcd(A)} \right).$$

Under the above assumptions, for given A, b and τ, such a sparse solution can be computed in polynomial time.

From the optimization perspective, Corollary 2 deals with the problem

$$\min \{\|x\|_0 \, : \, Ax = b, \, x \in \mathbb{Z}^n\}$$

of minimization of the ℓ_0-norm over the affine lattice $\{x \in \mathbb{Z}^n \, : \, Ax = b\}$.

1.2   Semigroups: Sparse Solutions in Integer Programming

Consider next the standard form of the feasibility constraints of integer linear programming

$$Ax = b, \, x \in \mathbb{Z}^n_{\geq 0}. \tag{3}$$

For a given matrix A, the set of all b such that (3) is feasible, is the *semigroup* $Sg(A) = \{Ax : x \in \mathbb{Z}^n_{\geq 0}\}$ generated by the columns of A.

If (3) has a solution, i.e., $b \in Sg(A)$, *how sparse can such a solution be?* In other words, we are interested in the ℓ_0-norm minimization problem

$$\min \{\|x\|_0 \, : \, Ax = b, \, x \in \mathbb{Z}^n_{\geq 0}\}. \tag{4}$$

It is clear that Problem (4) is NP-hard, because deciding the feasibility of (3) [23, §18.2] or even solving the relaxation of (4) with the condition $x \in \mathbb{Z}^n_{\geq 0}$ replaced by $x \in \mathbb{R}^n$ [19] is NP-hard.

Taking the NP-hardness of Problem (4) into account, our aim is to *estimate* the optimal value of (4) under the assumption that this problem is feasible. In [2, Theorem 1.1 (i)] (see also [1, Theorem 1]), it was shown that for any $b \in Sg(A)$, there exists a $x \in \mathbb{Z}^n$, such that $Ax = b$ and

$$\|x\|_0 \leq m + \left\lfloor \log_2 \left(\frac{\sqrt{\det(AA^\top)}}{\gcd(A)} \right) \right\rfloor. \tag{5}$$

In [1, Theorem 2], it was shown that (5) cannot be improved significantly, but nevertheless we show here how to improve it in some special cases. As a consequence of Theorem 1 we obtain the following.

Corollary 3. *Let $A \in \mathbb{Z}^{m \times n}$ be a matrix whose columns positively span $\mathbb{R}^m$ and let $b \in \mathbb{Z}^m$. Then $\mathcal{L}(A) = Sg(A)$. Furthermore, if $b \in \mathcal{L}(A)$, and $\tau \in \binom{[n]}{m}$ is a set, for which the matrix A_τ is non-singular, then there is a solution x of the integer-programming feasibility problem $Ax = b, x \in \mathbb{Z}^m_{\geq 0}$ that satisfies the sparsity bound*

$$\|x\|_0 \leq 2m + \Omega_m \left(\frac{|\det(A_\tau)|}{\gcd(A)} \right). \tag{6}$$

Under the above assumptions, for given A, b and τ, such a sparse solution x can be computed in polynomial time.

Note that for a fixed m, (6) is usually much tighter than (5), because the function $\Omega_m(z)$ is bounded from above by the logarithmic function $\log_2(z)$ and is much smaller than $\log_2(z)$ on the average. Furthermore, $|\det(A_\tau)| \leq \sqrt{\det(AA^\top)}$ in view of the Cauchy-Binet formula.

We take a closer look at the case $m = 1$ of a single equation and tighten the given bounds in this case. That is, we consider the *knapsack feasibility problem*

$$a^\top x = b, \ x \in \mathbb{Z}_{\geq 0}^n, \tag{7}$$

where $a \in \mathbb{Z}^n$ and $b \in \mathbb{Z}$. Without loss of generality we can assume that all components of the vector a are not equal to zero. It follows from (5) that a feasible problem (7) has a solution x with

$$\|x\|_0 \leq 1 + \left\lfloor \log_2\left(\frac{\|a\|_2}{\gcd(a)}\right) \right\rfloor. \tag{8}$$

If all components of a have the same sign, without loss of generality we can assume $a \in \mathbb{Z}_{>0}^n$. In this setting, Theorem 1.2 in [2] strengthens the bound (8) by replacing the ℓ_2-norm of the vector a with the ℓ_∞-norm. It was conjectured in [2, page 247] that a bound $\|x\|_0 \leq c + \lfloor \log_2(\|a\|_\infty/\gcd(a)) \rfloor$ with an absolute constant c holds for an *arbitrary* $a \in \mathbb{Z}^n$. We obtain the following result, which covers the case that has not been settled so far and yields a confirmation of this conjecture.

Corollary 4. *Let $a = (a_1, \ldots, a_n)^\top \in (\mathbb{Z} \setminus \{0\})^n$ be a vector that contains both positive and negative components. If the knapsack feasibility problem $a^\top x = b, \ x \in \mathbb{Z}_{\geq 0}^n$ has a solution, then there is a solution x satisfying the sparsity bound*

$$\|x\|_0 \leq 2 + \min\left\{\omega\left(\frac{|a_i|}{\gcd(a)}\right) : i \in [n]\right\}.$$

Under the above assumptions, for given a and b, such a sparse solution x can be computed in polynomial time.

Our next contribution is that, given additional structure on A, we can improve on [2, Theorem 1.1 (i)], which in turn also gives an improvement on [2, Theorem 1.2]. For $a_1, \ldots, a_n \in \mathbb{R}^m$, we denote by $\text{cone}(a_1, \ldots, a_n)$ the convex conic hull of the set $\{a_1, \ldots, a_n\}$. Now assume the matrix $A = (a_1, \ldots, a_n) \in \mathbb{Z}^{m \times n}$ with columns a_i satisfies the following conditions:

$$a_1, \ldots, a_n \in \mathbb{Z}^m \setminus \{0\}, \tag{9}$$

$$\text{cone}(a_1, \ldots, a_n) \text{ is an } m\text{-dimensional pointed cone}, \tag{10}$$

$$\text{cone}(a_1) \text{ is an extreme ray of } \text{cone}(a_1, \ldots, a_n). \tag{11}$$

Note that the previously best sparsity bound for the general case of the integer-programming feasibility problem is (5). Using the Cauchy-Binet formula, (5) can be written as

$$\|x\|_0 \leq m + \log_2 \frac{\sqrt{\sum_{I \in \binom{[n]}{m}} \det(A_I)^2}}{\gcd(A)}.$$

The following theorem improves this bound in the *"pointed cone case"* by removing a fraction of m/n of terms in the sum under the square root.

Theorem 5. *Let* $A = (a_1, \ldots, a_n) \in \mathbb{Z}^{m \times n}$ *satisfy* (9)–(11) *and, for* $b \in \mathbb{Z}^m$, *consider the integer-programming feasibility problem*

$$Ax = b, \quad x \in \mathbb{Z}_{\geq 0}^n. \tag{12}$$

If (12) *is feasible, then there is a feasible solution* x *satisfying the sparsity bound*

$$\|x\|_0 \leq m + \left\lfloor \log_2 \frac{q(A)}{\gcd(A)} \right\rfloor,$$

where

$$q(A) := \sqrt{\sum_{I \in \binom{[n]}{m} : 1 \in I} \det(A_I)^2}.$$

We omit the proof of this result due to the page limit for the IPCO proceedings. Instead we focus on the particularly interesting case $m = 1$. In this case, assumption (10) is equivalent to $a \in \mathbb{Z}_{>0}^n \cup \mathbb{Z}_{<0}^n$. Without loss of generality, one can assume $a \in \mathbb{Z}_{>0}^n$.

Theorem 6. *Let* $a = (a_1, \ldots, a_n)^\top \in \mathbb{Z}_{>0}^n$ *and* $b \in \mathbb{Z}_{\geq 0}$. *If the knapsack feasibility problem* $a^\top x = b$, $x \in \mathbb{Z}_{\geq 0}^n$ *has a solution, there is a solution* x *satisfying the sparsity bound*

$$\|x\|_0 \leq 1 + \left\lfloor \log_2 \left(\frac{\min\{a_1, \ldots, a_n\}}{\gcd(a)} \right) \right\rfloor.$$

When dealing with bounds for sparsity it would be interesting to understand *the worst case scenario among all members of the semigroup*, which is described by the function

$$\mathrm{ICR}(A) = \max_{b \in \mathcal{S}g(A)} \min\{\|x\|_0 : Ax = b, \ x \in \mathbb{Z}_{\geq 0}^n\}. \tag{13}$$

We call $\mathrm{ICR}(A)$ the *integer Carathéodory rank* in resemblance to the classical problem of finding the integer Carathéodory number for Hilbert bases [24]. Above results for the problem $Ax = b$, $x \in \mathbb{Z}_{\geq 0}^n$ can be phrased as upper bounds on $\mathrm{ICR}(A)$. We are interested in the complexity of computing $\mathrm{ICR}(A)$. The first question is: *can the integer Carathéodory rank of a matrix A be computed at all?* After all, remember that the semigroup has infinitely many elements and, despite the fact that $\text{ICR}(A)$ is a finite number, a direct usage of (13) would result into the determination of the sparsest representation $Ax = b$ for all of the infinitely many elements b of $\mathcal{S}g(A)$. It turns out that $\text{ICR}(A)$ is computable, as the inequality $\text{ICR}(A) \leq k$ can be expressed as the formula $\forall x \in \mathbb{Z}^n_{\geq 0} \exists y \in \mathbb{Z}^n_{\geq 0} : (Ax = Ay) \wedge (\|y\|_0 \leq k)$ in *Presburger arithmetic* [14]. Beyond this fact, the complexity status of computing $\text{ICR}(A)$ is largely open, even when A is just one row:

Problem 7. *Given the input* $a = (a_1, \ldots, a_n)^\top \in \mathbb{Z}^n$, *is it NP-hard to compute* $\text{ICR}(a^\top)$?

The *Frobenius number* $\max \mathbb{Z}_{\geq 0} \setminus \mathcal{S}g(a^\top)$, defined under the assumptions $a \in \mathbb{Z}^n_{>0}$ and $\gcd(a) = 1$, is yet another value associated to $\mathcal{S}g(a^\top)$. The Frobenius number can be computed in polynomial time when n is fixed [5,16] but is NP-hard to compute when n is not fixed [21]. It seems that there might be a connection between computing the Frobenius number and $\text{ICR}(a^\top)$.

2   Proofs of Theorem 1 and its consequences

The proof of Theorem 1 relies on the theory of finite Abelian groups. We write Abelian groups additively. An Abelian group G is said to be a *direct sum* of its finitely many subgroups $G_1, \ldots, G_m$, which is written as $G = \bigoplus_{i=1}^m G_i$, if every element $x \in G$ has a unique representation as $x = x_1 + \cdots + x_m$ with $x_i \in G_i$ for each $i \in [m]$. A *primary cyclic group* is a non-zero finite cyclic group whose order is a power of a prime number. We use G/H to denote the quotient of G modulo its subgroup H.

The fundamental theorem of finite Abelian groups states that every finite Abelian group G has a *primary decomposition*, which is essentially unique. This means, G is decomposable into a direct sum of its primary cyclic groups and that this decomposition is unique up to automorphisms of G. We denote by $\kappa(G)$ the number of direct summands in the primary decomposition of G.

For a subset S of a finite Abelian group G, we denote by $\langle S \rangle$ the subgroup of G generated by S. We call a subset S of G *non-redundant* if the subgroups $\langle T \rangle$ generated by proper subsets T of S are properly contained in $\langle S \rangle$. In other words, S is non-redundant if $\langle S \setminus \{x\} \rangle$ is a proper subgroup of $\langle S \rangle$ for every $x \in S$. The following result can be found in [13, Lemma A.6].

Theorem 8. *Let G be a finite Abelian group. Then the maximum cardinality of a non-redundant subset S of G is equal to $\kappa(G)$.*

We will also need the following lemmas, proved in the Appendix.

Lemma 1. *Let G be a finite Abelian group representable as a direct sum $G = \bigoplus_{j=1}^m G_j$ of $m \in \mathbb{Z}_{>0}$ cyclic groups. Then $\kappa(G) \leq \Omega_m(|G|)$.*

Lemma 2. *Let Λ be a sublattice of $\mathbb{Z}^m$ of rank $m \in \mathbb{Z}^m_{>0}$. Then $G = \mathbb{Z}^m/\Lambda$ is a finite Abelian group of order $\det(\Lambda)$ that can be represented as a direct sum of at most m cyclic groups.*

Proof (Theorem 1). Let $a_1, \ldots, a_n$ be the columns of A. Without loss of generality, let $\tau = [m]$. We use the notation $B := A_\tau$.

Reduction to the case $\gcd(A) = 1$. For a non-singular square matrix M, the columns of $M^{-1}A$ are representations of the columns of A in the basis of columns of M. In particular, for a matrix M whose columns form a basis of $\mathcal{L}(A)$, the matrix $M^{-1}A$ is integral and the $m \times m$ minors of $M^{-1}A$ are the respective $m \times m$ minors of A divided by $\det(M) = \gcd(A)$. Thus, replacing A by $M^{-1}A$, we pass from $\mathcal{L}(A)$ to $\mathcal{L}(M^{-1}A) = \{M^{-1}z : z \in \mathcal{L}(A)\}$, which corresponds to a change of a coordinate system in $\mathbb{R}^m$ and ensures that $\gcd(A) = 1$.

Sparsity bound (1). The matrix B gives rise to the lattice $\Lambda := \mathcal{L}(B)$ of rank m, while Λ determines the finite Abelian group $\mathbb{Z}^m / \Lambda$.

Consider the canonical homomorphism $\phi : \mathbb{Z}^m \to \mathbb{Z}^m / \Lambda$, sending an element of $\mathbb{Z}^m$ to its coset modulo Λ. Since $\gcd(A) = 1$, we have $\mathcal{L}(A) = \mathbb{Z}^m$, which implies $\langle T \rangle = \mathbb{Z}^m / \Lambda$ for $T := \{\phi(a_{m+1}), \ldots, \phi(a_n)\}$. For every non-redundant subset S of T, we have

$$|S| \leq \kappa(\mathbb{Z}^m / \Lambda) \qquad \text{(by Theorem 8)}$$
$$\leq \Omega_m(|\det(A_\tau)|) \qquad \text{(by Lemmas 1 and 2)}.$$

Fixing a set $I \subseteq \{m+1, \ldots, n\}$ that satisfies $|I| = |S|$ and $S = \{\phi(a_i) : i \in I\}$, we reformulate $\langle S \rangle = \mathbb{Z}^m / \Lambda$ as $\mathbb{Z}^m = \mathcal{L}(A_I) + \Lambda = \mathcal{L}(A_I) + \mathcal{L}(A_\tau) = \mathcal{L}(A_{I \cup \tau})$. Thus, (1) holds for $\gamma = I \cup \tau$.

Construction of γ *in polynomial time.* The matrix M used in the reduction to the case $\gcd(A) = 1$ can be constructed in polynomial time: one can obtain M from the Hermite Normal Form of A (with respect to the column transformations) by discarding zero columns. For the determination of γ, the set I that defines the non-redundant subset $S = \{\phi(a_i) : i \in I\}$ of $\mathbb{Z}^m / \Lambda$ needs to be determined. Start with $I = \{m+1, \ldots, n\}$ and iteratively check if some of the elements $\phi(a_i) \in \mathbb{Z}^m / \Lambda$, where $i \in I$, is in the group generated by the remaining elements. Suppose $j \in I$ and we want to check if $\phi(a_j)$ is in the group generated by all $\phi(a_i)$ with $i \in I \setminus \{j\}$. Since $\Lambda = \mathcal{L}(A_\tau)$, this is equivalent to checking $a_j \in \mathcal{L}(A_{I \setminus \{j\} \cup \tau})$ and is thus reduced to solving a system of linear Diophantine equations with the left-hand side matrix $A_{I \setminus \{j\} \cup \tau}$ and the right-hand side vector a_j. Thus, carrying the above procedure for every $j \in I$ and removing j from I whenever $a_j \in \mathcal{L}(A_{I \setminus \{j\} \cup \tau})$, we eventually arrive at a set I that determines a non-redundant subset S of $\mathbb{Z}^m / \Lambda$. This is done by solving at most $n - m$ linear Diophantine systems in total, where the matrix of each system is a sub-matrix of A and the right-hand vector of the system is a column of A.   □

Remark 1 (Optimality of the bounds). For a given $\Delta \in \mathbb{Z}_{\geq 2}$ let us consider matrices $A \in \mathbb{Z}^{m \times n}$ with $\Delta = |\det(A_\tau)| / \gcd(A)$. We construct a matrix A that shows the optimality of the bound (1). As in the proof of Theorem 1, we assume $\tau = [m]$ and use the notation $B = A_\tau$. Consider the prime factorization $\Delta = p_1^{n_1} \cdots p_s^{n_s}$. We will fix the matrix B to be a diagonal matrix with diagonal entries $d_1, \ldots, d_m \in \mathbb{Z}_{>0}$ so that $\det(B) = d_1 \cdots d_m = \Delta$.

The diagonal entries are defined by distributing the prime factors of Δ among the diagonal entries of B. If the multiplicity n_i of the prime p_i is less than m, we introduce p_i as a factor of multiplicity 1 in n_i of the m diagonal entries of B. If the multiplicity n_i is at least m, we are able distribute the factors p_i among *all* of the diagonal entries of B so that each diagonal entry contains the factor p_i with multiplicity at least 1.

The group $\mathbb{Z}^m/\Lambda = \mathbb{Z}^m/\mathcal{L}(B)$ is a direct sum of m cyclic groups $G_1, \ldots, G_m$ of orders $d_1, \ldots, d_m$, respectively. By the Chinese Remainder Theorem, these cyclic groups can be further decomposed into the direct sum of primary cyclic groups. By our construction, the prime factor p_i of the multiplicity $n_i < m$ generates a cyclic direct summand of order p_i in n_i of the subgroups $G_1, \ldots, G_m$. If $n_i \geq m$, then each of the groups $G_1, \ldots, G_m$ has a direct summand, which is a non-trivial cyclic group whose order is a power of p_i. Summarizing, we see that the decomposition of $\mathbb{Z}^m/\Lambda$ into primary cyclic groups contains n_i summands of order p_i, when $n_i < m$, and m summands, whose order is a power of p_i, when $n_i \geq m$. The total number of summands is thus $\sum_{i=1}^{s} \min\{m, n_i\} = \Omega_m(\Delta)$.

Now, fix $n = m + \Omega_m(\Delta)$ and choose columns $\boldsymbol{a}_{m+1}, \ldots, \boldsymbol{a}_n$ so that $\phi(\boldsymbol{a}_{m+1})$, $\ldots, \phi(\boldsymbol{a}_n)$ generate all direct summands in the decomposition of $\mathbb{Z}^m/\Lambda$ into primary cyclic groups. With this choice, $\phi(\boldsymbol{a}_{m+1}), \ldots, \phi(\boldsymbol{a}_n)$ generate $\mathbb{Z}^m/\Lambda$, which means that $\mathcal{L}(A) = \mathbb{Z}^m$ and implies $\gcd(A) = 1$. On the other hand, any proper subset $\{\phi(\boldsymbol{a}_{m+1}), \ldots, \phi(\boldsymbol{a}_n)\}$ generates a proper subgroup of $\mathbb{Z}^m/\Lambda$, as some of the direct summands in the decomposition of $\mathbb{Z}^m/\Lambda$ into primary cyclic groups will be missing. This means $\mathcal{L}(A_{[m]\cup I}) \subsetneq \mathbb{Z}^m$ for every $I \subsetneq \{m+1, \ldots, n\}$.

Proof (Corollary 2). Feasiblity of (2) can be expressed as $\boldsymbol{b} \in \mathcal{L}(A)$. Choose γ from the assertion of Theorem 1. One has $\boldsymbol{b} \in \mathcal{L}(A) = \mathcal{L}(A_\gamma)$ and so there exists a solution $\boldsymbol{x}$ of (2) whose support is a subset of γ. This sparse solution $\boldsymbol{x}$ can be computed by solving the Diophantine system with the left-hand side matrix A_γ and the right-hand side vector $\boldsymbol{b}$.

Proof (Corollary 3). Assume that the Diophantine system $A\boldsymbol{x} = \boldsymbol{b}$, $\boldsymbol{x} \in \mathbb{Z}^n$ has a solution. It suffices to show that, in this case, the integer-programming feasibility problem $A\boldsymbol{x} = \boldsymbol{b}$, $\boldsymbol{x} \in \mathbb{Z}^n_{\geq 0}$ has a solution, too, and that one can find a solution of the desired sparsity to the integer-programming feasibility problem in polynomial time.

One can determine γ as in Theorem 1 in polynomial time. Using γ, we can determine a solution $\boldsymbol{x}^* = (x_1^*, \ldots, x_n^*)^\top \in \mathbb{Z}^n$ of the Diophantine system $A\boldsymbol{x} = \boldsymbol{b}$, $\boldsymbol{x} \in \mathbb{Z}^n$ satisfying $x_i^* = 0$ for $i \in [n] \setminus \gamma$ in polynomial time, as described in the proof of Corollary 2.

Let $\boldsymbol{a}_1, \ldots, \boldsymbol{a}_n$ be the columns of A. Since the matrix A_τ is non-singular, the m vectors $\boldsymbol{a}_i$, where $i \in \tau$, together with the vector $\boldsymbol{v} = -\sum_{i\in\tau} \boldsymbol{a}_i$ positively span $\mathbb{R}^n$. Since all columns of A positive span $\mathbb{R}^n$, the conic version of the Carathéodory theorem implies the existence of a set $\beta \subseteq [m]$ with $|\beta| \leq m$, such that $\boldsymbol{v}$ is in the conic hull of $\{\boldsymbol{a}_i : i \in \beta\}$. Consequently, the set $\{\boldsymbol{a}_i : i \in \beta \cup \tau\}$ and by this also the larger set $\{\boldsymbol{a}_i : i \in \beta \cup \gamma\}$ positively span $\mathbb{R}^m$. Let $I = \beta \cup \gamma$. By construction, $|I| \leq |\beta| + |\gamma| \leq m + |\gamma|$.

Since the vectors $\boldsymbol{a}_i$ with $i \in I$ positively span $\mathbb{R}^m$, there exist a choice of rational coefficients $\lambda_i > 0$ $(i \in I)$ with $\sum_{i\in I} \lambda_i \boldsymbol{a}_i = 0$. After rescaling we can assume $\lambda_i \in \mathbb{Z}_{>0}$. Define $\boldsymbol{x}' = (x'_1, \ldots, x'_n)^\top \in \mathbb{Z}^n_{\geq 0}$ by setting $x'_i = \lambda_i$ for $i \in I$ and $x'_i = 0$ otherwise. The vector $\boldsymbol{x}'$ is a solution of $A\boldsymbol{x} = \boldsymbol{0}$. Choosing $N \in \mathbb{Z}_{>0}$ large enough, we can ensure that the vector $\boldsymbol{x}^* + N\boldsymbol{x}'$ has non-negative components. Hence, $\boldsymbol{x} = \boldsymbol{x}^* + N\boldsymbol{x}'$ is a solution of the system $A\boldsymbol{x} = \boldsymbol{b}$, $\boldsymbol{x} \in \mathbb{Z}^n_{\geq 0}$ satisfying the desired sparsity estimate. The coefficients λ_i and the number N can be computed in polynomial time.

Proof (Corollary 4). The assertion follows by applying Corollary 3 for $m = 1$ and all $\tau = \{i\}$ with $i \in [n]$.

3   Proof of Theorem 6

Lemma 3. *Let $a_1, \ldots, a_t \in \mathbb{Z}_{>0}$, where $t \in \mathbb{Z}_{>0}$. If $t > 1 + \log_2(a_1)$, then the system*

$$y_1 a_1 + \cdots + y_t a_t = 0,$$
$$y_1 \in \mathbb{Z}_{\geq 0}, \; y_2, \ldots, y_t \in \{-1, 0, 1\}.$$

in the unknowns $y_1, \ldots, y_t$ has a solution that is not identically equal to zero.

Proof. The proof is inspired by the approach in [3, §3.1] (used in a different context) that suggests to reformulate the underlying equation over integers as two strict inequalities and then use Minkowski's first theorem [4, Ch. VII, Sect. 3] from the geometry of numbers. Consider the convex set $Y \subseteq \mathbb{R}^t$ defined by $2t$ strict linear inequalities

$$-1 < y_1 a_1 + \cdots + y_t a_t < 1,$$
$$-2 < y_i < 2 \text{ for all } i \in \{2, \ldots, t\}.$$

Clearly, the set Y is the interior of a hyper-parallelepiped and can also be described as $Y = \{\boldsymbol{y} \in \mathbb{R}^t : \|M\boldsymbol{y}\|_\infty < 1\}$, where M is the upper triangular matrix

$$M = \begin{pmatrix} a_1 & a_2 & \cdots & a_t \\ & 1/2 & & \\ & & \ddots & \\ & & & 1/2 \end{pmatrix}.$$

It is easy to see that the t-dimensional volume $\mathrm{vol}(Y)$ of Y is

$$\mathrm{vol}(Y) = \mathrm{vol}(M^{-1}[-1, 1]^t) = \frac{1}{\det(M)} 2^t = \frac{4^t}{2a_1}.$$

The assumption $t > 1 + \log_2(a_1)$ implies that the volume of Y is strictly larger than 2^t. Thus, by Minkowski's first theorem, the set Y contains a non-zero integer vector $\boldsymbol{y} = (y_1, \ldots, y_t)^\top \in \mathbb{Z}^t$. Without loss of generality we can assume that $y_1 \geq 0$ (if the latter is not true, one can replace $\boldsymbol{y}$ by $-\boldsymbol{y}$). The vector $\boldsymbol{y}$ is a desired solution from the assertion of the lemma. ∎

Proof (Theorem 6). Without loss of generality we can assume that $\gcd(\boldsymbol{a}) = 1$. In fact, if b is divisible by $\gcd(\boldsymbol{a})$ we can convert $\boldsymbol{a}^\top \boldsymbol{x} = b$ to $\overline{\boldsymbol{a}}^\top \boldsymbol{x} = \overline{b}$ with $\overline{\boldsymbol{a}} = \frac{a}{\gcd(\boldsymbol{a})}$ and $\overline{b} = \frac{b}{\gcd(\boldsymbol{a})}$, and, if b is not divisible by $\gcd(\boldsymbol{a})$, the knapsack feasibility problem $\boldsymbol{a}^\top \boldsymbol{x} = b$, $\boldsymbol{x} \in \mathbb{Z}_{\geq 0}^n$ has no solution.

Without loss of generality, let $a_1 = \min\{a_1, \ldots, a_n\}$. We need to show the existence of solution of the knapsack feasibility problem satisfying $\|\boldsymbol{x}\|_0 \leq 1 + \log_2(a_1)$.

Choose a solution $\boldsymbol{x} = (x_1, \ldots, x_n)^\top$ of the knapsack feasibility problem with the property that the number of indices $i \in \{2, \ldots, n\}$ for which $x_i \neq 0$ is minimized. Without loss of generality we can assume that, for some $t \in \{2, \ldots, n\}$ one has $x_2 > 0, \ldots, x_t > 0, x_{t+1} = \cdots = x_n = 0$. Lemma 3 implies $t \leq 1 + \log_2(a_1)$. In fact, if the latter was not true, then a solution $\boldsymbol{y} \in \mathbb{R}^t$ of the system in Lemma 3 could be extended to a solution $\boldsymbol{y} \in \mathbb{R}^n$ by appending zero components. It is clear that some of the components $y_2, \ldots, y_t$ are negative, because $a_2 > 0, \ldots, a_t > 0$. It then turns out that, for an appropriate choice of $k \in \mathbb{Z}_{\geq 0}$, the vector $\boldsymbol{x}' = (x_1', \ldots, x_n')^\top = \boldsymbol{x} + k\boldsymbol{y}$ is a solution of the same knapsack feasibility problem satisfying $x_1' \geq 0, \ldots, x_t' \geq 0$, $x_{t+1}' = \cdots = x_n' = 0$ and $x_i' = 0$ for at least one $i \in \{2, \ldots, t\}$. Indeed, one can choose k to be the minimum among all a_i with $i \in \{2, \ldots, t\}$ and $y_i = -1$.

The existence of $\boldsymbol{x}'$ with at most $t - 1$ non-zero components x_i' with $i \in \{2, \ldots, n\}$ contradicts the choice of $\boldsymbol{x}$ and yields the assertion. $\square$

Acknowledgements. The second author is supported by the DFG (German Research Foundation) within the project number 413995221. The third author acknowledges partial support from NSF grant 1818969.

A   Appendix

Proof (Lemma 1). Consider the prime factorization $|G| = p_1^{n_1} \cdots p_s^{n_s}$. Then $|G_j| = p_1^{n_{i,j}} \cdots p_s^{n_{i,j}}$ with $0 \leq n_{i,j} \leq n_i$ and, by the Chinese Remainder Theorem, the cyclic group G_j can be represented as $G_j = \bigoplus_{i=1}^s G_{i,j}$, where $G_{i,j}$ is a cyclic group of order $p_i^{n_{i,j}}$. Consequently, $G = \bigoplus_{i=1}^s \bigoplus_{j=1}^m G_{i,j}$. This is a decomposition of G into a direct sum of primary cyclic groups and, possibly, some trivial summands $G_{i,j}$ equal to $\{0\}$. We can count the non-trivial direct summands whose order is a power of p_i, for a given $i \in [s]$. There is at most one summand like this for each of the groups G_j. So, there are at most m non-trivial summands in the decomposition whose order is a power of p_i. On the other hand, the direct sum of all non-trivial summands whose order is a power of p_i is a group of order $p_i^{n_{i,1} + \cdots + n_{i,s}} = p_i^{n_i}$ so that the total number of such summands is not larger than n_i, as every summand contributes the factor at least p_i to the power $p_i^{n_i}$. This shows that the total number of non-zero summands in the decomposition of G is at most $\sum_{i=1}^s \min\{m, n_i\} = \Omega_m(|G|)$. $\square$

Proof (Lemma 2). The proof relies on the relationship of finite Abelian groups and lattices, see [23, §4.4]. Fix a matrix $M \in \mathbb{Z}^{m \times m}$ whose columns form a basis of Λ. Then $|\det(M)| = \det(\Lambda)$. There exist unimodular matrices $U \in \mathbb{Z}^{m \times m}$

and $V \in \mathbb{Z}^{m \times m}$ such that $D := UMV$ is diagonal matrix with positive integer diagonal entries. For example, one can choose D to be the Smith Normal Form of M [23, §4.4]. Let $d_1, \ldots, d_m \in \mathbb{Z}_{>0}$ be the diagonal entries of D. Since U and V are unimodular, $d_1 \cdots d_m = \det(D) = \det(\Lambda)$.

We introduce the quotient group $G' := \mathbb{Z}^m/\Lambda' = (\mathbb{Z}/d_1\mathbb{Z}) \times \cdots \times (\mathbb{Z}/d_m\mathbb{Z})$ with respect to the lattice $\Lambda' := \mathcal{L}(D) = (d_1\mathbb{Z}) \times \cdots \times (d_m\mathbb{Z})$. The order of G' is $d_1 \cdots d_m = \det(D) = \det(\Lambda)$ and G' is a direct sum of at most m cyclic groups, as every $d_i > 1$ determines a non-trivial direct summand.

To conclude the proof, it suffices to show that G' is isomorphic to G. To see this, note that $\Lambda' = \mathcal{L}(D) = \mathcal{L}(UMV) = \mathcal{L}(UM) = \{Uz : z \in \Lambda\}$. Thus, the map $z \mapsto Uz$ is an automorphism of $\mathbb{Z}^m$ and an isomorphism from Λ to Λ'. Thus, $z \mapsto Uz$ induces an isomorphism from the group $G = \mathbb{Z}^m/\Lambda$ to the group $G' = \mathbb{Z}^m/\Lambda'$. $\square$

References

1. Aliev, I., De Loera, J.A., Eisenbrand, F., Oertel, T., Weismantel, R.: The support of integer optimal solutions. SIAM J. Optim. **28**(3), 2152–2157 (2018)
2. Aliev, I., De Loera, J.A., Oertel, T., O'Neill, C.: Sparse solutions of linear diophantine equations. SIAM J. Appl. Algebra Geom. **1**(1), 239–253 (2017)
3. Averkov, G.: On the size of lattice simplices with a single interior lattice point. SIAM J. Discrete Math. **26**(2), 515–526 (2012)
4. Barvinok, A.I.: A Course in Convexity. Graduate Studies in Mathematics, vol. 54. American Mathematical Society, Providence, RI (2002)
5. Barvinok, A.I., Woods, K.: Short rational generating functions for lattice point problems. J. AMS **16**(4), 957–979 (2003)
6. Boche, H., Calderbank, R., Kutyniok, G., Vybíral, J.: A survey of compressed sensing. In: Boche, H., Calderbank, R., Kutyniok, G., Vybíral, J. (eds.) Compressed Sensing and its Applications. ANHA, pp. 1–39. Springer, Cham (2015). https://doi.org/10.1007/978-3-319-16042-9_1
7. Candès, E., Rudelson, M., Tao, T., Vershynin, R.: Error correction via linear programming. In: 46th Annual IEEE Symposium on Foundations of Computer Science (FOCS 2005), pp. 668–681 (2005)
8. Candès, E.J., Romberg, J.K., Tao, T.: Stable signal recovery from incomplete and inaccurate measurements. Comm. Pure Appl. Math. **59**(8), 1207–1223 (2006)
9. Candès, E.J., Tao, T.: Decoding by linear programming. IEEE Trans. Inform. Theory **51**(12), 4203–4215 (2005)
10. Eisenbrand, F., Shmonin, G.: Carathéodory bounds for integer cones. Oper. Res. Lett. **34**(5), 564–568 (2006)
11. Flinth, A., Kutyniok, G.: PROMP: a sparse recovery approach to lattice-valued signals. Appl. Comput. Harmon. Anal. **45**(3), 668–708 (2018)
12. Fukshansky, L., Needell, D., Sudakov, B.: An algebraic perspective on integer sparse recovery. Appl. Math. Comput. **340**, 31–42 (2019)
13. Geroldinger, A., Halter-Koch, F.: Non-Unique Factorizations: Algebraic Combinatorial and Analytic Theory. Pure and Applied Mathematics. Chapman and Hall/CRC, Boca Raton (2006)
14. Haase, C.: A survival guide to Presburger arithmetic. ACM SIGLOG News **5**(3), 67–82 (2018)

15. Hardy, G.H., Wright, E.M., Heath-Brown, R., Silverman, J.: An Introduction to the Theory of Numbers. Oxford Mathematics. OUP, Oxford (2008)

16. Kannan, R.: Lattice translates of a polytope and the frobenius problem. Combinatorica **12**(2), 161–177 (1992)

17. Konyagin, S.V.: On the recovery of an integer vector from linear measurements. Mat. Zametki **104**(6), 863–871 (2018)

18. Lovász, L.: Matching structure and the matching lattice. J. Comb. Theory Ser. B **43**(2), 187–222 (1987)

19. Natarajan, B.K.: Sparse approximate solutions to linear systems. SIAM J. Comput. **24**(2), 227–234 (1995)

20. Oertel, T., Paat, J., Weismantel, R.: Sparsity of integer solutions in the average case. In: Lodi, A., Nagarajan, V. (eds.) IPCO 2019. LNCS, vol. 11480, pp. 341–353. Springer, Cham (2019). https://doi.org/10.1007/978-3-030-17953-3_26

21. Ramírez-Alfonsín, J.L.: Complexity of the Frobenius problem. Combinatorica **16**(1), 143–147 (1996)

22. Rossi, M., Haimovich, A.M., Eldar, Y.C.: Spatial compressive sensing for MIMO radar. IEEE Trans. Signal Process. **62**(2), 419–430 (2014)

23. Schrijver, A.: Theory of Linear and Integer Programming. Wiley-Interscience Series in Discrete Mathematics. John Wiley & Sons Ltd., Chichester (1986). A Wiley-Interscience Publication

24. Sebö, A.: Hilbert bases, Carathéodory's Theorem and combinatorial optimization. In: Proceedings of the 1st Integer Programming and Combinatorial Optimization Conference, Waterloo, Ont., Canada, pp. 431–455. University of Waterloo Press (1990)

A Technique for Obtaining True Approximations for k-Center with Covering Constraints

Georg Anegg, Haris Angelidakis, Adam Kurpisz, and Rico Zenklusen[✉]

Department of Mathematics, ETH Zurich, Zurich, Switzerland
{ganegg,angelidc,kurpisza,ricoz}@ethz.ch

Abstract. There has been a recent surge of interest in incorporating fairness aspects into classical clustering problems. Two recently introduced variants of the k-Center problem in this spirit are Colorful k-Center, introduced by Bandyapadhyay, Inamdar, Pai, and Varadarajan, and lottery models, such as the Fair Robust k-Center problem introduced by Harris, Pensyl, Srinivasan, and Trinh. To address fairness aspects, these models, compared to traditional k-Center, include additional covering constraints. Prior approximation results for these models require to relax some of the normally hard constraints, like the number of centers to be opened or the involved covering constraints, and therefore, only obtain constant-factor pseudo-approximations. In this paper, we introduce a new approach to deal with such covering constraints that leads to (true) approximations, including a 4-approximation for Colorful k-Center with constantly many colors—settling an open question raised by Bandyapadhyay, Inamdar, Pai, and Varadarajan—and a 4-approximation for Fair Robust k-Center, for which the existence of a (true) constant-factor approximation was also open.

We complement our results by showing that if one allows an unbounded number of colors, then Colorful k-Center admits no approximation algorithm with finite approximation guarantee, assuming that $P \neq NP$. Moreover, under the Exponential Time Hypothesis, the problem is inapproximable if the number of colors grows faster than logarithmic in the size of the ground set.

1   Introduction

Along with k-Median and k-Means, k-Center is one of the most fundamental and heavily studied clustering problems. In k-Center, we are given a finite metric

G. Anegg and R. Zenklusen—Research supported by Swiss National Science Foundation grant 200021_184622.

H. Angelidakis and R. Zenklusen—This project has received funding from the European Research Council (ERC) under the European Union's Horizon 2020 research and innovation programme (grant agreement No 817750).

A. Kurpisz—Research supported by Swiss National Science Foundation grant PZ00P2_174117.

D. Bienstock and G. Zambelli (Eds.): IPCO 2020, LNCS 12125, pp. 52–65, 2020.
https://doi.org/10.1007/978-3-030-45771-6_5

space (X, d) and an integer $k \in [\|X\|] := \{1, \ldots, |X|\}$, and the task is to find a set $C \subseteq X$ with $|C| \leq k$ minimizing the maximum distance of any point in X to its closest point in C. Equivalently, the problem can be phrased as covering X with k balls of radius as small as possible, i.e., finding the smallest radius $r \in \mathbb{R}_{\geq 0}$ together with a set $C \subseteq X$ with $|C| \leq k$ such that $X = B(C, r) := \bigcup_{c \in C} B(c, r)$, where $B(c, r) := \{u \in X : d(c, u) \leq r\}$ is the ball of radius r around c.

k-Center, like most clustering problems, is computationally hard; actually it is NP-hard to approximate to within any constant below 2 [18]. On the positive side, various 2-approximations [12,16] have been found, and thus, its approximability is settled. Many variations of k-Center have been studied, most of which are based on generalizations along one of the following two main axes:

(i) which sets of centers can be selected, and
(ii) which sets of points of X need to be covered.

The most prominent variations along (i) are variations where the set of centers is required to be in some down-closed family $\mathcal{F} \subseteq 2^X$. For example, if centers have non-negative opening costs and there is a global budget for opening centers, Knapsack Center is obtained. If $\mathcal{F}$ are the independent sets of a matroid, the problem is known as Matroid Center. The best-known problem type linked to (ii) is Robust k-Center. Here, an integer $m \in [\|X\|]$ is given, and one only needs to cover any m points of X with k balls of radius as small as possible. Research on k-Center variants along one or both of these axes has been very active and fruitful, see, e.g., [7,9,10,17]. In particular, a recent elegant framework of Chakrabarty and Negahbani [8] presents a unifying framework for designing best possible approximation algorithms for all above-mentioned variants.

All the above variants have in common that there is a single covering requirement; either all of X needs to be covered or a subset of it. Moreover, they come with different kinds of packing constraints on the centers to be opened as in Knapsack or Matroid Center. However, the desire to address fairness in clustering, which has received significant attention recently, naturally leads to multiple covering constraints. Here, existing techniques only lead to constant-factor pseudo-approximations that violate at least one constraint, like the number of centers to be opened. In this work, we present techniques for obtaining (true) approximations for two recent fairness-inspired generalizations of k-Center along axis (ii), namely

(i) γ-Colorful k-Center, as introduced by Bandyapadhyay et al. [3], and
(ii) Fair Robust k-Center, a lottery model introduced by Harris et al. [15].

γ-Colorful k-Center (γCkC) is a fairness-inspired k-Center model imposing covering constraints on subgroups. It is formally defined as follows:[1]

---

[1] The version introduced in [3] requires $X_1, \ldots, X_\gamma$ to partition X. However, γCkC readily reduces to the more restrictive model in [3] by replacing an element with q colors by q elements on the same location with each having a single color.

Definition 1 (γ-Colorful k-Center (γCkC) [3]). *Let $\gamma, k \in \mathbb{Z}_{\geq 1}$, (X, d) be a finite metric space, $X_\ell \subseteq X$ for $\ell \in [\gamma]$, and $m \in \mathbb{Z}_{\geq 0}^\gamma$. The γ-Colorful k-Center problem (γCkC) asks to find the smallest radius $r \in \mathbb{R}_{\geq 0}$ together with centers $C \subseteq X$, $|C| \leq k$ such that*

$$|B(C, r) \cap X_\ell| \geq m_\ell \quad \forall \ell \in [\gamma] .$$

Such a set of centers C is called a (γCkC) solution of radius r.

The name stems from interpreting each set X_ℓ for $\ell \in [\gamma]$ as a color assigned to the elements of X_ℓ. In particular, an element can have multiple colors or no color. In words, the task is to open k centers of smallest possible radius such that, for each color $\ell \in [\gamma]$, at least m_ℓ points of color ℓ are covered. Hence, for $\gamma = 1$, we recover the Robust k-Center problem.

We briefly contrast γCkC with related fairness models. A related class of models that has received significant attention also assumes that the ground set is colored, but requires that each cluster contains approximately the same number of points from each color. Such variants have been considered for k-Median, k-Means, and k-Center, e.g., see [2,4,5,11,23] and references therein. γCkC differentiates itself from the above notion of fairness by not requiring a per-cluster guarantee, but a global fairness guarantee. More precisely, each color can be thought of as representing a certain group of people (demographic), and a global covering requirement is given per demographic. Also notice the difference with the well-known Robust k-Center problem, where a feasible solution might, potentially, completely ignore a certain subgroup, resulting in a heavily unfair treatment. γCkC addresses this issue.

The presence of multiple covering constraints in γCkC, imposed by the colors, hinders the use of classical k-Center clustering techniques, which, as mentioned above, have mostly been developed for packing constraints on the centers to be opened. An elegant first step was done by Bandyapadhyay et al. [3]. They exploit sparsity of a well-chosen LP (in a similar spirit as in [15]) to obtain the following pseudo-approximation for γCkC: they efficiently compute a solution of twice the optimal radius by opening at most $k + \gamma - 1$ centers. Hence, up to $\gamma - 1$ more centers than allowed may have to be opened. Moreover, [3] shows that in the Euclidean plane, a significantly more involved extension of this technique allows for obtaining a $(17 + \varepsilon)$-approximation for $\gamma = O(1)$. Unfortunately, this approach is heavily problem-tailored and does not even extend to 3-dimensional Euclidean spaces. This naturally leads to the main open question raised in [3]:

Does γCkC with $\gamma = O(1)$ admit an $O(1)$-approximation, for any finite metric?

Here, we introduce a new approach that answers this question affirmatively.

Together with additional ingredients, our approach also applies to Fair Robust k-Center, which is a natural lottery model introduced by Harris et al. [15]. We introduce the following generalization thereof that can be handled with our techniques, which we name *Fair γ-Colorful k-Center problem (Fair γCkC)*. (The Fair Robust k-Center problem, as introduced in [15], corresponds to $\gamma = 1$.)

Definition 2 (Fair γ-Colorful k-Center (Fair γCkC)). *Given is a γCkC instance on a finite metric space (X, d) together with a vector $p \in [0, 1]^X$. The goal is to find the smallest radius $r \in \mathbb{R}_{\geq 0}$, together with an efficient procedure returning a random γCkC solution $C \subseteq X$ of radius r such that*

$$\Pr[u \in B(C, r)] \geq p(u) \quad \forall u \in X \ .$$

Hence, Fair γCkC is a generalization of γCkC, where each element $u \in X$ needs to be covered with a prescribed probability $p(u)$. The Fair Robust k-Center problem, i.e., Fair γCkC with $\gamma = 1$, is indeed a fairness-inspired generalization of Robust k-Center, since Robust k-Center is obtained by setting $p(u) = 0$ for $u \in X$. One example setting where the additional fairness aspect of Fair γCkC compared to γCkC is nicely illustrated, is when k-Center problems have to be solved repeatedly on the same metric space. The introduction of the probability requirements p allows for obtaining a distribution to draw from that needs to consider all elements of X (as prescribed by p), whereas classical Robust k-Center likely ignores a group of badly-placed elements. We refer to Harris et al. [15] for further motivation of the problem setting. They also discuss the Knapsack and Matroid Center problem under the same notion of fairness.

For Fair Robust k-Center, [15] presents a 2-pseudo-approximation that slightly violates both the number of points to be covered and the probability of covering each point. More precisely, for any constant $\varepsilon > 0$, only a $(1 - \varepsilon)$-fraction of the required number of elements are covered, and element $u \in X$ is covered only with probability $(1 - \varepsilon)p(u)$ instead of $p(u)$. It was left open in [15] whether a true approximation may exist for Fair Robust k-Center.

1.1 Our Results

Our main contribution is a method to obtain 4-approximations for variants of k-Center with unary encoded covering constraints on the points to be covered. We illustrate our technique in the context of γCkC, affirmatively resolving the open question of Bandyapadhyay et al. [3] about the existence of an $O(1)$-approximation for constantly many colors (without restrictions on the underlying metric space).

Theorem 3. *There is a 4-approximation for γCkC running in time $|X|^{O(\gamma)}$.*

In a second step we extend and generalize our technique to Fair γCkC, which, as mentioned, is a generalization of γCkC. We show that Fair γCkC admits a $O(1)$-approximation, which neither violates covering nor probabilistic constraints.

Theorem 4. *There is a 4-approximation for Fair γCkC running in time $|X|^{O(\gamma)}$.*

We complete our results by showing inapproximability for γCkC when γ is not bounded. This holds even on the real line (1-dimensional Euclidean space).

Theorem 5. *It is* NP-*hard to decide whether* γCkC *on the real line admits a solution of radius* 0. *Moreover, unless the Exponential Time Hypothesis fails, for any function* $f : \mathbb{Z}_{\geq 0} \to \mathbb{Z}_{\geq 0}$ *with* $f(n) = \omega(\log n)$, *no polynomial time algorithm can distinguish whether* γCkC *on the real line with* $\gamma = f(|X|)$ *admits a solution of radius* 0.

Hence, assuming the Exponential Time Hypothesis, γCkC is not approximable (with a polynomial-time algorithm) if the number of colors grows faster than logarithmic in the size of the ground set. Notice that, for a logarithmic number of colors, our procedures run in quasi-polynomial time.

1.2 Outline of Main Technical Contributions and Paper Organization

We introduce two main technical ingredients. The first is a method to deal with additional covering constraints in k-Center problems, which leads to Theorem 3. For this, we combine polyhedral sparsity-based arguments as used by Bandyapadhyay et al. [3], which by themselves only lead to pseudo-approximations, with dynamic programming to design a round-or-cut approach. Round-or-cut approaches, first used by Carr et al. [6], leverage the ellipsoid method in a clever way. In each ellipsoid iteration they either separate the current point from a well-defined polyhedron P, or round the current point to a good solution. The rounding step may happen even if the current point is not in P. Round-or-cut methods have found applications in numerous problem settings (see, e.g., [1,8,13,19–22]). The way we employ round-or-cut is inspired by a powerful round-or-cut approach of Chakrabarty and Negahbani [8] also developed in the context of k-Center. However, their approach is not applicable to k-center problems as soon as multiple covering constraints exist, like in γCkC.

Our second technical contribution first employs LP duality to transform lottery-type models, like Fair γCkC, into an auxiliary problem that corresponds to a weighted version of k-center with covering constraints. We then show how a certain type of approximate separation over the dual is possible, by leveraging the techniques we introduced in the context of γCkC, leading to a 4-approximation.

Even though Theorem 4 is a strictly stronger statement than Theorem 3, we first prove Theorem 3 in Sect. 2, because it allows us to give a significantly cleaner presentation of some of our main technical contributions. In Sect. 3, we then focus on the additional techniques needed to deal with Fair γCkC, by reducing it to a problem that can be tackled with the techniques introduced in Sect. 2.

Due to space constraints, various proofs are deferred to the full version of the paper, including the proof of our hardness result, Theorem 5.

2 A 4-approximation for γCkC for $\gamma = O(1)$

In this section, we prove Theorem 3, which implies a 4-approximation algorithm for γCkC with constantly many colors. We assume $\gamma \geq 2$, since $\gamma = 1$ corresponds to Robust k-Center, for which an (optimal) 2-approximation is known [7,15].

We present a procedure that for any $r \in \mathbb{R}_{\geq 0}$ returns a solution of radius $4r$ if a solution of radius r exists. This implies Theorem 3 because the optimal radius is a distance between two points. Hence, we can run the procedure for all possible pairwise distances r between points in X and return the best solution found. Hence, we fix $r \in \mathbb{R}_{\geq 0}$ in what follows. We denote by $\mathcal{P}$ the following canonical relaxation of $\gamma\mathrm{CkC}$ with radius r:

$$\mathcal{P} = \left\{ (x,y) \in [0,1]^X \times [0,1]^X \;\middle|\; \begin{array}{l} \sum_{v \in X} y(v) \leq k \\ \sum_{v \in B(u,r)} y(v) \geq x(u) \quad \forall u \in X \\ \sum_{u \in X_\ell} x(u) \geq m_\ell \quad \forall \ell \in [\gamma] \end{array} \right\}. \tag{1}$$

Integral points $(x,y) \in \mathcal{P}$ correspond to solutions of radius r, where y indicates the opened centers and x indicates the points that are covered. We denote by $\mathcal{P}_I := \mathrm{conv}\left(\mathcal{P} \cap (\{0,1\}^X \times \{0,1\}^X)\right)$ the integer hull of $\mathcal{P}$.

Our algorithm is based on the round-or-cut framework, first used in [6]. The main building block is a procedure that rounds a point $(x,y) \in \mathcal{P}$ to a radius $4r$ solution under certain conditions. It will turn out that these conditions are always satisfied if $(x,y) \in \mathcal{P}_I$. If they are not satisfied, then we can prove that $(x,y) \notin \mathcal{P}_I$ and generate in polynomial time a hyperplane separating (x,y) from $\mathcal{P}_I$. This separation step now becomes an iteration of the ellipsoid method, employed to find a point in $\mathcal{P}_I$, and we continue with a new candidate point (x,y). Schematically, the whole process is described in Fig. 1.

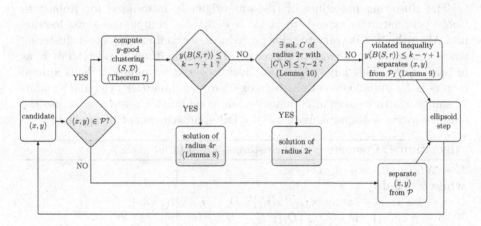

Fig. 1. An iteration of the ellipsoid method.

On a high level, we realize our round-or-cut procedure as follows. First, we check whether $(x,y) \in \mathcal{P}$ and return a violated constraint if this is not the case. If $(x,y) \in \mathcal{P}$, we partition the metric space, based on a natural greedy heuristic introduced by Harris et al. [15]. This gives a set of centers $S = \{s_1, \ldots, s_q\}$ with corresponding clusters $\mathcal{D} = \{D_1, \ldots D_q\}$. We now exploit a technique by Bandyapadhyay et al. [3], which implies that if $y(B(S,r)) \leq k - \gamma + 1$, then one can leverage sparsity arguments in a simplified LP to obtain a radius $4r$ solution that picks centers only within S. We then turn to the case where $y(B(S,r)) > k - \gamma + 1$. At this point, we show that one can efficiently check whether there exists a solution of radius $2r$ that opens at most $\gamma - 2$ centers outside of S. This is achieved by guessing $\gamma - 2$ centers and using dynamic programming to find the remaining $k - \gamma + 2$ centers in S. If no such radius $2r$ solution exists, we argue that any solution of radius r has at most $k - \gamma + 1$ centers in $B(S,r)$, proving that $y(B(S,r)) \leq k - \gamma + 1$ is an inequality separating (x,y) from $\mathcal{P}_I$.

We now give a formal treatment of each step of the algorithm described in Fig. 1. Given a point $(x,y) \in \mathbb{R}^X \times \mathbb{R}^X$, we first check whether $(x,y) \in \mathcal{P}$, and, if not, return a violated constraint of $\mathcal{P}$. Such a constraint separates (x,y) from $\mathcal{P}_I$ because $\mathcal{P}_I \subseteq \mathcal{P}$. Hence, we may assume that $(x,y) \in \mathcal{P}$.

We now use a clustering technique by Harris et al. [15] that, given $(x,y) \in \mathcal{P}$, allows for obtaining what we call a y-good clustering $(S, \mathcal{D})$, defined as follows.[2]

Definition 6 (y-good clustering). *Let $(x,y) \in \mathcal{P}$. A tuple $(S, \mathcal{D})$, where the family $\mathcal{D} = \{D_1, \ldots, D_q\}$ partitions X and $S = \{s_1, \ldots, s_q\} \subseteq X$ with $s_i \in D_i$ for $i \in [q]$, is a y-good clustering if:*

(i) $d(s_i, s_j) > 4r \quad \forall i, j \in [q], i \neq j$,
(ii) $D_i \subseteq B(s_i, 4r) \quad \forall i \in [q]$, and
(iii) $\sum_{i \in [q]} \min\{1, y(B(s_i, r))\} \cdot |D_i \cap X_\ell| \geq m_\ell \quad \forall \ell \in [\gamma]$.

The clustering procedure of [15] was originally introduced for Robust k-Center and naturally extends to γCkC (see [3]). For completeness, we describe it in Algorithm 1. Contrary to prior procedures, we compute a y-good clustering whose centers have pairwise distances of strictly more than $4r$ (instead of $2r$ as in prior work). This large separation avoids overlap of radius $2r$ balls around centers in S, and allows us to use dynamic programming (DP) to build a radius $2r$ solution with centers in S under certain conditions. However, it is also the reason why get a 4-approximation if the DP approach cannot be applied.

Algorithm 1: Compute y-good clustering, given $(x,y) \in \mathcal{P}$

$U \leftarrow X; \quad i \leftarrow 0; \quad S \leftarrow \emptyset; \quad \mathcal{D} \leftarrow \emptyset;$
while $U \neq \emptyset$ **do**
$\quad$ $i \leftarrow i + 1; \quad s_i \leftarrow \text{argmax}_{u \in U}\{x(u)\}; \quad D_i \leftarrow U \cap B(s_i, 4r);$
$\quad$ $S \leftarrow S \cup \{s_i\}; \quad \mathcal{D} \leftarrow \mathcal{D} \cup \{D_i\}; \quad U \leftarrow U \setminus B(s_i, 4r);$
end
return $(S, \mathcal{D})$

Theorem 7 ([3,15]). *For $(x,y) \in \mathcal{P}$, Algorithm 1 computes a y-good clustering $(S, \mathcal{D})$ in polynomial time.*

---

[2] As the name suggests, the properties of a y-good clustering do not depend on x. Hence, we could equivalently define the clustering for any $y \in \mathbb{R}^X$ that lies in the projection of $\mathcal{P}$ onto the last $|X|$ coordinates, i.e., the ones corresponding to y.

Theorem 7 as well as the following lemma follow from the results in [3].

Lemma 8 ([3]). *Let $(x,y) \in \mathcal{P}$ and $(S,\mathcal{D})$ be a y-good clustering. Then, if $y(B(S,r)) \leq k - \gamma + 1$, a solution of radius $4r$ can be found in polynomial time.*

We are left with the case $y(B(S,r)) > k - \gamma + 1$. If $(x,y) \in \mathcal{P}_I$, then there must exist a solution $C_1 \subseteq X$ of radius r with $|C_1 \cap B(S,r)| > k - \gamma + 1$. Hence, C_1 has at most $\gamma - 2$ centers outside of $B(S,r)$. We observe that if such solution C_1 exists, then there must be a solution C_2 of radius $2r$ with all centers being within S, except for $\gamma - 2$ many. This is formalized in the following lemma (which states the contrapositive of the mentioned implication because this form is slightly more convenient later).

Lemma 9. *Let $S \subseteq X$ with $d(s,s') > 4r$ for all $s \neq s' \in S$, $\beta \in \mathbb{Z}_{\geq 0}$. If no radius $2r$ solution $C_2 \subseteq X$ satisfies $|C_2 \setminus S| \leq \beta$, then $|C_1 \cap B(S,r)| \leq k - \beta - 1$ for any radius r solution C_1.*

Proof. Assume there is a solution C_1 of radius r where $|C_1 \cap B(S,r)| \geq k - \beta$. Let $A = C_1 \cap B(S,r)$. For each $p \in A$, let $\phi(p) \in S$ be the unique point in S such that $p \in B(\phi(p), r)$; $\phi(p)$ is well defined because $d(s,s') > 4r$ for every $s \neq s' \in S$. Let $C_2 = \phi(A) \cup (C_1 \setminus A)$. Then $|C_2| \leq |\phi(A)| + |C_1 \setminus A| \leq |A| + |C_1 \setminus A| \leq k$. Moreover, as $d(p, \phi(p)) \leq r$ for every $p \in A$, we conclude that $B(C_1, r) \subseteq B(C_2, 2r)$. Thus, C_2 is a feasible solution of radius $2r$. $\square$

Hence, if $y(B(S,r)) > k - \gamma + 1$ and $(x,y) \in \mathcal{P}_I$, then there is a solution C_2 of radius $2r$ with $|C_2 \setminus S| \leq \gamma - 2$. The motivation for considering solutions of radius $2r$ with all centers in S except for constantly many (if $\gamma = O(1)$) is that such solutions can be found efficiently via dynamic programming. This is possible because the centers in S are separated by distances strictly larger than $4r$, which implies that radius $2r$ balls centered at points in S do not overlap. Hence, there are no interactions between such balls. This is formalized below.

Lemma 10. *Let $S \subseteq X$ with $d(s,s') > 4r$ for all $s,s' \in S$ with $s \neq s'$, and $\beta \in \mathbb{Z}_{\geq 0}$. If a radius $2r$ solution $C \subseteq X$ with $|C \setminus S| \leq \beta$ exists, then we can find such a solution in time $|X|^{O(\beta+\gamma)}$.*

Proof. Suppose there is a solution $C \subseteq X$ of radius $2r$ with $|C \setminus S| \leq \beta$. The algorithm has two components. We first guess the set $Q := C \setminus S$. Because $|Q| \leq \beta$, there are $|X|^{O(\beta)}$ choices. Given Q, it remains to select at most $k - |Q|$ centers $W \subseteq S$ to fulfill the color requirements. Note that for any $W \subseteq S$, the number of points of color $\ell \in [\gamma]$ that $B(W, 2r)$ covers on top of those already covered by $B(Q, 2r)$ is $|(B(W, 2r) \setminus B(Q, 2r)) \cap X_\ell| = \sum_{w \in W} |(B(w, 2r) \setminus B(Q, 2r)) \cap X_\ell|$, where equality holds because centers in W are separated by distances strictly larger than $4r$, and thus $B(W, 2r)$ is the disjoint union of the sets $B(w, 2r)$ for $w \in W$. Hence, the task of finding a set $W \subseteq S$ with $|W| \leq k - |Q|$ such that $Q \cup W$ is a solution of radius $2r$ can be phrased as finding a feasible solution to the following binary program:

$$\sum_{s \in S} z(s) \cdot |(B(s,2r) \setminus B(Q,2r)) \cap X_\ell| \geq m_\ell - |B(Q,2r) \cap X_\ell| \quad \forall \ell \in [\gamma]$$

$$\sum_{s \in S} z(s) \leq k - |Q| \qquad (2)$$

$$z \in \{0,1\}^S.$$

As this is a binary linear program with $\gamma + 1$ constraints with coefficients in $\{0, \dots, |X|\}$, it can be solved by standard dynamic programming techniques in $|X|^{O(\gamma)}$ time.[3] As the dynamic program is run for $|X|^{O(\beta)}$ many guesses of Q, we obtain an overall running time of $|X|^{O(\beta+\gamma)}$, as claimed. $\square$

This completes the last ingredient for an iteration of our round-or-cut approach as shown in Fig. 1. In summary, assuming $y(B(S,r)) > k - \gamma + 1$ (for otherwise Lemma 8 leads to a solution of radius $4r$) we use Lemma 10 to check whether there is a radius $2r$ solution C_2 with $|C_2 \setminus S| \leq \gamma - 2$. If this is the case, we are done. If not, Lemma 9 implies that every radius r solution C_1 fulfills $|C_1 \cap B(S,r)| \leq k - \gamma + 1$. Hence, every point $(\overline{x}, \overline{y}) \in \mathcal{P}_I$ satisfies $\overline{y}(B(S,r)) \leq k - \gamma + 1$. However, this constraint is violated by (x,y), and so it separates (x,y) from $\mathcal{P}_I$. Thus, we proved that the process described in Fig. 1 is a valid round-or-cut procedure that runs in polynomial time.

Corollary 11. *There is a polynomial-time algorithm that, given a point $(x,y) \in \mathbb{R}^X \times \mathbb{R}^X$, either returns a γCkC solution of radius $4r$ or an inequality separating (x,y) from $\mathcal{P}_I$.*

We can now prove the main theorem.

Proof (of Theorem 3). We run the ellipsoid method on $\mathcal{P}_I$ for each of the $O(|X|^2)$ candidate radii r. For each r, the number of ellipsoid iterations is polynomially bounded as the separating hyperplanes have encoding length at most $O(|X|)$ (see Theorem 6.4.9 of [14]). To see this, note that all generated hyperplanes are either inequalities defining $\mathcal{P}$ or inequalities of the form $y(B(S,r)) \leq k - \gamma + 1$. For the correct guess of r, $\mathcal{P}_I$ is non-empty and the algorithm terminates by returning a radius $4r$ solution. Hence, if we return the best solution among those computed for all guesses of r, we have a 4-approximation. $\square$

3   The Lottery Model of Harris et al. [15]

Let (X,d) be a Fair γCkC instance, and let $\mathcal{F}(r)$ be the family of sets of centers satisfying the covering requirements with radius r, i.e.,

$$\mathcal{F}(r) := \left\{ C \subseteq X \, \middle| \, |C| \leq k \text{ and } |B(C,r) \cap X_\ell| \geq m_\ell \ \forall \ell \in [\gamma] \right\} .$$

Note that a radius r solution for Fair γCkC defines a distribution over the sets in $\mathcal{F}(r)$. Given r, such a distribution exists if and only if the following (exponential-size) linear program PLP(r) is feasible (with DLP(r) being its dual):

---

[3] For any ordering $S = \{s_1, \dots, s_q\}$ of the elements in S, the DP successively computes for any prefix $\{s_1, \dots, s_i\}$ all possible left-hand side values for the $\gamma + 1$ constraints that can be achieved if $\mathrm{supp}(z) := \{s \in S : z(s) > 0\} \subseteq \{s_1, \dots, s_i\}$. This is a straightforward extension of classical DPs for binary knapsack problems.

$$\text{PLP}(r) : \min 0 \qquad\qquad\qquad \text{DLP}(r) : \max \sum_{u \in X} p(u)\alpha(u) - \mu$$

$$\sum_{\substack{C \in \mathcal{F}(r): \\ u \in B(C,r)}} \lambda(C) \geq p(u) \quad \forall u \in X \qquad \sum_{u \in B(C,r)} \alpha(u) \leq \mu \qquad \forall C \in \mathcal{F}(r)$$

$$\sum_{C \in \mathcal{F}(r)} \lambda(C) = 1 \qquad\qquad\qquad\qquad \alpha \in \mathbb{R}_{\geq 0}^{X}$$

$$\lambda \in \mathbb{R}_{\geq 0}^{\mathcal{F}(r)} \qquad\qquad\qquad\qquad\qquad \mu \in \mathbb{R} \; .$$

Clearly, if $\text{PLP}(r)$ is feasible, then its optimal value is 0. Observe that $\text{DLP}(r)$ always has a feasible solution (the zero vector) of value 0. Thus, by strong duality, $\text{PLP}(r)$ is feasible if and only if the optimal value of $\text{DLP}(r)$ is 0. We note now that $\text{DLP}(r)$ is scale-invariant, meaning that if (α, μ) is feasible for $\text{DLP}(r)$ then so is $(t\alpha, t\mu)$ for $t \in \mathbb{R}_{\geq 0}$. Hence, $\text{DLP}(r)$ has a solution of strictly positive objective value if and only if $\text{DLP}(r)$ is unbounded. We thus define the following polyhedron $\mathcal{Q}(r)$, which contains all solutions of $\text{DLP}(r)$ of value at least 1:

$$\mathcal{Q}(r) := \left\{ (\alpha, \mu) \in \mathbb{R}_{\geq 0}^{X} \times \mathbb{R} \;\middle|\; \sum_{u \in X} p(u)\alpha(u) \geq \mu + 1, \; \sum_{u \in B(C,r)} \alpha(u) \leq \mu \; \forall C \in \mathcal{F}(r) \right\}.$$

As discussed, we have the following.

Lemma 12. $\mathcal{Q}(r)$ *is empty if and only if* $PLP(r)$ *is feasible.*

The main lemma that allows us to obtain our result is the following.

Lemma 13. *There is a polynomial-time algorithm that, given a point* $(\alpha, \mu) \in \mathbb{R}_{\geq 0}^{X} \times \mathbb{R}$ *satisfying* $\sum_{u \in X} p(u)\alpha(u) \geq \mu + 1$ *and a radius* $r \geq 0$, *either certifies that* $(\alpha, \mu) \in \mathcal{Q}(r)$, *or outputs a set* $C \in \mathcal{F}(4r)$ *with* $\sum_{u \in B(C,4r)} \alpha(u) > \mu$.

In words, Lemma 13 either certifies $(\alpha, \mu) \in \mathcal{Q}(r)$ or returns a hyperplane separating (α, μ) from $\mathcal{Q}(4r)$. Its proof leverages techniques introduced in Sect. 2, and we sketch it in Appendix A. Using Lemma 13, we can now prove Theorem 4.

Proof (of Theorem 4). As noted, there are polynomially many choices for the radius r, for each of which we run the ellipsoid method to check emptiness of $\mathcal{Q}(4r)$ as follows. Whenever there is a call to the separation oracle for a point $(\alpha, \mu) \in \mathbb{R}^{X} \times \mathbb{R}$, we first check whether $\alpha \geq 0$ and $\sum_{u \in X} p(u)\alpha(u) \geq \mu + 1$. If one of these constraints is violated, we return it as separating hyperplane. Otherwise, we invoke the algorithm of Lemma 13. The algorithm either returns a constraint in the inequality description of $\mathcal{Q}(4r)$ violated by (α, μ), which solves the separation problem, or certifies $(\alpha, \mu) \in \mathcal{Q}(r)$. If, at any iteration of the ellipsoid method, the separation oracle is called for a point (α, μ) for which Lemma 13 certifies $(\alpha, \mu) \in \mathcal{Q}(r)$, then Lemma 12 implies $\text{PLP}(r)$ is infeasible. Thus, there is no solution to the considered Fair γCkC instance of radius r. Hence, consider from now on that the separation oracle always returns a separating hyperplane, in which case the ellipsoid method certifies that $\mathcal{Q}(4r) = \emptyset$ as follows. Let $\mathcal{H} \subseteq \mathcal{F}(4r)$ be the family of all sets $C \in \mathcal{F}(4r)$ returned by Lemma 13 through calls to the separation oracle. Then, the following polyhedron:

$$\mathcal{Q}_{\mathcal{H}}(4r) = \left\{ (\alpha, \mu) \in \mathbb{R}_{\geq 0}^X \times \mathbb{R} \,\middle|\, \sum_{u \in X} p(u)\alpha(u) \geq \mu + 1, \; \sum_{u \in B(C, 4r)} \alpha(u) \leq \mu \; \forall C \in \mathcal{H} \right\},$$

containing $\mathcal{Q}(4r)$, is empty. As the encoding length of any constraint in the inequality description of $\mathcal{Q}(4r)$ is polynomially bounded in the input, the ellipsoid method runs in polynomial time (see Theorem 6.4.9 of [14]). In particular, the number of calls to the separation oracle, and thus $|\mathcal{H}|$, is polynomially bounded.

As $\mathcal{Q}(4r) \subseteq \mathcal{Q}_{\mathcal{H}}(4r) = \emptyset$, Lemma 12 implies that $PLP(4r)$ is feasible. More precisely, because $\mathcal{Q}_{\mathcal{H}}(4r) = \emptyset$, the linear program obtained from $DLP(4r)$ by replacing $\mathcal{F}(4r)$, which parameterizes the constraints in $DLP(4r)$, by $\mathcal{H}$, has optimal value equal to zero. Hence, its dual, which corresponds to $PLP(4r)$ where we replace $\mathcal{F}(4r)$ by $\mathcal{H}$ is feasible. As this feasible linear program has polynomial size, because $|\mathcal{H}|$ is polynomially bounded, we can solve it efficiently to obtain a distribution with the desired properties.   □

A   Appendix

In this appendix we discuss the proof of Lemma 13. Due to space constraints, we only sketch the proofs of some auxiliary lemmas; formal proofs are deferred to the full version.

The desired separation algorithm requires us to find a solution for a γCkC instance with an extra covering constraint; the procedure of Sect. 2 generalizes to handle this extra constraint. We follow similar steps as in Fig. 1.

Let $(\alpha, \mu) \in \mathbb{R}_{\geq 0}^X \times \mathbb{R}$ be a point satisfying $\sum_{u \in X} p(u)a(u) \geq \mu + 1$, let $r \geq 0$, and, moreover, let $\mathcal{F}^{\alpha,\mu}(r) := \{C \in \mathcal{F}(r) \mid \sum_{u \in B(C,r)} \alpha(u) > \mu\}$. We have to decide whether $\mathcal{F}^{\alpha,\mu}(4r) = \emptyset$ and, if not, find a set $C \in \mathcal{F}^{\alpha,\mu}(4r)$. We claim that there is a polynomially encoded $\varepsilon > 0$, such that this is equivalent to finding $C \in \mathcal{F}(4r)$ with $\sum_{u \in B(C,4r)} \alpha(u) \geq \mu + \varepsilon$, or deciding that no such C exists. The next standard result guarantees that such an $\varepsilon > 0$ can be computed efficiently.

Lemma 14. *Let* $(\alpha, \mu) \in \mathbb{R}_{\geq 0}^X \times \mathbb{R}$. *Then one can efficiently compute an* $\varepsilon > 0$ *with encoding length* $O(L)$, *where* L *is the encoding length of* (α, μ), *such that the following holds: For any* $C \in \mathcal{F}(r)$, *we have* $\sum_{u \in B(C,r)} \alpha(u) > \mu$ *if and only if* $\sum_{u \in B(C,r)} \alpha(u) \geq \mu + \varepsilon$.

Proof. The tuple (α, μ) consists of $|X| + 1$ rationals $\{p_i/q_i\}_{i \in [N]}$, with $p_i \in \mathbb{Z}$ and $q_i \in \mathbb{Z}_{>0}$. Let $\Pi = \prod_{i \in [N]} q_i$. Note that if $\sum_{u \in B(C,r)} \alpha(u) > \mu$, then $\sum_{u \in B(C,r)} \alpha(u) - \mu \geq \frac{1}{\Pi}$. Thus, we set $\varepsilon = 1/\Pi$. Moreover $\log \Pi = \sum_{i \in [N]} \log q_i$, and so the encoding length of ε is $O(L)$.   □

Let $\mathcal{P}^{\alpha,\mu}$ be the following modified relaxation of γCkC, defined for given $(\alpha,\mu) \in \mathbb{R}_{\geq 0}^X \times \mathbb{R}$, where the polytope $\mathcal{P}$ is defined for a fixed radius r, as in Sect. 2 (see (1)):

$$\mathcal{P}^{\alpha,\mu} := \left\{ (x,y) \in \mathcal{P} \,\middle|\, \sum_{u \in X} \alpha(u)x(u) \geq \mu + \varepsilon \right\} .$$

Let $\mathcal{P}_I^{\alpha,\mu} := \mathrm{conv}\left(\mathcal{P}^{\alpha,\mu} \cap (\{0,1\}^X \times \{0,1\}^X)\right)$ be the integer hull of $\mathcal{P}^{\alpha,\mu}$. We now state the following straightforward lemma, whose proof is an immediate consequence of the definitions of the corresponding polytopes and Lemma 14.

Lemma 15. *Let $(\alpha,\mu) \in \mathbb{R}_{\geq 0}^X \times \mathbb{R}$ be such that $\sum_{u \in X} p(u)\hat{\alpha}(u) \geq \mu + 1$ and $\mathcal{P}_I^{\alpha,\mu} = \emptyset$. Then $(\alpha,\mu) \in \mathcal{Q}(r)$.*

We now state Lemma 16, a modified version of Lemma 8. Its proof is analogous to the one of Lemma 8 and is deferred to the full version. The only difference is that the auxiliary polytope on which we exploit sparsity has one additional constraint, the one corresponding to $\sum_{u \in X} \alpha(u)x(u) \geq \mu + \varepsilon$, which explains the shift by one unit in the condition on $y(B(S,r))$.

Lemma 16. *Let $(\alpha,\mu) \in \mathbb{R}_{\geq 0}^X \times \mathbb{R}$, let $(x,y) \in \mathcal{P}^{\alpha,\mu}$, and let $(S,\mathcal{D})$ be a y-good clustering. If $y(B(S,r)) \leq k - \gamma$, a set $C \in \mathcal{F}^{\alpha,\mu}(4r)$ can be found in polynomial time.*

If $y(B(S,r)) \leq k - \gamma$, then Lemma 16 leads to a set $C \in \mathcal{F}(4r)$ that satisfies $\sum_{u \in B(S,4r)} \alpha(u) \geq \mu + \varepsilon$; this gives a constraint separating (α,μ) from $\mathcal{Q}(4r)$.

It remains to consider the case $y(B(S,r)) > k - \gamma$. As in Sect. 2, we can either find a set $C_2 \in \mathcal{F}^{\alpha,\mu}(2r)$ or certify that every $C_1 \in \mathcal{F}^{\alpha,\mu}(r)$ satisfies $|C_1 \cap B(S,r)| \leq k - \gamma$.

Lemma 17. *Let $(\alpha,\mu) \in \mathbb{R}_{\geq 0}^X \times \mathbb{R}$, $S \subseteq X$ with $d(s,s') > 4r$ for all $s,s' \in S$ with $s \neq s'$, and $\beta \in \mathbb{Z}_{\geq 0}$. If no $C_2 \in \mathcal{F}^{\alpha,\mu}(2r)$ satisfies $|C_2 \setminus S| \leq \beta$, then $|C_1 \cap B(S,r)| \leq k - \beta - 1$ for any $C_1 \in \mathcal{F}^{\alpha,\mu}(r)$.*

Proof (Sketch). As in Lemma 9, we prove that any $C_1 \in \mathcal{F}^{\alpha,\mu}(r)$ satisfying $|C_1 \cap B(S,r)| > k - \beta - 1$ can be transformed into a set $C_2 \in \mathcal{F}^{\alpha,\mu}(2r)$ with $|C_2 \setminus S| \leq \beta$. $\qquad\square$

Lemma 18. *Let $(\alpha,\mu) \in \mathbb{R}_{\geq 0}^X \times \mathbb{R}$, $S \subseteq X$ with $d(s,s') > 4r$ for all $s,s' \in S$ with $s \neq s'$, and $\beta \in \mathbb{Z}_{\geq 0}$. If there exists a set $C \in \mathcal{F}^{\alpha,\mu}(2r)$ with $|C \setminus S| \leq \beta$, then we can find such a set in time $|X|^{O(\beta+\gamma)}$.*

Proof (Sketch). As in the proof of Lemma 10, we first guess up to β centers $Q = X \setminus S$. For each of those guesses, we then consider the binary program (2) with objective function $\sum_{s \in S} z(s) \cdot \alpha(B(s,2r))$ to be maximized. For the guess $Q = C \setminus S$, the characteristic vector $\chi^{C \cap S}$ is feasible for this binary program, implying that the optimal centers $Z \subseteq S$ chosen by the binary program fulfill $Z \cup Q \in \mathcal{F}^{\alpha,\mu}(2r)$. $\qquad\square$

Corollary 19. *Let $(\alpha, \mu) \in \mathbb{R}^X_{\geq 0} \times \mathbb{R}$. There is a polynomial-time algorithm that, given $(x, y) \in \mathbb{R}^X \times \mathbb{R}^X$, either returns a set $C \in \mathcal{F}^{\alpha,\mu}(4r)$ or returns a hyperplane separating (x, y) from $\mathcal{P}^{\alpha,\mu}_I$.*

Proof. If $(x, y) \notin \mathcal{P}^{\alpha,\mu}$, we return a violated constraint separating (x, y) from $\mathcal{P}^{\alpha,\mu}_I$. Hence we may assume $(x, y) \in \mathcal{P}^{\alpha,\mu}$. Since $\mathcal{P}^{\alpha,\mu} \subseteq \mathcal{P}$, we can use Theorem 7 to get a y-good clustering $(S, \mathcal{D})$. If $y(B(S, r)) \leq k - \gamma$, Lemma 16 gives a set $C \in \mathcal{F}^{\alpha,\mu}(4r)$. So, assuming $y(B(S, r)) > k - \gamma$, we use Lemma 18 to check whether there is $C_2 \in \mathcal{F}^{\alpha,\mu}(2r)$ with $|C_2 \setminus S| \leq \gamma - 1$. If this is the case, we are done. If not, Lemma 17 implies that every $C_1 \in \mathcal{F}^{\alpha,\mu}(r)$ fulfills $|C_1 \cap B(S, r)| \leq k - \gamma$. Hence, every point $(\overline{x}, \overline{y}) \in \mathcal{P}^{\alpha,\mu}_I$ satisfies $\overline{y}(B(S, r)) \leq k - \gamma$. However, this constraint is violated by (x, y), and it thus separates (x, y) from $\mathcal{P}^{\alpha,\mu}_I$. $\qquad\square$

Proof (Proof of Lemma 13). We use the ellipsoid method to check emptiness of $\mathcal{P}^{\alpha,\mu}_I$. Whenever the separation oracle gets called for a point $(x, y) \in \mathbb{R}^X \times \mathbb{R}^X$, we invoke the algorithm of Corollary 19. If the algorithm returns at any point a set $C \in \mathcal{F}^{\alpha,\mu}(4r)$, then C corresponds to a constraint in the inequality description of $\mathcal{Q}(4r)$ violated by (α, μ). Otherwise, the ellipsoid method certifies that $\mathcal{P}^{\alpha,\mu}_I = \emptyset$, which implies $(\alpha, \mu) \in \mathcal{Q}(r)$ by Lemma 15. Note that the number of iterations of the ellipsoid method is polynomial as the hyperplanes used by the procedure above have encoding length $O(\text{poly}(|X|))$ (see Theorem 6.4.9 of [14]). $\qquad\square$

References

1. An, H.C., Singh, M., Svensson, O.: LP-based algorithms for capacitated facility location. SIAM J. Comput. **46**(1), 272–306 (2017)
2. Backurs, A., Indyk, P., Onak, K., Schieber, B., Vakilian, A., Wagner, T.: Scalable fair clustering. In: Proceedings of the 36th International Conference on Machine Learning (ICML), pp. 405–413 (2019)
3. Bandyapadhyay, S., Inamdar, T., Pai, S., Varadarajan, K.R.: A constant approximation for colorful k-center. In: Proceedings of the 27th Annual European Symposium on Algorithms (ESA), pp. 12:1–12:14 (2019)
4. Bera, S.K., Chakrabarty, D., Flores, N., Negahbani, M.: Fair algorithms for clustering. In: Proceedings of the 33rd International Conference on Neural Information Processing Systems (NeurIPS), pp. 4955–4966 (2019)
5. Bercea, I.O., et al.: On the cost of essentially fair clusterings. In: Proceedings of the 22nd International Conference on Approximation Algorithms for Combinatorial Optimization Problems (APPROX/RANDOM), pp. 18:1–18:22 (2019)
6. Carr, R.D., Fleischer, L.K., Leung, V.J., Phillips, C.A.: Strengthening integrality gaps for capacitated network design and covering problems. In: Proceedings of the 11th Annual ACM-SIAM Symposium on Discrete Algorithms (SODA), pp. 106–115 (2000)
7. Chakrabarty, D., Goyal, P., Krishnaswamy, R.: The non-uniform k-center problem. In: Proceedings of the 43rd International Colloquium on Automata, Languages, and Programming, (ICALP), pp. 67:1–67:15 (2016)
8. Chakrabarty, D., Negahbani, M.: Generalized center problems with outliers. ACM Trans. Algorithms **15**(3), 1–14 (2019)

9. Charikar, M., Khuller, S., Mount, D.M., Narasimhan, G.: Algorithms for facility location problems with outliers. In: Proceedings of the 12th Annual Symposium on Discrete Algorithms (SODA), pp. 642–651 (2001)
10. Chen, D.Z., Li, J., Liang, H., Wang, H.: Matroid and knapsack center problems. Algorithmica **75**(1), 27–52 (2016)
11. Chierichetti, F., Kumar, R., Lattanzi, S., Vassilvitskii, S.: Fair clustering through fairlets. In: Proceedings of the 31st International Conference on Neural Information Processing Systems (NIPS), pp. 5029–5037 (2017)
12. Gonzalez, T.F.: Clustering to minimize the maximum intercluster distance. Theor. Comput. Sci. **38**, 293–306 (1985)
13. Grandoni, F., Kalaitzis, C., Zenklusen, R.: Improved approximation for tree augmentation: saving by rewiring. In: Proceedings of the 50th ACM Symposium on Theory of Computing (STOC), pp. 632–645 (2018)
14. Grötschel, M., Lovász, L., Schrijver, A.: Geometric Algorithms and Combinatorial Optimization, vol. 2. Springer, Heidelberg (2012). https://doi.org/10.1007/978-3-642-78240-4
15. Harris, D.G., Pensyl, T., Srinivasan, A., Trinh, K.: A lottery model for center-type problems with outliers. ACM Trans. Algorithms **15**(3), 36:1–36:25 (2019)
16. Hochbaum, D.S., Shmoys, D.B.: A best possible heuristic for the k-center problem. Math. Oper. Res. **10**(2), 180–184 (1985)
17. Hochbaum, D.S., Shmoys, D.B.: A unified approach to approximation algorithms for bottleneck problems. J. ACM **33**(3), 533–550 (1986)
18. Hsu, W., Nemhauser, G.L.: Easy and hard bottleneck location problems. Discrete Appl. Math. **1**(3), 209–215 (1979)
19. Levi, R., Lodi, A., Sviridenko, M.: Approximation algorithms for the capacitated multi-item lot-sizing problem via flow-cover inequalities. Math. Oper. Res. **33**(2), 461–474 (2008)
20. Li, S.: Approximating capacitated k-median with $(1 + \epsilon)k$ open facilities. In: Proceedings of the 27th Annual ACM Symposium on Discrete Algorithms (SODA), pp. 786–796 (2016)
21. Li, S.: On uniform capacitated k-median beyond the natural LP relaxation. ACM Trans. Algorithms **13**(2), 22:1–22:18 (2017)
22. Nutov, Z.: On the tree augmentation problem. In: Proceedings of the 25th Annual Symposium on Algorithms (ESA), pp. 61:1–61:14 (2017)
23. Rösner, C., Schmidt, M.: Privacy preserving clustering with constraints. In: Proceedings of the 45th International Colloquium on Automata, Languages, and Programming (ICALP), pp. 96:1–96:14 (2018)

Tight Approximation Bounds
for Maximum Multi-coverage

Siddharth Barman[1(✉)], Omar Fawzi[2], Suprovat Ghoshal[1],
and Emirhan Gürpınar[2]

[1] Indian Institute of Science, Bangalore, India
barman@iisc.ac.in
[2] Univ Lyon, ENS Lyon, UCBL, CNRS, LIP, 69342 Lyon Cedex 07, France

Abstract. In the classic maximum coverage problem, we are given subsets $T_1, \ldots, T_m$ of a universe $[n]$ along with an integer k and the objective is to find a subset $S \subseteq [m]$ of size k that maximizes $C(S) := |\cup_{i \in S} T_i|$. It is well-known that the greedy algorithm for this problem achieves an approximation ratio of $(1 - e^{-1})$ and there is a matching inapproximability result. We note that in the maximum coverage problem if an element $e \in [n]$ is covered by several sets, it is still counted only once. By contrast, if we change the problem and count each element e as many times as it is covered, then we obtain a linear objective function, $C^{(\infty)}(S) = \sum_{i \in S} |T_i|$, which can be easily maximized under a cardinality constraint.

We study the maximum ℓ-multi-coverage problem which naturally interpolates between these two extremes. In this problem, an element can be counted up to ℓ times but no more; hence, we consider maximizing the function $C^{(\ell)}(S) = \sum_{e \in [n]} \min\{\ell, |\{i \in S : e \in T_i\}|\}$, subject to the constraint $|S| \leq k$. Note that the case of $\ell = 1$ corresponds to the standard maximum coverage setting and $\ell = \infty$ gives us a linear objective.

We develop an efficient approximation algorithm that achieves an approximation ratio of $1 - \frac{\ell^{\ell} e^{-\ell}}{\ell!}$ for the ℓ-multi-coverage problem. In particular, when $\ell = 2$, this factor is $1 - 2e^{-2} \approx 0.73$ and as ℓ grows the approximation ratio behaves as $1 - \frac{1}{\sqrt{2\pi\ell}}$. We also prove that this approximation ratio is tight, i.e., establish a matching hardness-of-approximation result, under the Unique Games Conjecture.

This problem is motivated by the question of finding a code that optimizes the list-decoding success probability for a given noisy channel. We show how the multi-coverage problem can be relevant in other contexts, such as combinatorial auctions.

A full version of the paper containing the proofs can be found in [3]. This research is supported by the French ANR project ANR-18-CE47-0011 (ACOM). Siddharth Barman gratefully acknowledges the support of a Ramanujan Fellowship (SERB - SB/S2/RJN-128/2015) and a Pratiksha Trust Young Investigator Award. Part of this work was conducted during the first author's visit to École Normale Supérieure de Lyon and was supported by the Administration de la recherche (ADRE), France.

D. Bienstock and G. Zambelli (Eds.): IPCO 2020, LNCS 12125, pp. 66–77, 2020.
https://doi.org/10.1007/978-3-030-45771-6_6

Keywords: Approximation algorithms · Covering problems · Submodular optimization

1   Introduction

Coverage problems lie at the core of combinatorial optimization and have been extensively studied in computer science. A quintessential example of such problems is the *maximum coverage* problem wherein we are given subsets $T_1, \ldots, T_m$ of a universe $[n]$ along with an integer $k \in \mathbb{Z}_+$, and the objective is to find a size-k set $S \subseteq [m]$ that maximizes the covering function $C(S) := |\cup_{i \in S} T_i|$. It is well-known that a natural greedy algorithm achieves an approximation ratio of $1 - e^{-1}$ for this problem (see, e.g., [16]). Furthermore, the work of Feige [13] shows that this approximation factor is tight, under the assumption that $P \neq NP$. Over the years, a large body of work has been directed towards extending these fundamental results and, more generally, coverage problems have been studied across multiple fields, such as operations research [8], machine learning [15], and algorithmic game theory [12].

In this paper, we study the ℓ-multi-coverage (ℓ-coverage for short) problem, which is a natural generalization of the classic maximum coverage problem. Here, we are given a universe of elements $[n]$ and a collection of subsets $\mathcal{F} = \{T_i \subseteq [n]\}_{i=1}^m$. For any integer $\ell \in \mathbb{Z}_+$ and a choice of index set $S \subseteq [m]$, we define the ℓ-coverage of an element e to be $C_e^{(\ell)}(S) := \min\{\ell, |\{i \in S : e \in T_i\}|\}$, i.e., $C_e^{(\ell)}(S)$ counts—up to ℓ—how many times element e is covered by the subsets indexed in S. We extend this definition to that of ℓ-coverage of all the elements, $C^{(\ell)}(S) := \sum_{e \in [n]} C_e^{(\ell)}(S)$.

The ℓ-multi-coverage problem is defined as follows: given a universe of elements $[n]$, a collection $\mathcal{F}$ of subsets of $[n]$ and an integer $k \leq m$, find a size-k subset $S \subseteq [m]$ which maximizes $C^{(\ell)}(S)$. For $\ell = 1$, it is easy to see that this reduces to the standard maximum coverage problem.

1.1   Our Results and Techniques

Our main result is a polynomial-time algorithm that achieves a tight approximation ratio for the ℓ-multi-coverage problem, with any $\ell \geq 1$.

Theorem 1. *Let ℓ be a positive integer. There exists a polynomial-time algorithm that takes as input an integer n, a set system $\mathcal{F} = \{T_i \subseteq [n]\}_{i=1}^m$ along with an integer $k \leq m$ and outputs a size-k set $S \subseteq [m]$ (i.e., identifies k subsets $\{T_i\}_{i \in S}$ from $\mathcal{F}$) such that*

$$C^{(\ell)}(S) \geq \left(1 - \frac{\ell^\ell e^{-\ell}}{\ell!}\right) \max_{S' \in \binom{[m]}{k}} C^{(\ell)}(S').$$

One way to interpret this approximation ratio $\rho_\ell := \left(1 - \frac{\ell^\ell e^{-\ell}}{\ell!}\right)$ is that $\rho_\ell = \mathbb{E}\left[\frac{1}{\ell}\min\{\ell, \text{Poi}(\ell)\}\right]$, where $\text{Poi}(\ell)$ denotes a Poisson random variable with rate parameter ℓ.

We complement Theorem 1 by proving that the achieved approximation guarantee is tight, under the Unique Games Conjecture. Formally,

Theorem 2. *Assuming the Unique Games Conjecture, it is NP-hard to approximate the maximum ℓ-multi-coverage problem to within a factor greater than $\left(1 - \frac{\ell^\ell}{\ell!} e^{-\ell} + \varepsilon\right)$, for any constant of $\varepsilon > 0$.*

The Approximation Algorithm. We first observe that for the maximum multi-coverage problem the standard greedy algorithm fails: the approximation guarantee does not improve with ℓ. The greedy algorithm starts with $S_0 = \emptyset$ and repeats the following k times for $j = 0$ to $k - 1$. Let i_j be a maximizer of the function $i \mapsto C^{(\ell)}(S_j \cup \{i\})$ and let $S_{j+1} \leftarrow S_j \cup \{i_j\}$. The greedy algorithm outputs S_k. As the function $C^{(\ell)}$ is a monotone, submodular function, the greedy algorithm will certainly still achieve an approximation ratio of $1 - e^{-1}$. However, it is simple to construct instances wherein exactly this ratio is achieved. In fact, if $\mathcal{F}$ is a collection of distinct subsets, let $\mathcal{F}^{(\ell)}$ contain the same subsets as $\mathcal{F}$ but each one appearing ℓ times. Then, it is easy to see that the greedy algorithm, when applied to $\mathcal{F}^{(\ell)}$, will simply choose ℓ times the sets chosen by the algorithm on input $\mathcal{F}$. So the greedy algorithm is not able to take advantage when we have $\ell > 1$.

Instead, we use another algorithmic idea, which is standard in the context of covering problems. We consider the natural linear programming (LP) relaxation of the problem to obtain a fractional, optimal solution and apply pipage rounding [1,5,27]. Pipage rounding is a method that maps a fractional solution $x \in [0,1]^m$ into an integral one $x^{\text{int}} \in \{0,1\}^m$, in a way that does not decrease the expected value of the objective function; here the fractional solution $x \in [0,1]^m$ is viewed as a (product) distribution over the index set $[m]$.

Hence, the core of the analysis of this algorithm is to compute the expected ℓ-coverage, $\mathbb{E}_{S \sim x}\left[C^{(\ell)}(S)\right]$, and relate it to the optimal value of the linear program (which, of course, upper bounds the value of an integral, optimal solution). With a careful use of convexity, one can establish that the analytic form of this expectation corresponds to the expected value of a binomial random variable truncated at ℓ.

To obtain the claimed approximation ratio (which, as mentioned above, has a Poisson interpretation), one would like to use the well-known Poisson approximation for binomial distributions. However, this convergence statement is only asymptotic and thus will lead to an error term that will depend on the size of the problem instance and on the value of ℓ. One can alternatively try to compare the two distributions using the natural notion of stochastic domination. It turns out that indeed a binomial distribution can be stochastically dominated by a Poisson distribution, but this again cannot be used in our setting for two reasons: there is a loss in terms of the underlying parameters (and, hence, this cannot lead to a tight approximation factor) and more importantly the inequality goes in the wrong direction.[1]

---

[1] Here, the Poisson distribution stochastically dominates the binomial. Hence, instead of a lower bound, we obtain an upper bound.

The right tool for us turns out to be the notion of *convex order* between distributions. It expresses the property that one distribution is more "spread" than the other. While this notion has found several applications in statistics, economics, and other related fields (see [25] and references therein), to the best of our knowledge, its use in the context of analyzing approximation algorithms is novel. In particular, it leads to tight comparison inequalities between binomial and Poisson distributions, even in non-asymptotic regimes (see Lemma 1). Overall, using this tool we are able to obtain optimal approximation guarantees for all values of ℓ.

We also note that our algorithmic result directly generalizes to the weighted version of maximum ℓ-coverage and we can replace the constraint $|S| \leq k$ by a matroid constraint $S \in \mathcal{M}$; here, $\mathcal{M}$ is any matroid that admits an efficient, optimization algorithm (equivalently, any matroid whose basis polytope admits an efficient separation oracle). To keep the exposition simple, we conform to the unweighted case and to the cardinality constraint $|S| \leq k$.

Hardness Result. We now give a brief description of our hardness result and the techniques used to establish it. In [13], the $(1 - 1/e)$ inapproximability of the standard maximum coverage problem was shown using the tight $\ln n$ inapproximability of the *set cover* problem, which in turn was obtained via a reduction from a variant of MAX-3-SAT. However, in our setting, one cannot hope to show tight inapproximability for the maximum ℓ-multi-coverage problem by a similar sequence of reductions. This is because, as detailed in Sect. 1.2, the multi-coverage analogue of the set cover problem is as inapproximable as the usual set cover problem. Therefore, one cannot hope to directly reuse the arguments from Feige's reduction in order to get tight inapproximability for the maximum ℓ-multi-coverage problem. We bypass this by developing a direct reduction to the maximum ℓ-multi-coverage problem without going through the set cover variant.

Our reduction is from an h-ary hypergraph variant of UNIQUEGAMES [19,20], which we call h-ARYUNIQUEGAMES. Here the constraints are given by h-uniform hyperedges on a vertex set V with a label set Σ. A salient feature of the h-ARYUNIQUEGAMES, which is crucially used in our reduction, is that it involves two distinct notions of satisfied hyperedges, namely *strongly* and *weakly* satisfied hyperedges. A labeling $\sigma : V \mapsto \Sigma$ strongly satisfies a hyperedge $e = (v_i)_{i \in [h]}$ if all the labels project to the same alphabet, i.e., $\pi_{e,v_1}(\sigma(v_1)) = \pi_{e,v_2}(\sigma(v_2)) = \cdots = \pi_{e,v_3}(\sigma(v_h))$. We say a labelling σ weakly satisfies the hyperedge e if at least two of the projected labels match, i.e., $\pi_{e,v_i}(\sigma(v_i)) = \pi_{e,v_j}(\sigma(v_j))$ for some $i, j \in [h], i \neq j$. For these instances, the equivalent form of the Unique Games Conjecture is the following: *It is NP-Hard to distinguish between whether (YES): most hyperedges can be strongly satisfied, or (NO): even a small fraction of hyperedges cannot be weakly satisfied.*

We employ the above variant of UNIQUEGAMES with a generalization of Feige's partitioning gadget, which has been tailored to work with the ℓ-coverage objective $C^{(\ell)}(\cdot)$. This gadget is essentially a collection of s set families $\mathcal{P}_1, \mathcal{P}_2, \ldots, \mathcal{P}_s$ over a universe $[\widehat{n}]$ satisfying (i) Each family $\mathcal{P}_i$ is a collection of sets such that each element in $[\widehat{n}]$ is covered exactly ℓ-times, i.e., it has (normalized) ℓ-coverage 1 (ii) Any choice of sets $S_1, S_2, \ldots, S_h$ from distinct families has ℓ-coverage at most ρ_ℓ (the target approximation ratio). We combine the h-ARYUNIQUEGAMES instance with the partitioning gadget by associating each hyperedge constraint with a disjoint copy of the gadget. The construction of the set family in our reduction ensures that sets corresponding to strongly satisfied edges use property (i), whereas sets corresponding to not even weakly satisfied hyperedges use property (ii). Since in the YES case, most hyperedges can be strongly satisfied, we get that there exists a choice of sets for which the normalized ℓ-multi-coverage is close to 1. On the other hand, in the NO case, since most hyperedges are not even weakly satisfied, for any choice of sets, the normalized ℓ-multi-coverage will be at most ρ_ℓ. Combining the two cases gives us the desired inapproximability. Due to space limitations, we do not discuss further the hardness result here and we refer to the full version [3] for the proof.

1.2   Related Covering Problems and Submodular Function Maximization

Another fundamental problem in the covering context is the *set cover* problem: given subsets $T_1, \ldots, T_m$ of a universe $[n]$, the objective is to find a set $S \subseteq [m]$ of minimum cardinality that covers all of n, i.e., $C(S) = \cup_{i \in S} T_i = [n]$. This is one of the first problems for which approximation algorithms were studied: Johnson [18] showed that the natural greedy algorithm achieves an approximation ratio of $1 + \ln n$ and much later Feige [13], building on a long line of works, established a matching inapproximability result.

Along the lines of maximum coverage, one can also consider the ℓ-version of set cover. In this version, the goal is to find the smallest set S such that $C^{(\ell)}(S) = \ell n$ (this corresponds to every element being covered at least ℓ times). Here, with $\ell > 1$, we observe an interesting dichotomy: while one achieves improved approximation guarantees for the maximum ℓ-multi-coverage, this is not the case for set ℓ-cover. In particular, set ℓ-cover is essentially as hard as to approximate as the standard set cover problem. To see this, consider the instance where $\mathcal{F}^{(\ell)}$ is obtained from $\mathcal{F}$ by adding $\ell - 1$ copies of the whole set $[n]$. Then, we have that $[n]$ can be 1-covered with k sets in $\mathcal{F}$ if and only if $[n]$ can be ℓ-covered with $k + \ell - 1$ sets in $\mathcal{F}^{(\ell)}$.

A well-studied generalization of the set cover problem, called the set multi-cover problem, requires element $e \in [n]$ to be covered at least d_e times, where the demand d_e is part of the input. The greedy algorithm was shown to also achieve $1 + \ln n$ approximation for this problem as well [9, 24]. Even though there has been extensive research on set multicover, its variants, and applications (see e.g., [4, 6, 17] and references therein), we are not aware of any previous work that considers the maximum multi-coverage problem.

The problem of maximum coverage fits within the larger framework of *submodular function maximization* [22]. In fact, the covering function $C : 2^{[n]} \to \mathbb{R}$ is submodular in the sense that it satisfies a diminishing-returns property: $C(S \cup \{i\}) - C(S) \geq C(S' \cup \{i\}) - C(S')$ for any $S \subseteq S'$ and $i \notin S'$. Nemhauser et al. [22] showed that the greedy algorithm achieves the ratio $1 - e^{-1}$ not only for the coverage function C but for any (monotone) submodular function. Submodular functions are a central object of study in combinatorial optimization and appear in a wide variety of applications; we refer the reader to [21] for a textbook treatment of this topic. Here, an important thread of research is that of maximizing submodular functions that have an additional structure which render them closer to linear functions. Specifically, the notion of curvature of a function was introduced by [7]. The curvature of a monotone submodular function $f : 2^{[m]} \to \mathbb{R}$ is a parameter $c \in [0,1]$ such that for any $S \subset [m]$ and $j \notin S$, we have $f(S \cup \{j\}) - f(S) \geq (1-c)f(\{j\})$. Note that if $c = 0$, this means that f is a linear function and if $c = 1$, the condition is mute.

Conforti and Cornuéjols [7] have shown that when the greedy algorithm is applied to a function with curvature c, the approximation guarantee is $\frac{1}{c}(1-e^{-c})$. Using a different algorithm, this was later improved by Sviridenko et al. [26] to a factor of approximately $1 - \frac{c}{e}$. This notion of curvature does have applications in some settings (see e.g., [26] and references therein), but the requirement is too strong and does not apply to the ℓ-coverage function $C^{(\ell)}$. In fact, if S is such that the sets T_i for $i \in S$ cover all the universe at least ℓ times, then adding another set T_j will not change the function $C^{(\ell)}$. Another way to see that this condition is not adapted to our ℓ-coverage problem is that we know that the greedy algorithm will not be able to beat the factor $1 - e^{-1}$ for any value of ℓ. We hope that this work will help in establishing a more operational way of interpolating between general submodular functions and linear functions.

1.3 Applications

We now briefly discuss some applications of the ℓ-coverage problem, the main message being that for most settings where coverage is used, ℓ-coverage has a very natural and meaningful interpretation as well. We leave the more detailed discussion of such applications for future work.

Our initial motivation for studying the maximum multi-coverage problem was in understanding the complexity of finding the code for which the list-decoding success probability is optimal. More precisely, consider a noisy channel with input set X and output set Y that maps an input $x \in X$ to $y \in Y$ with probability $W(y|x)$. To simplify the discussion, assume that for any input x, the output is uniform on a set T_x of size t, i.e., $W(y|x) = \frac{1}{t}$ if $y \in T_x$ and 0 otherwise. We would like to send a message m belonging to the set $\{1,\ldots,k\}$ using this noisy channel in such a way to maximize the probability of successfully decoding the message m. It is elementary to see that this problem can be written as one of maximizing the quantity $\frac{1}{tk}|\cup_{x \in S} T_x|$ over codes $S \subseteq X$ of size k [2]. Thus, the problem of finding the optimal code can be written as a covering problem, and handling general noisy channels corresponds to a weighted covering problem. This connection was exploited in [2] to prove tightness of the bound known as the *meta-converse* in the information theory literature [23] and to give limitations on the effect of quantum entanglement to decrease the communication errors. Suppose we now consider the list-decoding success probability, i.e., the receiver now decodes y into a list of size ℓ and we deem the decoding successful if m is in this list. Then the success probability can be written as: $\frac{1}{tk}\sum_{y \in Y} \min\{\ell, |\{x \in S : y \in T_x\}|\}$, i.e., an ℓ-coverage function. Our main result thus shows that the code with the maximum list-decoding success probability can be approximated to a factor of $\rho_\ell = 1 - \frac{\ell^\ell e^{-\ell}}{\ell!}$ and it shows that the meta-converse for list-decoding is tight within the factor ρ_ℓ.

The applicability of the multi-coverage can also be observed in game-theoretic settings in which the (standard) covering function is used to represent valuations of agents; see, e.g., works on combinatorial auctions [10,11]. As a stylized instantiation, consider a setup wherein the elements in the ground set represent types of goods and the given subsets correspond to bundles of goods (of different types). Assuming that, for each agent, goods of a single type are perfect substitutes of each other, one obtains valuations (defined over the bundles) that correspond to covering functions. In this context, the ℓ-multi-coverage formulation provides an operationally-useful generalization: additional copies (of the same type of the good) are valued, till a threshold ℓ. Indeed, our algorithmic result shows that if the diminishing-returns property does not come into effect right away, then better (compared to $1 - e^{-1}$) approximation guarantees can be obtained.

2 Approximation Algorithm for the ℓ-Multi-coverage Problem

The algorithm we analyze is simple and composed of two steps (relax and round): First, we solve the natural linear programming relaxation (see Eq. (1)) obtaining a fractional, optimal solution $x^* \in [0,1]^m$, which satisfies $\sum_{i \in [m]} x_i^* = k$. The second step is to use *pipage rounding* to find an *integral* vector $x^{\text{int}} \in \{0,1\}^m$ with the property that $\sum_{i \in [m]} x_i^{\text{int}} = k$. This is the size-$k$ set returned by the algorithm, $S = \{i \in [m] : x_i^{\text{int}} = 1\}$. These two steps are detailed below.

Step 1. Solve the Linear Programming Relaxation: Specifically, we consider the following linear programming relaxation of the ℓ-multi-coverage problem. Here, with the given collection of sets $\mathcal{F} = \{T_1, T_2, \ldots, T_m\}$, the set $\Gamma_e := \{i \in [m] : e \in T_i\}$ denotes the indices of T_is that contain the element e.

$$
\begin{aligned}
\max_{x,c} \quad & \sum_{e \in [n]} c_e \\
\text{subject to} \quad & c_e \leq \ell \quad \forall e \in [n] \\
& c_e \leq \sum_{i \in \Gamma_e} x_i \quad \forall e \in [n] \\
& 0 \leq x_i \leq 1 \quad \forall i \in [m] \\
& \sum_{i=1}^{m} x_i = k.
\end{aligned}
\tag{1}
$$

In this linear program (LP), the number of variables is $n + m$ and the number of constraints is $O(n + m)$ and, hence, an optimal solution can be found in polynomial time.

Step 2. Round the fractional, optimal solution: We round the computed fractional solution x^* by considering the *multilinear extension* of the objective, and applying pipage rounding [1,5,27] on it. Formally, given any function $f : \{0,1\}^m \to \mathbb{R}$, one can define the multilinear extension $F : [0,1]^m \to \mathbb{R}$ by $F(x_1,\dots,x_m) := \mathbb{E}[f(X_1,\dots,X_m)]$, where $X_1,\dots,X_m \in \{0,1\}$ are independent random variables with $\Pr[X_i = 1] = x_i$.

For a submodular function f, one can use pipage rounding to transform, in polynomial time, any fractional solution $x \in [0,1]^m$ satisfying $\sum_{i \in [m]} x_i = k$ into an integral vector $x^{\text{int}} \in \{0,1\}^m$ such that $\sum_{i \in [m]} x_i^{\text{int}} = k$ and $F(x^{\text{int}}) \geq F(x)$. We apply this strategy for the ℓ-coverage function and the fractional, optimal solution x^* of the LP relaxation (1). It is simple to check that the ℓ-coverage function $C^{(\ell)}$ is submodular. We thus get the following lower bound for the ℓ-coverage value of the set returned by the algorithm:[2]

$$C^{(\ell)}(x^{\text{int}}) = \mathbb{E}_{(X_1,\dots,X_m) \sim (x_1^{\text{int}},\dots,x_m^{\text{int}})}\left[C^{(\ell)}(X_1,\dots,X_m)\right]$$
$$\geq \mathbb{E}_{(X_1,\dots,X_m) \sim (x_1^*,\dots,x_m^*)}\left[C^{(\ell)}(X_1,\dots,X_m)\right].$$

To conclude it suffices to relate $\mathbb{E}_{(X_1,\dots,X_m) \sim (x_1^*,\dots,x_m^*)}\left[C^{(\ell)}(X_1,\dots,X_m)\right]$ to the value taken by the LP at the optimal solution $x^* = (x_1^*,\dots,x_m^*)$. In particular, Theorem 1 directly follows from the following result (Theorem 3), which provides a lower bound in terms of the value achieved by the LP relaxation.

Indeed, this deterministic algorithm is quite direct: it simply solves a linear program and applies pipage rounding. We consider this as a positive aspect of the work and note that our key technical contribution lies in the underlying analysis; specifically, Theorem 3.

Theorem 3. *Let $x \in [0,1]^m$ and $c \in [0,1]^n$ constitute a feasible solution of the LP relaxation (1). Then we have,*

$$\mathbb{E}_{(X_1,\dots,X_m) \sim (x_1,\dots,x_m)}\left[C^{(\ell)}(X_1,\dots,X_m)\right] \geq \rho_\ell \sum_{e \in [n]} c_e$$

where ρ_ℓ is defined by

$$\rho_\ell := 1 - \frac{\ell^\ell}{\ell!}e^{-\ell}. \tag{2}$$

2.1 Proof Sketch of Theorem 3

To prepare for the proof, we need to establish some useful properties of the quantity ρ_ℓ. As mentioned previously, ρ_ℓ has an interpretation as the expectation of some function applied to a Poisson random variable. However, for the analysis of the algorithm, we need to relate ρ_ℓ to the expectation of this function applied to a Binomial random variable. In order to do this we use the following lemma.

---

[2] That is, a lower bound for the ℓ-coverage value of the size-k set $\{i \in [m] : x_i^{\text{int}} = 1\}$.

Lemma 1. *For any convex function f, any integer $N \geq 1$ and parameter $p \in [0, 1]$, we have*

$$\mathbb{E}\big[f(\mathrm{Bin}(N, p))\big] \leq \mathbb{E}\big[f(\mathrm{Poi}(Np))\big] . \tag{3}$$

Proof. The notion of convex order between two distributions is defined as follows. If X and Y are random variables, we say that $X \leq_{\mathrm{cvx}} Y$ whenever $\mathbb{E}\big[f(X)\big] \leq \mathbb{E}\big[f(Y)\big]$ holds for any convex function $f : \mathbb{R} \to \mathbb{R}$. We refer the reader to [25, Section 3.A] for more information and properties of this order.

As a result, the lemma will follow once we show that

$$\mathrm{Bin}(N, p) \leq_{\mathrm{cvx}} \mathrm{Poi}(Np). \tag{4}$$

First, we note that it suffices to prove this inequality for $N = 1$; this is a direct consequence of the fact that the convex order is closed under convolution [25, Theorem 3.A.12] (i.e., it is closed under the addition of independent random variables).[3]

Now, using [25, Theorem 3.A.2], we have that Eq. (4) for $N = 1$ is equivalent to showing that $\mathbb{E}\big[|\mathrm{Ber}(p) - a|\big] \leq \mathbb{E}\big[|\mathrm{Poi}(p) - a|\big]$ for any $a \in \mathbb{R}$. This follows from a simple case analysis.   □

We now state the main lemma and give a sketch for its proof. We refer to the full version for the complete argument.

Lemma 2. *Let $x \in [0, 1]^m$ and $c \in [0, 1]^n$ constitute a feasible solution of the linear program (1). Then, we have for any $e \in [n]$:*

$$\mathbb{E}_{(X_1, \ldots, X_m) \sim (x_1, \ldots, x_m)}\big[C_e^{(\ell)}(X_1, \ldots, X_m)\big] \geq \rho_\ell c_e.$$

Proof (sketch). To make the notation lighter, we write $C_e^{(\ell)} = C_e^{(\ell)}(X_1, \ldots, X_m)$ with indicators $X_i \sim \mathrm{Ber}(x_i)$. Recall that $C_e^{(\ell)} = \min\{\ell, \sum_{i \in \Gamma_e} X_i\}$, where $\Gamma_e := \{i \in [m] : e \in T_i\}$ denotes the indices of all the given subsets $T_i \in \mathcal{F}$ that contain the element e.

The tail-sum formula gives us

$$\mathbb{E}\big[C_e^{(\ell)}\big] = \sum_{a=1}^{\ell} \Pr\Big[\sum_{i \in \Gamma_e} X_i \geq a\Big]$$

$$= \ell - \sum_{a=0}^{\ell-1} (\ell - a) \Pr\Big[\sum_{i \in \Gamma_e} X_i = a\Big].$$

Now we can apply a standard lemma in the context of approximation algorithms (see e.g., [14]) and get that the expression for $\mathbb{E}\big[C_e^{(\ell)}\big]$ is minimum when for all

---

[3] Recall that if X and Y are independent, Poisson random variables with rate parameters λ_1 and λ_2, respectively, then $X + Y$ is Poisson-distributed with parameter $\lambda_1 + \lambda_2$.

$i \in \Gamma_e$, $x_i \in \{0, 1, q\}$ for some $q \in (0, 1)$. We now assume x has this form. Let $\bar{\ell}$ be the number of elements $i \in \Gamma_e$ such that $x_i = 1$. In this proof sketch, we assume $\bar{\ell} = 0$. We can write

$$\mathbb{E}\big[C_e^{(\ell)}\big] = \ell - \sum_{a=0}^{\ell-1}(\ell - a)\Pr\Big[\sum_{i \in \Gamma_e} X_i = a\Big].$$

We write t for the number of elements $i \in \Gamma_e$ such that $x_i = q$; hence, $\sum_{i \in \Gamma_e} x_i = qt$.

Note that $\sum_{i \in \Gamma_e} X_i$ has a binomial distribution with parameters t and q. We can then write

$$\mathbb{E}\big[C_e^{(\ell)}\big] = \ell - \sum_{a=0}^{\ell-1}(\ell - a)\binom{t}{a}q^a(1-q)^{t-a}$$

$$\geq \ell - \sum_{a=0}^{\ell-1}(\ell - a)\binom{t}{a}\Big(\frac{c_e}{t}\Big)^a\Big(1 - \frac{c_e}{t}\Big)^{t-a},$$

where we used the fact that this expression is increasing in q together with the inequality $c_e \leq qt$ (this follows from the linear program (1)). Using the fact that the function $x \mapsto \sum_{a=0}^{\ell-1}(\ell - a)\binom{t}{a}\big(\frac{x}{t}\big)^a\big(1 - \frac{x}{t}\big)^{t-a}$ is convex in the interval $[0, t]$ together with properties of the quantity ρ_ℓ that follow from Lemma 1, we finally get

$$\mathbb{E}\big[C_e^{(\ell)}\big] \geq \rho_\ell c_e.$$

3   Concluding Remarks

Generalizing the results of this paper, it would be interesting to consider the general setting where c copies have a value $\varphi(c)$ for some function φ, and identify what properties of φ lead to better approximation bounds. More generally, we believe that our work paves the way towards an operationally motivated notion of submodularity that interpolates between linear functions and completely general submodular functions. The previously mentioned notion of curvature studied in [7, 26] does this interpolation but the definition is unfortunately too restrictive and thus difficult to interpret operationally.

Another interesting question is whether there exists combinatorial algorithms that achieve the approximation ratio ρ_ℓ for maximum ℓ-coverage for $\ell \geq 2$. For $\ell = 1$, the simple greedy algorithm does the job, but as we observed previously, the greedy algorithm continues to provide a $1 - e^{-1}$ approximation ratio even for $\ell \geq 2$. Hence, an interesting direction of future work is to understand if the greedy algorithm can be generalized to obtain an approximation ratio better than $1 - e^{-1}$.

References

1. Ageev, A.A., Sviridenko, M.I.: Pipage rounding: a new method of constructing algorithms with proven performance guarantee. J. Comb. Optim. **8**(3), 307–328 (2004)
2. Barman, S., Fawzi, O.: Algorithmic aspects of optimal channel coding. IEEE Trans. Inform. Theory (2018). arXiv:1508.04095
3. Barman, S., Fawzi, O., Ghoshal, S., Gürpınar, E.: Tight approximation bounds for maximum multi-coverage (2019). arXiv:1905.00640
4. Berman, P., DasGupta, B., Sontag, E.: Randomized approximation algorithms for set multicover problems with applications to reverse engineering of protein and gene networks. Discrete Appl. Math. **155**(6), 733–749 (2007). https://doi.org/10.1016/j.dam.2004.11.009. http://www.sciencedirect.com/science/article/pii/S0166218X06003659. Computational Molecular Biology Series, Issue V
5. Calinescu, G., Chekuri, C., Pál, M., Vondrák, J.: Maximizing a monotone submodular function subject to a matroid constraint. SIAM J. Comput. **40**(6), 1740–1766 (2011)
6. Chekuri, C., Clarkson, K.L., Har-Peled, S.: On the set multicover problem in geometric settings. ACM Trans. Algorithms (TALG) **9**(1), 9 (2012)
7. Conforti, M., Cornuéjols, G.: Submodular set functions, matroids and the greedy algorithm: tight worst-case bounds and some generalizations of the Rado-Edmonds theorem. Discrete Appl. Math. **7**(3), 251–274 (1984)
8. Cornuejols, G., Fisher, M.L., Nemhauser, G.L.: Location of bank accounts to optimize float: an analytic study of exact and approximate algorithms. Manag. Sci. **23**(8), 789–810 (1977). https://doi.org/10.1287/mnsc.23.8.789
9. Dobson, G.: Worst-case analysis of greedy heuristics for integer programming with nonnegative data. Math. Oper. Res. **7**(4), 515–531 (1982)
10. Dobzinski, S., Schapira, M.: An improved approximation algorithm for combinatorial auctions with submodular bidders. In: Proceedings of the Seventeenth Annual ACM-SIAM Symposium on Discrete Algorithm, pp. 1064–1073. Society for Industrial and Applied Mathematics (2006)
11. Dughmi, S., Roughgarden, T., Yan, Q.: Optimal mechanisms for combinatorial auctions and combinatorial public projects via convex rounding. J. ACM (JACM) **63**(4), 30 (2016)
12. Dughmi, S., Vondrák, J.: Limitations of randomized mechanisms for combinatorial auctions. In: Proceedings of the 2011 IEEE 52nd Annual Symposium on Foundations of Computer Science (FOCS 2011), pp. 502–511. IEEE Computer Society, Washington, DC (2011). https://doi.org/10.1109/FOCS.2011.64
13. Feige, U.: A threshold of ln n for approximating set cover. J. ACM **45**(4), 634–652 (1998)
14. Feige, U.: On maximizing welfare when utility functions are subadditive. SIAM J. Comput. **39**(1), 122–142 (2009)
15. Feldman, V., Kothari, P.: Learning coverage functions and private release of marginals. In: Conference on Learning Theory, pp. 679–702 (2014)
16. Hochbaum, D.S.: Approximating covering and packing problems: set cover, vertex cover, independent set, and related problems. In: Approximation Algorithms for NP-Hard Problem, pp. 94–143. PWS Pub. (1997)
17. Hua, Q.-S., Yu, D., Lau, F.C.M., Wang, Y.: Exact algorithms for set multicover and multiset multicover problems. In: Dong, Y., Du, D.-Z., Ibarra, O. (eds.) ISAAC 2009. LNCS, vol. 5878, pp. 34–44. Springer, Heidelberg (2009). https://doi.org/10.1007/978-3-642-10631-6_6

18. Johnson, D.S.: Approximation algorithms for combinatorial problems. J. Comput. Syst. Sci. **9**(3), 256–278 (1974)
19. Khot, S.: On the power of unique 2-prover 1-round games. In: Proceedings of the Thirty-Fourth Annual ACM Symposium on Theory of Computing, pp. 767–775. ACM (2002)
20. Khot, S., Vishnoi, N.K.: On the unique games conjecture. In: FOCS, vol. 5, p. 3 (2005)
21. Krause, A., Golovin, D.: Submodular function maximization. In: Tractability: Practical Approaches to Hard Problems, vol. 3, p. 19 (2012)
22. Nemhauser, G.L., Wolsey, L.A., Fisher, M.L.: An analysis of approximations for maximizing submodular set functions. Math. Program. **14**(1), 265–294 (1978)
23. Polyanskiy, Y., Poor, H.V., Verdú, S.: Channel coding rate in the finite blocklength regime. IEEE Trans. Inf. Theory **56**(5), 2307–2359 (2010)
24. Rajagopalan, S., Vazirani, V.V.: Primal-dual RNC approximation algorithms for set cover and covering integer programs. SIAM J. Comput. **28**(2), 525–540 (1998)
25. Shaked, M., Shanthikumar, J.G.: Stochastic Orders. Springer, New York (2007). https://doi.org/10.1007/978-0-387-34675-5
26. Sviridenko, M., Vondrák, J., Ward, J.: Optimal approximation for submodular and supermodular optimization with bounded curvature. Math. Oper. Res. **42**(4), 1197–1218 (2017)
27. Vondrák, J.: Submodularity in combinatorial optimization (2007)

Implementing Automatic Benders Decomposition in a Modern MIP Solver

Pierre Bonami[1]([⊠]), Domenico Salvagnin[2], and Andrea Tramontani[3]

[1] CPLEX Optimization, IBM, Madrid, Spain
pierre.bonami@es.ibm.com
[2] DEI, Via Gradenigo, 6/B, 35131 Padova, Italy
[3] CPLEX Optimization, IBM, Bologna, Italy

Abstract. We describe the automatic Benders decomposition implemented in the commercial solver IBM CPLEX. We propose several improvements to the state-of-the-art along two lines: making a numerically robust method able to deal with the general case and improving the efficiency of the method on models amenable to decomposition. For the former, we deal with: unboundedness, failures in generating cuts and scaling of the artificial variable representing the objective. For the latter, we propose a new technique to handle so-called generalized bound constraints and we use different types of normalization conditions in the Cut Generating LPs. We present computational experiments aimed at assessing the importance of the various enhancements. In particular, on our test bed of models amenable to a decomposition, our implementation is approximately 5 times faster than CPLEX default branch-and-cut. A remarkable result is that, on the same test bed, default branch-and-cut is faster than a Benders decomposition that doesn't implement our improvements.

1 Introduction

Benders decomposition was originally proposed in [5] to solve Mixed Integer Programs (MIP). The decomposition consists in splitting the original problem between a *master problem*, that consists of the integer variables of the original problem and possibly some additional continuous variables, and a *Cut Generating Linear Program (CGLP)* formulated in the space of the remaining variables. As originally stated, Benders method is iterative. At each step, the master is first solved to optimality. The CGLP is then constructed using the solution of master and solved: from its solution cuts for the master are derived (Benders cuts herein). This process is repeated until no cuts are found by the CGLP. The theorem of Benders guarantees that at this point the original problem is solved to optimality.

Benders decomposition is more often applied to MIPs where the CGLP constraint matrix has a block diagonal structure and can be further decomposed into smaller problems. Among the numerous applications the most notable are Facility Location [11,12], Network Design [15] and Stochastic Optimization [6].

© Springer Nature Switzerland AG 2020
D. Bienstock and G. Zambelli (Eds.): IPCO 2020, LNCS 12125, pp. 78–90, 2020.
https://doi.org/10.1007/978-3-030-45771-6_7

In the last decades, a vast body of research has examined every step of Benders decomposition. A recent and comprehensive survey can be found in [24]. Here, we overview the literature most relevant to our work. First, a feasible solution of the master problem is not needed to generate a Benders cut. An initial good set of cuts can usually be found by solving the initial LP relaxation of the problem by Benders decomposition [21]. More generally, in branch-and-cut, Benders cuts can be separated at any node of the search tree. Second, several computational studies (e.g., [4,11]) have shown that a simple stabilization mechanism for the cutting plane loop allows to significantly improve the effectiveness of the method. Finally, most of the research has been devoted to separating non-dominated or even facet defining Benders cuts. For the case where the CGLP is feasible and bounded, [20] proposes to solve two LPs to guarantee non-domination. In [23], this approach was improved and it was shown that non-domination can be obtained with a single LP. For the infeasible case, in [14] a normalization was proposed based on the concept of *minimal infeasible subsystem*.

For the applications mentioned above, Benders decomposition is often the only method able to solve problems of realistic size. Most implementations in the scientific literature or in the industry are ad-hoc implementations for a specific class of problems. In this paper, we report on the automatic Benders decomposition solver implemented in CPLEX [17]. Our goal is to have a numerically robust implementation that can be used as a black-box on any MIP, and that is competitive on the classes of problems where Benders decomposition has been reported to be useful in practice.

To achieve this goal our main contributions are: applying a generic stabilization procedure to solve the initial Benders cut loop efficiently, dealing with cases of unboundedness, dealing with numerical stability of Benders cuts and artificial variables, handling of linking constraints with special structure to simplify the CGLP, and applying normalizations to find "good" Benders cuts when the CGLP is infeasible. Our computational results show that the resulting algorithm is considerably faster than default branch-and-cut on models amenable to decomposition, and that the algorithmic enhancements we propose have a dramatic effect.

The outline of the paper is as follows: in Sect. 2, we outline the overall Benders decomposition algorithm, setting up the required notation. In Sect. 3, we detail enhancements to the construction of the master problem and overall numerical stability of the procedure. In Sect. 4, we describe the improvements to solving the CGLP mentioned above. Finally, in Sect. 5 we computationally evaluate our algorithm and we analyze the effect of the different algorithmic ideas we propose.

2   Benders Decomposition

In this section we briefly outline the Benders decomposition algorithm. Most textbooks use the so-called dual space of the CGLP, which has the advantage of directly expressing coefficients for the Benders cuts. However, as noted in the original paper [5], it is often computationally more convenient to work in the primal space, i.e., the space in which the problem is originally formulated. The implementation in CPLEX is entirely done in the primal space, and we believe it is also a simpler way to view the method. We will use this point of view in the remainder. We consider a MIP of the form:

$$\min cx + dy \tag{1}$$
$$Ax \geq b, \tag{2}$$
$$Tx + Qy \geq r, \tag{3}$$
$$x, y \geq 0 \text{ and } x \in \mathbb{Z}^n, \tag{4}$$

where $c \in \mathbb{Q}^n$, $d \in \mathbb{Q}^p$, $A \in \mathbb{Q}^{m_1 \times n}$, $T \in \mathbb{Q}^{m_2 \times n}$, and $Q \in \mathbb{Q}^{m_2 \times p}$. The decomposition starts by creating the master problem. It involves the x variables and an additional variable η representing the contribution to the objective of the y variables:

$$\min cx + \eta \tag{5}$$
$$Ax \geq b, \tag{6}$$
$$<\text{Benders cuts}>, \tag{7}$$
$$x \in \mathbb{Z}^n_+, \eta \text{ free}. \tag{8}$$

Some of the x variables could be continuous without significantly changing the method. Note that the set of Benders cuts (7) is initially empty. Also, at any point we can assume that (5)–(8) is feasible, otherwise the original problem is proven infeasible and the method is stopped.

Given a solution (x^*, η^*) to the master problem, the CGLP tries to find a feasible y satisfying the linking constraints (3) for fixed $x = x^*$. The following lemma details how the CGLP is defined and how it is used to either derive a Benders cut, or conclude optimality.

Lemma 1 (Benders Theorem [5]**).** *Let (x^*, η^*) be an optimal solution of (5)–(8) and define the CGLP:*

$$\min dy \tag{9}$$
$$Tx^* + Qy \geq r \tag{10}$$
$$y \geq 0. \tag{11}$$

(i) If (9)–(11) is infeasible, then there exists $\pi \in \mathbb{Q}^{m_2}_+$ such that

$$\pi Tx \geq \pi r \tag{12}$$

   is valid and $\pi Tx^* < \pi r$.

(ii) If (9)–(11) has a finite optimum y^*, then: Either $dy^* \leq \eta^*$ and (x^*, y^*) is optimal for (1)–(4). Otherwise, there exists $\pi \in \mathbb{Q}^{m_2}_+$ such that

$$\eta + \pi Tx \geq \pi r \tag{13}$$

   is valid and cuts off (x^*, η^*).

Although the result is well known, we give a short proof in the appendix.

The two cuts (12) and (13) defined in Lemma 1 are the so-called Benders cuts. The former is the *feasibility cut* and the latter the *optimality cut*. After the CGLP is solved, if a cut has been derived, it can be added to the master problem, and the method is iterated. Otherwise, (1)–(4) is solved. Note that we have neglected the particular case where the master problem is unbounded. We will deal with it in Sect. 3.

Before proceeding to our implementation of this procedure, two remarks are in order. First, as noted in the introduction, the CGLP is often itself decomposable (Q is block diagonal). In this case, introducing an η variable for each block, the CGLP can be split into smaller problems and each one may give a cut. Second, the procedure outlined doesn't require solving (5)–(8) to optimality before solving the CGLP. Instead, in a branch-and-cut algorithm, the CGLP can be solved every time an integer feasible solution for (5)–(8) is encountered. Either this solution is cut and the search can proceed or it is not cut and it is indeed also feasible for (1)–(4).

3   The Master Problem

The overall Benders decomposition algorithm implemented in CPLEX follows the algorithm described in the previous section. In this section, we detail in particular how the master problem is constructed. Here is the main workflow of the method:

1. A decomposition of the complete model (1)–(4) is detected. If the user provided a decomposition (via annotations [17]), then CPLEX will decompose the model according to it. Otherwise, it performs an automatic split, in which integer variables are assigned to the master, while continuous variables are assigned to the CGLP. The latter is eventually split into several independent CGLPs if a block decomposition can be detected. This detection is very efficient, as it is linear in the number of nonzeros of the CGLP matrix.
2. The complete model (1)–(4) is presolved using CPLEX regular MIP presolve (only reductions that may invalidate the decomposition previously identified are disabled). This step is applied before actually decomposing the problem because more reductions are typically found on the complete model and presolve is usually cheap enough that it can be done on the full model.
3. The presolved model is decomposed according to the decomposition identified at Step 1.
4. An initial stabilized Benders cut loop is executed on the LP relaxation of the problem. The rationale is to warm start the Benders search with a tighter approximation of the projection of the complete model, so that the subsequent search can benefit from it. We detail the stabilization procedure below.
5. Once the initial Benders cut loop is over, a regular branch-and-cut is started from the current master. This is pretty much a regular MIP solve, but in which Benders cuts are separated on the fly as lazy constraints.

To deal with all challenges in maintaining a numerically stable master problem, several additional ingredients are needed. We detail them in the next paragraphs.

Stabilization of Initial Cut Loop. Stabilization is obtained using an in-out strategy [4,11]. Briefly, this consists in not trying to separate the optimal solution of the LP relaxation (x^*, η^*) of (5)–(8), but using instead a suitable convex combination of (x^*, η^*) and a point (x^0, η^0) that is in the relative interior of the projection of the feasible region of (1)–(4) onto the variables (x, η). The point (x^0, η^0) is called the *core point*. The in-out strategy performs a binary search on the line segment joining (x^0, η^0) and (x^*, η^*), until a violated cut is found. In applications, the corepoint is usually obtained by exploiting the specific structure of the problem. In our generic framework, we compute it by solving the LP relaxation of the complete model without the objective and using the barrier algorithm without crossover. This is to attempt to get a point close to the analytic center of the LP relaxation of (1)–(4).

Cut Violation. A key decision for the soundness (and numerical stability) of the overall decomposition is the strategy used to decide when a master solution (x^*, η^*) is violated or not by a Benders cut. As CPLEX is based on floating point arithmetic, tolerances are needed. For optimality cuts, the violation of a cut has a natural interpretation: it is the amount by which the artificial variable η^* underestimates the contribution of the CGLP variables to the objective function. As such, for optimality cuts we use the regular optimality tolerance used by the solver. For feasibility cuts, there is no such natural interpretation, and any scaling of feasibility cuts is arbitrary. Therefore, we consider a master solution (x^*, η^*) to be violated if and only if the corresponding CGLP is infeasible (according to the regular feasibility tolerances) regardless of whether we are actually able to derive a sufficiently violated feasibility cut out of it. This is also consistent with what a regular B&C algorithm would have done on the complete model.

Another key aspect is what to do if we fail to derive a cut. When the issue is on the CGLP side, sometimes it can pay off to resolve the CGLP from scratch (possibly forcing a different LP algorithm), and reconstruct the cut. However, in some cases the cut is inherently bad, e.g., when the cut returned by the CGLP is violated but its dynamism is such that it cannot be added to master. In this case, the current master solution is still flagged as invalid, and we act as follows. If the master solution comes from a heuristic, we just discard the solution. If it comes from a node, we have no choice but to branch on a non-fractional variable, unless we already reached a leaf of the enumeration tree. If we are at a leaf of the tree, and there are no continuous variables (except the artificial η) in master, we can still prune the node. However, this would not be correct if there are structural continuous variables in master, so in this latter case we have no choice but to abort with a numerical failure.

Scaling of the Variable η in Optimality Cuts. Another aspect where the method may fail for numerical reasons is the scaling of η in optimality cuts. Note that in (13) the coefficient of η is 1 by construction. Depending on the contribution of the y variables to the objective this can pose severe problems in the numerical behavior of the method. If the coefficients πT for the variables x in the cuts are very large or very small, an LP solver using floating point arithmetic might not be able to handle them correctly. In such a case it is possible to scale η to get more numerically stable cuts. In particular, we can define η so that instead of being equal to dy, it is a fraction of it, i.e., $\alpha\eta = dy$. The derivation of a cut is the same as before, except that in the aggregation used to construct the optimality cut we now use the inequality $dy \leq \alpha\eta$ leading to the cut $\alpha\eta + \pi Tx \geq \pi r$. Of course all optimality cuts must share the same scaling for η while the coefficients for x may be vastly different among them. Therefore choosing an appropriate value for α is not trivial. In our implementation, we define α to be equal to the largest coefficient for an x variable in the first optimality cut derived. Note that α could be dynamically rescaled in the procedure but our simple attempt did not find any advantage to it.

Cutting a Ray. Finally, a case that we left out in Sect. 2 is the separation of cuts when master is unbounded. In most of the literature this is excluded by construction, but CPLEX has to deal with the general case. Denote with P and R the continuous relaxations of (1)–(4) and (5)–(8), respectively. When R is unbounded, the next lemma shows that a variant of the CGLP can be used to conclude that P is unbounded as well (and hence (1)–(4) is either infeasible or unbounded), or to separate a cut to truncate an unbounded ray of R.

Lemma 2. *Let (u^*, u_0^*) be an unbounded ray of R and consider the modified CGLP*

$$\min ds \tag{14}$$
$$Tu^* + Qs \geq 0 \tag{15}$$
$$s \geq 0. \tag{16}$$

(i) *If (14)–(16) is unbounded, then P is unbounded.*
(ii) *If (14)–(16) is infeasible, then there exists $\pi \in \mathbb{Q}_+^{m_2}$ such that $\pi Tx \geq \pi r$ is valid and $\pi Tu^* < 0$.*
(iii) *If (14)–(16) has a finite optimum s^*, then: If $ds^* \leq u_0^*$, P is unbounded. Otherwise, $\exists \pi \in \mathbb{Q}_+^{m_2}$ such that $\eta + \pi Tx \geq \pi r$ is valid and $u_0^* + \pi Tu^* < 0$.*

The proof is in the appendix. Note that the only differences w.r.t. the case in which we cut a point are (i) fix the master variables to the values in the ray (rather than a point), and (ii) zero out the right hand side of the constraints[1].

4   CGLP Improvements

In this section we describe several improvements to the CGLP: exploiting linking constraints with special structure in Sect. 4.1, and normalization conditions to separate "good" feasibility cuts in Sect. 4.2.

---

[1] This also applies to bounds: bounded y variables turn into directions s fixed to zero.

4.1   Generalized Bound Constraints

Benders decomposition can be effective when the problem structurally simplifies after fixing variables x. The most common case is when Q is block diagonal. However, the simplification can be significant in other cases as well. A relevant case arises when many linking constraints (3) involve only one y variable. We denote those constraints as Generalized Bound Constraints (GBCs) since in the CGLP, with the x variables fixed, they boil down to simple bound constraints on the y variables. A prime example of GBCs is the one of variable bound constraints like, e.g., constraints of the form $y_j \leq x_i$ that are prevalent in facility location problems. However, other and more complex GBCs, involving several x variables at a time, can also arise in practice, as, for instance, in the case of partial set covering location problems [9].

Translating GBCs to simple bound constraints is key to solving the CGLP faster. However, fully exploiting the presence of GBCs requires some dedicated machinery. Blindly fixing the x variable in the CGLP and having LP presolve do the necessary simplifications potentially destroys the warm-starting capabilities of the simplex method. On the other hand, disabling presolve is of course not an option either as in this case GBCs would not be turned into simple bounds anymore, negating all the benefits of the method. For this reason, we treat GBCs explicitly when we set up the CGLP: we don't add them to the CGLP formulation, but rather compute on the fly the corresponding simple bounds and directly change those. Note that we need to keep track of which GBC (if any) is active for each CGLP variable y_j, as we need to multiply it with the corresponding dual multiplier when computing the cut coefficients.

4.2   CGLP Normalization

It was noted in [14] that the textbook implementation of the Benders CGLP gives little control on which feasibility cut is returned. Actually, any dual feasible solution is optimal and any arbitrary unbounded dual ray will be returned by an LP solver. To select "good" feasibility cuts, we need to add a normalization condition that truncates the dual cone: clearly, the choice of the normalization is critical. In the following, we will describe two such normalizations. Note that, by adding the objective as a constraint, we can always reduce ourselves to the case where the CGLP is infeasible, and treat feasibility and optimality cuts in a unified way. However, this has several drawbacks: the objective is often dense, numerically shaky and badly scaled w.r.t. the other constraints in the model. A preliminary implementation of this unified approach indicated it is not effective. For this reason, we adopt a two-stage approach. We first solve (9)–(11): if it is feasible, an optimality cut is derived. Otherwise, we temporarily remove the objective and add a normalization. Once the cut is obtained, the normalization is removed and the objective restored[2]. It is important to note that the addition

---

[2] There is a notable exception to this strategy: if the CGLP has no objective (i.e. $d = 0$) we never remove the normalization, as any violated Benders cut will be a feasibility one by construction.

of the normalization can be done in both cases without hindering the warm-start capabilities of the simplex method. Finally, we note that GBCs (see Sect. 4.1) do not simplify to simple bounds if they are used in the normalization. Therefore, we do not apply the normalization to those constraints. As a result cuts are potentially weaker, but separation would be orders of magnitude slower on some models classes otherwise.

L^1 **Normalization.** Assume that (9)–(11) is infeasible. A normalization is simply introduced by adding a penalty variable z_0 as follows:

$$\min z_0 \tag{17}$$
$$Tx^* + Qy + z_0 \geq r \tag{18}$$
$$y, z_0 \geq 0 \tag{19}$$

Note that the addition of z_0 acts as normalization condition in the dual space, specifically the L^1-norm of the dual multipliers is constrained to be 1. This normalization is the one used in [14] but expressed in the primal space. It was originally proposed in the context of lift-and-project cuts in [2] and, as shown in [13], it has nice numerical properties as it favors the separation of cuts with a sparse support.

As we assumed that (9)–(11) is infeasible, the optimal solution of (17)–(19) has $z_0^* > 0$. Using the vector π of optimal dual multipliers we can obtain (12). By strong duality it holds that $z_0^* = \pi(r - Tx^*)$, hence the cut is violated by x^*.

CW Normalization. The L^1 normalization is known to have nice numerical properties, but it does not give any theoretical guarantee on the strength of the feasibility cuts. A better approach in this sense is the one proposed by Conforti and Wolsey in [8] and recently implemented in [7,9]. Let x^0 be the core point defined in Sect. 3. The geometric idea is to find the point on the line segment (x^0, x^*) which is feasible for (9)–(11) and further away from x^0. It is shown in [8] that this approach separates facet defining inequalities with probability 1. Defining the convex combination as $x^* + \lambda(x^0 - x^*)$, after introducing the variable λ, we can write the CGLP as:

$$\min \lambda$$
$$Tx^* + Qy + \lambda[T(x^0 - x^*)] \geq r$$
$$y \geq 0$$
$$0 \leq \lambda \leq 1 \tag{20}$$

Note that the dual constraint associated with λ reads $\pi T(x^0 - x^*) = 1$, which is the well known Balas–Perregaard normalization on the polar space [3]. Given the optimal dual multipliers π, a feasibility cut is derived as previously.

Although the CW normalization is theoretically stronger than the L^1 normalization, (20) is typically harder to solve than (17)–(19). Therefore, CPLEX chooses at runtime which normalization to apply, with the rationale of trying to use the CW normalization when feasibility cuts appear to be important w.r.t. optimality cuts.

5 Computational Results

In this section we report on some computational experiments aimed at assessing the importance of the algorithmic components previously described. In particular, we focus on (i) in-out techniques to stabilize and accelerate the convergence of the initial Benders cut loop, (ii) CW normalization to separate stronger feasibility cuts, and (iii) special handling of GBCs in the CGLP.

To this end, we considered a benchmark test bed of instances that are suitable for Benders decomposition. Specifically, we collected 209 two-stage stochastic models from various applications (capacitated facility location [6,19], network interdiction [6,22], fixed charge multi-commodity network design [10], chance-constrained programs [18], and others from CPLEX internal library) and 166 non-stochastic models also coming from different applications (capacitated and uncapacitated facility location [11,12,16,25], network expansion [1], partial set covering location [9], and others from CPLEX internal library), for a total of 375 benchmark instances on which Benders decomposition is expected to be effective. All tests were conducted by running CPLEX 12.10 [17] on a cluster of identical 12 core Intel Xeon CPU E5430 machines running at 2.66 GHz and equipped with 24 GB of RAM. A time limit of 10,000 s was enforced on each run.

Table 1 compares the default CPLEX branch-and-cut ("B&C") to two versions of CPLEX automatic Benders search: the default method ("Benders default") and the much weaker variant where we disabled in-out, CW normalization[3], and the special handling of GBCs ("Benders no all"). The table reports aggregated results on all 375 instances which are grouped in each row based on the *hardness of the models*. First, the set "all" consists of all the models for which no method had a failure and all methods gave consistent objective values. Then, the set "all" is subdivided in classes "[n, 10k]" ($n \in \{0, 1, 10, 100\}$), containing the models for which at least one of the methods took at least n seconds and that were solved to optimality within the time limit by at least one. For each set, we report: the number of models ("#models"), the number of time limit hit by each method ("#tilim"), then for the two methods "Benders default" and "Benders no all", the ratio of the shifted geometric means with respect to the reference "B&C" for solution times ("time") and number of nodes to optimality ("nodes")[4] (a value $t < 1$ indicates that the specific method is faster than the reference one by a factor of $1/t$).

The results reported in Table 1 clearly show that, on a test bed of instances amenable to Benders decomposition, the default variant of CPLEX Benders significantly outperforms regular branch-and-cut. In particular, considering the instances in the class [0,10k], the number of timeouts is reduced from 117 to 24 and the Benders solver is around 5.26 times faster. However, the table also shows that advanced algorithmic components are crucial to achieve good performance. Indeed, by disabling in-out techniques, the CW normalization and the special

---

[3] Note that, when the CW normalization is disabled, the L^1 normalization is used instead, and thus feasibility cuts are still separated using some normalization.

[4] The shift applied is of 1 s for "time" and 10 nodes for "nodes".

Table 1. Comparison between regular B&C and Benders decomposition.

Class	#models	B&C #tilim	Benders default #tilim	time	nodes	Benders no all #tilim	time	nodes
All	361	165	72	0.23	58.6	179	1.44	149.
[0,10k]	313	117	24	0.19	44.8	131	1.53	162.
[1,10k]	310	117	24	0.18	45.0	131	1.53	166.
[10,10k]	304	117	24	0.18	47.8	131	1.51	160.
[100,10k]	285	117	24	0.16	55.5	131	1.55	192.

handling of GBCs, performance dramatically deteriorates. In particular, still considering the instances in [0,10k], the number of timeouts increases to 131 and the Benders solver becomes 1.53 times slower than regular branch-and-cut.

In order to better assess the performance impact of the individual algorithmic components highlighted in Table 1, we conducted a set of experiments in which we disabled each of them individually. The outcome of these experiments is summarized in Table 2. Each row compares a variant of the CPLEX Benders solver, obtained by disabling one or more features, against the default Benders solver. For each comparison we report only the results for the instances in the class [0,10k]. We remark that each row is independent of the others and the number of models varies a little. The structure of the table is similar to Table 1. We add three columns under the header "affected" to report results only on the models on which the specific solver in the comparison is at least 10% slower or faster than the reference solver (i.e., default Benders decomposition).

Table 2. Impact of the individual features in the Benders solver on the [0,10k] bracket.

Feature	#models	Default #tilim	All models #tilim	time	nodes	Affected #models	time	nodes
No InOut	298	1	41	1.31	0.99	243	1.40	0.96
No CW-norm	301	0	0	1.14	1.40	71	1.83	4.36
No InOut and CW-norm	297	0	76	3.35	3.52	266	3.85	4.06
No GBCs	305	3	20	2.31	0.83	214	3.29	0.79
No all	300	6	113	9.03	3.87	275	11.00	4.45

The results reported in Table 2 lead to the following observations:

1. In-out appears to be the most important feature, as it affects 82% of the models and disabling it leads to 40 additional timeouts.
2. The CW normalization seems to be less important than in-out, as it affects only 24% of the models and no timeouts are introduced by disabling it. However, by comparing "No InOut" and "No InOut and CW-norm", we can clearly see that CW normalization becomes fundamental if in-out is disabled.

Intuitively, the two techniques are related as they both use the segment joining x^* to x^0 to separate deeper cuts. In this sense, the CW normalization is theoretically superior, but our experiments show that in-out is also essential.

3. Handling of GBCs affects 70% of the models and allows to solve 17 additional instances. Also, it appears to be the most important single feature in terms of overall improvement of computing time.

Acknowledgement. We thank Daniel Junglas and Roland Wunderling for many discussions and helping out with the implementation of some of the ideas. We also thank Michele Conforti for the many discussions on normalizations in Benders CGLPs.

A   Appendix

A.1   Proof of Lemma 1

(i) Assume that (9)–(11) is infeasible. Then, Farkas lemma implies that there exists a ray $\pi \geq 0$ such that $\pi Q \leq 0$ and $\pi(r - Tx^*) > 0$. Multiplying (3) by π, and eliminating the y variables from the resulting constraint using $\pi Q \leq 0$ and $y \geq 0$, we get the inequality (12). This inequality is violated by x^* by definition of π.

(ii) Suppose now that (9)–(11) has a finite optimal value and let y^* be an optimal solution. If $dy^* \leq \eta^*$ we claim that (x^*, y^*) is optimal for (1)–(4). Indeed, it is feasible, and since (5)–(8) is a relaxation of (1)–(4) and $cx^* + dy^* \leq cx^* + \eta^*$, (x^*, y^*) is optimal. Otherwise, let's consider the optimal dual vector π. It satisfies the conditions $\pi Q \leq d$, $\pi(r - Tx^*) = dy^*$ and $\pi \geq 0$. Multiplying again (3) by π and using $\pi Q \leq d$, $y \geq 0$ and $dy \leq \eta$, we can eliminate the y variables from the resulting constraint and we obtain the inequality (13). This inequality is violated by the point (x^*, η^*) by strong duality.

A.2   Proof of Lemma 2

By definition, (u, u_0) is an unbounded ray of R if

$$u \geq 0, Au \geq 0, cu + u_0 < 0,$$

and (u, s) is an unbounded ray of P if

$$u \geq 0, Au \geq 0, s \geq 0, Tu + Qs \geq 0, cu + ds < 0. \tag{21}$$

Given an unbounded ray (u^*, u_0^*) of R, we want to check whether it can be turned into an unbounded ray (u^*, s^*) of P, meaning that P itself is unbounded, or find a cut that truncates R along (u^*, u_0^*).

(i) Assume (14)–(16) is unbounded, and consider an unbounded ray s^*. By definition, $s^* \geq 0$, $Qs^* \geq 0$ and $ds^* < 0$. Thus, there exists a scalar $\lambda > 0$ such that $(u^*, \lambda s^*)$ satisfies (21). This proves that P is unbounded.

(ii) Assume that (14)–(16) is infeasible. Then by Farkas lemma $\exists \pi \geq 0$ such that $\pi Q \leq 0$ and $\pi T u^* < 0$. Multiplying (3) by π, and eliminating the y variables from the resulting constraint using $\pi Q \leq 0$ and $y \geq 0$, we get the inequality (12). By definition of π we have $\pi T u^* < 0$ and thus the inequality truncates R along (u^*, u_0^*).

(iii) Finally, suppose that (14)–(16) is feasible and bounded and let s^* be an optimal solution. If $ds^* \leq u_0^*$, then (u^*, s^*) satisfies (21) and P is unbounded. Now suppose $ds^* > u_0^*$, and consider the optimal dual vector $\pi \geq 0$ which satisfies the conditions $\pi Q \leq d$ and $\pi T u^* = -ds^*$. Multiplying again (3) by π and using $\pi Q \leq d$, $y \geq 0$ and $dy \leq \eta$, we can eliminate the y variables from the resulting constraint and we obtain the inequality (13). From $\pi T u^* = -ds^*$ and $ds^* > u_0^*$ we get $\pi T u^* + u_0^* < 0$ and thus the inequality truncates R along (u^*, u_0^*).

References

1. Atamtürk, A., Nemhauser, G.L., Savelsbergh, M.W.P.: Valid inequalities for problems with additive variable upper bounds. Math. Program. **91**, 145–162 (2001)
2. Balas, E.: A modified lift-and-project procedure. Math. Program. **79**, 19–31 (1997)
3. Balas, E., Perregaard, M.: Lift-and-project for mixed 0–1 programming: recent progress. Discret. Appl. Math. **123**, 129–154 (2002)
4. Ben-Ameur, W., Neto, J.: Acceleration of cutting-plane and column generation algorithms: applications to network design. Networks: Int. J. **49**(1), 3–17 (2007)
5. Benders, J.F.: Partitioning procedures for solving mixed-variables programming problems. Numer. Math. **4**, 238–252 (1962)
6. Bodur, M., Dash, S., Günlük, O., Luedtke, J.: Strengthened benders cuts for stochastic integer programs with continuous recourse. INFORMS J. Comput. **29**(1), 77–91 (2017)
7. Bucarey, V., Elloumi, S., Labbé, M., Plein, F.: Models and algorithms for the product pricing with single-minded customers requesting bundles. Technical report hal-02056763 (2019)
8. Conforti, M., Wolsey, L.A.: Facet separation with one linear program. Math. Program. **178**(1–2), 361–380 (2019). https://doi.org/10.1007/s10107-018-1299-8D
9. Cordeau, J.F., Furini, F., Ljubić, I.: Benders decomposition for very large scale partial set covering and maximal covering location problems. Eur. J. Oper. Res. **275**(3), 882–896 (2019)
10. Crainic, T.G., Hewitt, M., Rei, W.: Partial decomposition strategies for two-stage stochastic integer programs. Technical report 13, CIRRELT (2014)
11. Fischetti, M., Ljubic, I., Sinnl, M.: Benders decomposition without separability: a computational study for capacitated facility location problems. Eur. J. Oper. Res. **253**(3), 557–569 (2016)
12. Fischetti, M., Ljubić, I., Sinnl, M.: Redesigning benders decomposition for large-scale facility location. Manag. Sci. **63**(7), 2146–2162 (2017)
13. Fischetti, M., Lodi, A., Tramontani, A.: On the separation of disjunctive cuts. Math. Program. **128**, 205–230 (2011). https://doi.org/10.1007/s10107-009-0300-y
14. Fischetti, M., Salvagnin, D., Zanette, A.: A note on the selection of Benders' cuts. Math. Program. B **124**, 175–182 (2010). https://doi.org/10.1007/s10107-010-0365-7

15. Geoffrion, A.M., Graves, G.W.: Multicommodity distribution system design by benders decomposition. Manag. Sci. **20**(5), 822–844 (1974)
16. Görtz, S., Klose, A.: A simple but usually fast branch-and-bound algorithm for the capacitated facility location problem. INFORMS J. Comput. **24**(4), 597–610 (2012)
17. IBM CPLEX Optimizer: CPLEX user's manual (2019). https://www.ibm.com/analytics/cplex-optimizer
18. Liu, X., Küçükyavuz, S., Luedtke, J.: Decomposition algorithms for two-stage chance-constrained programs. Math. Program. **157**(1), 219–243 (2014). https://doi.org/10.1007/s10107-014-0832-7
19. Louveaux, F.V.: Discrete stochastic location models. Ann. Oper. Res. **6**(2), 21–34 (1986). https://doi.org/10.1007/BF02027380
20. Magnanti, T., Wong, R.: Accelerating Benders decomposition: algorithmic enhancement and model selection criteria. Oper. Res. **29**, 464–484 (1981)
21. McDaniel, D., Devine, M.: A modified Benders' partitioning algorithm for mixed integer programming. Manag. Sci. **4**, 312–319 (1977)
22. Pan, F., Morton, D.P.: Minimizing a stochastic maximum-reliability path. Networks: Int. J. **52**(3), 111–119 (2008)
23. Papadakos, N.: Practical enhancements to the Magnanti-Wong method. Oper. Res. Lett. **36**(4), 444–449 (2008)
24. Rahmaniani, R., Crainic, T.G., Gendreau, M., Rei, W.: The benders decomposition algorithm: a literature review. Eur. J. Oper. Res. **259**(3), 801–817 (2017)
25. UflLib: Uncapacitated facility location library. http://resources.mpi-inf.mpg.de/departments/d1/projects/benchmarks/UflLib/packages.html

Improved Approximation Algorithms
for Inventory Problems

Thomas Bosman[1] and Neil Olver[2(✉)]

[1] Booking.com, Amsterdam, The Netherlands
tbosman@gmail.com
[2] Department of Mathematics,
London School of Economics and Political Science, London, UK
N.Olver@lse.ac.uk

Abstract. We give new approximation algorithms for the submodular joint replenishment problem and the inventory routing problem, using an iterative rounding approach. In both problems, we are given a set of N items and a discrete time horizon of T days in which given demands for the items must be satisfied. Ordering a set of items incurs a cost according to a set function, with properties depending on the problem under consideration. Demand for an item at time t can be satisfied by an order on any day prior to t, but a holding cost is charged for storing the items during the intermediate period; the goal is to minimize the sum of the ordering and holding cost.

Our approximation factor for both problems is $O(\log \log \min(N, T))$; this improves exponentially on the previous best results.

Keywords: Approximation algorithms · Iterative rounding · Inventory

1 Introduction

The inventory problem studied in this paper captures a number of related models studied in the supply chain literature. One of the simplest is the dynamic economic lot size model [14]. Here we have varying demand for a single item over T time units. Demand at time t or later can be satisfied by an order at time t (but not vice versa). For each day, there is a per-unit cost for holding the items in storage; there is also a fixed setup cost for ordering any positive quantity of the item, which is the same for each day. We want to decide on how many items to order on each day so as to minimize the total ordering and holding cost.

The *joint replenishment problem (JRP)* generalizes this problem to multiple items. We now have a unique holding cost for each day and each item, and a per item setup cost for ordering any quantity of that item. Furthermore, there is a general setup cost for ordering any items at all. This setup cost structure is called *additive*. While having limited expressive power in comparison to

N. Olver—Supported by Dutch Science Foundation (NWO) Vidi grant 016.Vidi. 189.087.

some of the more complex generalizations that have been studied, the additive joint replenishment problem is long known to be NP-hard [1]. This problem has attracted considerable attention from the theory community in the past, resulting in a line of progressively stronger approximation algorithms [11,12], the best of which gives an approximation ratio of 1.791 [2].

A more general version of this problem uses an ordering cost structure in which the setup cost is a submodular function of the items ordered. This model, introduced by Cheung et al. [6], is called the *submodular joint replenishment problem (SJRP)* and captures both the additive cost structure as well as other sensible models. In the same work, the authors give constant factor approximation algorithms for special cases of submodular cost functions, such as tree cost, laminar cost (i.e., coverage functions where the sets form a laminar family) and cardinality cost (where the cost is a concave function of the number of distinct items). For the general case, they provide an $O(\log NT)$ approximation algorithm.

In the *inventory routing problem (IRP)* setup costs are routing costs in a metric space. There is a root node and every item corresponds to a point in the metric. The setup cost for a given item set is then given by the length of the shortest tour visiting the depot and every item in the set. The usual interpretation of the model is that the root node represents a central depot and every other point in the metric represents a warehouse to be supplied from the depot. (To streamline terminology with the joint replenishment problem, we will keep using the term "item" rather than "location".)

The IRP has been extensively studied in the past three decades (see [7] for a recent survey), but primarily from a computational perspective. But very little is known about its approximability. Fukunaga et al. [9] presented a constant factor approximation under the restriction that orders for a given item must be scheduled *periodically*. This restriction appears to significantly simplify the construction of an approximation algorithm, as prior to this work the best known polynomial time approximation algorithms gave (incomparable) $O(\log N)$ [9] and $O(\log T)$ [13] performance guarantees.

Nagarajan and Shi [13] break the logarithmic barrier for both IRP and SJRP, under the condition that holding costs grow as a fixed degree monomial. This is a very natural restriction; in particular it captures the case where holding an item incurs a fixed rate per unit per day, depending only on the item. Building on their approach, we improve exponentially on their $O(\log T/\log\log T)$ approximation factor. We also provide some general techniques to turn (sub)logarithmic approximation algorithms in terms of T into equivalent algorithms in terms of N; and we are able to obtain results without restriction on the holding costs. Our main contributions are summarized in the following theorems.

Theorem 1. *There is a polynomial time $O(\log\log\min(N, T))$-approximation algorithm for the inventory routing problem.*

Theorem 2. *There is a polynomial time $O(\log\log\min(N, T))$-approximation algorithm for the submodular joint replenishment problem.*

We also mention the works on submodular partitioning problems [4,5,8]. In these problems, a ground set V must be partitioned across k different sets to minimize a submodular cost function. They use rounding of a relaxation based on the Lovász extension to unify and improve several prior results. Their approach inspired our use of the Lovász extension in the rounding algorithm for SJRP.

2   Preliminaries, Model and Technical Overview

We use log for the base 2 logarithm. We write $[k]$ for $\{1, \ldots, k\}$, and $[k, \ell]$ for $\{k, k+1, \ldots, \ell\}$, for any integers $k \leq \ell$.

The general framework of the inventory problems we investigate is defined by a set of items V of size N, ordering cost function $f : 2^V \to \mathbb{R}_{\geq 0}$ and a time horizon $[T] = \{1, \ldots, T\}$. We will assume throughout this paper that f is monotone and subadditive, with $f(\emptyset) = 0$. We will typically refer to the atomic time units as *days*.

For each item $v \in V$ and day t, there is a demand $d_{vt} \geq 0$. The collection of item-day pairs for which there is positive demand is denoted $D := \{(v, t) : d_{vt} > 0\}$. Demand for day t can be satisfied on or before day t. If we satisfy demand for item i on day t using an order on some day $s < t$, we need to store the items in the intervening days, and we pay a holding cost of h_{st}^v per unit we store. The magnitude of the demand only plays a role in the holding cost; the ordering cost is determined by the unique items ordered and is independent of how many units are ordered.

Given these inputs, we need to place an order for items to be delivered on each day so as to minimize the total ordering cost plus the holding cost. Since the cost of delivering an item does not depend on the size of the order and we want to store items as briefly as possible, it is always optimal to deliver just enough units of an item to satisfy demand until the next order for that item is scheduled. Hence, once we decide which items to order on which days, the optimal schedule is completely determined.

The inventory routing problem is the special case where we have a metric on V, some distinguished root node $r \in V$, and the ordering cost $f(S)$ of a set $S \subseteq V$ is the minimum possible length of a tree containing $S \cup \{r\}$. Here we differ from the usual definition, where $f(S)$ is defined to be the length of a shortest tour on $S \cup \{r\}$; but as is well known, these two definitions differ only by a factor of at most 2, which will not concern us. The submodular joint replenishment problem is the special case where f is submodular (in addition to the required properties already listed).

An integer programming formulation for this problem is given in (1). Here the variable y_t^S indicates whether item set S is ordered on day t, and x_{st}^v indicates whether the demand for item v on day t is satisfied by an order on day s.

$$\begin{array}{ll}
\text{minimize} & \sum_{t\in[T]}\sum_{S\subseteq V}f(S)y_t^S + \sum_{t\in[T]}\sum_{v\in V}\sum_{s\leq t}d_{vt}h_{st}^v x_{st}^v \\
\text{subject to} & x_{st}^v \leq \sum_{S:v\in S}y_s^S \qquad \forall(v,t)\in D, s\leq t \\
& \sum_{s\leq t}x_{st}^v = 1 \qquad \forall(v,t)\in D \\
& y_t^S, x_{st}^v \in \{0,1\} \qquad \forall v\in V, s\leq t, S\subseteq V
\end{array} \qquad (1)$$

Let (LP) denote the LP relaxation obtained by replacing the integrality constraints of ILP (1) by nonnegativity constraints; this LP has an exponential number of variables. To efficiently solve (LP), it suffices to provide an efficient separation oracle for the dual. This can be done for both SJRP and IRP (in the latter case, in an approximate sense); see [13].

Nagarajan and Shi [13] show that in order to round (LP), it suffices to round an associated covering problem. Essentially, given a solution $(\hat{x},\hat{y})$ to (LP), we require that each demand $(v,t)\in D$ is served within an interval $[s'(v,t),t]$, where $s'(v,t)$ is the median of the distribution $(\hat{x}_{st}^v)_{s\leq t}$. Serving (v,t) anywhere within this interval will incur cost at most twice what the fractional solution pays; and moreover, they show that enforcing this restriction cannot make the optimal solution more than a constant factor more expensive. The holding costs can then be dropped from the objective function. All in all, we obtain an instance of the following *subadditive cover over time* problem: for each item $v\in V$ we are given an associated set of *demand windows* $\mathcal{W}_v \subseteq \{[s,t] : s\leq t\in[T]\}$. We must choose a subset $S_t\subseteq V$ for each day $t\in[T]$ such that every item $v\in V$ is covered in each of its demand windows—that is, $v\in S_r$ for some $r\in[s,t]$, for each $[s,t]\in\mathcal{W}_v$. The goal is to find a feasible solution minimizing the total cost $\sum_{t\in[T]}f(S_t)$.

We also associate the canonical LP (2) with the subadditive cover over time problem.

$$\begin{array}{ll}
\text{minimize} & \sum_{t\in[T]}\sum_{S\subseteq V}f(S)y_t^S \\
\text{subject to} & \sum_{r\in[s,t]}\sum_{S:v\in S}y_r^S \geq 1 \qquad \forall[s,t]\in\mathcal{W}_v, \quad \forall v\in V \\
& y_t^S \geq 0 \qquad \forall t\in[T], S\subseteq V
\end{array} \qquad (2)$$

Our goal, given a fractional solution to this LP, is to round it to an integral solution. Note that the instance of subadditive cover over time is constructed from a solution $(\hat{x},\hat{y})$ to (LP) in such a way that $\hat{y}$ is already a feasible fractional solution to (2).

We now come to our first contribution. We show that this reduction can be taken much further: we can reduce to covering problems where the set of intervals have a very special structure. This structure, which we call *left aligned*, shares many of the benefits of a laminar family. For example, they have a natural notion of depth, which is always logarithmically bounded by T; the approximation factors of our algorithms are essentially logarithmic in this depth. We describe this reduction, which is rather general and applies identically to both SJRP and IRP, in Sect. 3. We also show, again generally, how the time horizon T can be polynomially bounded in terms of the number of items N.

So to obtain our main theorems, it suffices to give $O(\log \log T)$-factor approximation algorithms for these well-structured covering variants of SJRP and IRP. Here the approaches diverge; we give rather different algorithms for these two problems, albeit based on iterative rounding [10] in both cases.

For IRP, the algorithm uses randomized iterative rounding [3] of a certain path-based relaxation. We can show that after sampling every path in the support of the relaxation $O(\log \log T)$ times, we can remove a constant fraction of the edges and reorder the remaining paths such that we retain a feasible solution. Details are given in Sect. 4.

For SJRP, by contrast, the iterative rounding approach is naturally deterministic in nature. Instead of randomly rounding item sets in the support of a relaxation, we carefully try to pick a set for each day such that we win a constant fraction of its cost back in the subsequent reduction of the cost of the relaxation. If such a set cannot be found, we show that we can shrink the time horizon T by merging some adjacent time units (or put differently, we are able to remove the bottom "leaf" layer of the left-aligned family); we can then recurse on the smaller instance. We discuss this further in Sect. 5.

3   Reducing to Structured Covering Problems

The results of this section will not use any properties of the ordering cost function f that differ between IRP and SJRP. All that we will need, other than the general properties of f assumed at the start of Sect. 2, is that we are given an (approximately) optimal solution to the LP relaxation of (1).

Let $\mathcal{D} = \{[k2^i+1, (k+1)2^i] : i, k \in \mathbb{Z}_{\geq 0}\}$ denote the family of dyadic intervals over the nonnegative integers; the value of i for one of these intervals in $\mathcal{D}$ we call the *level* of that interval.

Definition 1. A family of intervals $\mathcal{F} \subseteq \{[s, t] : s, t \in \mathbb{Z}_{\geq 0}\}$ is called

- *left aligned* if for all $[s, t] \in \mathcal{F}$ there exists $[s, t'] \in \mathcal{D}$ with $t' \geq t$,
- *right aligned* if for all $[s, t] \in \mathcal{F}$ there exists $[s', t] \in \mathcal{D}$ with $s' \leq s$.

The *level* of an interval $[s, t] \in \mathcal{F}$ is the level of the minimal interval of $\mathcal{D}$ containing $[s, t]$.

We will call an instance of subadditive cover over time *left (right) aligned* if $\bigcup_{v \in V} \mathcal{W}_v$ is left (right) aligned.

Theorem 3. *At the loss of a constant factor, we can reduce an instance of the subadditive cover over time problem to a pair of instances, one left aligned and the other right aligned.*

The proof is given in the appendix. Right-aligned instances can be handled identically to left-aligned ones (by simply reversing the time indexing in the instance), so we consider only left-aligned instances in the sequel.

3.1 Bounding the Time Horizon, and Further Simplifications

Due to space constraints, we defer the proof of the following to the full version.

Theorem 4. *At the loss of a constant factor we can reduce a left-aligned sub-additive cover over time problem to a polynomial-sized collection of left-aligned subadditive cover over time problems with time horizons equal to N^2.*

This theorem could already be used to improve the dependence of the Nagarajan-Shi algorithm from $O\left(\frac{\log T}{\log \log T}\right)$ to $O\left(\frac{\log \min\{T,N\}}{\log \log \min\{T,N\}}\right)$. It also allows us to make the simplifying assumption that each item has positive demand on exactly one day, by making a copy of an item for each day in which it has a positive demand (with $T \leq N^2$, this only increases the number of items by a polynomial factor). Finally, we also assume that $T = 2^{2^k}$ for some $k \in \mathbb{N}$; if not, simply round up. Call an instance with all these properties (including being left aligned) *nice*; we will assume throughout the remainder of the paper that we are working with a nice instance.

4 Steiner Tree over Time

So in order to prove Theorem 1, we need to give an $O(\log \log T)$-approximation algorithm to nice instances of subadditive cover over time, for the appropriate class of order functions f. More precisely, V is the set of nodes of a semimetric space with distance function $c : V \times V \to \mathbb{R}_{\geq 0}$; a root node $r \in V$ is specified, and for all other nodes $v \in V \setminus \{r\}$, a time window $F_v = [a_v, b_v] \subseteq [T]$ is given. Since $f(S)$ denotes the cost of a cheapest tree containing $S \cup \{r\}$, we will consider a solution to be described by a collection of trees $(\mathcal{T}_t)_{t \in T}$, all containing the root. To be feasible, each node $v \neq r$ must be contained in $\mathcal{T}_t$ for some $t \in F_v$. The cost of a tree $\mathcal{T}$ (i.e., the sum of the length of its edges) is denoted $c(\mathcal{T})$; the objective is to minimize the total cost $\sum_t c(\mathcal{T}_t)$. We will call this the *Steiner tree over time problem.*

    The main part of our result works by iteratively massaging a specific type of fractional solution until it becomes integral. We now describe this type of solution.

    We let $\mathcal{P}$ denote the collection of directed paths in V. For each such directed path $P \in \mathcal{P}$, let $P \odot_t v$ signify that P connects v to $\mathcal{T}_t$, i.e., $P \odot_t v$ if there is a directed subpath on P from v to a node in the tree $\mathcal{T}_t$ containing the root on day t. If $v \in \mathcal{T}_t$ we let $P \odot_t v$ hold for all P by convention.

Definition 2. *A fractional path solution* (FPS) *is a pair* $(\mathcal{T}, w)$, *giving for each day t a tree $\mathcal{T}_t$ rooted at r and weights $w_t : \mathcal{P} \to [0, 1]$, with the property that*

$$\sum_{t \in F_v} \sum_{P \in \mathcal{P} : P \odot_t v} w_t(P) \geq 1 \qquad \forall v \in V \setminus \{r\}.$$

The cost of the fractional path solution is given by the sum of the cost of the trees and the cost of the paths weighted by w:

$$\sum_{t \in [T]} \sum_{P \in \mathcal{P}} w_t(P) c(P) + \sum_{t \in [T]} c(\mathcal{T}_t).$$

Note that a fractional path solution with $w_t(P) = 0$ for all P and t corresponds to a feasible integral solution to the Steiner tree over time problem. Moreover, we can start with a solution y to (2) and convert it to a fractional path solution at the loss of a constant (or more directly, we can solve a compact LP corresponding to fractional path solution). To do this, start by initializing all trees T_t to contain only the root. Then for each day t and set S in supp(y_t), construct a minimum spanning tree on $S \cup \{r\}$, and use this to define a path P to r that contains S and has cost at most twice the cost of this spanning tree (simply shortcut the doubled tree). Add P to the solution with weight $w_t(P) = y_t^S$.

Hence, we focus on turning a fractional path solution into one where all path weights are zero, without losing too much in the cost of the solution.

We need some preliminary notation and definitions. Given a directed path P, we use $V(P)$ and $E(P)$ to denote the node set and edge set, respectively. We similarly define $V(T)$ and $E(T)$ for a tree T. The head and tail of P are denoted head(P) and tail(P) respectively. Given a tree T and path P with head(P) in T, *adding* P to T (which we may write as $T + P$) results in a spanning tree of the union of T and P. In particular, $T + P$ is a tree, costing no more than $c(T) + c(P)$, and spanning $V(T) \cup V(P)$. We associate with each node v a *level* $\ell(v)$ in the natural way, namely as the level of F_v in the left aligned family $\bigcup_{v \in V \setminus \{r\}} F_v$.

The rounding algorithm will consist of a number of iterations. Each iteration will increase the size of the integral part $(T_t)_{t \in [T]}$, while reducing the size of the fractional part $(w_t)_{t \in [T]}$, until the solution is entirely integral. We will ensure that the cost increase in the integral part is an $O(\log \log T)$ factor times the cost decrease in the fractional part, which clearly yields the required approximation guarantee.

Each iteration of the rounding scheme involves two steps. The first step is, for each $t \in [T]$, to independently sample the paths according to the weights $w_t(P)$, upscaled by a factor $K \log \log T$; K is a fixed constant we will choose later. These paths are added (one by one) to T_t. The total cost increase due to this step is $O(C \log \log T)$, where $C = \sum_{t \in [T]} \sum_{P \in \mathcal{P}} w_t(P) c(P)$ is the total cost of the fractional part. Our goal will now be to adjust the fractional paths in a way that reduces the total fractional cost by a constant factor with high probability, while maintaining feasibility. This will clearly lead to our desired approximation ratio: each iteration we pay $O(\log \log T)$ times the decrease in the total fractional cost, and once the total fractional cost reaches zero, we have an integral solution.

The main operation that the algorithm will perform in order to achieve this is a "split and shift" operation. Let (T, w) be the fractional path solution after the above sampling step. Let P be some path in supp(w_t). Our goal will be to

- **(split)** remove some edges from P, resulting in a collection of subpaths $P_1, P_2, \ldots, P_q$, which may *not* have their heads in T_t; and then
- **(shift)** for each one of these paths P_j, assign its weight to some day t_j, in such a way that now head(P_j) is in T_{t_j}, and t_j is still in the time windows of all the nodes in P_j.

This would ensure feasibility, while reducing the fractional cost by $w(P)$ times the total cost of the removed edges. If each edge is removed with constant probability, we obtain the required cost decrease.

In order to make this work, we need some control of the interaction of the different time windows of the nodes on a given path. The most important fact, immediate from the left aligned structure, is the following.

Lemma 1. *If the time windows of v and w overlap, with $\ell(w) \geq \ell(v)$, then $a_w \leq a_v$.*

This means that if we consider a path P' with head v' (which might be a subpath of a path in supp(w_t) say) where $\ell(v')$ is minimal amongst all nodes in P', then any $t' \in F_{v'}$ with $t' \leq t$ will be in F_v for all $v \in V(P')$.

The following definition will aid us in shifting always to earlier days, so that the above can be applied.

Definition 3. Given a fractional path solution $(\mathcal{T}, w)$, for each $v \in V \setminus \{r\}$, let m_v be maximal such that

$$\sum_{t=m_v}^{b_v} \sum_{P \in \mathcal{P}: P \odot_t v} w_t(P) \geq \tfrac{1}{2}.$$

Then for any $v \in V$, we call $[a_v, m_v]$ the *sow* phase of v and $[m_v, b_v]$ the *reap* phase of v. *(Note that the sow and reap phases both contain m_v.)*

Let S_v and R_v denote the sow and reap phases of $v \in V$, at the start of this iteration. We first double all weights; let $w' = 2w$. This ensures that

$$\sum_{t \in R_v} \sum_{P \in \mathcal{P}: P \odot_t v} w'_t(P) \geq 1 \quad \text{and} \quad \sum_{t \in S_v} \sum_{P \in \mathcal{P}: P \odot_t v} w'_t(P) \geq 1. \tag{3}$$

Our goal will be to show that every edge on a path can be removed with probability $\tfrac{3}{4}$; this counteracts the doubling of the weights and ensures that the expected fractional cost at the end of the iteration is at most half its initial value.

Consider each day $t \in [T]$ and path $P \in \text{supp}(w'_t)$ separately. Let us say that a node $v \in V(P)$ is *serviced* by P if $t \in R_v$. Assume that tail(P) is serviced by P; otherwise replace P in w'_t by the subpath of P from the first serviced node, retaining feasibility by (3). We say that a node v serviced by P has *germinated* if it lies in $V(\mathcal{T}_t)$ for some $t \in S_v$. Let $v_1, v_2, \ldots, v_k$ be the nodes serviced by P, in order from the tail to the head of P; so $v_1 = \text{tail}(P)$ and $v_k = \text{head}(P)$. We consider each node v_j (with $j < k$) in turn, and check if (i) v_j has germinated, and (ii) $\ell(v_j)$ is minimal amongst $\ell(v_1), \ell(v_2), \ldots, \ell(v_j)$. If this is the case, we proceed as follows:

- Let $P^{(1)}$, $P^{(2)}$ and $P^{(3)}$ be the subpaths of P from tail(P) to v_j, from v_j to v_{j+1}, and from v_{j+1} to head(P), respectively.
- Shift $P^{(1)}$ to a day t_j witnessing that v_j germinated, and delete all edges of $P^{(2)}$. Thus, we modify w' by increasing both $w'_{t_j}(P^{(1)})$ and $w'_t(P^{(3)})$ by $w'_t(P)$, and then setting $w'_t(P)$ to zero.

Since $t_j \in S_{v_j}$, and $t \in R_{v_j}$, we know that $t_j \leq t$; thus by the condition (ii) and Lemma 1, t_j lies in the sow phases of $v_1, v_2, \ldots, v_j$. Thus, this modification retains feasibility of $(\mathcal{T}, w')$. We then repeat this process, continuing with path $P^{(3)}$ instead of P (resulting in a new sequence of nodes serviced by $P^{(3)}$; $v_1, \ldots, v_j$ will not be part of this sequence). The iteration ends when this process has been completed for all fractional paths on all days.

As observed, $(\mathcal{T}, w')$ is still feasible at the end of this process. All that remains is to show that indeed each edge is removed with the desired $\frac{3}{4}$ probability.

So fix a day $t \in [T]$ and a path $P \in \operatorname{supp}(w_t)$ (thus, a path in the support of the solution before this process began). As before, let $v_1, \ldots, v_k$ be the ordered sequence of nodes serviced by P. Fix now also an edge $e \in E(P)$, and let v_j be the last node in the sequence that lies in the subpath of P from the tail of P to the tail of e. We have the following characterization of when e can be deleted.

Lemma 2. *Edge e will be deleted in the described split-and-shift procedure if the following condition holds:*

For each layer i, the last (furthest from tail(P)) node from $v_1, \ldots, v_j$ with layer at most i has germinated.

The proof can be found in the appendix.

The probability that e is deleted is now easily controlled. Fix a level i, and consider the last node u from $v_1, \ldots, v_j$ of level at most i. Since

$$\sum_{t \in S_u} \sum_{P \in \mathcal{P}: P \odot_t u} w_t(P) \geq \tfrac{1}{2},$$

standard calculations imply that the probability that u has not germinated is at most $e^{-K \log \log T / 2} = \epsilon / \log T$, where $\epsilon = \log(-K/2)$. A union bound over all levels, and an appropriate choice of K, gives us the desired result (with high probability).

To complete the proof of Theorem 1, we should observe that the number of iterations is polynomial (in expectation). This is fairly clear, and we omit the details in this extended abstract.

5 Submodular Cover over Time

Here we consider the subadditive cover over time problem where in addition to the previously required properties (in particular, that f is monotone with $f(\emptyset) = 0$), f is submodular. We assume throughout that we have a nice instance, and use F_v to denote the single time window for $v \in V$.

First, we observe that the LP relaxation (2) has an equivalent convex formulation in terms of the Lovász extension $\hat{f}$ of f.

$$\begin{aligned}
\min \quad & \sum_{t \in [T]} \hat{f}(x^t) \\
\text{s.t.} \quad & \sum_{t \in F_v} x_v^t = 1 && \forall v \in V \\
& x \geq 0
\end{aligned} \tag{4}$$

For $x \in [0,1]^V$ and $\theta \in [0,1]$, we define the *level set* $L_\theta(x) = \{v \in V : x_v \geq \theta\}$. Then $\hat{f}(x) = \mathbb{E}_\theta[f(L_\theta(x))]$, where $\theta \sim \text{Uniform}(0,1)$. Define the truncation $x^{|\theta}$ by $x_v^{|\theta} = \min\{x_v, \theta\}$.

Definition 4. Given $\theta \in [0,1]$ and $x \in [0,1]^V$, we say that the set $L_\theta(x)$ is *α-supported* if:

$$\hat{f}(x) - \hat{f}(x^{|\theta}) \geq \alpha f(L_\theta(x)). \tag{5}$$

We will provide some intuition for this definition later, but first we describe the algorithm.

Input: A solution x to (4).
Output: A solution $(S_t)_{t \in [T]}$ to the submodular cover over time problem.
1: $S_t \leftarrow \emptyset$ for all $t \in [T]$.
2: **for** $i = 1, \ldots, \log T$ **do**
3:     **for** $t \in [T]$ **do**
4:         **if** there exists $\theta \in [0,1]$ such that $L_\theta(x^t)$ is $\frac{1}{32 \log \log T}$-supported **then**
5:             Choose such a θ and set $S_t \leftarrow S_t \cup L_\theta(x^t)$, $x^t \leftarrow x^{t|\theta}$.
6:         **else**
7:             $S_t \leftarrow S_t \cup L_1(x^t)$.
8:     Merge time periods by setting, for all $t \in \{k2^i : k = 0, 1, 2, \ldots\}$,
        $x^{t+1} \leftarrow x^{t+1} + x^{t+2^{i-1}+1}$ and $x^{t+2^{i-1}+1} \leftarrow 0$.
9: **return** $(S_t)_{t \in [T]}$.

Feasibility. The only steps where the coverage $\sum_{t \in F_v} x_v^t$ for an item v could possibly decrease are steps 5 and 8. In step 5, v is added to S^t, so this is clearly no problem. In step 8, we are shifting weight from some time $t + 2^{i-1} + 1$ to $t + 1$. This cannot leave the time window of v unless $\ell(v) < i$. But if v has not been covered by the end of iteration $\ell(v)$, all of the fractional coverage for v will have been merged into the left endpoint of its time window $t' = \min F_v$. This means that $L_\theta(x^{t'})$ contains v for any θ, ensuring v will be added to $S_{t'}$ in step 3 of iteration $\ell(v) + 1$.

Cost analysis. This is where the key insights lie. The main driver is the following technical lemma (the proof is postponed to the appendix).

Lemma 3. *For any $x \in [0,1]^V$ and $\alpha \in (0,1]$, either there exists $\theta \in [0,1]$ such that $L_\theta(x)$ is $\frac{\alpha}{32}$-supported, or otherwise $2^{1/\alpha} f(L_1(x)) \leq \hat{f}(x)$.*

The intuition for this lemma is that if no θ with the desired property exists, it can be shown that $f(L_\theta(x))$ decreases quickly everywhere, and consequently $f(L_1(x))$ is small compared to $\hat{f}(x) = \mathbb{E}_\theta[f(L_\theta(x))]$.

Consider now some iteration i of the algorithm, and a particular choice of t. If step 5 is executed, then $\hat{f}(x^t)$ decreases by at least a $\frac{1}{32 \log \log T}$ fraction of $f(L_\theta(x^t))$, which is an upper bound on the cost increase of the current solution by subadditivity. Otherwise, by Lemma 3 (with $\alpha = \frac{1}{\log \log T}$),

$f(L_1(x^t)) \le \hat{f}(x^t)/\log T$. The total cost of sets chosen in step 7 in a single iteration is thus at most $\sum_{t \in [T]} \hat{f}(x^t)/\log T$. So over all $\log T$ iterations, this incurs a total cost not more than the original objective value of the relaxation.

Again, to complete the proof of Theorem 2, we should argue that the algorithm runs in polynomial time; we postpone the straightforward details to the full version.

A   Some Omitted Proofs

Proof (Theorem 3). Let y be a solution to (2). We will first generate two new instances of the subadditive cover over time problem, one being left aligned and the other right aligned.

Given an interval $[s, t]$, define the *right-aligned part* $R([s, t])$ and the *left-aligned part* $L([s, t])$ by

$$R([s, t]) = [s, k2^i] \quad \text{and} \quad L([s, t]) = [k2^i + 1, t],$$

where i, k are integers such that $k2^i \in [s, t]$ and i is maximal. If $k2^i = t$, then $L([s, t]) = \emptyset$, and if $k2^i + 1 = s$ then $R([s, t]) = \emptyset$ by convention. It is clear from this definition that $\{L([s, t]) : v \in V, [s, t] \in \mathcal{W}_v\}$ forms a left-aligned family, and similarly the right-aligned parts form a right-aligned family.

Any LP solution must cover every item by at least half in either the right-aligned or left-aligned part of its demand window. For each $v \in V$ and demand window $[s, t] \in \mathcal{W}_v$, if $L([s, t])$ receives half a unit of coverage under y, add $L([s, t])$ as a time window for v in the left-aligned instance; otherwise put $R([s, t])$ in the right-aligned instance.

It is immediate from the way in which we constructed the two instances that $2y$ is a feasible solution to each. Hence the combined cost of the optimal solutions to the LP relaxations of the generated instances is at most 4 times that of the original instance. Furthermore, we can translate integral solutions to the left and right aligned instances back to one for the original instance by adding them together, which does not increase the cost by subadditivity of f.   □

Proof (Lemma 2). We proceed by induction on j, the number of serviced nodes on the subpath of P from the tail of P until before edge e. The claim is clearly true if $j = 1$, since the condition ensures that v_1 germinated, in which case all edges from v_1 to v_2 will be deleted by the procedure. (The claim is trivial if $j = 0$, in which case edge e is always deleted).

Suppose $j > 1$, and that that the condition holds. First, by considering level $\ell(v_j)$, it follows that v_j germinated. Next, consider level $i = \ell(v_j) - 1$. If none of $v_1, v_2, \ldots, v_j$ have level i or less, then the procedure will clearly remove the edges between v_j and v_{j+1}, irrespective of what edges have already been removed from P. Otherwise, let q be chosen maximally from $\{1, 2, \ldots, j - 1\}$ so that $\ell(v_q) \le i$. Then the condition of the lemma holds for an edge between v_q and v_{q+1}; hence by our inductive assumption, these edges were removed by the split-and-shift procedure. So at the point that the current edge e is being considered for removal, the subpath of P remaining contains only the services nodes $v_{q+1}, \ldots, v_k$. Since v_j has the smallest level amongst $v_{q+1}, \ldots, v_j$ and has germinated, e will be removed. This completes the induction.   □

Proof (Lemma 3). Let $k \in \mathbb{N}$ be such that $\frac{1}{16}\alpha \le \frac{1}{k} \le \frac{1}{8}\alpha$. Note that this implies $k \ge 8$. Let $z = f(L_1(x))$.

Claim. *If $2^{1/\alpha}z > \hat{f}(x)$, there exists $m \in \mathbb{N}$ such that*

$$\dot{f}(L_{\frac{k-m}{k}}(x)) < 2^m z. \tag{6}$$

Before we prove the claim, let's see that it implies the lemma. Suppose that $2^{1/\alpha}f(L_1(x)) > \hat{f}(x)$, since otherwise we are done. The condition of the claim then holds, so take the smallest m that satisfies (6), and let $\theta = \frac{k-m}{k}$. We claim that

$$\frac{\alpha}{32}f(L_\theta(x)) \le \hat{f}(x) - \hat{f}(x^{|\theta}).$$

To see this we first rewrite the right hand side as an integral.

$$\hat{f}(x) - \hat{f}(x^{|\theta}) = \int_0^1 f(L_\eta(x))\, d\eta - \int_0^1 f(L_\eta(x^{|\theta}))\, d\eta \tag{7}$$

$$= \int_0^1 f(L_\eta(x))\, d\eta - \int_0^\theta f(L_\eta(x))\, d\eta = \int_\theta^1 f(L_\eta(x))\, d\eta.$$

Recall that $f(L_\eta(x))$ is monotonically decreasing and that $m \ge 1$ so that $\theta + \frac{1}{k} = \frac{k-m+1}{k} \le 1$. Then

$$\int_\theta^1 f(L_\eta(x))\, d\eta \ge \int_\theta^{\theta+\frac{1}{k}} f(L_\eta(x))\, d\eta \ge \frac{1}{k}f(L_{\theta+\frac{1}{k}}(x)). \tag{8}$$

Finally, we use the fact that m is minimal, which implies that $f(L_{\frac{k-m+1}{k}}(x)) \ge 2^{m-1}z$, together with (7) and (8):

$$\hat{f}(x) - \hat{f}(x^{|\theta}) \ge \frac{1}{k}2^{m-1}z = \frac{1}{2k}2^m z > \frac{\alpha}{32}f(L_\theta(x)). \tag{9}$$

In the final inequality of (9) we use that the fact that we chose m to satisfy $2^m z > f(L_{\frac{k-m}{k}}(x))$ and $\frac{1}{k} \ge \frac{1}{16}\alpha$.

Now we proceed to prove the claim. Suppose for contradiction that the condition of the claim holds but no m satisfies inequality (6). Then, in particular it must hold that $f(L_{\frac{1}{k}}(x)) \ge 2^{k-1}z$ and therefore we obtain

$$\hat{f}(x) \ge \int_0^{\frac{1}{k}} f(L_\eta(x))\, d\eta \ge \frac{1}{k}f(L_{\frac{1}{k}}(x)) \ge \frac{1}{k}2^{k-1}z.$$

Since $k \ge 8$, $\frac{1}{k}2^{k-1}z \ge 2^{k/2}z$. Since also $\frac{1}{k} \le \frac{1}{8}\alpha$, we deduce

$$\hat{f}(x) \ge 2^{k/2}z \ge 2^{4/\alpha}z \ge 2^{1/\alpha}z,$$

contradicting that $2^{1/\alpha}z > \hat{f}(x)$. This proves the claim, and hence the lemma.   □

References

1. Arkin, E., Joneja, D., Roundy, R.: Computational complexity of uncapacitated multi-echelon production planning problems. Oper. Res. Lett. **8**(2), 61–66 (1989)
2. Bienkowski, M., Byrka, J., Chrobak, M., Jeż, Ł., Nogneng, D., Sgall, J.: Better approximation bounds for the joint replenishment problem. In: Proceedings of the 25th Annual ACM-SIAM Symposium on Discrete Algorithms (SODA), pp. 42–54 (2014)
3. Byrka, J., Grandoni, F., Rothvoß, T., Sanità, L.: Steiner tree approximation via iterative randomized rounding. J. ACM **60**(1), 6:1–6:33 (2013)
4. Chekuri, C., Ene, A.: Approximation algorithms for submodular multiway partition. In: Proceedings of the 52nd Annual IEEE Symposium on Foundations of Computer Science (FOCS), pp. 807–816 (2011)
5. Chekuri, C., Ene, A.: Submodular cost allocation problem and applications. In: Aceto, L., Henzinger, M., Sgall, J. (eds.) ICALP 2011. LNCS, vol. 6755, pp. 354–366. Springer, Heidelberg (2011). https://doi.org/10.1007/978-3-642-22006-7_30
6. Cheung, M., Elmachtoub, A.N., Levi, R., Shmoys, D.B.: The submodular joint replenishment problem. Math. Programm. **158**(1–2), 207–233 (2016)
7. Coelho, L.C., Cordeau, J.-F., Laporte, G.: Thirty years of inventory routing. Transp. Sci. **48**(1), 1–19 (2013)
8. Ene, A., Vondrák, J., Wu, Y.: Local distribution and the symmetry gap: approximability of multiway partitioning problems. In: Proceedings of the 24th Annual ACM-SIAM Symposium on Discrete Algorithms (SODA), pp. 306–325 (2013)
9. Fukunaga, T., Nikzad, A., Ravi, R.: Deliver or hold: approximation algorithms for the periodic inventory routing problem. In: Proceedings of APPROX/RANDOM (2014)
10. Jain, K.: A factor 2 approximation algorithm for the generalized steiner network problem. Combinatorica **21**(1), 39–60 (2001)
11. Levi, R., Roundy, R., Shmoys, D., Sviridenko, M.: A constant approximation algorithm for the one-warehouse multiretailer problem. Manag. Sci. **54**(4), 763–776 (2008)
12. Levi, R., Roundy, R.O., Shmoys, D.B.: Primal-dual algorithms for deterministic inventory problems. Math. Oper. Res. **31**(2), 267–284 (2006)
13. Nagarajan, V., Shi, C.: Approximation algorithms for inventory problems with submodular or routing costs. Math. Program. **160**(1–2), 225–244 (2016)
14. Wagner, H.M., Whitin, T.M.: Dynamic version of the economic lot size model. Manag. Sci. **5**(1), 89–96 (1958)

Extended Formulations for Stable Set Polytopes of Graphs Without Two Disjoint Odd Cycles

Michele Conforti[1], Samuel Fiorini[2], Tony Huynh[3], and Stefan Weltge[4(✉)]

[1] Dipartimento di Matematica, Università degli Studi di Padova, Padova, Italy
conforti@math.unipd.it
[2] Département de Mathématique, Université libre de Bruxelles, Brussels, Belgium
sfiorini@ulb.ac.be
[3] School of Mathematics, Monash University, Melbourne, Australia
tony.bourbaki@gmail.com
[4] Fakultät für Mathematik, Technische Universität München, Munich, Germany
weltge@tum.de

Abstract. Let G be an n-node graph without two disjoint odd cycles. The algorithm of Artmann, Weismantel and Zenklusen (STOC'17) for bimodular integer programs can be used to find a maximum weight stable set in G in strongly polynomial time. Building on structural results characterizing sufficiently connected graphs without two disjoint odd cycles, we construct a size-$O(n^2)$ extended formulation for the stable set polytope of G.

1 Introduction

It is a classic result that integer programs with a totally unimodular constraint matrix A are solvable in strongly polynomial time. Very recently, Artmann, Weismantel and Zenklusen [1] generalized this to *bimodular* matrices A. These include all matrices with all subdeterminants in $\{-2, -1, 0, 1, 2\}$. As noted in [1], this has consequences for the maximum weight stable set problem in graphs as follows.

Let $\mathrm{STAB}(G)$ be the *stable set polytope* of a graph G and note that

$$\mathrm{STAB}(G) = \mathrm{conv}\{x \in \{0,1\}^{V(G)} \mid Mx \leqslant \mathbf{1}\};$$

where $M \in \{0,1\}^{E(G) \times V(G)}$ is the edge-node incidence matrix of G. It is well-known that the maximum absolute value of a subdeterminant of M is equal to $2^{\mathrm{ocp}(G)}$, where $\mathrm{ocp}(G)$ is the maximum number of (node-)disjoint odd cycles of G (see [12]). Therefore, the bimodular algorithm of [1] can be used to efficiently compute a maximum weight stable set in a graph without two disjoint odd cycles.

Although the bimodular algorithm is extremely powerful, it provides limited insight on which properties of graphs with $\mathrm{ocp}(G) \leqslant 1$ are relevant to derive

---

Extended abstract. For the full version, see [8].

© Springer Nature Switzerland AG 2020
D. Bienstock and G. Zambelli (Eds.): IPCO 2020, LNCS 12125, pp. 104–116, 2020.
https://doi.org/10.1007/978-3-030-45771-6_9

efficient algorithms for graphs with *higher* odd cycle packing number. Indeed, in light of recent work linking the complexity and structural properties of integer programs to the magnitude of its subdeterminants [1,4,10,11,14,17,18], it is tempting to believe that integer programs with bounded subdeterminants can be solved in polynomial time. This would imply in particular that the stable set problem on graphs with $\mathrm{ocp}(G) \leqslant k$ is polynomial for every fixed k. Conforti, Fiorini, Huynh, Joret, and Weltge [7] recently proved this is true under the additional assumption that G has bounded (Euler) genus.[1]

Furthermore, by itself the bimodular algorithm does not imply any linear description of the stable set polytope of graphs G with $\mathrm{ocp}(G) = 1$. It turns out that for such graphs, $\mathrm{STAB}(G)$ may have many facets with high coefficients that do not seem to allow a "nice" combinatorial description in the original space. While stable set polytopes have been studied for several classes of graphs, very little is known about $\mathrm{STAB}(G)$ when $\mathrm{ocp}(G) = 1$. Our main result is to show that every such stable set polytope admits a compact description in an "extended" space.

To this end, we say that an *extended formulation* of a polyhedron P is a description of the form $P = \{x \mid \exists y : Ax + By \leqslant b\}$ whose *size* is the number of inequalities in $Ax + By \leqslant b$. The *extension complexity* of P, denoted $\mathrm{xc}(P)$, is the minimum size of an extended formulation of P. Our main result is the following.

Theorem 1. *For every n-node graph G with $\mathrm{ocp}(G) \leqslant 1$, $\mathrm{STAB}(G)$ admits a size-$O(n^2)$ extended formulation. Moreover, this extended formulation can be constructed in polynomial time.*

Note that this does not follow from the main result of [1]. As noted in [5, Thm. 5.4], integer hulls of bimodular integer programs can have exponential extension complexity. Moreover, Theorem 1 does also not follow from [7] since here we are dealing with *arbitrary* graphs G with $\mathrm{ocp}(G) \leqslant 1$.

On the one hand, our proof uses a characterization of graphs with $\mathrm{ocp}(G) \leqslant 1$ due to Lovász (see Seymour [15]). Kawarabayashi and Ozeki [13] later gave a short, purely graph-theoretical proof of the same result. Before stating Lovász' theorem, we need a few more definitions. The *odd cycle transversal number* of a graph G, denoted $\mathrm{oct}(G)$, is the minimum size of a set of nodes X such that $G - X$ is bipartite. The *projective plane* is the surface obtained from a closed disk by identifying antipodal points on its boundary. An embedding of a graph G in a surface is an *even-face embedding* if every face of G is an open disk bounded by an even cycle of G.

Theorem 2 (Lovász, cited in [15]). *Let G be a 4-connected graph with $\mathrm{ocp}(G) \leqslant 1$. Then*

---

[1] The *Euler genus* of graph G is the minimum of $|E(G)| - |V(G)| - |F(G)| - 2$, taken over all embeddings of G in a (orientable or non-orientable) surface, where $F(G)$ denotes the set of faces of G with respect to the embedding.

(i) $\mathrm{oct}(G) \leqslant 3$, *or*

(ii) G has an even-face embedding in the projective plane.

Note that if a graph G satisfies (i) of Theorem 2, then $\mathrm{STAB}(G)$ has a compact extended formulation since it is the convex hull of the union of at most eight polytopes described by nonnegativity and edge constraints. As a special case of [7, Theorem 3], $\mathrm{STAB}(G)$ also has a compact extended formulation if G satisfies (ii) of Theorem 2.

Theorem 3. *Let G be an n-node graph that is even-face embedded in the projective plane. Then* $\mathrm{STAB}(G)$ *has a size-$O(n^2)$ extended formulation.*

However, the decomposition portion of our proof is non-trivial since Theorem 2 is stated for 4-connected graphs. Hence, we have to deal with the polyhedral aspects of performing 2- and 3-sums, using the properties of graphs without two disjoint odd cycles. In general graphs, performing multiple k-sums does not preserve small extended formulations for the respective stable set polytopes, even for $k = 2$.

On the other hand, our polyhedral analysis crucially relies on new insights about the structure of facets of stable set polytopes (see Lemma 12) and a transformation of stable set polytopes into the edge space (see Appendix A and Sect. 4). We believe that this perspective can be equally beneficial for other future investigations of (general) stable set polytopes.

Finally, we remark that our proof also can be turned into a direct, purely graph-theoretic strongly polynomial time algorithm for the stable set problem in graphs G with $\mathrm{ocp}(G) \leqslant 1$.

We conclude this introduction with a brief outline of the paper. In Sect. 2, we build on Theorem 2 and its signed version due to Slilaty [16] to describe the structure of graphs without two disjoint odd cycles. Roughly, we prove that each such graph G either has $\mathrm{oct}(G) \leqslant 3$ or can be obtained from a graph H_0 having an even-face embedding in the projective plane by gluing internally disjoint bipartite graphs $T_1, \ldots, T_\ell$ "around" H_0 in a certain way. Appendix A gives a short account of the known compact extended formulation for $\mathrm{STAB}(G)$ for graphs G admitting an even-face embedding in the projective plane, see [7]. This is our base case. The general case is treated in Sects. 3 and 4 by a delicate argument using certain gadgets $H_1, \ldots, H_\ell$ "simulating" the bipartite graphs $T_1, \ldots, T_\ell$.

Due to length restrictions, some material has been deleted from this extended abstract. For a full version, see [8].

2   The Structure of Graphs Without Two Disjoint Odd Cycles

In this section we show that every graph without two disjoint odd cycles either has a small odd cycle transversal or has a structure that we will exploit later.

For this purpose we use the notion of separations. A k-*separation* of a graph G is an ordered pair (G_0, G_1) of edge-disjoint subgraphs of G with $G = G_0 \cup G_1$, $|V(G_0) \cap V(G_1)| = k$, and $E(G_0), E(G_1), V(G_1) \smallsetminus V(G_0), V(G_0) \smallsetminus V(G_1)$ all non-empty. We say that a k-separation is *linked* if for every two distinct nodes of $V(G_0) \cap V(G_1)$ there exists a u–v path in G_1 whose internal nodes are disjoint from G_0.

Definition 4. *A comb structure of a graph G is a set of subgraphs $H_0, T_1, \ldots, T_\ell$ of G such that for all $i \in [\ell]$: T_i is bipartite, $(H_0 \cup_{j \neq i} T_j, T_i)$ is a linked k-separation of G with $k \leq 3$, and $V(T_i) \cap V(T_j) \subseteq V(H_0)$ for all $j \neq i$ (Fig. 1).*

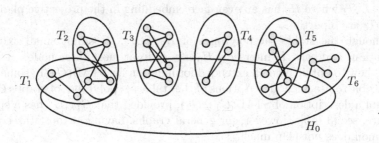

Fig. 1. A comb structure.

For our structural result we will also use the notion of signed graphs. A *signed graph* is a pair (G, Σ) where G is a graph and $\Sigma \subseteq E(G)$. A subgraph of G is said to be Σ-*odd* if it contains an odd number of edges in Σ, and is Σ-*even* otherwise. The *odd cycle packing number* of a signed graph (G, Σ) is the maximum number of disjoint Σ-odd cycles in (G, Σ), and is denoted by $\mathrm{ocp}(G, \Sigma)$. A signed graph (G, Σ) is *balanced* if $\mathrm{ocp}(G, \Sigma) = 0$. The *odd cycle transversal number* of (G, Σ) is the minimum number of nodes in (G, Σ) intersecting every Σ-odd cycle in (G, Σ), and is denoted by $\mathrm{oct}(G, \Sigma)$. An embedding of a signed graph (G, Σ) in a surface is an *even-face embedding* if every face of (G, Σ) is an open disk bounded by a Σ-even cycle of (G, Σ). Graphs in this section may have parallel edges.

In the definition below, $\uplus$ is used to denote the edge-disjoint union of graphs.

Definition 5. *Let G be a graph with comb structure $H_0, T_1, \ldots, T_\ell$. For each $i \in [\ell]$, let $S_i := V(H_0) \cap V(T_i)$ and note that there is a signed clique (K_i, Σ_i) with $V(K_i) = S_i$ such that $(K_i \uplus T_i, \Sigma_i \uplus E(T_i))$ is balanced. The signed graph (H^+, Σ) is defined via $H^+ := H_0 \uplus K_1 \uplus \cdots \uplus K_\ell$ and $\Sigma := E(H_0) \uplus \Sigma_1 \uplus \cdots \uplus \Sigma_\ell$.*

Our structural result is as follows. The proof can be found in the full version of the paper. It uses a finer version of Theorem 2 that is suited for signed graphs, due to Slilaty [16].

Theorem 6. *Let G be a graph with $\mathrm{ocp}(G) = 1$ and $\mathrm{oct}(G) \geq 4$. Then G admits a comb structure $H_0, T_1, \ldots, T_\ell$, such that $S_1, \ldots, S_\ell$ and (H^+, Σ) from Definition 5 have the following properties:*

- S_i is not a subset of S_j for all distinct $i, j \in [\ell]$,
- (H^+, Σ) has an even-face embedding in the projective plane, and
- the nodes of each S_i are on the boundary of some face of the embedding.

3   Constructing a Compact Extended Formulation

Here, we describe how Theorem 1 can be proven using Theorems 6 and 3. Let G be a graph with $\mathrm{ocp}(G) = 1$. If $\mathrm{oct}(G) \leqslant 3$, then $\mathrm{STAB}(G)$ has a $O(n^2)$-size extended formulation by Balas' theorem [2]. Otherwise, $\mathrm{oct}(G) \geqslant 4$ and G can be decomposed as in Theorem 6. In particular, G is the union of graphs $H_0, T_1, \ldots, T_\ell$ where H_0 has an even-face embedding in the projective plane and $T_1, \ldots, T_\ell$ are bipartite.

Although the stable set polytopes of $H_0, T_1, \ldots, T_\ell$ admit small extended formulations and each T_i intersects $H_0 \cup_{j \neq i} T_j$ in at most three nodes, it is not obvious how to obtain a small extended formulation for $\mathrm{STAB}(G)$. In some cases it is possible to use linear descriptions of the stable set polytopes of graphs G_1, G_2 to obtain a description of $\mathrm{STAB}(G_1 \cup G_2)$, provided that $G_1 \cap G_2$ has a specific structure, see [3,6,9]. However, for general graphs, having at most three nodes in common does not help much.

With this in mind, recall that not only H_0 but also the signed graph (H^+, Σ) has an even-face embedding in the projective plane. We will replace each signed clique used to define (H^+, Σ) by a constant size gadget H_i corresponding to each T_i in a way that the resulting graph $G^{(\ell)} := H_0 \cup H_1 \cup \cdots \cup H_\ell$ (the "core") still has an even-face embedding in the projective plane. Moreover, each $T_i' := T_i \cup H_i$ will still be bipartite. In this way G is obtained from $G^{(\ell)}$ by iteratively performing k-sums ($k \leqslant 3$) with $T_1', \ldots, T_\ell'$ along $H_1, \ldots, H_\ell$. In each such operation, the specific choice of the gadget will allow us to relate the extension complexities of the stable set polytopes of the participating graphs in a controlled way. Let us start with describing the gadgets that will be used.

Definition 7. *A gadget is a graph isomorphic to P_3, P_4, $S_{2,2,2}$ or $S_{2,3,3}$, see Fig. 2. Let G be a graph with a linked k-separation (G_0, G_1) such that $k \in \{2, 3\}$ and G_1 is bipartite. We say that a gadget H is attachable to G_1 (with respect to separation (G_0, G_1)) if its set of leaf nodes equals $V(G_0) \cap V(G_1)$, its set of non-leaf nodes is disjoint from $V(G)$, and $G_1' := G_1 \cup H$ is bipartite.*

Note that if G is a graph with a linked k-separation (G_0, G_1) such that $k \in \{2, 3\}$ and G_1 is bipartite, then there is a unique gadget $H \in \{P_3, P_4, S_{2,2,2}, S_{2,3,3}\}$ that is attachable to G_1.

Next, let us formally describe how the signed cliques used to define (H^+, Σ) are replaced by gadgets in order to obtain the core.

Definition 8. *Let G be a 2-connected graph with comb structure $H_0, T_1, \ldots, T_\ell$. For each $i \in [\ell]$, pick a gadget H_i that is attachable to T_i with respect to the separation $(H_0 \cup \bigcup_{j \neq i} T_j, T_i)$. (We always assume that the set of non-leaf nodes of the gadgets H_i, $i \in [\ell]$ are mutually disjoint.) We call the graph $H_0 \cup H_1 \cup \cdots \cup H_\ell$ the core.*

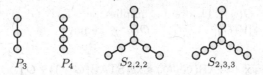

Fig. 2. Gadgets and their names.

Lemma 9. *Every 2-connected graph G with $\mathrm{ocp}(G) = 1$ and $\mathrm{oct}(G) \geqslant 4$ admits a comb structure whose core has an even-face embedding in the projective plane.*

Proof. The proof is immediate by choosing a comb structure that satisfies Theorem 6.   □

The remaining ingredient for our proof of Theorem 1 will be the following result. To this end, let (G_0, G_1) be a separation of graph G. Below, for $i \in \{0,1\}$, we call a vertex *internal* if it belongs to $V(G_i) \smallsetminus V(G_{1-i})$ and an edge of G_i *internal* if at least one of its endnodes is not in G_{1-i}.

Theorem 10. *Let G be a 2-connected, non-bipartite graph. Assume that G has a k-separation (G_0, G_1) such that G_1 is bipartite, and $k \in \{2,3\}$. Let μ_1 denote the number of internal vertices and edges of G_1. Let H be a gadget that is attachable to G_1, and let $G_0' := G_0 \cup H$. Then $\mathrm{xc}(\mathrm{STAB}(G)) \leqslant \mathrm{xc}(\mathrm{STAB}(G_0')) + O(\mu_1)$.*

Before we continue with the proof of Theorem 10 in the next section, let us see how this yields a proof of our main result.

Proof of Theorem 1. By induction on the number of nodes n, we may assume that G is 2-connected. Indeed, suppose that G has a k-separation (G_0, G_1) with $k \in \{0,1\}$. For $i \in \{0,1\}$, let $n_i := |V(G_i)|$. Thus $n = n_0 + n_1 - k$. If c is any constant such that $\mathrm{xc}(\mathrm{STAB}(G_i)) \leqslant c \cdot n_i^2$ for $i \in \{0,1\}$, we get

$$\mathrm{xc}(\mathrm{STAB}(G)) \leqslant \mathrm{xc}(\mathrm{STAB}(G_0)) + \mathrm{xc}(\mathrm{STAB}(G_1)) \leqslant c \cdot n_0^2 + c \cdot n_1^2 \leqslant c \cdot n^2,$$

where the first inequality is due to [6, Thm. 4.1].

As observed above, if $\mathrm{oct}(G) \leqslant 3$ then $\mathrm{STAB}(G)$ trivially has a size-$O(n^2)$ extended formulation. Now assume that $\mathrm{ocp}(G) = 1$ and $\mathrm{oct}(G) \geqslant 4$. Let $H_0, T_1, \dots, T_\ell$ be a comb structure of G as in Lemma 9. Since G is 2-connected, each separation $(H_0 \cup_{j \neq i} T_j, T_i)$ is either a 2- or a 3-separation. For each $i \in [\ell]$, we consider the graph

$$G^{(i)} := H_0 \cup H_1 \cup \cdots \cup H_i \cup T_{i+1} \cup \cdots \cup T_\ell .$$

where H_i denotes a gadget attachable to T_i. For $i \in [\ell]$, let μ_i denote the number of internal vertices and edges of T_i. Notice that $G^{(\ell)}$ is the core, and thus by Lemma 9 has an even-face embedding in the projective plane. By Theorem 10,

$$\mathrm{xc}(\mathrm{STAB}(G^{(i-1)})) \leqslant \mathrm{xc}(\mathrm{STAB}(G^{(i)})) + O(\mu_i) .$$

Since $|V(G^{(\ell)})| = O(n)$, Theorem 3 implies $\mathrm{xc}(\mathrm{STAB}(G^{(\ell)})) = O(n^2)$. Since moreover $\sum_{i=1}^{\ell} \mu_i \leqslant |V(G)| + |E(G)| = O(n^2)$, we have

$$\mathrm{xc}(\mathrm{STAB}(G)) = \mathrm{xc}(\mathrm{STAB}(G^{(0)})) \leqslant \mathrm{xc}(\mathrm{STAB}(G^{(\ell)})) + O\left(\sum_{i=1}^{\ell} \mu_i\right) = O(n^2).$$

□

4 Dealing with Small Separations

In this section we describe an extended formulation that yields the bound claimed in Theorem 10. Given a stable set S in a graph G, we say that an edge is *slack* if none of its endnodes is in S. We denote by $\sigma(S)$ the set of slack edges. An edge is said to be *tight* if it is not slack.

Lemma 11. *Let G, G_0, G_0', G_1 and H be as in Theorem 10. Letting $\overline{\mathrm{STAB}}(G_1')$ denote the convex hull of characteristic vectors of stable sets S in $G_1' := G_1 \cup H$ having at most one slack edge in H, we have*

$$\mathrm{STAB}(G) = \{(x^0, x^1, x^{01}) \in \mathbb{R}^{V(G)} \mid \exists x^H : (x^0, x^{01}, x^H) \in \mathrm{STAB}(G_0'), \qquad (1)$$
$$(x^1, x^{01}, x^H) \in \overline{\mathrm{STAB}}(G_1')\}.$$

where $x^0 \in \mathbb{R}^{V(G_0) \setminus V(G_1)}$, $x^1 \in \mathbb{R}^{V(G_1) \setminus V(G_0)}$, $x^{01} \in \mathbb{R}^{V(G_0) \cap V(G_1)}$ and $x^H \in \mathbb{R}^{V(H) \setminus V(G)}$.

Let us first verify that Lemma 11 indeed implies Theorem 10.

Proof of Theorem 10. By Lemma 11, we have $\mathrm{xc}(\mathrm{STAB}(G)) \leqslant \mathrm{xc}(\mathrm{STAB}(G_0')) + \mathrm{xc}(\overline{\mathrm{STAB}}(G_1'))$. Since gadget H has constant size, $\overline{\mathrm{STAB}}(G_1')$ is the convex hull of the union of a constant number of faces of $\mathrm{STAB}(G_1')$ in which the coordinates of the nodes in H are fixed. Hence by Balas' union of polytopes [2], we obtain $\mathrm{xc}(\overline{\mathrm{STAB}}(G_1')) = O(\mathrm{xc}(\mathrm{STAB}(G_1))) = O(|V(G_1)| + |E(G_1)|)$. Since $|V(G_1)| + |E(G_1)| \leqslant 6 + \mu_1$ and $\mu_1 \geqslant 1$, we conclude $\mathrm{xc}(\overline{\mathrm{STAB}}(G_1')) = O(\mu_1)$. This proves the claim.   □

In the proof of Lemma 11, we exploit a special property of the weight functions that are facet-defining for stable set polytopes, which we describe next.

4.1 Reducing to Edge-Induced Weights

We call a weight function $w : V(G) \to \mathbb{R}$ on the nodes of G *edge-induced* if there is a nonnegative cost function $c : E(G) \to \mathbb{R}_{\geqslant 0}$ such that $w(v) = c(\delta(v))$ for all $v \in V(G)$. For a given node-weighted graph (G, w) we let $\alpha(G, w)$ denote the maximum weight of a stable set. The proof of the next lemma is given in the full version.

Lemma 12. *Let $G = (V, E)$ be a graph without isolated nodes and let $w : V \to \mathbb{R}$ be a weight function. There exists an edge-induced weight function $w' : V \to \mathbb{R}$ such that $w(v) \leqslant w'(v)$ for all nodes v and $\alpha(G, w) = \alpha(G, w')$. In particular, the node weights of every non-trivial facet-defining inequality of $\mathrm{STAB}(G)$ are edge-induced.*

For $c \colon E(G) \to \mathbb{R}_{\geqslant 0}$, we let

$$\beta(G, c) := \min \left\{ \sum_{e \in E(G)} c(e) y_e \mid y \in \sigma(\mathrm{STAB}(G)) \right\}.$$

In fact, in [7, Propositions 11 and 14] it is shown that one can optimize over $Q(G) = \sigma(P(G))$ instead of $\sigma(\mathrm{STAB}(G))$ without changing the optimum. However, we will not need this here. Our last lemma follows from [7, Observation 13].

Lemma 13. *Let $G = (V, E)$ be a graph. If $w : V(G) \to \mathbb{R}$ is induced by $c : E(G) \to \mathbb{R}_{\geqslant 0}$, then $\alpha(G, w) = c(E(G)) - \beta(G, c)$.*

4.2   Correctness of the Extended Formulation

In this section we prove Lemma 11. To this end, let $R(G)$ denote the right-hand side of (1). We leave to the reader to verify that for each stable set S of G, there exists a stable set S' of $G' := G \cup H$ such that $S' \cap V(G) = S$ and moreover at most one edge of H is slack with respect to S'. The inclusion $\mathrm{STAB}(G) \subseteq R(G)$ follows directly from this.

In order to prove the reverse inclusion $R(G) \subseteq \mathrm{STAB}(G)$, first observe that $R(G) \subseteq \mathbb{R}_{\geqslant 0}^{V(G)}$. Thus, by Lemma 12 it suffices to show that, for all edge-induced node weights $w : V(G) \to \mathbb{R}$, the inequality

$$\sum_{v \in V(G)} w(v) x_v \leqslant \alpha(G, w) \tag{2}$$

is valid for all $x \in R(G)$. As in Sect. A it will be convenient to work in the edge space instead of the node space. To this end, let $c \colon E(G) \to \mathbb{R}_+$ be non-negative edge costs, and let $w(v) := c(\delta(v))$ for every node v. By Lemma 13 we see that (2) is valid for $R(G)$ if and only if

$$\sum_{e \in E(G)} c(e) y_e \geqslant \beta(G, c) \tag{3}$$

is satisfied by all points $y \in \sigma(R(G))$. Our proof strategy to derive (3) is to seek additional edge costs $c^H : E(H) \to \mathbb{R}_{\geqslant 0}$ on the edges of gadget H such that

$$\sum_{e \in E(G_0)} c(e) y_e^0 + \sum_{e \in E(H)} c^H(e) y_e^H \geqslant \beta(G, c) \tag{4}$$

is valid for all $(y^0, y^H) \in \sigma(\mathrm{STAB}(G_0'))$ and

$$\sum_{e \in E(G_1)} c(e)y_e^1 - \sum_{e \in E(H)} c^H(e)y_e^H \geqslant 0 \tag{5}$$

is valid for all $(y^1, y^H) \in \sigma(\overline{\mathrm{STAB}}(G_1'))$. Then (3) is obtained as the sum of (4) and (5).

Let us first focus on Inequality (5). Independently of how the edge costs c^H are defined, in order to prove that (5) holds for all $(y^0, y^H) \in \sigma(\overline{\mathrm{STAB}}(G_1'))$, we may assume that y^H is a 0/1-vector with at most one nonzero entry. The general case follows by convexity. Since the case $y^H = \mathbf{0}$ is trivial, assume that $y^H = \chi^{\{f\}}$ for some $f \in E(H)$. Hence (5) can be rewritten as

$$\sum_{e \in E(G_1)} c(e)y_e^1 \geqslant c^H(f). \tag{6}$$

This suggests the following definition of c^H. For $F \subseteq E(H)$, we let

$$\gamma(F) := \min\left\{c\big(\sigma(S) \cap E(G_1)\big) \mid S \text{ stable set of } G_1', \ \sigma(S) \cap E(H) = F\right\}.$$

Observe that $\gamma(F) \in \mathbb{R}_{\geqslant 0} \cup \{\infty\}$. We say that F is *feasible* if $\gamma(F)$ is finite, that is, there exists a stable set S of G_1' such that $\sigma(S) \cap E(H) = F$. Notice that $F := \{f\}$ is feasible for all $f \in E(H)$. By setting $c^H(f) := \gamma(\{f\}) \in \mathbb{R}_{\geqslant 0}$ for each $f \in E(H)$ we clearly satisfy (6), and hence (5) is valid for all $(y^1, y^H) \in \sigma(\overline{\mathrm{STAB}}(G_1'))$ for this choice of c^H.

It remains to prove that (4) is valid for all $(y^0, y^H) \in \sigma(\mathrm{STAB}(G_0'))$. To this end, we will need the following observation, see Lemma 14 below. The geometric intuition behind Lemma 14 is that $Q(G) = \sigma(P(G))$ is a polyhedral cone whenever G is bipartite.

Lemma 14. *If $F \subseteq E(H)$ is feasible and the disjoint union of A and B, then $\gamma(F) \leqslant \gamma(A) + \gamma(B)$.*

To prove that the inequality in (4) is valid for all $(y^0, y^H) \in \sigma(\mathrm{STAB}(G_0'))$, it suffices to consider any vertex (y^0, y^H) of $\sigma(\mathrm{STAB}(G_0'))$ minimizing the left-hand size of (4). We may even assume that (y^0, y^H) minimizes $\|y^H\|_1$ among all such vertices.

Let S^0 denote the stable set of G_0' corresponding to (y^0, y^H) and let $F := \sigma(S^0) \cap E(H)$. Note that $y^H = \chi^F$. Observe that S^0 is not properly contained in another stable set, since this would contradict the minimality of y. Moreover, we claim that F has at most one edge. In order to prove the claim, we consider only the case where $H = S_{2,2,2}$, see Fig. 2. The other cases are easier or similar, and we leave the details to the reader.

Let us assume that F contains at least two edges, that is, $\|y^H\|_1 \geqslant 2$. We will replace y^H by a new vector $\bar{y}^H \in \{0,1\}^{E(H)}$ such that $(y^0, \bar{y}^H) \in \sigma(\mathrm{STAB}(G_0'))$ with smaller ℓ_1-norm in such a way that the cost of $(y^0, \bar{y}^H)$ is not higher than that of (y^0, y^H), arriving at a contradiction. In order to prove that $(y^0, \bar{y}^H) \in \sigma(\mathrm{STAB}(G_0'))$ we will explain how to obtain the corresponding stable set $\bar{S}^0$ from stable set S^0 in each case. To guarantee that the cost of $(y^0, \bar{y}^H)$ does not exceed that of (y^0, y^H), we will mainly rely on Lemma 14.

To distinguish the different cases, let v_1, v_2 and v_3 denote the leaves of H and v_0 denote its degree-3 node. For $i, j \in \{0, 1, 2, 3\}$ we let P_{ij} denote the v_i–v_j path in H. For $i \in [3]$, let v_{0i} denote the middle vertex of P_{ij} and let e_i and f_i denote the edges of the path P_{0i} incident to v_i and v_0 respectively. The relevant cases and the replacements are listed in Fig. 3. We treat each of them below. Notice that the case $|S^0 \cap \{v_1, v_2, v_3\}| = 3$ cannot arise since this would contradict the maximality of S^0.

Case 1: $|S^0 \cap \{v_1, v_2, v_3\}| = 0$. In this case we set $\bar{y}^H := \mathbf{0}$, which corresponds to letting $\bar{S}^0 := (S^0 \cup \{v_{01}, v_{02}, v_{03}\}) \smallsetminus \{v_0\}$. In this case it is clear that the cost of $(y^0, \bar{y}^H)$ is at most the cost of (y^0, y^H).

Case 2: $|S^0 \cap \{v_1, v_2, v_3\}| = 1$. We may assume that $S^0 \cap \{v_1, v_2, v_3\} = \{v_3\}$. Since $|F| \geqslant 2$ and S^0 is maximal, we must have $S^0 \cap V(H) = \{v_0, v_3\}$ and hence $y^H = \chi^{\{e_1, e_2\}}$.

We let $\bar{y}^H := \chi^{\{f_3\}}$, which corresponds to letting $\bar{S}^0 := S^0 \smallsetminus \{v_0\} \cup \{v_{01}, v_{02}\}$. The cost of $(y^0, \bar{y}^H)$ equals the cost of (y^0, y^H) minus $\gamma(\{e_1\}) + \gamma(\{e_2\}) - \gamma(\{f_3\}) = \gamma(\{e_1\}) + \gamma(\{e_2\}) - \gamma(\{e_1, e_2\}) \geqslant 0$. The equality follows from the fact that stable sets S of G_1' such that $\sigma(S) \cap E(H) = \{f_3\}$ and stable sets S of G_1' such that $\sigma(S) \cap E(H) = \{e_1, e_2\}$ have the same intersection with the leaves of H. The inequality follows from Lemma 14.

Case 3: $|S^0 \cap \{v_1, v_2, v_3\}| = 2$. We may assume that $S^0 \cap \{v_1, v_2, v_3\} = \{v_1, v_2\}$. Again, since $|F| \geqslant 2$ and S^0 is maximal, we must have $S^0 \cap V(H) = \{v_1, v_2, v_{03}\}$ and hence $y^H = \chi^{\{f_1, f_2\}}$. We let $\bar{y}^H := \chi^{\{e_3\}}$, which corresponds to letting $\bar{S}^0 := S^0 \smallsetminus \{v_{03}\} \cup \{v_{01}, v_{02}\}$. Similar to the previous case, we obtain that the cost of $(y^0, \bar{y}^H)$ equals the cost of (y^0, y^H) minus $\gamma(\{f_1\}) + \gamma(\{f_2\}) - \gamma(\{e_3\}) = \gamma(\{f_1\}) + \gamma(\{f_2\}) - \gamma(\{f_1, f_2\}) \geqslant 0$.

Thus, F has indeed at most one edge. There exists a stable set S^1 of G_1' that is a minimizer for $\gamma(F)$ such that $S^1 \cap V(G) \cap V(H) = S^0 \cap V(G) \cap V(H)$. Hence, $S := S^1 \cup S^0$ is a stable set of G. Let (y^0, y^1) denote the characteristic vector of $\sigma(S)$, so that $(y^0, y^1) \in \sigma(\mathrm{STAB}(G))$. We get

$$\sum_{e \in E(G_0)} c(e) y_e^0 + \sum_{e \in E(H)} c^H(e) y_e^H = \sum_{e \in E(G_0)} c(e) y_e^0 + \gamma(F)$$

$$= \sum_{e \in E(G_0)} c(e) y_e^0 + \sum_{e \in E(G_1)} c(e) y_e^1 \geqslant \beta(G, c).$$

Above, the first equality comes from the fact that F has at most one edge, the definition of $c^H(f)$ for $f \in E(H)$ and $\gamma(\varnothing) = 0$. The second equality follows from the hypothesis that S^1 is a minimizer for $\gamma(F)$. Finally, the inequality is due to the validity of (3) for $\sigma(\mathrm{STAB}(G))$. This shows that (4) is indeed valid for $(y^0, y^H) \in \sigma(\mathrm{STAB}(G_0'))$, which concludes the proof of Lemma 11.

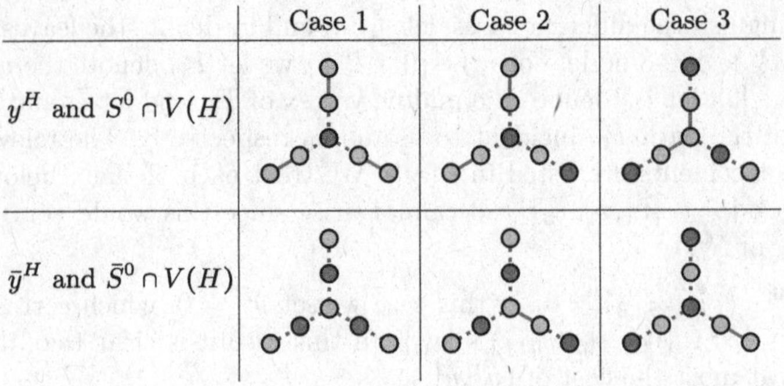

	Case 1	Case 2	Case 3
y^H and $S^0 \cap V(H)$			
$\bar{y}^H$ and $\bar{S}^0 \cap V(H)$			

Fig. 3. Replacements in the proof of Lemma 11 (top row: before, bottom row: after). Red thick edges are slack. Blue thick, dotted edges are tight. Red nodes are in the stable set, blue nodes are not. (Color figure online)

Acknowledgements. This paper was supported by ERC Consolidator Grant 615640-ForEFront and FNRS PDR Funding T008720F "Structural and Algorithmic Aspects of Graphs with Few Disjoint Odd Cycles". We would like to thank the IPCO reviewers for their comments. Also, we would like to mention that the total unimodularity test of Truemper & Walter [19] provided valuable insights in the early stages of our project, especially for gaining a better understanding of the algorithm in [1].

A   Review of the Projective Planar Case

In this section, we briefly review the compact extended formulation from [7] for STAB(G), when k and g are fixed constants, where $k = \mathrm{ocp}(G)$ and g denotes the Euler genus of G. Since here we are only interested in the case $k = g = 1$, the extended formulation is much easier to describe. Our starting point is the unbounded polyhedron

$$P(G) := \mathrm{conv}\{x \in \mathbb{Z}^{V(G)} \mid Mx \leqslant \mathbf{1}\},$$

where M is the edge-node incidence matrix of G. Its relationship to STAB(G) is as follows.

Lemma 15 ([7, Prop. 49]). *For every graph G,* STAB(G) $= P(G) \cap [0,1]^{V(G)}$.

Thus, it suffices to study $P(G)$ instead of STAB(G). To this end, it is convenient to switch from the node space of G to the edge space of G by considering the affine map $\sigma : \mathbb{R}^{V(G)} \to \mathbb{R}^{E(G)}$ defined via

$$\sigma(x) := \mathbf{1} - Mx.$$

Under σ, a vector $x \in \mathbb{R}^{V(G)}$ is mapped to $y = \sigma(x) \in \mathbb{R}^{E(G)}$ where $y_{vw} = 1 - x_v - x_w$ for every edge $vw \in E(G)$. Since σ is invertible if and only if G has no bipartite component, we can focus on $Q(G) := \sigma(P(G))$.

We provide an extended formulation for $Q(G)$, assuming that G is even-face embedded in the projective plane. Let G^* be the dual graph of G. An orientation D of the edges of G^* is called *alternating* if in the local cyclic ordering of the edges incident to each dual node f, the edges alternatively leave and enter f. We say that a graph G satisfies the *standard assumptions* if it is non-bipartite, 2-connected, and even-face embedded in the projective plane.

Lemma 16 ([7, Lem. 17]). *Let G be a graph satisfying the standard assumptions. Then the dual graph G^* of G has an alternating orientation.*

Let G be even-face embedded in the projective plane and D be an alternating orientation of G^*. Note that there is a bijection between the edges of G and the arcs of D. Therefore, we may regard a vector $y \in \mathbb{R}^{E(G)}$ as a vector in $\mathbb{R}^{A(D)}$, and vice versa. With this identification, $Q(G)$ turns out to be the convex hull of all non-negative integer circulations of D that satisfy one additional constraint.

Lemma 17 ([7, Lem. 18]). *Let G be a graph satisfying the standard assumptions, D be an alternating orientation of G^*, and C be an arbitrary odd cycle in G. Then*

$$Q(G) = \mathrm{conv}\{y \in \mathbb{Z}_{\geq 0}^{E(G)} \mid y \text{ is a circulation in } D \text{ and } y(E(C)) \text{ is odd}\}.$$

Motivated by Lemma 17, we now introduce an auxiliary directed graph to design an extended formulation for $Q(G)$. Let G be a graph satisfying the standard assumptions, D be an alternating orientation of G^*, and C be an odd cycle in G. The *cover graph* of D is the directed graph $\overline{D}$ with node set $\{(f,p) \mid f \in V(D), \, p \in \{0,1\}\}$ and an arc from (f_1,p_1) to (f_2,p_2) if and only if $(f_1,f_2) \in A(D)$ and $p_1+p_2 \equiv \chi_e^{E(C)} \pmod 2$, where e is the edge of G corresponding to the arc (f_1,f_2). For each node $f \in V(D)$ we let Q_f be the polyhedron of all (uncapacitated) unit flows from $(f,0)$ to $(f,1)$ in $\overline{D}$. Finally, we let $\overline{Q}(G)$ denote the convex hull of the union of all polyhedra Q_f for $f \in V(D)$.

By [7, Sec. 12.3], $\overline{Q}(G)$ is an extension[2] of $Q(G)$. Moreover, each Q_f has $O(|A(\overline{D})|) = O(|E(G)|) = O(|V(G)|)$ facets. Finally, by applying Balas' theorem [2], we obtain a quadratic size extended formulation for $Q(G)$, and thus for $\mathrm{STAB}(G)$ in case G satisfies the standard assumptions. By [7, Sec. 12.1], this result extends to all graphs that are even-face embedded in the projective plane. This proves Theorem 3.

References

1. Artmann, S., Weismantel, R., Zenklusen, R.: A strongly polynomial algorithm for bimodular integer linear programming. In: STOC 2017–Proceedings of the 49th Annual ACM SIGACT Symposium on Theory of Computing, pp. 1206–1219. ACM, New York (2017)

---

[2] If $P = \{x \mid Ax + By \leqslant b\}$ is an extended formulation of some polyhedron, then the polyhedron $Q := \{(x,y) \mid Ax + By \leqslant b\}$ is called an *extension* of P.

2. Balas, E.: Disjunctive programming. Ann. Discrete Math. **5**, 3–51 (1979). discrete optimization (Proc. Adv. Res. Inst. Discrete Optimization and Systems Appl., Banff, Alta., 1977), II

3. Barahona, F., Mahjoub, A.R.: Compositions of graphs and polyhedra ii: stable sets. SIAM J. Discrete Math. **7**(3), 359–371 (1994)

4. Bonifas, N., Di Summa, M., Eisenbrand, F., Hähnle, N., Niemeier, M.: On subdeterminants and the diameter of polyhedra. Discrete Comput. Geom. **52**(1), 102–115 (2014)

5. Cevallos, A., Weltge, S., Zenklusen, R.: Lifting linear extension complexity bounds to the mixed-integer setting. In: Proceedings of the Twenty-Ninth Annual ACM-SIAM Symposium on Discrete Algorithms, SODA '18, Society for Industrial and Applied Mathematics, Philadelphia, PA, USA, pp. 788–807 (2018). http://dl.acm.org/citation.cfm?id=3174304.3175321

6. Chvátal, V.: On certain polytopes associated with graphs. J. Comb. Theory Ser. B **18**(2), 138–154 (1975)

7. Conforti, M., Fiorini, S., Huynh, T., Joret, G., Weltge, S.: The stable set problem in graphs with bounded genus and bounded odd cycle packing number (2020). https://arxiv.org/abs/1908.06300. to appear in SODA '20

8. Conforti, M., Fiorini, S., Huynh, T., Weltge, S.: Extended formulations for stable set polytopes of graphs without two disjoint odd cycles (2019). http://arxiv.org/abs/1911.12179

9. Conforti, M., Gerards, B., Pashkovich, K.: Stable sets and graphs with no even holes. Math. Program. **153**(1), 13–39 (2015). https://doi.org/10.1007/s10107-015-0912-3

10. Dyer, M., Frieze, A.: Random walks, totally unimodular matrices, and a randomised dual simplex algorithm. Math. Program. **64**(1–3), 1–16 (1994)

11. Eisenbrand, F., Vempala, S.: Geometric random edge. Math. Program. **164**(1–2), 325–339 (2017)

12. Grossman, J.W., Kulkarni, D.M., Schochetman, I.E.: On the minors of an incidence matrix and its smith normal form. Linear Algebra Appl. **218**, 213–224 (1995)

13. Kawarabayashi, K.I., Ozeki, K.: A simpler proof for the two disjoint odd cycles theorem. J. Comb. Theory Ser. B **103**(3), 313–319 (2013). https://doi.org/10.1016/j.jctb.2012.11.004

14. Paat, J., Schlöter, M., Weismantel, R.: Most IPs with bounded determinants can be solved in polynomial time (2019). http://arxiv.org/abs/1904.06874

15. Seymour, P.D.: Matroid minors. In: Handbook of Combinatorics, vol. 1, no. 2, pp. 527–550. Elsevier Sci. B. V., Amsterdam (1995)

16. Slilaty, D.: Projective-planar signed graphs and tangled signed graphs. J. Comb. Theory Ser. B **97**(5), 693–717 (2007). https://doi.org/10.1016/j.jctb.2006.10.002

17. Tardos, E.: A strongly polynomial algorithm to solve combinatorial linear programs. Oper. Res. **34**(2), 250–256 (1986)

18. Veselov, S.I., Chirkov, A.J.: Integer program with bimodular matrix. Discrete Optim. **6**(2), 220–222 (2009)

19. Walter, M., Truemper, K.: Implementation of a unimodularity test. Math. Program. Ser. C **5**(1), 57–73 (2013). https://doi.org/10.1007/s12532-012-0048-x

On a Generalization
of the Chvátal-Gomory Closure

Sanjeeb Dash[1], Oktay Günlük[1], and Dabeen Lee[2(✉)]

[1] IBM Research, Yorktown Heights, USA
{sanjeebd,gunluk}@us.ibm.com
[2] Discrete Mathematics Group, Institute for Basic Science (IBS),
Daejeon, Republic of Korea
dabeenl@ibs.re.kr

Abstract. Many practical integer programming problems involve variables with one or two-sided bounds. Dunkel and Schulz (2012) considered a strengthened version of Chvátal-Gomory (CG) inequalities that use 0–1 bounds on variables, and showed that the set of points in a rational polytope that satisfy all these strengthened inequalities is a polytope. Recently, we generalized this result by considering strengthened CG inequalities that use all variable bounds. In this paper, we generalize further by considering not just variable bounds, but general linear constraints on variables. We show that all points in a rational polyhedron that satisfy such strengthened CG inequalities form a rational polyhedron.

Keywords: Integer programming · Cutting planes · Chvátal-Gomory cuts

1 Introduction

Let $S \subseteq \mathbb{Z}^n$, and let P be a rational polyhedron. Let $\alpha x \leq \beta$ be a valid inequality for P and assume that $\alpha \in \mathbb{Z}^n$. Assume further that S has a point satisfying $\alpha x \leq \beta$. Let

$$\lfloor \beta \rfloor_{S,\alpha} = \max\{\alpha x : x \in S, \alpha x \leq \beta\}.$$

We call the inequality $\alpha x \leq \lfloor \beta \rfloor_{S,\alpha}$ an *S-Chvátal-Gomory cut* for P (or *S-CG cut*, for short). This inequality is valid for $P \cap S$. If $\alpha x \leq \beta$ is valid for P, but S does not contain a point satisfying this inequality, then $P \cap S$ is clearly empty, and we say that $0x \leq -1$ is an S-CG cut for P derived from $\alpha x \leq \beta$. In a similar manner, we define

$$\lceil \beta \rceil_{S,\alpha} = \min\{\alpha x : x \in S, \alpha x \geq \beta\},$$

assuming that S has a point satisfying $\alpha x \geq \beta$. Then we say that $\alpha x \geq \lceil \beta \rceil_{S,\alpha}$ is the S-CG cut obtained from $\alpha x \geq \beta$. We define the S-CG closure of a polyhedron

© Springer Nature Switzerland AG 2020
D. Bienstock and G. Zambelli (Eds.): IPCO 2020, LNCS 12125, pp. 117–129, 2020.
https://doi.org/10.1007/978-3-030-45771-6_10

P to be the set of all points in P that satisfy all S-CG cuts for P, and we denote this set by P_S.

For the case $S = \mathbb{Z}^n$, S-CG cuts for a polyhedron P are essentially the same as Chvátal-Gomory (CG) cuts [4,11] for P, an important class of cutting planes for integer programming problems. More precisely, let $\alpha x \leq \beta$ be a valid inequality for P such that α is an integral vector. Then $\alpha x \leq \lfloor \beta \rfloor$ is called a CG inequality for P. We also have

$$\lfloor \beta \rfloor \geq \max\{\alpha x : x \in \mathbb{Z}^n, \alpha x \leq \beta\} = \lfloor \beta \rfloor_{\mathbb{Z}^n, \alpha}.$$

The above inequality becomes an equality if the coefficients of α are coprime integers. Therefore, when $S = \mathbb{Z}^n$, P_S is equal to the *Chvátal closure* of P. The Chvátal closure is known to be a rational polyhedron when P is a rational polyhedron (see Schrijver [14]). There exist other classes of convex sets whose Chvátal closure is a rational polyhedron [2,5,9]. For general S, the hyperplane $\{x \in \mathbb{R}^n : \alpha x = \beta\}$ is moved until it hits a point in S; the resulting hyperplane is given by $\{x \in \mathbb{R}^n : \alpha x = \lfloor \beta \rfloor_{S,\alpha}\}$. Here, the gap $\beta - \lfloor \beta \rfloor_{S,\alpha}$ can be larger than 1 when $S \neq \mathbb{Z}^n$. In this case, S-CG cuts can be viewed as a special case of *wide split cuts*, where each cut coincides with one side of a *wide split disjunction* (introduced by Bonami, Lodi, Tramontani, and Wiese [1]). Since there is no constant bound on the gap $\beta - \lfloor \beta \rfloor_{S,\alpha}$, standard arguments for proving the polyhedrality of the Chvátal closure cannot be directly applied to the S-CG closure when $S \neq \mathbb{Z}^n$.

Dunkel and Schulz [8] studied S-CG cuts and P_S for the case $P \subseteq [0,1]^n$ and $S = \{0,1\}^n$. Clearly, the inequality $\alpha x \leq \lfloor \beta \rfloor_{S,\alpha}$ dominates the inequality $\alpha x \leq \lfloor \beta \rfloor$ in this case. They proved that when $S = \{0,1\}^n$, the set P_S is a rational polyhedron. The inequalities studied by Dunkel and Schulz are valid for the 0–1 knapsack set $\{x \in \{0,1\}^n : \alpha x \leq \beta\}$; valid inequalities for such knapsack sets are used to solve practical problem instances in Crowder, Johnson, and Padberg [3], and an associated closure operation is defined by Fischetti and Lodi [10]. Pashkovich, Poirrier, and Pulyassary [12] showed that the aggregation closure – which is the set of points satisfying valid inequalities for all knapsack sets $\{x \in \mathbb{Z}^n : x \geq 0, \alpha x \leq \beta\}$ where $\alpha x \leq \beta$ is valid for P and $\alpha \leq 0$ or $\alpha \geq 0$ – is polyhedral for packing and covering polyhedra. For packing polyhedra, Del Pia, Linderoth, and Zhu [13] independently proved the same result.

In [6], we showed that P_S is a rational polyhedron when $P \subseteq \text{conv}(S)$ and S is the set of integral points that satisfy an arbitrary collection of single variable bounds. In this paper, we prove that even if S is the set of integer points contained in an arbitrary rational polyhedron, P_S is a rational polyhedron.

Theorem 1. *Let $S = R \cap \mathbb{Z}^n$ for some rational polyhedron R and $P \subseteq \text{conv}(S)$ be a rational polyhedron. Then P_S is a rational polyhedron.*

The proof outline is as follows. We first prove that the result holds when R is a rational cylinder, via a unimodular mapping of R to the set $T \times \mathbb{R}^l$ where $l \leq n$ and $T \subseteq \mathbb{R}^{n-l}$ is a polytope. The case $R = T \times \mathbb{R}^l$ is already covered in [6]. We then consider the case when R is pointed, and we show, via a translation, that R

can be assumed to be contained in its recession cone. The hardest case in [6] is the case when $S = \mathbb{Z}_+^n$ and P is a packing or covering polyhedron contained in $\mathbb{R}_+^n$. Similarly, the case when R is a pointed polyhedron and P behaves like a packing or covering polyhedron with respect to R is the hardest case in this paper. The main technical difference between a pointed polyhedron R and $\text{conv}(\mathbb{Z}_+^n) = \mathbb{R}_+^n$ is that R can have more than n extreme rays and, in particular, the extreme rays can be linearly dependent. Nevertheless, this case can be dealt with by an argument generalizing the argument in [6] for the case $S = \mathbb{Z}_+^n$. Essentially, we prove that given a valid inequality for P (and the associated hyperplane) that yields a nonredundant S-CG cut, the points at which the hyperplane intersects the rays of the recession cone of R are bounded.

1.1 Preliminaries

Given a rational polyhedron $P = \{x \in \mathbb{R}^n : Ax \leq b\}$ where $A \in \mathbb{Z}^{m \times n}$ and $b \in \mathbb{Z}^m$, we define Π_P as the set of all coefficient vectors that define valid, supporting inequalities for P with integral left-hand-side coefficients:

$$\Pi_P = \left\{ (\lambda A, \lambda b) \in \mathbb{Z}^n \times \mathbb{R} : \lambda \in \mathbb{R}_+^m, \ \lambda b = \max\{\lambda Ax : x \in P\} \right\}.$$

Finally, for $\Omega \subseteq \Pi_P$, we define $P_{S,\Omega}$ as follows:

$$P_{S,\Omega} = \bigcap_{(\alpha,\beta) \in \Omega} \{x \in \mathbb{R}^n : \alpha x \leq \lfloor \beta \rfloor_{S,\alpha}\}.$$

P_S, the S-CG closure of P, can be formally defined as P_{S,Π_P}. Notice that if $\Gamma \subseteq \Omega \subseteq \Pi_P$, then $P_S \subseteq P_{S,\Omega} \subseteq P_{S,\Gamma}$. Also, if $S \subseteq T$ for some $T \subseteq \mathbb{Z}^n$, then $P_S \subseteq P_T$. Likewise, for any $\Omega \subseteq \Pi_P$, we have $P_{S,\Omega} \subseteq P_{T,\Omega}$ if $S \subseteq T$.

2 Integer Points in a General Cylinder

In [6], Dash, Günlük, and Lee proved the following:

Theorem 2 ([6, Theorem 3.4]). *Let $S = F \times \mathbb{Z}^l$ for some finite $F \subseteq \mathbb{Z}^{n-l}$ and $P \subseteq \mathbb{R}^n$ be a rational polyhedron. Then P_S is a rational polyhedron.*

We say that a rational polyhedron R is a *rational cylinder* if the recession cone and lineality space of R are the same. For instance, the convex hull of $F \times \mathbb{Z}^l$ for some finite $F \subseteq \mathbb{Z}^{n-l}$ is a rational cylinder. In this section, we consider the case where $S = R \cap \mathbb{Z}^n$ for some rational cylinder R.

Remember that a unimodular transformation is a mapping $\tau : \mathbb{R}^n \to \mathbb{R}^n$ that maps $x \in \mathbb{R}^n$ to $Ux + v \in \mathbb{R}^n$ for some unimodular matrix $U \in \mathbb{Z}^{n \times n}$ and some integral vector $v \in \mathbb{Z}^n$. Note that the inverse mapping $\tau^{-1}(x) = U^{-1}x - U^{-1}v$ is also a unimodular transformation. For $X \subseteq \mathbb{R}^n$, we denote by $\tau(X)$ the image of X under τ. For $\Pi \subseteq \Pi_P$, although Π is not in the space of $\mathbb{R}^n$, we abuse our notation and define $\tau(\Pi)$ as $\{(\alpha U^{-1}, \beta + \alpha U^{-1}v) : (\alpha, \beta) \in \Pi\} \subseteq \Pi_{\tau(P)}$.

Lemma 1 (Unimodular mapping lemma [6]**).** *Let* $S \subseteq \mathbb{Z}^n$ *and* $P \subseteq \text{conv}(S)$ *be a rational polyhedron. Then* $\tau(P) \subseteq \text{conv}(\tau(S))$, *and for any* $\Pi \subseteq \Pi_P$, $\tau(P_{S,\Pi}) = \tau(P)_{\tau(S),\tau(\Pi)}$. *In particular,* $\tau(P_S) = \tau(P)_{\tau(S)}$.

Essentially, we will argue that there is a unimodular transformation mapping S to $F \times \mathbb{Z}^l$ for some finite $F \subseteq \mathbb{Z}^{n-l}$.

Theorem 3. *Let* $S = R \cap \mathbb{Z}^n$ *for some rational cylinder* R *and* $P \subseteq \text{conv}(S)$ *be a rational polyhedron. Then* P_S *is a rational polyhedron.*

Proof. Since $S = R \cap \mathbb{Z}^n$ and R is a rational cylinder, we have $\text{conv}(S) \cap \mathbb{Z}^n = S$ and, for some integer $g \geq 0$, there exist integer vectors $v^1, \ldots, v^g$ such that $\text{conv}(S) = \text{conv}\left\{v^1, \ldots, v^g\right\} + \mathcal{L}$ where $\mathcal{L}$ is the lineality space of $\text{conv}(S)$. Let $P \subseteq \text{conv}(S)$ be a rational polyhedron. There exists a unimodular transformation τ such that $\tau(\mathcal{L}) = \{\mathbf{0}\} \times \mathbb{R}^l$ and $\tau(\text{conv}(S)) = \tau\left(\text{conv}\left\{v^1, \ldots, v^g\right\}\right) + \tau(\mathcal{L})$ where $0 \leq l \leq n$ is the dimension of $\mathcal{L}$. This implies that $\tau(S) = F \times \mathbb{Z}^l$ for some finite $F \subseteq \mathbb{Z}^{n-l}$. Then Theorem 2 implies that $\tau(P)_{\tau(S)}$ is a rational polyhedron, and by Lemma 1, P_S is a rational polyhedron, as required. ☐

3   Integer Points in a Pointed Polyhedron

In this section, we consider the case when

$$S = R \cap \mathbb{Z}^n \quad \text{where } R \text{ is a rational pointed polyhedron.}$$

Then $\text{conv}(S) \cap \mathbb{Z}^n = S$ and, for some integers $g, h \geq 0$, there exist integer vectors $v^1, \ldots, v^g, r^1, \ldots, r^h \in \mathbb{Z}^n$ such that $\text{conv}(S)$ can be rewritten as

$$\text{conv}(S) = \text{conv}\left\{v^1, \ldots, v^g\right\} + \text{cone}\left\{r^1, \ldots, r^h\right\}. \tag{1}$$

Since R is pointed, $\text{cone}\left\{r^1, \ldots, r^h\right\}$ has to be pointed as well. Let $N_v = \{1, \ldots, g\}$ and $N_r = \{1, \ldots, h\}$. We will show that the S-CG closure of a rational polyhedron $P \subseteq \text{conv}(S)$ is again a rational polyhedron. To simplify the proof, we will reduce this setting to a more restricted setting with additional assumptions on S and P, and we will see that these assumptions make the structure of S and that of P easier to deal with. The first part of Sect. 3 explains the reduction, and Sects. 3.1 and 3.2 consider the narrower case of S and P obtained after the reduction.

The first assumption we make is the following:

$$\text{rec}\left(\text{conv}(S)\right) \subseteq \{\mathbf{0}\} \times \mathbb{R}^{n_2}, \quad \text{conv}(S) \subseteq \text{cone}\left\{e^1, \ldots, e^{n_1}, r^1, \ldots, r^h\right\} \tag{2}$$

where n_2 is the dimension of $\text{rec}\left(\text{conv}(S)\right)$, $n_1 = n - n_2$, and $e^1, \ldots, e^{n_1}$ are unit vectors in $\mathbb{R}^{n_1} \times \{\mathbf{0}\}$. The following lemma justifies the assumption (2).

Lemma 2. *Let* $S = R \cap \mathbb{Z}^n$ *for some rational pointed polyhedron* R. *Then there is a unimodular transformation* τ *so that* $T := \tau(S)$ *has the property that* $\text{conv}(T) \cap \mathbb{Z}^n = T$ *and* $\text{conv}(T)$ *is of the form* (1) *satisfying* (2).

Proof. As R is pointed, $\text{conv}(S)$ is also pointed and $\text{conv}(S) \cap \mathbb{Z}^n = S$. Let n_2 be the dimension of $\text{rec}(\text{conv}(S))$. Then there is a unimodular transformation u such that $u(\text{rec}(\text{conv}(S))) = \text{rec}(\text{conv}(u(S))) \subseteq \{\mathbf{0}\} \times \mathbb{R}^{n_2}$. Let $\text{rec}(\text{conv}(u(S))) = \text{cone}\{r^1, \ldots, r^h\}$. Then it follows that $e^1, \ldots, e^{n_1}, r^1, \ldots, r^h$ span $\mathbb{R}^n$. Therefore, there exists a sufficiently large integer M such that $v + M(\sum_{i=1}^{n_1} e^i + \sum_{j=1}^{h} r^j) \in \text{cone}\{e^1, \ldots, e^{n_1}, r^1, \ldots, r^h\}$ for all vertices v of $\text{conv}(u(S))$. Let ν be the undimodular transformation defined by $\nu(x) := x + M(\sum_{i=1}^{n_1} e^i + \sum_{j=1}^{h} r^j)$ for $x \in \mathbb{R}^n$. Then $\text{conv}(\nu(u(S))) \subseteq \text{cone}\{e^1, \ldots, e^{n_1}, r^1, \ldots, r^h\}$, and since ν is just a translation, the recession cone of $\text{conv}(\nu(u(S)))$ remains the same as that of $\text{conv}(u(S))$. So, $\tau = \nu \circ u$ is the desired unimodular transformation. $\qquad\square$

By Lemma 1, P_S is a rational polyhedron if and only if $\tau(P)_{\tau(S)}$ is a rational polyhedron, so we may assume that S satisfies (2).

The second assumption is on the structure of the polyhedron P. Let P^1 and P^2 be defined as follows:

$$P^1 := P + \text{cone}\{e^1, \ldots, e^{n_1}, r^1, \ldots, r^h\}, P^2 := P - \text{cone}\{e^1, \ldots, e^{n_1}, r^1, \ldots, r^h\}.$$

Since $P \subseteq \text{conv}(S) \subseteq \text{cone}\{e^1, \ldots, e^{n_1}, r^1, \ldots, r^h\}$, P^1 is pointed and the extreme points of P^1 are contained in $\text{conv}(S)$. Moreover, P^1 can be written as $P^1 = \{x \in \mathbb{R}^n : Ax \geq b\}$ where $A \in \mathbb{Z}^{m \times n}$, $b \in \mathbb{Z}^m$ are matrices satisfying

$$Ax \geq \mathbf{0} \text{ for all } x \in \{e^1, \ldots, e^{n_1}, r^1, \ldots, r^h\} \quad \text{and} \quad b \geq \mathbf{0}. \qquad (3)$$

Similarly, P^2 can be written as $P^2 = \{x \in \mathbb{R}^n : Ax \leq b\}$ for some $A \in \mathbb{Z}^{m \times n}$, $b \in \mathbb{Z}^m$ satisfying (3). Essentially, we can focus on the polyhedra of the form $P^\uparrow$ or $P^\downarrow$:

$$P^\uparrow = \{x \in \mathbb{R}^n : Ax \geq b\} \quad \text{or} \quad P^\downarrow = \{x \in \mathbb{R}^n : Ax \leq b\} \qquad (4)$$

for some $A \in \mathbb{Z}^{m \times n}$, $b \in \mathbb{Z}^m$ satisfying (3). In Sects. 3.1 and 3.2, we prove that the following holds:

($\star$) Let $Q \subseteq \mathbb{R}^n$ be a rational polyhedron of the form $P^\uparrow$ or $P^\downarrow$ as in (4) for some $A \in \mathbb{Z}^{m \times n}$, $b \in \mathbb{Z}^m$ satisfying (3). We further assume that if $Q = P^\uparrow$, then $P^\uparrow \subseteq \text{cone}\{e^1, \ldots, e^{n_1}, r^1, \ldots, r^h\}$ and the extreme points of $P^\uparrow$ are contained in $\text{conv}(S)$. Then Q_S is a rational polyhedron.

The following Lemma implies that proving ($\star$) is sufficient to prove the result for $\text{conv}(S)$ being pointed polyhedra.

Lemma 3. *Let $S \subseteq \mathbb{Z}^n$ be such that $\text{conv}(S) \cap \mathbb{Z}^n = S$, $\text{conv}(S)$ is of the form (1) satisfying (2). Let $P \subseteq \text{conv}(S)$ be a rational polyhedron. Assume that ($\star$) holds. Then P_S is a rational polyhedron.*

The basic idea for proving Lemma 3 is to come up with a series of unimodular transformations so that the setting in Lemma 3 is reduced to the narrow case in ($\star$). We defer the formal proof of Lemma 3 to a full version of this paper.

3.1 Covering Polyhedra

In Sects. 3.1 and 3.2, we assume that $\text{conv}(S) \cap \mathbb{Z}^n = S$ and $\text{conv}(S)$ is of the form (1) satisfying (2). In this section, we consider polyhedra of the form $P^\uparrow$ as in (4) where $A \in \mathbb{Z}^{m \times n}$ and $b \in \mathbb{Z}^m$ satisfy (3). We will prove that if $P^\uparrow \subseteq \text{cone}\left\{e^1, \ldots, e^{n_1}, r^1, \ldots, r^h\right\}$ and the extreme points of $P^\uparrow$ are contained in $\text{conv}(S)$, then $P^\uparrow_S$ is a rational polyhedron. Notice that every valid inequality for $P^\uparrow$ is of the form

$$\alpha x \geq \beta \quad \text{where } \alpha x \geq 0 \text{ for all } x \in \left\{e^1, \ldots, e^{n_1}, r^1, \ldots, r^h\right\} \text{ and } \beta \geq 0. \quad (5)$$

As we will be dealing with inequalities of the greater or equal to form in this section, we will abuse notation and define $\Pi_{P^\uparrow}$ as follows:

$$\Pi_{P^\uparrow} = \left\{(\lambda A, \lambda b) \in \mathbb{Z}^n \times \mathbb{R} : \lambda \in \mathbb{R}^m_+, \; \lambda b = \min\{\lambda A x : x \in P^\uparrow\}\right\}.$$

Given $(\alpha, \beta) \in \Pi_{P^\uparrow}$, the S-CG cut obtained from $\alpha x \geq \beta$ is $\alpha x \geq \lceil \beta \rceil_{S, \alpha}$.

To prove that $P^\uparrow_S$ is a rational polyhedron, we define the notion of "ray-support". Given a vector $\alpha \in \mathbb{R}^n$, we define the *ray-support* of α, denoted $r\text{-supp}(\alpha)$, as follows:

$$r\text{-supp}(\alpha) := \left\{j \in N_r : \alpha r^j > 0\right\}.$$

If $(\alpha, \beta) \in \Pi_{P^\uparrow}$ and $j \in r\text{-supp}(\alpha)$, then $\alpha r^j \geq 1$. For $j \in r\text{-supp}(\alpha)$, the ray generated by r^j always intersects the hyperplane $\{x \in \mathbb{R}^n : \alpha x = \beta\}$, and $(\beta/\alpha r^j) r^j$ is the intersection point. Henceforth, $\beta/\alpha r^j$ is referred to as an "intercept" for convenience for $j \in r\text{-supp}(\alpha)$. Lemma 4 implies that if every nondominated S-CG cut for $P^\uparrow$ has bounded intercepts, then $P^\uparrow_S$ is a rational polyhedron. The following proposition will be useful:

Proposition 1 ([6, Proposition 2.9]). *Let S be a finite subset of $\mathbb{Z}^n$ and $P \subseteq \mathbb{R}^n$ be a rational polyhedron. Let $H \subseteq \mathbb{R}^n \times \mathbb{R}$ be a rational polyhedron that is contained in its recession cone $\text{rec}(H)$ and let $\Omega = \Pi_P \cap H$. Then, $P_{S,\Omega}$ is a rational polyhedron.*

Lemma 4. *Let M^* be a positive integer, and let*

$$\Pi = \left\{(\alpha, \beta) \in \Pi_{P^\uparrow} : \beta/\alpha r^j \leq M^* \text{ for all } j \in r\text{-supp}(\alpha)\right\}. \quad (6)$$

Then $P^\uparrow_{S,\Pi}$ is a rational polyhedron.

Proof. Recall that $\text{conv}(S) = \text{conv}\left\{v^1, \ldots, v^g\right\} + \text{cone}\left\{r^1, \ldots, r^h\right\}$. Let S^* be a finite subset of S defined as

$$S^* := S \cap \left(\text{conv}\left\{v^1, \ldots, v^g\right\} + \left\{\mu_1 r^1 + \cdots + \mu_h r^h : 0 \leq \mu_j \leq M^* \text{ for } j \in N_r\right\}\right).$$

As $S^* \subseteq S$, we have $P^\uparrow_{S^*,\Pi} \subseteq P^\uparrow_{S,\Pi}$. In fact, we can show that $P^\uparrow_{S^*,\Pi} = P^\uparrow_{S,\Pi}$. It suffices to argue that $\lceil \beta \rceil_{S^*,\alpha} = \lceil \beta \rceil_{S,\alpha}$ for every $(\alpha, \beta) \in \Pi$. The details of this are deferred to a full version of this paper. So, it remains to show that

$P^\uparrow_{S^*,\Pi}$ is a rational polyhedron. Note that we can write $\Pi = \bigcup_{I \subseteq N_r} \Pi(I)$ where $\Pi(I) = \{(\alpha, \beta) \in \Pi : r\text{-supp}(\alpha) = I\}$. Therefore, $\Pi(I) = \Pi_{P^\uparrow} \cap H(I)$ where

$$H(I) = \left\{(\alpha, \beta) \in \mathbb{R}^n \times \mathbb{R} : \begin{array}{l} \alpha r^j \geq 1 \text{ for } j \in I, \ \alpha r^j = 0 \text{ for } j \in N_r \setminus I, \\ M^* \alpha r^j \geq \beta \text{ for } j \in I \end{array}\right\}.$$

Notice that $H(I) \subseteq \text{rec}(H(I))$. So, by Proposition 1, $P^\uparrow_{S^*,\Pi(I)}$ is a rational polyhedron. As $P^\uparrow_{S^*,\Pi} = \bigcap_{I \subseteq N_r} P^\uparrow_{S^*,\Pi(I)}$, the proof is complete. $\qquad\square$

By Lemma 4, it is enough to argue that all nondominated S-CG cut for $P^\uparrow$ have "bounded" intercepts, in the sense that these inequalities belong to Π defined in (6). In the end, we will prove that $P^\uparrow_S = P^\uparrow_{S,\Pi}$. Recall that $P^\uparrow$ is described by the system $Ax \geq b$ consisting of m inequalities, denoted $a_1 x \geq b_1, \ldots, a_m x \geq b_m$. Since $P^\uparrow$ is pointed, we know that $m \geq 1$. By (3), for every $i = 1, \ldots, m$, we have $a_i r^j \geq 0$ for $j \in N_r$. Then for any $\lambda \in \mathbb{R}^m_+$, it follows that $r\text{-supp}(\lambda_i a_i) \subseteq r\text{-supp}(\lambda A)$.

Definition 1. Let $\lambda \in \mathbb{R}^m_+ \setminus \{0\}$, and $\lambda_1 \geq \lambda_2 \geq \cdots \geq \lambda_m$. The tilting ratio of λ with respect to A is defined as

$$r(\lambda, A) = \lambda_1 / \lambda_{t(\lambda, A)}$$

where $t(\lambda, A) = \min \left\{ j \in \{1, \ldots, m\} : \bigcup_{i=1}^j r\text{-supp}(a_i) = r\text{-supp}(\lambda A) \right\}$. In particular, $\lambda_1 \ldots, \lambda_{t(\lambda,A)} > 0$ and $r(\lambda, A) > 0$.

Definition 2. Let $B = \max_{1 \leq i \leq m} \{b_i\}$ and $D = \sum_{i=1}^m a_i \left(\sum_{i=1}^{n_1} e^i + \sum_{j=1}^h r^j \right)$. We define $M_1 = 2(mB + 2D)$ and $M = \prod_{i=1}^{m-1} M_i$ where

$$M_i = \left(2mB \prod_{j=1}^{i-1} M_j\right)^{i-1} M_1 \quad \text{for } i = 2, \ldots, m-1.$$

In particular, $M = 1$ if $m = 1$ and $M \geq M_1 \geq 4$ if $m \geq 2$.

Moreover, $(M_i/M_1)^{1/(i-1)} \geq 4$, and thus, $(M_1/M_i)^{1/(i-1)} \leq 1/4$ for all $i \geq 2$.

We will show in Lemma 5 that if $\lambda \in \mathbb{R}^m_+ \setminus \{0\}$ has tilting ratio $r(\lambda, A) > M$, then there exists a $\mu \in \mathbb{R}^m_+ \setminus \{0\}$ that defines an S-CG cut dominating the one defined by λ, but with $\|\mu\|_1 \leq \|\lambda\|_1 - 1$. We will need a result of Dirichlet:

Theorem 4 (Simultaneous Diophantine Approximation Theorem [7]). Let k be a positive integer. Given any real numbers $r_1, \ldots, r_k$ and $0 < \varepsilon < 1$, there exist integers $p_1, \ldots, p_k$ and q such that $\left| r_i - \frac{p_i}{q} \right| < \frac{\varepsilon}{q}$ for $i = 1, \ldots, k$ and $1 \leq q \leq \varepsilon^{-k}$.

The following technical lemma generalizes Lemma 4.10 in [6]:

Lemma 5. *Let* $\lambda \in \mathbb{R}_+^m \setminus \{0\}$ *be such that* $(\lambda A, \lambda b) \in \Pi_{P\uparrow}$. *If* $r(\lambda, A) > M$, *then there exists* $\mu \in \mathbb{R}_+^m \setminus \{0\}$ *that satisfies the following: (i)* $\|\mu\|_1 \leq \|\lambda\|_1 - 1$, *(ii)* $(\mu A, \mu b) \in \Pi_{P\uparrow}$, *and (iii)* $\mu Ax \geq \lceil \mu b \rceil_{S,\mu A}$ *dominates* $\lambda Ax \geq \lceil \lambda b \rceil_{S,\lambda A}$.

Proof. After relabeling the rows of $Ax \geq b$, we may assume that $\lambda_1 \geq \cdots \geq \lambda_m$. Let t stand for $t(\lambda, A)$. If $t = 1$, we have $r(\lambda, A) = 1 \leq M$, a contradiction to our assumption. This implies that $t \geq 2$, and thus, $m \geq 2$. Let Δ be defined as

$$\Delta = \min \left\{ \lambda A r^j : j \in r\text{-supp}(\lambda A) \right\}, \tag{7}$$

and let

$$k = \operatorname{argmin} \left\{ \lambda A r^j : j \in r\text{-supp}(\lambda A) \setminus \bigcup_{i=1}^{t-1} r\text{-supp}(a_i) \right\}. \tag{8}$$

By the definition of t, it follows that $r\text{-supp}(\lambda A) \setminus \bigcup_{i=1}^{t-1} r\text{-supp}(a_i)$ is not empty, and therefore, k is a well-defined index. Moreover, we obtain

$$\Delta \leq \lambda A r^k = \sum_{i=t}^{m} \lambda_i a_i r^k \leq \lambda_t \sum_{i=t}^{m} a_i r^k \leq D \lambda_t \tag{9}$$

where the first inequality is due to (7), the equality holds due to (8), the second inequality follows from the assumption that $\lambda_1 \geq \cdots \geq \lambda_m$, and the last inequality follows from the definition of D given in Definition 2. Notice that as $r(\lambda, A) = \frac{\lambda_1}{\lambda_t} = \frac{\lambda_1}{\lambda_2} \times \cdots \times \frac{\lambda_{t-1}}{\lambda_t} > M \geq M_1 \times \cdots \times M_{t-1}$, there exists $\ell \in \{1, \dots, t-1\}$ such that

$$\lambda_i / \lambda_{i+1} \leq M_i \text{ for all } i \in \{1, \dots, \ell-1\} \quad \text{and} \quad \lambda_\ell / \lambda_{\ell+1} > M_\ell. \tag{10}$$

We now construct the vector $\mu \in \mathbb{R}^m \setminus \{0\}$. We consider the case $\ell \geq 2$ first. It follows from the Simultaneous Diophantine Approximation Theorem (with $k = \ell - 1$ and $r_i = \lambda_i / \lambda_\ell$ for $i \in \{1, \dots, \ell-1\}$) that there exist positive integers $p_1, \dots, p_\ell$ satisfying

$$\left| \frac{\lambda_i}{\lambda_\ell} - \frac{p_i}{p_\ell} \right| < \frac{\varepsilon}{p_\ell}, \ i \in \{1, \dots, \ell\} \quad \text{and} \quad p_\ell \leq \varepsilon^{-(\ell-1)} \tag{11}$$

where $\varepsilon = (M_1 / M_\ell)^{1/(\ell-1)}$. Moreover, for all $i \in \{1, \dots, \ell-1\}$, we can assume that $p_i \geq p_{i+1} \geq p_\ell$, as $\lambda_i \geq \lambda_{i+1}$. If $p_i < p_{i+1}$ for some $i \in \{1, \dots, \ell-1\}$, then increasing p_i to p_{i+1} can only reduce $|\lambda_i / \lambda_\ell - p_i / p_\ell|$. Now we define $\mu_1, \dots, \mu_m$ as follows:

$$\mu_i = \begin{cases} \lambda_i - p_i \Delta & \text{for } i \in \{1, \dots, \ell\}, \\ \lambda_i & \text{otherwise.} \end{cases} \tag{12}$$

If, on the other hand, $\ell = 1$, we define μ as in (12) with $p_1 = 1$.

We can show that $\mu \geq 0$ (see Claim 1 in Appendix A), implying in turn that $\|\mu\|_1 \leq \|\lambda\|_1 - 1$. Then we can prove that $\mu Ax \geq \lceil \mu b \rceil_{S,\mu A}$ dominates $\lambda Ax \geq \lceil \lambda b \rceil_{S,\lambda A}$ (See Claim 4 in Appendix A). □

Now we are ready to prove that $P^\uparrow{}_S$ is a rational polyhedron. The following theorem extends Theorem 4.11 in [6]:

Theorem 5. *Let*

$$\Pi = \left\{ (\alpha, \beta) \in \Pi_{P^\uparrow} : \beta/\alpha r^j \le M^* \ \text{for all } j \in r-\text{supp}(\alpha) \right\}.$$

where $M^* = mBM$. If $P^\uparrow \subseteq \text{cone} \left\{ e^1, \ldots, e^{n_1}, r^1, \ldots, r^h \right\}$ and the extreme points of $P^\uparrow$ are contained in $\text{conv}(S)$, then $P^\uparrow{}_S = P^\uparrow{}_{S,\Pi}$, and in particular, $P^\uparrow{}_S$ is a rational polyhedron.

Proof. As $\Pi \subseteq \Pi_{P^\uparrow}$, we have $P^\uparrow{}_S \subseteq P^\uparrow{}_{S,\Pi}$. We will show that $P^\uparrow{}_S = P^\uparrow{}_{S,\Pi}$ by arguing that for each $(\alpha, \beta) \in \Pi_{P^\uparrow}$, there is an $(\alpha', \beta') \in \Pi$ such that the S-CG cut derived from (α', β') dominates the S-CG cut derived from (α, β) on $P^\uparrow$.

Let $\lambda \in \mathbb{R}^m_+ \setminus \{0\}$ be such that $(\lambda A, \lambda b) \in \Pi_{P^\uparrow}$, and set $(\alpha, \beta) = (\alpha A, \beta b)$. If $\beta/\alpha r^j \le M^*$ for all $j \in r$-supp(α), then $(\alpha, \beta) \in \Pi$ as desired. Otherwise, consider an arbitrary $j \in r$-supp(α) such that $\beta/\alpha r^j > M^*$. Let t stand for $t(\lambda, A)$ and note that

$$M^* < \frac{\beta}{\alpha r^j} = \frac{\sum_{i=1}^m \lambda_i b_i}{\sum_{i=1}^m \lambda_i a_i r^j} \le \frac{\lambda_1 \sum_{i=1}^m b_i}{\lambda_t \sum_{i=1}^t a_i r^j} = r(\lambda, A) \frac{\sum_{i=1}^m b_i}{\sum_{i=1}^t a_i r^j} \le mB\, r(\lambda, A),$$

where the last inequality follows from the fact that $b_i \le B$ for all $i \in \{1, \ldots, m\}$ and the fact that $\sum_{i=1}^t a_i r^j \ge 1$ as $\bigcup_{i=1}^t r$-supp $(a_i) = r$-supp (λA).

As $M^* = mBM$, we have $r(\lambda, A) > M$. Then, by Lemma 5, there exists a $\mu \in \mathbb{R}^m_+ \setminus \{0\}$ such that $\|\mu\|_1 \le \|\lambda\|_1 - 1$ and the S-CG cut generated by μ dominates the S-CG cut generated by λ for $P^\uparrow$. If necessary, we can repeat this argument and construct a sequence of vectors $\mu^1, \mu^2, \ldots$, with decreasing norms such that each vector in the sequence defines an S-CG cut that dominates the previous one. Therefore, after at most $\|\lambda\|_1$ iterations, we must obtain a vector $\hat{\mu} \in \mathbb{R}^m_+ \setminus \{0\}$ such that $r(\hat{\mu}, A) \le M$ and $(\hat{\mu}A, \hat{\mu}b) \in \Pi$. As $(\hat{\mu}A, \hat{\mu}b) \in \Pi$ and the S-CG cut generated by $\hat{\mu}$ dominates the S-CG cut generated by λ for $P^\uparrow$, we conclude that $P^\uparrow{}_S = P^\uparrow{}_{S,\Pi}$. Moreover, as $P^\uparrow{}_{S,\Pi}$ is a rational polyhedron by Lemma 4, it follows that $P^\uparrow{}_S$ is a rational polyhedron, as desired.   □

3.2   Packing Polyhedra and General Pointed Polyhedra

Similarly, we can show that $P^\downarrow{}_S$ is a rational polyhedron, where $P^\downarrow$ is defined as in (4) for some $A \in \mathbb{Z}^{m \times n}$ and $b \in \mathbb{Z}^m$ satisfying (3). Unlike $P^\uparrow$, $P^\downarrow$ is not necessarily pointed. Moreover, we do not assume that the extreme points of $P^\downarrow$ are contained in $\text{conv}(S)$. We may assume that $m \ge 1$. Otherwise, $P^\downarrow = \mathbb{R}^n$, and therefore, $P^\downarrow{}_S = \mathbb{R}^n$ is trivially a rational polyhedron. We defer the proof of the following theorem to a full version of this paper.

Theorem 6. *$P^\downarrow{}_S$ is a rational polyhedron.*

By Theorems 5 and 6, it follows that $(\star)$, stated before Lemma 3, holds. As a consequence of Lemmas 2 and 3, we obtain the main theorem of Sect. 3.

Theorem 7. *Let $S = R \cap \mathbb{Z}^n$ for some rational pointed polyhedron R and $P \subseteq \text{conv}(S)$ be a rational polyhedron. Then P_S is a rational polyhedron.*

4   The General Case

In this section, we get back to the most general case:

$$S = R \cap \mathbb{Z}^n \quad \text{where } R \text{ is a rational polyhedron}$$

and R is not necessarily pointed. Then $\text{conv}(S) \cap \mathbb{Z}^n = S$ and $\text{conv}(S)$ can be written as

$$\text{conv}(S) = \mathcal{P} + \mathcal{R} + \mathcal{L}$$

where $\mathcal{L}$ is the lineality space of $\text{conv}(S)$, $\mathcal{P} + \mathcal{R}$ is the pointed polyhedron $\text{conv}(S) \cap \mathcal{L}^\perp$ whose recession cone is $\mathcal{R}$, and $\mathcal{P}$ is a polytope. Let $S_0 \subseteq \mathbb{Z}^n$ be the set of integer points such that $\text{conv}(S_0) \cap \mathbb{Z}^n = S_0$ and

$$\text{conv}(S_0) = \mathcal{P} + \text{lin}(\mathcal{R}) + \mathcal{L}$$

where $\text{lin}(\mathcal{R})$ is the linear hull of $\mathcal{R}$ or $\mathcal{R} + (-\mathcal{R})$. By definition, $S \subseteq S_0$ and $\text{conv}(S_0)$ is a relaxation of $\text{conv}(S)$. Moreover, $\text{conv}(S_0)$ is a rational cylinder.

Lemma 6. *If $P \subseteq \text{conv}(S)$ is a rational polyhedron, then*

$$P_S = P_{S_0} \cap P_{S,\Pi} \quad \text{where} \quad \Pi := \{(\alpha, \beta) \in \Pi_P : \alpha \ell = 0 \text{ for } \ell \in \mathcal{L}\}. \tag{13}$$

Due to the space limit, we defer the proof of Lemma 6 to a full version of this paper. By this lemma, it is sufficient to show that $P_{S,\Pi}$ is a rational polyhedron, and the following lemma will be useful for that:

Lemma 7 (Projection lemma [6]). *Let F, S and P be defined as*

$$S = F \times \mathbb{Z}^{n_2} \text{ for some } F \subseteq \mathbb{Z}^{n_1} \quad \text{and} \quad P = \{(x,y) \in \mathbb{R}^{n_1} \times \mathbb{R}^{n_2} : Ax + Cy \leq b\}$$

where the matrices A, C, b have integral components and $n_1, n_2, 1$ columns, respectively. Let $\Omega \subseteq \{(\alpha, \beta) \in \Pi_P : \alpha = (\phi, \mathbf{0}) \in \mathbb{R}^{n_1} \times \mathbb{R}^{n_2}\}$, and let $\Phi = \{(\phi, \beta) \in \mathbb{R}^{n_1} \times \mathbb{R} : (\phi, \mathbf{0}) = \alpha, (\alpha, \beta) \in \Omega\}$. If $Q = \text{proj}_x(P)$, then $P_{S,\Omega} = P \cap (Q_{F,\Phi} \times \mathbb{R}^{n_2})$.

Now we are ready to prove the main result of this paper:

Proof of Theorem 1. Let Π be defined as in (13). By Lemma 6, we know that $P_S = P_{S_0} \cap P_{S,\Pi}$. Since $\text{conv}(S_0)$ is a rational cylinder, Theorem 3 implies that P_{S_0} is a rational polyhedron. So, it is sufficient to show that $P_{S,\Pi}$ is a rational polyhedron. Since $\mathcal{P} + \mathcal{R} = \text{conv}(S) \cap \mathcal{L}^\perp$, there exists a unimodular transformation τ such that $\tau(\mathcal{L}) = \{\mathbf{0}\} \times \mathbb{R}^{n_2}$ and $\tau(\mathcal{P} + \mathcal{R}) \subseteq \mathbb{R}^{n_1} \times \{\mathbf{0}\}$. Let $Q = \tau(P)$ and $T = \tau(S)$. Then Lemma 1 implies that

$$\tau(P_{S,\Pi}) = Q_{T,\Omega} \quad \text{where} \quad \Omega := \{(\alpha, \beta) \in \Pi_Q : \alpha \ell = 0 \text{ for } \ell \in \tau(\mathcal{L})\}.$$

Since $\tau(\mathcal{L}) = \{\mathbf{0}\} \times \mathbb{R}^{n_2}$, we have $\Omega = \{(\alpha, \beta) \in \Pi_Q : \alpha_{n_1+1} = \cdots = \alpha_{n_1+n_2} = 0\}$. Moreover, T can be written as $T = T_C \times \mathbb{Z}^{n_2}$ where $\text{conv}(T_C) \subseteq \mathbb{R}^{n_1}$ is a pointed polyhedron and $T_C = \text{conv}(T_C) \cap \mathbb{Z}^{n_1}$. Let

$$\Phi = \{(\phi, \beta) \in \mathbb{R}^{n_1} \times \mathbb{R} : (\phi, \mathbf{0}) = \alpha, (\alpha, \beta) \in \Omega\}.$$

Let $\hat{Q}$ denote the projection of Q onto the $\mathbb{R}^{n_1}$-space. As $Q \subseteq \mathrm{conv}(S)$, we have $\hat{Q} \subseteq \mathrm{conv}(T_C)$ and $\Phi = \Pi_{\hat{Q}}$. Since $\mathrm{conv}(T_C)$ is pointed, we know from Theorem 7 that $\hat{Q}_{T_C,\Phi}$ is a rational polyhedron. Since $Q_{T,\Omega} = Q \cap \left(\hat{Q}_{T_C,\Phi} \times \mathbb{R}^{n_2}\right)$ by Lemma 7, it follows that $Q_{T,\Omega}$ is a rational polyhedron. Therefore, P_S is a rational polyhedron, as required. $\qquad\square$

Acknowledgments. This research is supported, in part, by the Institute for Basic Science (IBS-R029-C1).

A   Proof of Lemma 5

First of all, Claims 1 and 2 below are proved inside the proof of Lemma 4.10 in [6].

Claim 1. $\mu \geq 0$ and $supp(\mu) = supp(\lambda)$.

Claim 2. $\mu b = \min\{\mu A x : x \in P^\uparrow\}$ and therefore $(\mu A, \mu b) \in \Pi_{P^\uparrow}$.

Since $supp(\mu) = supp(\lambda)$ by Claim 1 and $Ar^j \geq 0$ for all $j \in N_r$, it follows that $r\text{-}supp(\mu A) = r\text{-}supp(\lambda A)$, and therefore, $t(\mu, A) = t(\lambda, A)$.

The next claim extends Claim 3 of Lemma 4.10 in [6].

Claim 3. Let $Q = \left\{x \in \mathrm{cone}\left\{e^1, \ldots, e^{n_1}; r^1, \ldots, r^{n_r}\right\} : \mu b \leq \mu A x \leq \mu b + \Delta\right\}$. There is no point $x \in Q$ that satisfies

$$\sum_{i=1}^{\ell} p_i a_i x \geq 1 + \sum_{i=1}^{\ell} p_i b_i. \tag{14}$$

Proof. Suppose for a contradiction that there exists $\tilde{x} \in Q$ satisfying (14). Recall that for the index k defined in (8), the inequality $\mu A r^k > 0$ holds. Let $v = \frac{\mu b}{\mu A r^k} r^k$. Then $\mu A v = \mu b$ and $v \in Q$. In addition, for the index ℓ defined in (10), we have $\sum_{i=1}^{\ell} p_i a_i v = 0$ since $k \notin \bigcup_{i=1}^{t-1} r\text{-}supp(a_i)$ and $a_i r^k = 0$ for $i \leq t-1$. As $\tilde{x} \in Q$ satisfies (14) and $v \in Q$ satisfies $\sum_{i=1}^{\ell} p_i a_i v = 0$, we can take a convex combination of these points to get a point $\bar{x} \in Q$ such that

$$\sum_{i=1}^{\ell} p_i a_i \bar{x} = 1 + \sum_{i=1}^{\ell} p_i b_i \quad \Rightarrow \quad \sum_{i=1}^{\ell} p_i(a_i \bar{x} - b_i) = 1. \tag{15}$$

As $\mu A \bar{x} \leq \mu b + \Delta$, we have

$$\sum_{i=1}^{\ell} \mu_i(a_i \bar{x} - b_i) \leq - \sum_{j=\ell+1}^{m} \mu_j(a_j \bar{x} - b_j) + \Delta. \tag{16}$$

As in [6], we can rewrite (16) as:

$$\frac{\lambda_\ell}{p_\ell}\left(1 + \sum_{i=1}^{\ell} \varepsilon_i(a_i \bar{x} - b_i)\right) \leq - \sum_{j=\ell+1}^{m} \mu_j(a_j \bar{x} - b_j) + 2\Delta$$

$$\leq \sum_{j=\ell+1}^{m} \mu_j b_j + 2\Delta \leq \lambda_{\ell+1}(mB + 2D) = \frac{1}{2}\lambda_{\ell+1} M_1 \tag{17}$$

where the second inequality in (17) follows from the assumption that $A \in \mathbb{Z}^{m \times n}$ and $b \in \mathbb{Z}^m$ satisfy (3), the third inequality follows from the fact that $\mu_i = \lambda_i \leq \lambda_{\ell+1}$ for $i = \ell+1, \ldots, m$ by (12) and that $b_j \leq B$ by Definition 2, and the last equality simply follows from the definition of M_1.

Next, we obtain a lower bound on the first term in (17). As $a_i \bar{x} \geq 0$, $b_i \geq 0$, and $\varepsilon_i \in [-\varepsilon, \varepsilon]$, we have

$$\sum_{i=1}^{\ell} \varepsilon_i (a_i \bar{x} - b_i) = \sum_{i=1}^{\ell} \varepsilon_i a_i \bar{x} - \sum_{i=1}^{\ell} \varepsilon_i b_i \geq -\varepsilon \sum_{i=1}^{\ell} (a_i \bar{x} + b_i). \tag{18}$$

Following the same argument in Claim 3 of Lemma 4.10 in [6], we can show that $-\varepsilon \sum_{i=1}^{\ell} (a_i \bar{x} + b_i) \geq -\frac{1}{2}$. Then it follows from (18) that $\sum_{i=1}^{\ell} \varepsilon_i (a_i \bar{x} - b_i) \geq -1/2$. So, the left hand side of (17) is lower bounded by $\lambda_\ell / 2p_\ell$.

Since the first term in (17) is at least $\lambda_\ell / 2p_\ell$, we obtain $\lambda_\ell \leq p_\ell \lambda_{\ell+1} M_1$ from (17), implying in turn that $M_\ell < p_\ell M_1$ as we assumed that $\lambda_\ell > M_\ell \lambda_{\ell+1}$ as in (10). However, (11) implies that $M_\ell \geq p_\ell M_1$, a contradiction.   □

Claim 4. $\mu A x \geq \lceil \mu b \rceil_{S, \mu A}$ *dominates* $\lambda A x \geq \lceil \lambda b \rceil_{S, \lambda A}$.

Proof. We will first show that

$$\mu b \leq \lceil \mu b \rceil_{S, \mu A} \leq \mu b + \Delta \tag{19}$$

holds. Set $(\alpha, \beta) = (\mu A, \mu b)$. By Claim 2, we have that $\beta = \min\{\alpha x : x \in P^\uparrow\}$. As the extreme points of $P^\uparrow$ are contained in $\text{conv}(S)$, it follows that $\beta \geq \min\{\alpha z : z \in S\}$. If $\beta = \min\{\alpha z : z \in S\}$, then $\beta = \lceil \beta \rceil_{S, \alpha}$. Thus we may assume that $\beta > \min\{\alpha z : z \in S\}$, so there exists $z' \in S$ such that $\beta > \alpha z'$. Remember that by (7), $\Delta = \min\{\lambda A r^j : j \in r\text{-supp}(\lambda A)\}$, and let j be such that $\lambda A r^j = \Delta$. As $r\text{-supp}(\lambda A) = r\text{-supp}(\mu A)$, we have $\alpha r^j > 0$ and $\kappa = (\beta - \alpha z')/\alpha r^j > 0$. Therefore $z'' = z' + \lceil \kappa \rceil r^j \in S$. Observe that

$$\beta = \alpha z' + (\beta - \alpha z') = \alpha \left(z' + \kappa r^j \right) \leq \alpha \left(z' + \lceil \kappa \rceil r^j \right) = \beta + \alpha r^j (\lceil \kappa \rceil - \kappa) \leq \beta + \alpha r^j.$$

As $\lambda \geq \mu$, we have $\Delta \geq \alpha r^j$ implying $\beta \leq \alpha z'' \leq \beta + \Delta$ and (19) holds, as desired.

Using (19), we will show that $\mu A x \geq \lceil \mu b \rceil_{S, \mu A}$ dominates $\lambda A x \geq \lceil \lambda b \rceil_{S, \lambda A}$. Let $z \in S$ be such that $\mu A z = \lceil \mu b \rceil_{S, \mu A}$. As z is integral and $\mu b \leq \lceil \mu b \rceil_{S, \mu A} \leq \mu b + \Delta$ by (19), Claim 3 implies that $\sum_{i=1}^{\ell} p_i a_i z < 1 + \sum_{i=1}^{\ell} p_i b_i$, and therefore, $\sum_{i=1}^{\ell} p_i a_i z = \sum_{i=1}^{\ell} p_i b_i - f$ for some integer $f \in [0, \sum_{i=1}^{\ell} p_i b_i]$. Consider $z + f r^j \in S$ and observe that

$$\lambda A \left(z + f r^j \right) = \lambda A z + f \lambda A r^j = \left(\mu A + \Delta \sum_{i=1}^{\ell} p_i a_i \right) z + \Delta \sum_{i=1}^{\ell} p_i (b_i - a_i z)$$

$$= \lceil \mu b \rceil_{S, \mu A} + \Delta \sum_{i=1}^{\ell} p_i b_i.$$

Since $\lceil \mu b \rceil_{S,\mu A} \geq \mu b$, we must have $\lceil \mu b \rceil_{S,\mu A} + \Delta \sum_{i=1}^{\ell} p_i b_i \geq \mu b + \Delta \sum_{i=1}^{\ell} p_i b_i = \lambda b$. Then $\lceil \mu b \rceil_{S,\mu A} + \Delta \sum_{i=1}^{\ell} p_i b_i \geq \lceil \lambda b \rceil_{S,\lambda A}$. Then the inequality $\lambda A x \geq \lceil \lambda b \rceil_{S,\lambda A}$ is dominated by $\mu A x \geq \lceil \mu b \rceil_{S,\mu A}$, as the former is implied by the latter and a nonnegative combination of the inequalities in $Ax \geq b$, as required.

$\square$

References

1. Bonami, P., Lodi, A., Tramontani, A., Wiese, S.: Cutting planes from wide split disjunctions. In: Eisenbrand, F., Koenemann, J. (eds.) IPCO 2017. LNCS, vol. 10328, pp. 99–110. Springer, Cham (2017). https://doi.org/10.1007/978-3-319-59250-3_9
2. Braun, G., Pokutta, S.: A short proof for the polyhedrality of the Chvátal-Gomory closure of a compact convex set. Oper. Res. Lett. **42**, 307–310 (2014)
3. Crowder, H., Johnson, E., Padberg, M.: Solving large-scale zero-one linear programming problems. Oper. Res. **31**, 803–834 (1983)
4. Chvátal, V.: Edmonds polytopes and a hierarchy of combinatorial problems. Discret. Math. **4**, 305–337 (1973)
5. Dadush, D., Dey, S.S., Vielma, J.P.: On the Chvátal-Gomory closure of a compact convex set. Math. Program. **145**, 327–348 (2014)
6. Dash, S., Günlük, O., Lee, D.: Generalized Chvátal-Gomory closures for integer programs with bounds on variables, June 2019. http://www.optimization-online.org/DB_HTML/2019/06/7245.html
7. Dirichlet, G.L.: Verallgemeinerung eines Satzes aus der Lehre von den Kettenbrichen nebst einigen Anwendungen auf die Theorie der Zahlen, Bericht iiber die zur Bekanntmachung geeigneten Verhandlungen der Königlich Preussischen Akademie der Wissenschaften zu Berlin, pp. 93–95 (1842). (Reprinted in: L. Kronecker (ed.), G. L. Dirichlet's Werke, vol. I, G. Reimer, Berlin 1889 (reprinted: Chelsea, New York 1969), pp. 635–638)
8. Dunkel, J., Schulz, A.S.: A refined Gomory-Chvátal closure for polytopes in the unit cube, Technical report, March 2012. http://www.optimization-online.org/DB_HTML/2012/03/3404.html
9. Dunkel, J., Schulz, A.S.: The Gomory-Chvátal closure of a nonrational polytope is a rational polytope. Math. Oper. Res. **38**, 63–91 (2013)
10. Fischetti, M., Lodi, A.: On the Knapsack closure of 0-1 integer linear programs. Electron. Notes Discret. Math. **36**, 799–804 (2010)
11. Gomory, R.E.: Outline of an algorithm for integer solutions to linear programs. Bull. Am. Math. Soc. **64**, 275–278 (1958)
12. Pashkovich, K., Poirrier, L., Pulyassary, H.: The aggregation closure is polyhedral for packing and covering integer programs arXiv:1910.03404 (2019)
13. Del Pia, A., Linderoth, J., Zhu, H.: Integer packing sets form a well-quasi-ordering arXiv:1911.12841 (2019)
14. Schrijver, A.: On cutting planes. Ann. Discret. Math. **9**, 291–296 (1980)

Algorithms for Flows over Time
with Scheduling Costs

Dario Frascaria[1(✉)] and Neil Olver[2,3]

[1] Department of Econometrics and Operations Research,
Vrije Universiteit Amsterdam, Amsterdam, The Netherlands
d.frascaria@vu.nl
[2] Department of Mathematics,
London School of Economics and Political Science, London, UK
N.Olver@lse.ac.uk
[3] CWI, Amsterdam, The Netherlands

Abstract. Flows over time have received substantial attention from both an optimization and (more recently) a game-theoretic perspective. In this model, each arc has an associated delay for traversing the arc, and a bound on the rate of flow entering the arc; flows are time-varying. We consider a setting which is very standard within the transportation economic literature, but has received little attention from an algorithmic perspective. The flow consists of users who are able to choose their route but also their departure time, and who desire to arrive at their destination at a particular time, incurring a *scheduling cost* if they arrive earlier or later. The total cost of a user is then a combination of the time they spend commuting, and the scheduling cost they incur. We present a combinatorial algorithm for the natural optimization problem, that of minimizing the average total cost of all users (i.e., maximizing the social welfare). Based on this, we also show how to set tolls so that this optimal flow is induced as an equilibrium of the underlying game.

Keywords: Flows over time · Tolls · Traffic

1 Introduction

The study of *flows over time* is a classical one in combinatorial optimization; it began already with the work of Ford and Fulkerson [9] in the 50s. It is a natural extension of static flows, which associates a single numerical value, representing a total quantity or rate of flow on the arc. In a flow over time, a second value associated with each arc represents the time it takes for flow to traverse it; the flow is then described by a function on each arc, representing the rate of flow entering the arc as a function of time.

Partially supported by NWO TOP grant 614.001.510 and NWO Vidi grant 016.Vidi.189.087. A version of this manuscript containing omitted proofs can be found on arXiv:1912.00082.

D. Bienstock and G. Zambelli (Eds.): IPCO 2020, LNCS 12125, pp. 130–143, 2020.
https://doi.org/10.1007/978-3-030-45771-6_11

Classical optimization problems involving static flows have natural analogs in the flow over time setting (see the surveys [17,24]). For example (restricting the discussion to single commodity flows), the *maximum flow over time* problem asks to send as much flow as possible, departing from the source starting from time 0 and arriving to the sink by a given time horizon T; this can be solved in polynomial time [8–10]. A *quickest flow* asks, conversely, for the shortest time horizon necessary to send a given amount of flow. Of particular importance for us is the notion of an *earliest arrival flow*: this has the very strong property that simultaneously for all $T' \leq T$, the amount of flow arriving by time T' is as large as possible [12]. Such a flow can also be characterized as minimizing the average arrival time [15]. Earliest arrival flows can be "complicated", in that they can require exponential space (in the input size) to describe [29], and determining the average arrival time of an earliest arrival flow is NP-hard [7]. But they can be constructed in time strongly polynomial in the sum of the input and output size [2].

Another important aspect of many settings where flow-over-time models are applicable—such as traffic—involves game theoretic considerations. In traffic settings, the flow is made up of a large number of individuals making their own routing choices, and aiming to maximize their own utility rather than the overall social welfare (e.g., average journey time). *Dynamic equilibria*, which is the flow over time equivalent of Wardrop equilibria for static flows, are key objects of study. Existence, uniqueness, structural and algorithmic issues, and much more have been receiving increasing recent interest from the optimization community [3–6,16,22,23].

Traffic, being such a relevant and important topic, has received attention from many different communities, each with their own perspective. Within the transportation economic literature, modelling other aspects of user choice besides route choice has been considered particularly important. A very standard setting, motivated by morning rush-hour traffic, is the following [1,26]. Users are able to choose not only their route, but also their *departure time*. They are then concerned not only with their journey time, but also their *arrival time* at the destination. This is captured in a *scheduling cost function* which we will denote by ρ: a user arriving at time θ will experience a scheduling cost of $\rho(\theta)$. The total disutility of a user is then the sum of their scheduling cost and their journey time (scaled by some factor $\alpha > 0$ representing their value for time spent commuting). A very standard choice of ρ is

$$\rho(\theta) = \begin{cases} -\beta\theta & \text{if } \theta \leq 0 \\ \gamma\theta & \text{if } \theta > 0 \end{cases}, \tag{1}$$

where $\beta < \alpha < \gamma$ (it is very bad to be late, but time spent in the office early is better than time spent in traffic). Our approach can handle essentially general scheduling cost functions, but we will restrict our discussion to strongly unimodal cost functions; these are the most relevant, and this avoids some distracting technical details.

Two very natural questions can be posed at this point. The first is a purely optimization question, with no attention paid to the decentralized nature of traffic.

Problem 1. How can one compute a flow over time minimizing the average total cost paid by users, i.e., maximizing the social welfare?

From now on, we will call a solution to this problem simply an *optimal flow*.

It is well understood that users will typically not coordinate their actions to induce a flow that minimizes total disutility. There is a huge body of literature (particularly in the setting of static flows [20]) investigating this phenomenon. In the traffic setting, the relevance of an optimal flow represented by an answer to this question comes primarily via the possibility of *pricing*. By putting appropriate tolls on roads, we can influence the behaviour of users and the resulting dynamic equilibrium. Thus:

Problem 2. How can one set tolls (possibly time-varying) on the arcs of a given instance so that an optimal flow is obtained in dynamic equilibrium?

One subtlety is that since dynamic equilibria need not be precisely unique, there is a distinction between tolls that induce an optimal flow as *an* equilibrium, compared to tolls for which *all* dynamic equilibria are optimal. (This is called *weak* and *strong* enforcement by Harks [14] in a general pricing setting). We will return to this subtlety shortly.

Questions like these are of great interest to transportation economists. However, most work in that community has focused on obtaining a fine-grained understanding of very restricted topologies (such as a single link, or multiple parallel links); see [25] for a survey.

Both of these questions (for general network topologies) were considered by Yang and Meng [28] in a discrete time setting, by exploiting the notion of *time-expanded graphs*. This is a standard tool in the area of flows over time; discrete versions all of the optimization questions concerning flows over time mentioned earlier can (in a sense) be dealt with in this way. A node v in the graph is expanded to a collection (v, i) of nodes, for $i \in \mathbb{Z}$ in a suitable interval, and an arc vw of delay τ_{vw} becomes a collection of arcs $((v, i), (w, i + \tau_{vw}))$ (this assumes a scaling so that τ_{vw} is a length in multiples of the chosen discrete timesteps). Scheduling costs are encoded by appropriately setting arc costs from (t, i) to a supersink t' for each i, and the problem can be solved by a minimum cost static flow computation. A primary disadvantage of this approach (and in the use of time-expanded graphs more generally) is that the running time of the algorithm depends polynomially on the number of time steps, which can be very large. Further, it cannot be used to exactly solve the continuous time version (our interest in this paper); by discretizing time, it can be used to approximate it, but the size of the time-expanded graph is inversely proportional to the step size of the discretization. In the same work [28], the authors also observe that in the discrete setting, an answer to the second question can be obtained from the time-expanded graph as well. Taking the LP describing the minimum cost flow problem on the time-expanded graph, the optimal dual solution to this LP provides the necessary tolls to enforce (weakly) an optimal flow. (This is no big surprise—dual variables can frequently be interpreted as prices.)

An Assumption on ρ. Suppose we consider ρ in the standard form given in (1), but with $\beta > \alpha$. This means that commuting is considered to be less unpleasant than arriving early. A user arriving earlier than time 0 at the sink would be better off "waiting" at the sink before leaving, in order to pay a scheduling cost of 0. Whether waiting in this way is allowed or not depends on the precise way one specifies the model, but it is most natural (and convenient) to allow this. If we do so, then it is clear that a scheduling cost function ρ can be replaced by

$$\hat{\rho}(\theta) := \min_{\xi \geq \theta} \rho(\xi) + \alpha(\xi - \theta)$$

without changing the optimal flow (except there is no longer any incentive to wait at the sink, and we need not even allow it). Then $\theta \rightarrow \hat{\rho}(\theta) + \alpha\theta$ is nondecreasing. From now on, we always assume that ρ satisfies this; we will call it the *growth bound* on ρ.

Our Results. We give a combinatorial algorithm to compute an optimal flow. Similarly to the case of earliest arrival flows, this flow can be necessarily complicated, and involves a description length that is exponential in the input size.

The algorithm is also similar to that for computing an earliest arrival flow. It is based on the (possibly exponentially sized) path decomposition of a minimum cost flow into *successive shortest paths*. In particular, suppose we choose the scheduling cost function to be as in (1), with $\beta = \alpha$ and $\gamma = \infty$. Then the disutility a user experiences is precisely described by how much before time 0 they depart; all users must arrive by time 0 to ensure finite cost. This is precisely the reversal (both in time and direction of all arcs) of an earliest arrival flow, from the sink to the source. Our algorithm will be the same as the earliest arrival flow in this case. This also shows that it may be the case that all optimal solutions to Problem 1 require exponential size (as a function of the input encoding length), since this is the case for earliest arrival flows.

Despite the close relation to earliest arrival flows, the proof of optimality of our algorithm is rather different. A key reason for this is the following. As mentioned, earliest arrival flows have the strong property that the amount of flow arriving before a given deadline T' is the maximum possible, *simultaneously for all choices of T'* (up to some maximum depending on the total amount of flow being sent). This implies that an earliest arrival flow certainly minimizes the average arrival time amongst all possible flows [15], but is a substantially stronger property. A natural analog of this stronger property in our setting would be to ask for a flow for which, simultaneously for any given cost horizon $C' \leq C$, the amount of flow consisting of agents experiencing disutility at most C' is as large as possible. Unfortunately, in general no such flow exists. The example is too involved to discuss here, but it relates to some questions on the behaviour of dynamic equilibria in this model that are investigated in a parallel manuscript [11].

Since the proofs for earliest arrival flows [2,12,19,27] show this stronger property which does not generalize, we take a different approach. Our proof is based on duality (of an infinite dimensional LP, though we do not require any technical results on such LPs). The main technical challenge in our work comes from determining the correct ansatz for the dual solution, as well as exploiting properties of the residual networks obtained from the successive shortest paths algorithm in precisely the right way to demonstrate certain complementary slackness conditions. As was the case with the time-expanded graph approach, the optimal dual solution immediately provides us with the tolls. However, we obtain an explicit formula for the optimal tolls, in terms of the successive shortest paths of the graph (see Sect. 3). This may be useful in obtaining a better structural understanding of optimal tolls, beyond just their computation. We also remark that a corollary of our result is that there is always an optimal solution without waiting (except at the source).

Consider for a moment the model where users cannot choose their departure time, but instead are released from the source at a fixed rate u_0, and simply wish to reach the destination as early as possible. This is the game-theoretic model that has received the most attention from the flow-over-time perspective [3,5, 6,16,22]. Our construction of optimal tolls is applicable to this model as well. Reverse all arcs, as well as the role of the source and sink (thus making s the new sink), and also introduce a replacement sink s' and arc ss' of capacity u_0 in the original instance. Then by choosing ρ as described in (1) with $\beta = \alpha$, $\gamma = \infty$, the optimal flow is an earliest arrival flow, and the tolls we construct will induce it in the original instance (after appropriate time reversal).

We now return to the subtlety alluded to earlier: the distinction between strongly enforcing an optimal flow, and only weakly enforcing it. Consider the simple instance shown. Suppose that the outflow of arc e is larger than 1 for some period in the optimum flow, due to the choice of scheduling cost function. In this period, one unit of flow would take the

bottom arc g, and the rest will be routed on f. Since the total cost (including tolls) of all users is the same in a tolled dynamic equilibrium, a toll of cost equivalent to a unit delay on arc g is needed in this period to induce the optimal flow. But then it will also be an equilibrium to send *all* flow in this period along f.

To strongly enforce an optimal flow, we need more flexible tolls. One way that we can do it is by "tolling lanes". If we are allowed to dynamically divide up the capacity of an arc into "lanes" (say a "fast lane" and a "slow lane"), and then separately set time-varying tolls on each lane, then we *can* strongly enforce any optimal flow. We discuss this further in Sect. 5. We are not aware of settings where this phenomenon has been previously observed, and it would be interesting to explore this further in a more applied context.

Outline of the Paper. We introduce some basic notation and notions, as well as formally define our model, in Sect. 2. In Sect. 3, we describe our algorithm, and show that it returns a feasible flow over time; we restrict ourselves to the most relevant case of a strictly unimodal scheduling cost function. In Sect. 4 we show optimality of this algorithm, and in Sect. 5 we derive optimal tolls from this analysis.

2 Model and Preliminaries

The notation $(z)^+$ is used to denote the nonnegative part of z, i.e., $(z)^+ = \max\{z, 0\}$. Given $v : X \to \mathbb{R}$ and $A \subseteq X$, we will use the shorthand notation $v(A) := \sum_{a \in A} v(a)$. We will not distinguish between a map $v : X \to \mathbb{R}$ and a vector in $\mathbb{R}^X$, and so the notation v_a and $v(a)$ is interchangeable. All graphs will be directed and (purely for notational convenience) simple and without digons.

Static Flows. Let $G = (V, E)$ be a directed graph, with source node $s \in V$ and sink node $t \in V$. Each arc $e \in E$ has a *capacity* ν_e and a *delay* τ_e (both nonnegative). We use $\delta^+(v)$ to denote the set of arcs in E with tail v, and $\delta^-(v)$ the set of arcs with head v.

Consider some $f : E \to \mathbb{R}_+$ (which we will equivalently view as a vector in $\mathbb{R}_+^E$). We use ∇f_v to denote the net flow into $v \in V$; a (static) s-t-flow satisfies the usual flow conservation conditions. Given an s-t-flow f, its *residual network* $G^f = (V, E^f)$ is defined by

$$E^f = \{vw : vw \in E \text{ and } f_{vw} < \nu_{vw}\} \cup \{vw : wv \in E \text{ and } f_{wv} > 0\}.$$

Call arcs in $E^f \cap E$ *forward arcs* and arcs in $E^f \setminus E$ *backwards arcs*. The *residual capacity* ν_e^f of an arc $e \in E^f$ is then $\nu_{vw}^f = \nu_{vw} - f_{vw}$ for vw a forward arc, and $\nu_{vw}^f = f_{wv}$ for vw a backwards arc. We also define $\tau_{vw} = -\tau_{wv}$ for all backwards arcs vw.

Given a subset $F \subseteq E$, we use $\chi(F)$ to denote the characteristic vector of F. We make the definitions $\overleftarrow{E} := \{wv : vw \in E\}$ and $\overleftrightarrow{E} := E \cup \overleftarrow{E}$. Given $f, g \in \mathbb{R}_+^E$, we define $f + g$ in the obvious way, and also define $f - g \in \mathbb{R}_+^{\overleftrightarrow{E}}$, by interpreting a negative value on vw instead as a positive value on wv.

Flows over Time. Consider some $f : E \times \mathbb{R} \to \mathbb{R}_+$. We will generally write $f_e(\theta)$ rather than $f(e, \theta)$. Define the *net flow into* v *at time* θ by

$$\nabla f_v(\theta) := \sum_{e \in \delta^-(v)} f_e(\theta - \tau_e) - \sum_{e \in \delta^+(v)} f_e(\theta).$$

Note that $f_e(\theta)$ represents the flow *entering* arc e at time θ; this flow will exit the arc at time $\theta + \tau_e$ (explaining the asymmetry between the terms for flow entering and flow leaving in the above).

We say that f is a *flow over time of value Q* if the following hold.

(i) For each $e \in E$, f_e is integrable and has compact support.
(ii) $\int_{-\infty}^{\infty} \nabla f_v(\theta) d\theta = Q(\mathbf{1}_{v=t} - \mathbf{1}_{v=s})$ for all $v \in V$.
(iii) $\int_{-\infty}^{\xi} \nabla f_v(\theta) d\theta \geq 0$ for all $v \in V \setminus \{s\}$ and $\xi \in \mathbb{R}$.
(iv) $f_e(\theta) \leq \nu_e$ for all $e \in E$ and $\theta \in \mathbb{R}$.

Note that this definition allows for flow to wait at a node; to disallow this and consider only *flows over time without waiting*, we would replace (iii) with the condition that $\nabla f_v(\theta) = 0$ for all $v \in V \setminus \{s, t\}$ and $\theta \in \mathbb{R}$.

We also have a natural notion of a residual network in the flow over time setting. Define, for any flow over time f and $\theta \in \mathbb{R}$,

$$E^f(\theta) = \{vw : vw \in E \text{ and } f_{vw}(\theta) < \nu_{vw}\} \cup \{vw : wv \in E \text{ and } f_{wv}(\theta - \tau_{wv}) > 0\}.$$

Minimizing Scheduling Cost. We are concerned with the following optimization problem. Given a *scheduling cost function* $\rho : \mathbb{R} \to \mathbb{R}_+$, as well as a value $\alpha > 0$, determine a flow over time f of value Q that minimizes the sum of the *commute cost* $\alpha \sum_{e \in E} \tau_e \cdot \int_{\mathbb{R}} f_e(\theta) d\theta$ and the *scheduling cost* $\int_{\mathbb{R}} \nabla f_t(\theta) \cdot \rho(\theta) d\theta$. As already discussed, we assume that ρ satisfies the growth bound, i.e., that $\theta \to \rho(\theta) + \alpha\theta$ is nondecreasing. This ensures that waiting at t is not needed, which is in fact disallowed by our definition[1], and makes various arguments cleaner. We will also make the assumption that ρ is strongly unimodal[2]. We then assume w.l.o.g. that the minimizer of ρ is at 0, and that $\rho(0) = 0$. For further technical convenience, by adjusting ρ on a set of measure zero we take ρ to be lower semi-continuous.

The unimodal assumption is not necessary; the algorithm and analysis can be extended to essentially general ρ, under some very weak technical conditions. We postpone discussion to the full version of the paper; no major new technical ideas are needed.

We also assume that we are able to query $\rho^{-1}(y)$ for a given rational $y > 0$, obtaining a pair of solutions (one positive, one negative) of moderate bit complexity.

3   A Combinatorial Algorithm

In this section we present an algorithm that computes an optimal flow over time, assuming that ρ is strongly unimodal. The proof of optimality is discussed in Sect. 4.

We begin by recalling the *successive shortest paths (SSP)* algorithm for computing a minimum cost static flow. It is not a polynomial time algorithm, so it is deficient as an algorithm for static flows, but it provides a structure that is relevant for flows over time. This is of course well known from its role in constructing earliest arrival flows, which we will briefly detail.

The SSP algorithm construct a sequence of paths $(P_1, P_2, \ldots)$ and associated amounts $(x_1, x_2, \ldots)$ inductively as follows. Suppose $P_1, \ldots, P_j$ and $x_1, \ldots, x_j$ have been defined. Let $f^{(j)} = \sum_{i=1}^{j} x_i \chi(P_i)$, and let G_j denote the residual graph of $f^{(j)}$ (G_0 being the original network). Also let $d_j(v, w)$ denote the length (w.r.t. arc delays τ in G_j) of a shortest path from v to w in G_j (this may be infinite). By construction, G_j will contain no negative cost cycles, so that d_j

---

[1] Were this really needed, one could simply add a dummy arc tt' to a new sink t'.

[2] I.e., (strictly) decreasing until some moment, and then (strictly) increasing.

is computable. If $d_j(s,t) = \infty$, we are done; set $m := j$. Otherwise, define P_{j+1} to be any shortest s-t-path in G_j, and x_{j+1} the minimum capacity in G_j of an arc in P_{j+1}. It can be shown that $\sum_{j=1}^{r} \tilde{x}_j \chi(P_j)$, with r and $\tilde{x}$ defined such that $\tilde{x}_j = x_j$ for $j < r$, $0 \le \tilde{x}_r \le x_r$ and $\sum_{j=1}^{r} \tilde{x}_j = M$, is a minimum cost flow of value M, as long as M is not larger than the value of a maximum flow.

To construct an earliest arrival flow of value Q and time horizon T, we (informally) send flow at rate x_j along path P_j for the time interval $[0, T - \tau(P_j)]$, for each $j \in [m]$ (if $\tau(P_j) > T$, we send no flow along the path). By this, we mean that for each $e = vw \in P_j$, we increase by x_j the value of $f_e(\theta)$ for $\theta \in [d_{j-1}(s,v), T - d_{j-1}(v,t)]$ (or if e is a backwards arc, we instead decrease $f_{wv}(\theta - \tau_{wv})$). An argument is needed to show that this defines a valid flow, since we must not violate the capacity constraints, and moreover, P_j may contain reverse arcs not present in G.

We are now ready to describe our algorithm for minimizing the disutility, which is a natural variation on the earliest arrival flow algorithm. It is also constructed from the successive shortest paths, but using a *cost horizon* rather than a *time horizon*. For now, consider C to be a given value (it will be the "cost horizon"). For each $j \in [m]$ with $\alpha d_{j-1}(s,t) \le C$, we send flow at rate x_j along path P_j for the time interval $[a_j, b_j]$ chosen maximally so that $\rho(\xi + d_{j-1}(s,t)) \le C - \alpha d_{j-1}(s,t)$ for all $\xi \in [a_j, b_j]$. (If ρ is continuous, then of course $\rho(a_j + d_{j-1}(s,t)) = \rho(b_j + d_{j-1}(s,t)) = C - \alpha d_{j-1}(s,t)$.) Note that a user leaving at time a_j or b_j and using path P_j, without waiting at any moment, incurs disutility C; whereas a user leaving at some time $\theta \in (a_j, b_j)$ and using path P_j will incur a strictly smaller total cost.

As we will shortly argue, this results in a feasible flow over time f. Given this, its value will be $\sum_{j=1}^{m} x_j(b_j - a_j)$. It is easy to see that this value changes continuously and monotonically with C (here we use the strong unimodality). Thus a bisection search can be used to determine the correct choice of C for a given value Q. Alternatively, bisection search can be avoided by using Megiddo's parametric search technique [18]; this will ensure a strongly polynomial running time, if queries to ρ^{-1} are considered to be of unit cost.

Feasibility. Given a vertex $v \in V$, a time $\theta \in \mathbb{R}$ and $j \in [m]$, let

$$c_j(v, \theta) = \alpha d_{j-1}(s,t) + \rho(\theta + d_{j-1}(v,t)).$$

If $v \in P_j$ then $c_j(v, \theta)$ is the travel cost of a user that utilizes path P_j and passes through node v at time θ; there does not seem to be a simple interpretation if $v \notin P_j$ however. Now define

$$J(v, \theta) = \max\{j \in [m] : c_j(v, \theta) \le C\}, \tag{2}$$

with the convention that the maximum over the empty set is 0. The motivation for this definition comes from the following theorem, which completely characterizes f.

Theorem 1. $f_{vw}(\theta) = f_{vw}^{(J(v,\theta))}$ *for any* $vw \in E$ *and* $\theta \in \mathbb{R}$.

Since f has value Q and satisfies flow conservation by construction, the feasibility of f is an immediate corollary of this theorem. We sketch the proof in the appendix.

4   Optimality

Duality-Based Certificates of Optimality. We can write the problem we are interested in as a (doubly) infinite linear program as follows:

$$
\begin{aligned}
\min \quad & \int_{-\infty}^{\infty} \rho(\theta) \nabla f_t(\theta) d\theta + \alpha \sum_{e \in E} \tau_e \int_{-\infty}^{\infty} f_e(\theta) d\theta + \alpha \sum_{v \in V \setminus \{s,t\}} \int_{-\infty}^{\infty} z_v(\theta) d\theta \\
\text{s.t.} \quad & -\int_{-\infty}^{\infty} \nabla f_s(\theta) d\theta = \int_{-\infty}^{\infty} \nabla f_t(\theta) d\theta = Q \\
& \int_{-\infty}^{\theta} \nabla f_v(\xi) d\xi = z_v(\theta) \qquad \forall v \in V \setminus \{s,t\}, \theta \in \mathbb{R} \\
& f_e(\theta) \leq \nu_e \qquad\qquad\quad \forall e \in E, \theta \in \mathbb{R} \\
& z, f \geq 0
\end{aligned}
$$

$$(3)$$

Here, $z_v(\theta)$ represents the amount of flow waiting at node v at time θ (which must always be nonnegative). The travel cost is captured on a per-arc basis, including waiting time as well.

The following theorem provides a certificate of optimality of a feasible solution to (3).

Theorem 2. *Let* f *be a flow over time with value* Q, *and suppose that* $\pi :$ $V \times \mathbb{R} \to \mathbb{R}$ *satisfies the following, for some choice of* C:

(i) $\theta \to \pi_v(\theta) - \alpha\theta$ *is nonincreasing.*
(ii) $\pi_w(\theta + \tau_{vw}) \leq \pi_v(\theta) + \alpha\tau_{vw}$ *for all* $\theta \in \mathbb{R}, vw \in E^f(\theta)$.
(iii) $\pi_s(\theta) = 0$ *for all* $\theta \in \mathbb{R}$.
(iv) $\pi_t(\theta) = (C - \rho(\theta))^+$ *for all* $\theta \in \mathbb{R}$, *and* $\nabla f_t(\theta) = 0$ *whenever* $\rho(\theta) > C$.

Then f *is an optimal solution.*

Essentially, $\pi_v(\theta)$ are dual variables, and the assumptions of the theorem are that f and π satisfy the complementary slackness conditions. There are many extensions of LP duality theory to infinite dimensional settings, e.g., [13, 21]; however the situation is subtle, since strong duality and even weak duality can fail [21]. We prefer to avoid technicalities and derive it directly (the proof is given in the full version).

The Dual Prescription. We now give a certificate of optimality $\pi : V \times \mathbb{R} \to \mathbb{R}$ for (3) that satisfies the conditions of the above LP. Given a vertex $v \in V$ and a time $\theta \in \mathbb{R}$ let

$$\pi_v(\theta) = \max\{\pi'_v(\theta), \bar{\pi}_v(\theta), 0\}$$

where
$$\pi'_v(\theta) = -\alpha d_{J(v,\theta)}(v, s),$$
$$\bar{\pi}_v(\theta) = C - \alpha d_{J(v,\theta)}(v, t) - \rho(\theta + d_{J(v,\theta)}(v, t)).$$

Notice that $\overset{*}{\pi}_s(\theta) = 0$ and $\pi_t(\theta) = \max\{C - \rho(\theta), 0\}$ for all $\theta \in \mathbb{R}$ and thus conditions (iii) and (iv) of Theorem 2 hold. The bulk of the technical work is in showing the remaining conditions; we sketch some part of the proof in the appendix.

5 Optimal Tolls

Tolls $\mu : E \times \mathbb{R} \to \mathbb{R}_+$ are per-arc, time-varying and nonnegative. The value $\mu_e(\xi)$ represents the toll a user is charged upon entering the link at time ξ.

We have the following theorem.

Theorem 3. *Let (f, π) be an optimal primal-dual solution to (3) (as constructed in Sects. 3 and 4) and define, for each $vw \in E$,*

$$\mu_{vw}(\theta) = (\pi_w(\theta + \tau_{vw}) - \pi_v(\theta) - \alpha\tau_{vw})^+.$$

Then f is a dynamic equilibrium under tolls μ.

Of course, to make sense of this theorem we must know what is meant by a dynamic equilibrium under tolls. A precise definition requires introducing the full game-theoretic fluid queueing model (also known as the Vickrey bottleneck model) [16, 26]. Tolls and departure time choice can be introduced into the definition of a dynamic equilibrium discussed in these works. Rather than going this route, we will show that the tolls satisfy a strong property that very clearly ensures the equilibrium property.

We show (in the full version—it is straightforward) that the following holds. A user starting from some $v \in V$ at some time $\theta \in \mathbb{R}$ cannot incur a total cost (including scheduling cost, and tolls and commuting cost from this point forward) less than $C - \pi_v(\theta)$. This is even allowing the user to take any link at any time, as if no other users were present in the network. Since the flow represents a solution where all users incur a total cost of precisely C, this must certainly be an equilibrium.

As already discussed, we cannot in general strongly enforce an optimal flow. The following shows that the "lane tolling" approach suffices to do this.

Theorem 4. *Let f, π and μ be as in the previous theorem, and suppose g is any dynamic equilibrium satisfying $g_e(\theta) \leq f_e(\theta)$ for all $e \in E$, $\theta \in \mathbb{R}$. Then $g = f$.*

Essentially, being able to dynamically split and separately toll the capacity of a link allows us to easily rule out all other potential equilibria just by using tolls to artificially constrict the capacities (in addition to choosing tolls that weakly enforce the desired flow, which is still needed). Tolling in this way seems quite distant from what could be imaginable in realistic traffic scenarios. But it does raise the interesting question of whether there is a tolling scheme which can strongly enforce an optimum flow, but which is more restricted (and more plausible) than fully dynamic lane tolling. Another natural question would be to determine if an optimum flow can be strongly enforced using lane tolling only on certain specified edges. We leave these as open questions.

A   Some Omitted Proofs

Proof (Theorem 1). The key ingredient is the following observation.

Lemma 1. $c_j(v, \theta)$ *is nondecreasing with j for any $\theta \in \mathbb{R}$.*

Proof. Consider any $j \in [m - 1]$; we show that $c_{j+1}(v, \theta) \geq c_j(v, \theta)$. Suppose Q is a shortest v-t-path in G_{j-1}, so $\tau(Q) = d_{j-1}(v, t)$. Consider the unit v-t flow $g = \chi(P_{j+1}) - \chi(P_j) + \chi(Q)$ in $\overleftarrow{E}$. Now observe that the support of g is contained in G_j: P_{j+1} and $\overleftarrow{P_j}$ are certainly contained in G_j; and if $e \in Q \cap (E_{j-1} \backslash E_j)$, then $e \in P_j$. Since G_j contains no negative cost cycles, the cost of g is at least that of a shortest v-t-path in G_j, and so $d_j(v, t) \leq \tau(P_{j+1}) - \tau(P_j) + \tau(Q)$. Finally, we can conclude

$$
\begin{aligned}
c_j(v, \theta) &= \alpha d_j(s, t) + \rho(\theta + d_{j-1}(v, t)) - \rho(\theta + d_{j-1}(v, t)) + \rho(\theta + d_j(v, t)) \\
&\geq \alpha d_j(s, t) + \rho(\theta + d_{j-1}(v, t)) - \alpha(d_j(v, t) - d_{j-1}(v, t)) \\
&\geq \alpha d_{j-1}(s, t) + \rho(\theta + d_{j-1}(v, t)),
\end{aligned}
$$

where the first inequality follows from the growth assumption, using $d_j(v, t) \geq d_{j-1}(v, t)$. $\qquad \square$

Fix some $vw \in E$ and $\theta \in \mathbb{R}$. Consider now any P_j (with $\alpha \tau(P_j) \leq C$, so that it is used for a nontrivial interval), with $vw \in P_j$. Since P_j is a shortest path in G_{j-1}, if we send flow along this path starting from some time ξ, it will arrive at v at time $\xi + d_{j-1}(s, v)$. Considering the definition of the interval $[a_j, b_j]$, we see that P_j contributes flow to vw at time θ if $c_j(v, \theta) \leq C$. By Lemma 1, this occurs precisely if $j \leq J(v, \theta)$. Considering in similar fashion paths P_j with $wv \in P_j$ (and noting that $J(w, \theta + \tau_{vw}) = J(v, \theta)$), the claim follows. $\qquad \square$

Optimality. We give the proof that π satisfies property (ii) in Theorem 2. The proof of property (i), while differing in the details, has a very similar flavour.

We first state a technical lemma involving distances in the residual graphs G_j; we omit the proof.

Lemma 2.

(a) For all $v \in V$, $d_j(v, s)$ is nonincreasing with j.
(b) For all $v \in V$ and $j \in [m]$, $d_{j-1}(v, t) - d_{j-1}(s, t) = d_j(v, s)$.

Lemma 3. *If $vw \in E^f(\theta)$, then $\pi_w(\theta + \tau_{vw}) \leq \pi_v(\theta) + \alpha \tau_{vw}$.*

Proof. Let $j := J(v, \theta)$ and $\ell := J(w, \theta + \tau_{vw})$. Note that since $vw \in E^f(\theta)$, Theorem 1 implies that $vw \in E_j$.

– **Case 1:** $\pi_w(\theta + \tau_{vw}) = -\alpha d_\ell(w, s)$.
   If $\ell \leq j$, then

$$
\begin{aligned}
\pi_v(\theta) &\geq -\alpha d_j(v, s) \\
&\geq -\alpha \tau_{vw} - \alpha d_j(w, s) && \text{since } vw \in E_j \\
&\geq -\alpha \tau_{vw} - \alpha d_\ell(w, s) && \text{by Lemma 2(a)} \\
&= \pi_w(\theta + \tau_{vw}) - \alpha \tau_{vw}.
\end{aligned}
$$

So suppose $\ell > j$. By the definition of $J(w, \theta + \tau_{vw})$ we know that

$$\alpha d_{\ell-1}(s,t) + \rho(\theta + \tau_{vw} + d_{\ell-1}(w,t)) \leq C. \tag{4}$$

Since $vw \in E_j$ and $d_j(w,t) \leq d_{\ell-1}(w,t)$, we also have

$$\theta + d_j(v,t) \leq \theta + \tau_{vw} + d_{\ell-1}(w,t). \tag{5}$$

Thus

$$
\begin{aligned}
\pi_v(\theta) &\geq \bar{\pi}_v(\theta) \\
&= C - \alpha d_j(v,t) - \rho(\theta + d_j(v,t)) \\
&\geq C - \alpha d_j(v,t) - \rho(\theta + \tau_{vw} + d_{\ell-1}(w,t)) - \alpha(\tau_{vw} + d_{\ell-1}(w,t) - d_j(v,t)) \\
&\geq \alpha d_{\ell-1}(s,t) - \alpha\tau_{vw} - \alpha d_{\ell-1}(w,t) && \text{by (4)} \\
&= -\alpha\tau_{vw} - \alpha d_\ell(w,s) && \text{by Lemma 2(b)} \\
&= \pi_w(\theta + \tau_{vw}) - \alpha\tau_{vw}
\end{aligned}
$$

where the second inequality follows from the growth assumption and (5).

Case 2: $\pi_w(\theta + \tau_{vw}) = C - \alpha d_\ell(w,t) - \rho(\theta + \tau_{vw} + d_\ell(w,t))$.
If $\ell \geq j$, since $vw \in E_j$ and $d_j(w,t) \leq d_\ell(w,t)$, we have that

$$\theta + d_j(v,t) \leq \theta + \tau_{vw} + d_\ell(w,t). \tag{6}$$

As a consequence, exploiting also the growth assumption, we have

$$
\begin{aligned}
\pi_v(\theta) &\geq C - \alpha d_j(v,t) - \rho(\theta + d_j(v,t)) \\
&\geq C - \alpha d_j(v,t) - \rho(\theta + \tau_{vw} + d_\ell(w,t)) - \alpha(\tau_{vw} + d_\ell(w,t) - d_j(v,t)) \\
&= C - \rho(\theta + \tau_{vw} + d_\ell(w,t)) - \alpha\tau_{vw} - \alpha d_\ell(w,t) \\
&= \pi_w(\theta + \tau_{vw}) - \alpha\tau_{vw}.
\end{aligned}
$$

If $\ell < j$, by definition of $J(w, \theta + \tau_{vw})$ we have that

$$\alpha d_\ell(s,t) + \rho(\theta + \tau_{vw} + d_\ell(w,t)) > C. \tag{7}$$

Thus

$$
\begin{aligned}
\pi_v(\theta) &\geq -\alpha d_j(v,s) \\
&\geq -\alpha d_j(w,s) - \alpha\tau_{vw} && \text{as } vw \in E_j \\
&\geq -\alpha d_{\ell+1}(w,s) - \alpha\tau_{vw} && \text{by Lemma 2(a)} \\
&> C - \alpha d_\ell(s,t) - \rho(\theta + \tau_{vw} + d_\ell(w,t)) \\
&\quad - \alpha d_{\ell+1}(w,s) - \alpha\tau_{vw} && \text{by (7)} \\
&= C - \alpha d_\ell(w,t) - \rho(\theta + \tau_{vw} + d_\ell(w,t)) - \alpha\tau_{vw} && \text{by Lemma 2(b)} \\
&= \pi_w(\theta + \tau_{vw}) - \alpha\tau_{vw}.
\end{aligned}
$$

$\square$

References

1. Arnott, R., de Palma, A., Lindsey, R.: Economics of a bottleneck. J. Urban Econ. **27**(1), 111–130 (1990)
2. Baumann, N., Skutella, M.: Solving evacuation problems efficiently-earliest arrival flows with multiple sources. In: Proceedings of the 47th Annual IEEE Symposium on Foundations of Computer Science (FOCS), pp. 399–410 (2006)
3. Bhaskar, U., Fleischer, L., Anshelevich, E.: A Stackelberg strategy for routing flow over time. In: Proceedings of the Twenty-Second Annual ACM-SIAM Symposium on Discrete Algorithms, (SODA), pp. 192–201 (2011)
4. Cominetti, R., Correa, J., Larré, O.: Dynamic equilibria in fluid queueing networks. Oper. Res. **63**(1), 21–34 (2015)
5. Cominetti, R., Correa, J., Olver, N.: Long term behavior of dynamic equilibria in fluid queuing networks. In: Eisenbrand, F., Koenemann, J. (eds.) IPCO 2017. LNCS, vol. 10328, pp. 161–172. Springer, Cham (2017). https://doi.org/10.1007/978-3-319-59250-3_14
6. Correa, J.R., Cristi, A., Oosterwijk, T.: On the price of anarchy for flows over time. In: Proceedings of the 2019 ACM Conference on Economics and Computation, (EC), pp. 559–577 (2019)
7. Disser, Y., Skutella, M.: The simplex algorithm is NP-mighty. In: Proceedings of the Twenty-Sixth Annual ACM-SIAM Symposium on Discrete Algorithms, (SODA), pp. 858–872 (2015)
8. Fleischer, L., Tardos, E.: Efficient continuous-time dynamic network flow algorithms. Oper. Res. Lett. **23**(3), 71–80 (1998)
9. Ford, L.R., Fulkerson, D.R.: Constructing maximal dynamic flows from static flows. Oper. Res. **6**(3), 419–433 (1958)
10. Ford, L.R., Fulkerson, D.R.: Flows in Networks. Princeton University Press, Princeton (1962)
11. Frascaria, D., Olver, N., Verhoef, E.T.: Emergent hypercongestion in Vickrey bottleneck networks. Preprint, Tinbergen Institute Discussion Paper TI 2020-002/VIII (2020). https://papers.tinbergen.nl/20002.pdf
12. Gale, D.: Transient flows in networks. Mich. Math. J. **6**(1), 59–63 (1959)
13. Grinold, R.: Infinite horizon programs. Manage. Sci. **18**, 157–170 (1971)
14. Harks, T.: Pricing in resource allocation games based on duality gaps. Preprint, arXiv:1907.01976 (2019). http://arxiv.org/abs/1907.01976
15. Jarvis, J.J., Ratliff, H.D.: Some equivalent objectives for dynamic network flow problems. Manage. Sci. **28**(1), 106–109 (1982)
16. Koch, R., Skutella, M.: Nash equilibria and the price of anarchy for flows over time. Theory Comput. Syst. **49**(1), 71–97 (2011)
17. Köhler, E., Möhring, R.H., Skutella, M.: Traffic networks and flows over time. In: Lerner, J., Wagner, D., Zweig, K.A. (eds.) Algorithmics of Large and Complex Networks. LNCS, vol. 5515, pp. 166–196. Springer, Heidelberg (2009). https://doi.org/10.1007/978-3-642-02094-0_9
18. Megiddo, N.: Combinatorial optimization with rational objective functions. In: Proceedings of the 10th Annual ACM Symposium on Theory of Computing, (STOC), pp. 1–12 (1978)
19. Minieka, E.: Maximal, lexicographic, and dynamic network flows. Oper. Res. **21**(2), 517–527 (1973)
20. Nisan, N., Roughgarden, T., Tardos, É., Vazirani, V.V. (eds.): Algorithmic Game Theory. Cambridge University Press, Cambridge (2007)

21. Romeijn, H.E., Smith, R.L., Bean, J.C.: Duality in infinite dimensional linear programming. Math. Program. **53**, 79–97 (1992)
22. Sering, L., Koch, L.V.: Nash flows over time with spillback. In: Proceedings of the Thirtieth Annual ACM-SIAM Symposium on Discrete Algorithms, (SODA), pp. 935–945 (2019)
23. Sering, L., Skutella, M.: Multi-source multi-sink Nash flows over time. In: 18th Workshop on Algorithmic Approaches for Transportation Modelling, Optimization, and Systems, (ATMOS), pp. 12:1–12:20 (2018)
24. Skutella, M.: An introduction to network flows over time. In: Cook, W., Lovász, L., Vygen, J. (eds.) Research Trends in Combinatorial Optimization, pp. 451–482. Springer, Heidelberg (2009). https://doi.org/10.1007/978-3-540-76796-1_21
25. Small, K.A.: The bottleneck model: an assessment and interpretation. Econ. Transp. **4**(1), 110–117 (2015)
26. Vickrey, W.: Congestion theory and transport investment. Am. Econ. Rev. **59**(2), 251–60 (1969)
27. Wilkinson, W.L.: An algorithm for universal maximal dynamic flows in a network. Oper. Res. **19**(7), 1602–1612 (1971)
28. Yang, H., Meng, Q.: Departure time, route choice and congestion toll in a queuing network with elastic demand. Transp. Res. Part B: Methodol. **32**(4), 247–260 (1998)
29. Zadeh, N.: A bad network problem for the simplex method and other minimum cost flow algorithms. Math. Program. **5**, 255–266 (1973)

Integer Plane Multiflow Maximisation: Flow-Cut Gap and One-Quarter-Approximation

Naveen Garg[1(✉)], Nikhil Kumar[1], and András Sebő[2]

[1] Indian Institute of Technology Delhi, Delhi, India
{naveen,nikhil}@cse.iitd.ac.in
[2] CNRS, Laboratoire G-SCOP, Univ. Grenoble Alpes, Grenoble, France
andras.sebo@grenoble-inp.fr

Abstract. In this paper, we bound the integrality gap and the approximation ratio for maximum plane multiflow problems and deduce bounds on the flow-cut-gap. Planarity means here that the union of the supply and demand graph is planar. We first prove that there exists a multiflow of value at least half of the capacity of a minimum multicut. We then show how to convert any multiflow into a half-integer one of value at least half of the original multiflow. Finally, we round any half-integer multiflow into an integer multiflow, losing again at most half of the value, in polynomial time, achieving a 1/4-approximation algorithm for maximum integer multiflows in the plane, and an integer-flow-cut gap of 8.

Keywords: Multicommodity flow · Multiflow · Multicut · Network design · Planar graphs · Flow-cut · Integrality gap · Approximation algorithm

1 Introduction

Given an undirected graph $G = (V, E)$ with edge capacities $c : E \to \mathbb{R}_+$, and some pairs of vertices given as edges of the graph $H = (V, F)$, the *maximum-multiflow problem* with input (G, H, c), asks for the maximum flow that can be routed in G, simultaneously between the endpoints of edges in F, respecting the capacities c.

This is one of many widely studied variants of the multiflow problem. Other popular variants include demand flows, all or nothing flows, unsplittable flows etc. In this paper, we are mainly interested in the integer version of this problem, and the half-integer or fractional versions also occur as tools. When capacities are 1, the capacity constraint specialises to edge-disjointness, whence the *maximum edge disjoint paths problem (MEDP)* between a given set of pairs is a special case; MEDP is NP-Hard to compute for general graphs, even in very restricted settings like when G is a tree [8].

The edges in F are called *demand edges* (sometimes commodities), those in E are called *supply edges*; accordingly, $H = (V, F)$ is the *demand graph*, and

D. Bienstock and G. Zambelli (Eds.): IPCO 2020, LNCS 12125, pp. 144–157, 2020.
https://doi.org/10.1007/978-3-030-45771-6_12

$G = (V, E)$ is the *supply graph*. If $G + H = (V, E \cup F)$ is planar we call the problem a *plane* multiflow problem. Plane multiflows have been studied for the past forty years, starting with Seymour [22].

Let $\mathcal{P}_e$ ($e \in F$) be the set of paths in G between the endpoints of e, and $\mathcal{P} := \cup_{e \in F} \mathcal{P}_e$. For $P \in \mathcal{P}_e$, the edge e is said to be the *demand-edge of P*, denoted by e_P. A *multiflow*, or for simplicity a *flow* in this paper is a function $f : \mathcal{P} \to \mathbb{R}^+$. The flow f is called *feasible*, if $\sum_{\{P \in \mathcal{P} : e \in P\}} f(P) \leq c(e)$ for all $e \in E$. The *value* of a flow f is defined as $|f| := \sum_{P \in \mathcal{P}} f(P)$.

For a path $P \in \mathcal{P}$, we refer to $f(P)$ as the *flow on P*. If the flow on every path is integer or half-integer, we say that the flow is integer or half-integer, respectively. A *circuit* is a connected subgraph with all degrees equal to two.

Multiflows (without restrictions on integrality) can be maximised in general, in (strongly) polynomial time [20, 70.6, page 1225] using a linear programming algorithm.

A *multicut* for (G, H) is a set of edges $M \subseteq E$ such that every $P \in \mathcal{P}$ contains at least one edge in M.[1] A multicut is the simplest possible and most natural dual to the maximum multiflow problem.

It is easy to see that the value of any feasible multiflow is smaller than or equal to the capacity of any multicut. Klein, Mathieu and Zhou [11] prove that computing the minimum multicut is NP-hard if $G + H$ is planar and also provide a PTAS.

There is a rich literature on the maximum ratio of a minimum capacity of a multicut over the maximum multiflow. With an abuse of terminology we will call this the *(possibly integer or half-integer) flow-cut gap*, sometimes also restricted to subclasses of problems. This is not to be confused with the same term used for demand problems. The integer flow-cut gap is 1 when G is a path and 2 when G is a tree. For arbitrary (G, H), the flow-cut gap is $\theta(\log |F|)$ [7]. Building on decomposition theorems from Klein, Plotkin and Rao [10], Tardos and Vazirani [23] showed a flow-cut gap of $O(r^3)$ for graphs which do not contain a $K_{r,r}$ minor; note that for $r = 3$ this includes the class of planar graphs. A long line of impressive work, culminated in [21] proving a constant approximation ratio for maximum half-integer flows, which together with [23] implies a constant half-integer flow-cut gap for planar supply graphs. A simple topological obstruction proves that the integer flow-cut gap for planar supply graphs, even when all demand edges are on one face of the graph, also called *Okamura-Seymour instances*, is $\Omega(|F|)$ [8].

These results are often bounded by the *integrality gap* of the multiflow problems for the problem classes, which is the infimum of $\mathrm{MAX}/|f|$ over all instances (G, H, c) of the problem class, and where MAX is the maximum value of an integer multiflow, and f is a multiflow for this same input, and the *half-integrality gap* is defined similarly by replacing "integer" by "half-integer" in the nominator. A ρ-approximation algorithm ($\rho \in \mathbb{R}$) for a maximisation problem is a

---

[1] Given a partition of V so that all edges in F join different classes, those of E joining different classes form a multicut, and inclusionwise minimal multicuts are like this.

polynomial algorithm which outputs a solution of value at least ρ times the optimum; ρ is also called the *approximation ratio* (or guarantee).

Our first result (Sect. 3, Theorem 1) is an upper bound of 2 for the flow-cut gap (i.e. multicut/multiflow ratio) for plane instances, the missing relation we mentioned. We prove this by relating multicuts to 2-edge-connectivity-augmentation in the plane-dual, and applying a bound of Williamson, Goemans, Mihail and Vazirani [25] for this problem.

We next show (Sect. 4, Theorem 2) how to obtain a half-integer flow from a given (fractional) flow in plane instances, reducing the problem to a linear program with a particular combinatorial structure, and solving it.

Finally, given any feasible half-integer flow, we show how to extract an integer flow of value at least half of the original, in polynomial time, using an algorithm that 4-colors planar graphs efficiently [18].

These results imply an integrality gap of $1/2$ for maximum half-integer flows, $1/4$ for maximum integer flows, and the same approximation ratios for each. In turn, the flow-cut gap of 2 implies a half-integer flow-cut gap of 4 and an integer flow-cut gap of 8 for plane instances.

A summary of the results completed by lower bounds and open problems are stated in Sect. 7. These are the first constant approximation ratios and gaps proved for the studied problems. Approximately at the same time as our submission to this volume, another manuscript was published on Arxiv, proving constant ratios for the same problems (https://arxiv.org/abs/2001.01715).

2   Preliminaries

We detail here some notions, notations, terminology, facts and tools we use, including some preceding results of influence.

Demands, the Cut Condition and Plane Duality

We describe the notion of demand flows, which is closely related to multi-flows defined in the previous section. The problem is defined by the quadruple (G, H, c, d), where G, H, c are as before, *demands* $d : F \to \mathbb{Z}_+$ are given, and we are looking for a feasible (sometimes in addition integer or half-integer) flow f satisfying $\sum_{\{P \in \mathcal{P} : e \in P\}} = d(e)$ for all $e \in F$.

A *cut* in a graph $G = (V, E)$ is a set of edges of the form $\delta(S) = \delta_E(S) := \{e \in E : e$ has exactly one endpoint in $S\}$ for all $S \subseteq V$. Note that $\delta(S) = \delta(V \setminus S)$; $S, V \setminus S$ are called the *shores* of the cut. For a subset $E' \subseteq E$ we use the notation $c(E') := \sum_{e \in E'} c(e)$, and we adopt this usual way of extending a function on single elements to subsets. For instance, $d(F') := \sum_{e \in F'} d(e)$ is the *demand* of the set $F' \subset F$.

A necessary condition for the existence of a feasible multiflow is the so called *Cut Condition*: for every $S \subseteq V, c(\delta_E(S)) \geq d(\delta_F(S))$, that is, the capacity of each cut must be at least as large as its demand. The cut condition is not sufficient for a flow in general, but Seymour [22] showed that it is sufficient for an integer flow, provided $G + H$ is planar, the capacities and demands are integer, and their sum is even on the edges incident to any vertex. A half-integer flow follows then for arbitrary integer capacities for such *plane* instances (see Sect. 1). The same holds for Okamura-Seymour instances [17]. There are more examples, unrelated to planarity, where the cut condition is sufficient to satisfy all demands, and with an integer flow, for instance when all demand edges can be covered by at most two vertices.

Seymour's theorem [22] on the sufficiency of the cut condition and about the existence of integer flows has promoted plane multiflow problems to become one of the targets of investigations. This paper is devoted to plane multiflow maximisation. Seymour's proof is based on a nice correspondence to other combinatorial problems through plane duality that we also adopt.

Middendorf and Pfeiffer [16] showed that the plane multiflow problem (edge disjoint paths already) is NP-Hard. The cut condition can be checked in polynomial time[2] so the difficult question to decide is the existence or not of an integer flow when the cut condition is satisfied. As far as multicuts are concerned, we show that their minimisation in plane instances is equivalent to a 2-edge-connectivity augmentation problem in planar graphs.

Following Schrijver [20, p. 27] we denote the dual of the planar graph G by $G^* = (V^*, E^*)$, where V^* is the set of faces of G and each edge $e \in E$ corresponds to an edge $e^* \in E^*$ joining the two faces that share e. For $X \subseteq E$ denote $X^* := \{e^* : e \in X\}$. An important fact we need and use about dualisation: *C is a minimal cut in G if and only if C^* is a circuit in G^*.*

Two Lemmas on Laminar Families

This correspondence by plane duality allows to transform any fact on cuts to circuits in the dual and vice versa. For example fractional, half-integer or integer packings of cuts in G^*, where each cut contains exactly one edge of F^* correspond to a fractional, half-integer or integer multiflow in G. We provide now some related definitions, notations and facts. We do this directly on the graph $(G+H)^*$ where they will be used; we denote by V^* the vertices of this graph, that is the faces of $G + H$. So $(G + H)^* = (V^*, E^* \cup F^*)$. Let $\delta(A) = \delta_{E^* \cup F^*}(A)$, $\delta(B) = \delta_{E^* \cup F^*}(B)$ $(A, B \subseteq V^*)$ be two *crossing* cuts, i.e. A and B are neither disjoint nor contain one another, each of which contains exactly one edge of F^*. Then they can be replaced by $\delta(A^* \cap B^*)$ and $\delta(A^* \cup B^*)$ or $\delta(A^* \setminus B^*)$ and $\delta(B^* \setminus A^*)$(in the plane dual). It is easy to check that every edge is contained in at most as many of these two cuts after the replacement as before, and if both cuts contain exactly one edge of F^* then this also holds for the replacing cuts. Doing this iteratively and using plane duality, we can convert any feasible flow into another one (without changing the value of the flow) in which no two flow paths cross. We formalise this below.

A family of subsets of V^* is called *laminar* if any two of its members are disjoint or one of them contains the other. If for any two members one of them

---

[2] Seymour's correspondence through dualisation reduces this problem to checking whether F^* is a minimum weight "T_{F^*}-join" (see e.g. [20]) in $(G + H)^*$ with weights defined by c and d, where T_{F^*} is the set of odd degree vertices of F^*. This also provides a polynomial separation algorithm for maximising the sum of (not necessarily integer) demands satisfying the cut condition.

contains the other we say that the family is a *chain*. Given a laminar family $\mathcal{L} \subseteq 2^{V^*}$, a chain $\mathcal{C} \subseteq \mathcal{L}$ is *full* (in $\mathcal{L}$) if $X, Y, Z \in \mathcal{L}, X \subseteq Y \subseteq Z$ and $X, Z \in \mathcal{C}$ implies $Y \in \mathcal{C}$. We call a multiflow f *laminar*, if $\{C^* : C = P \cup \{e_P\}, P \in \mathcal{P}, f(P) > 0\} = \{\delta_{E^* \cup F^*}(L) : L \in \mathcal{L}\}$, where $\mathcal{L} \subseteq 2^{V^*}$ is laminar. We state the following well-known Lemma without proof (see the full version of this paper (on ArXiv) for explanations and references).

Lemma 1. *For every feasible multiflow f there exists a laminar feasible multiflow f' so that $|f'| = |f|$, and f' can be found in polynomial time.*

The following useful properties are easy to check:

For a family $\mathcal{L}$ of subsets of V^* and $a \neq b \in V^*$, denote $\mathcal{L}(a) := \{L \in \mathcal{L} : a \in L\}$; $\mathcal{L}(a, b) := \{L \in \mathcal{L} : a \in L, b \notin L\}$.

Lemma 2. *Let $\mathcal{L}$ be a laminar family of subsets of V^*. Then*

a. $|\mathcal{L}| \leq 2(|V^*| - 1) = O(|V|)$.
b. $\mathcal{L}(a, b) \subseteq \mathcal{L}(a) \subseteq V^*$ *both form full chains of subsets of V^*.* ☐

Integrality in Demand Flows and Stable Sets
We first compute a half integer flow of value at least half the fractional flow and then convert this into an integer flow. We now describe an instance which illustrates the difficulty in finding an integer flow. Consider a planar graph with a perfect matching without any nontrivial *tight cut*[3] to be $(G + H)^*$, and F^* to be any perfect matching in it. Let all capacities be 1. Upper bound the demands by 1, by replacing each demand edge by two edges in series, a demand and a supply edge, the latter of capacity 1. Then multiflows can use only the dual edge-sets of stars of vertices in $(G + H)^*$, so an integer multiflow of value k corresponds exactly to a *stable set*, that is, a set of vertices not inducing any edge, of size k in $(G + H)^*$. This indicates that in order to find a large integer flow, we need to find large stable sets in planar graphs. Despite these restricted multiflows, the gap between integer and half-integer flow is at least $1/2$ for these graphs: it follows from the 4-color theorem [1] that a stable set of size at least $|V^*|/4$ exists while the maximum half integer flow cannot exceed $|F| = |V^*|/2$. We will be able to reach this ratio in general (see Sect. 5, Theorem 4), and K_4 with a matching F shows that this cannot be improved.

In order to reach this integrality gap of 2 in general, we will actually need to find a stable-set of size $|V^*|/4$ in $(G + H)^*$. The maximum stable set problem is NP-hard, but there is a PTAS for it in planar graphs [2], which, combined with the 4-color theorem [1] provides a stable-set of size $|V^*|/4$; the 4-coloring algorithm of Robertson, Sanders, Seymour and Thomas [18] directly provides a 4-coloring of a planar graph in polynomial time as well, and the largest color class is clearly of size at least $|V^*|/4$. Either of these algorithms can be used as a black-box-tool for rounding half-integer flows, so we state the result:

---

[3] i.e. a cut with both shores containing more than one vertex, and meeting every perfect matching in exactly one edge. Lovász characterised graphs without nontrivial tight cuts as "bicritical 3-connected graphs" [15]. Such graphs may have arbitrarily many vertices, even under the planarity constraint.

Lemma 3. *In a planar graph on n vertices, a stable set of size $n/4$ can be found in polynomial time.*

A similar rounding argument appeared in the work of Fiorini, Hardy, Reed and Vetta [5] in the somewhat different context of proving an upper bound of Král and Voss [14] for the ratio between "minimum size of an odd cycle edge transversal" versus the "maximum odd cycle edge packing" using the 4-color theorem [1] non-algorithmically. Our procedure in Sect. 5, occurs to be more general in that it is starting from an *arbitrary, not necessary optimal, half-integer multiflow* for an arbitrary capacity function. However, similar difficulties approached independently with the 4-color theorem confirm that the 4-color theorem may be unavoidable for plane multiflows.

3   Multicuts Versus Multiflows via 2-Connectors

We show in this section that the flow-cut gap is at most two for plane instances, via a reduction to the 2-edge-connectivity augmentation problem.

Given $G = (V, E)$, $H = (V, F)$, a *2-connector for H in G* is a set of edges $Q \subseteq E$ such that none of the edges $e \in F$ is a cut edge of $(V, Q \cup F)$; equivalently, Q is a 2-connector if and only if each $e \in F$ is contained in a circuit of $Q \cup F$. The *2-edge-connectivity Augmentation Problem (2ECAP)* is to find, for given edge costs $c : E \to \mathbb{Z}_+$ on E, a minimum cost 2-connector.

Let (G, H, c), $G = (V, E)$, $H = (V, F)$ be the input of a plane multiflow maximisation problem, and $(G + H)^* = (V^*, E^* \cup F^*)$, where V^* is the set of faces of $G + H$. Define $c(e^*) := c(e)$ $(e \in E)$.

Lemma 4. *The edge-set $Q \subseteq E$ is a multicut for (G, H) if and only if Q^* is a 2-connector for (V^*, F^*) in (V^*, E^*).*

Proof. The edge-set $Q \subseteq E$ forms a multicut in G if and only if the endpoints u, v of any edge $e \in F$ are in different components of $(V, E \setminus Q)$, that is, if and only if for all $e \in F$ there exists an inclusionwise minimal cut $C \subseteq Q \cup F$ of $G + H$ such that $e \in C$. But we saw among the preliminaries concerning duality that C is an inclusionwise minimal cut in $G + H$ if and only if C^* is a circuit in $Q^* \cup F^*$. Summarizing, $Q \subseteq E$ forms a multicut in G, if and only if for all $e^* \in F^*$ there exists a circuit C^* in $Q^* \cup F^*$ such that $e^* \in C^*$. This means exactly that $Q^* \subseteq E^*$ is a 2-connector for (V^*, F^*), in (V^*, E^*), finishing the proof.   □

Let (G, H, c) be an instance of multiflow problem with $G + H$ planar. Let $p : 2^{V^*} \to \{0, 1\}$ with $p(S) = 1$ if and only if $|\delta(S) \cap F^*| = 1$, otherwise $p(S) = 0$ $(S \subseteq V^*)$. The following linear program specialises the one investigated in [25]:

$$\text{minimise} \quad \sum_{e \in E^*} c(e)x(e),$$
$$\text{subject to}$$
$$\sum_{e \in \delta(S)} x(e) \geq p(S), \qquad S \subseteq V^*; \qquad \text{(2ECAP)}$$
$$x(e) \geq 0 \qquad e \in E^*.$$

Since p is $\{0,1\}$-valued so are the coordinatewise minimal integer solutions including the integer optima of (2ECAP). *The $\{0,1\}$-solutions are exactly the (incidence vectors of) 2-connectors of (V^*, F^*) in (V^*, E^*).* In the dual linear program of (2ECAP), we have a variable $y(S)$ for all $S \subseteq V^*$, constraints $\sum_{S:e\in\delta(S)} y(S) \leq c_e$ for all $e \in E^*$ and $y(S) \geq 0$ for all $S \subseteq V^*$. The objective is to maximize $\sum_{S \subseteq V^*} p(S)y(S)$. Williamson, Goemans, Mihail and Vazirani [25] developed a primal-dual algorithm finding for given input $(G + H)^*$ and c, an *integer primal solution x_{WGMV} to a class of linear programs including (2ECAP), together with a (not necessarily integer) dual solution y_{WGMV}, in polynomial time,* proving the following *WGMV inequality* [25, Lemma 2.1], see also [13, Section 20.4]:

$$\text{LIN} \leq OPT \leq \sum_{e\in E^*} c(e)x_{\text{WGMV}}(e) \leq 2 \sum_{S\subseteq V^*, p(S)=1} y_{\text{WGMV}}(S) \leq 2\,\text{LIN} \leq 2\,\text{OPT},$$

where OPT is the minimum cost of a 2-connector, and LIN is the optimum of (2ECAP). We will refer to this algorithm as the *WGMV algorithm*.

Note that the *WGMV* algorithm works for the class of weakly supermodular functions. A function $h : 2^V \to \{0,1\}$ is called *weakly supermodular* if $h(V) = 0$ and for any $A, B \subseteq V$, $h(A) + h(B) \leq \max\{h(A \cap B) + h(A \cup B), h(A \setminus B) + h(B \setminus A)\}$. It can be verified that p defined above is weakly supermodular.

Theorem 1. *Let (G, H, c) be a plane multiflow problem. Then there exists a feasible multiflow f and a multicut Q, such that $c(Q) \leq 2|f|$, where both f and Q can be computed in polynomial time.*

Proof. Recall that the WGMV algorithm finds x_{WGMV} and y_{WGMV} satisfying the WGMV inequality, where x_{WGMV} is the incidence vector of a 2-connector of (V^*, F^*) in (V^*, E^*), let us denote its plane dual set in $G + H$ by Q_{WGMV}. By Lemma 4 Q_{WGMV} is a multicut.

Consider a set S with $p(S) = 1$, that is, $|\delta(S) \cap F^*| = |\delta(S)^* \cap F| = 1$, all the other sets can be supposed to be absent from the inequalities. Then $\delta(S)^*$ contains a circuit C of $G + H$ containing the unique edge of $|\delta(S)^* \cap F|$. Therefore $C \setminus F$ is a path in G, denote it by P_S. Define a multiflow f in $G + H$ by $f(P_S) = y_{\text{WGMV}}(S)$. The dual feasibility of y_{WGMV} means exactly that the multiflow f is feasible. By our construction and the WGMV inequality we have

$$c(Q_{\text{WGMV}}) = \sum_{e\in E^*} c(e)x_{\text{WGMV}}(e) \leq 2 \sum_{S\subseteq V^*, p(S)=1} y_{\text{WGMV}}(S) = 2|f|.$$

So the multicut $Q := Q_{\text{WGMV}}$ and the multiflow f satisfy the claimed inequality; all operations, including the WGMV algorithm run in polynomial time.   □

Note that if y_{WGMV} is half-integer (assuming integer edge-costs), the obtained multiflow is half-integer and a half-integer flow-cut gap of 2 directly follows. There are examples where the WGMV algorithm does not produce a half-integer dual solution, but we do not know of an instance where half-integer flow-cut gap is larger than 2.

4   From Fractional to Half-Integer

We show here how to convert a flow to a half-integer one, loosing at most half of the flow value, provided the capacities are integers.

Theorem 2. *Let (G, H, c) be a plane multiflow problem, where $c : E \to \mathbb{Z}_+$ is an integer capacity function. Given a feasible multiflow f, there exists a feasible half-integer multiflow f', such that $|f'| \geq |f|/2$, and it can be computed in polynomial time.*

Proof. By Lemma 1 we can suppose that the given feasible multiflow f is laminar and can be found in polynomial time. Let $\mathcal{L} \subseteq 2^{V^*}$ be the laminar family of cuts in $(G + H)^*$, with $\{C^* : C = P \cup \{e_P\}, f(P) > 0\} = \{\delta_{E^* \cup F^*}(L) : L \in \mathcal{L}\}$ (see Sect. 2, just above Lemma 1). Denote $f_L := f(\delta(L)^* \setminus F)$, $(L \in \mathcal{L})$. The feasibility of the multiflow f means $\sum_{L \in \mathcal{L}, e \in \delta_{E^*}(L)} f_L \leq c(e)$, that is, $x_L = f_L \in \mathbb{N}_+$ satisfies

$$\sum_{L \in \mathcal{L}} x_L = |f|, \text{ and} \sum_{L \in \mathcal{L}, e \in \delta_{E^*}(L)} x_L \leq c(e), \text{for all } e \in E, \; x_L \geq 0, \; (L \in \mathcal{L}). \quad (1)$$

Clearly, the edge $e = uv$ is contained in exactly those sets $\delta(L)$ $(L \in \mathcal{L})$ for which $L \in L(u,v) \cup L(v,u)$, where $\mathcal{L}(a,b) := \{L \in \mathcal{L} : a \in L, b \notin L\}$ $(a, b \in V^*)$, and both $\mathcal{L}(u,v)$ and $\mathcal{L}(v,u)$ form full chains (Lemma 2). So the linear program

$$\max \quad \sum_{L \in \mathcal{L}} x_L$$

$$\text{subject to} \quad \sum_{L \in \mathcal{L}(u,v)} x_L \leq c(e) \quad \text{for all } (u,v) \in V^* \times V^*, uv = e^* \in E^* \quad (2)$$

$$x_L \geq 0 \quad \text{for all } L \in \mathcal{L},$$

is a relaxation of (1): for each $u, v \in V^*, uv = e^* \in E^*$ the coefficient vector of (1) associated with e^* is the sum of the coefficient vectors, one for each of (u, v) and (v, u), of the two inequalities associated with e^* in (2). Both of these ($\mathcal{L}(u, v)$ and $\mathcal{L}(v, u)$) correspond to full chains in the laminar family $\mathcal{L}$, and the right hand side $c(e)$ is repeated for both.

Denote $M = M(f)$ the $2|E| \times |\mathcal{L}|$ coefficient matrix of (2) (without the non-negativity constraints).

According to Edmonds and Giles [4], $\mathcal{L}$ has a rooted tree (arborescence) representation in which the full chains correspond to subpaths of paths from the root, so M is a network matrix. As such, it is totally unimodular by Tutte [24] and (2) has an integer optimum x, computable in polynomial time by [9], [19, Theorem 19.3 (ii), p. 269].

To finish the proof now, note that on the one hand, f_L $(L \in \mathcal{L})$ is a solution to (1), and therefore it is also a feasible solution of the relaxation (2). Since x_L $(L \in \mathcal{L})$ is the maximum of (2), $\sum_{L \in \mathcal{L}} x_L \geq \sum_{L \in \mathcal{L}} f_L = |f|$. According to the two inequalities of (2) associated to e^*, the sum of coefficients of the paths containing any given edge $e \in E$ is at most $c(e) + c(e) = 2c(e)$, so $f' = x/2$ defines a half-integer feasible flow in (G, H, c) (by assigning the flow value $f'(L)$ to the path $\delta(L)^* \setminus F$), so $|f'| \geq |f|/2$, finishing the proof. □

For more explanations and references showing that M is a network matrix, and an alternative direct combinatorial solution of the integer linear program with a simple greedy-type algorithm, see in the full version on ArXiv.

This proof does not fully exploit the possibilities of totally unimodular matrices: instead of putting $c(e)$ as right hand side for both inequalities of (2) associated with e^* we can put everywhere the smallest integer capacities satisfied by the fractional flow. Since the fractional values on the mentioned two inequalities sum to at most $c(e)$ and not $2c(e)$ we get a sharper result this way. The proof works if we replace the capacities by the rounded up fractional flow, leading to an error of only 1 compared to the original capacities, due to the round-up. Let us denote by $\underline{1}$ the all 1 function on E, and check this precisely:

Theorem 3. *Let (G, H, c) be a plane multiflow problem, where $c : E \to \mathbb{Z}_+$. Given a feasible multiflow f, there exists a feasible integer multiflow f', computable in polynomial time, feasible for the capacity function $c + \underline{1}$, and $|f'| \geq |f|$.*

Proof. Let $f \in \mathbb{R}^{\mathcal{L}}$ be a feasible multiflow, and $M = M(f)$ the $2|E| \times |\mathcal{L}|$ coefficient matrix defined in the proof of Theorem 2. Define $d := Mf \in \mathbb{R}^{2|E|}$, and consider the linear program

$$\max \qquad \sum_{L \in \mathcal{L}} x_L$$

$$\text{subject to} \qquad \sum_{L \in \mathcal{L}(u,v)} x_L \leq \lceil d(u,v) \rceil \quad \text{for all } (u,v) \in V^* \times V^*, uv = e^* \in E^*$$

$$x_L \geq 0 \qquad\qquad\qquad\qquad\qquad \text{for all } L \in \mathcal{L},$$
(3)

where $d(u, v) := \sum_{L \in \mathcal{L}(u,v)} f_{\delta(L)^* \setminus F}$. In words, (3) has exactly the same coefficients as (2), but the right hand sides $d(u, v)$ and $d(v, u)$ of the two inequalities associated with $e \in E$, $e^* = uv$ are defined with the sum of flow values on $\delta(L)^* \setminus F$, for $L \in \mathcal{L}(u, v)$ and $L \in \mathcal{L}(v, u)$ respectively.

Since f is a feasible flow for (G, H, c), $d(u, v) + d(v, u) \leq c(e)$, so $\lceil d(u,v) \rceil + \lceil d(v,u) \rceil \leq c(e) + 1$, and f is feasible for (3) since the capacities have been defined by rounding up the flow values. On the other hand, the coefficient matrix is totally unimodular, so by [9], [19, Theorem 19.3 (ii), p. 269] the linear program (3) has an integer maximum solution f', again computable in polynomial time, and $f' \leq c + \underline{1}$, $|f'| \geq |f|$. □

Theorem 2 is an immediate consequence of Theorem 3:

For each $k \in \mathbb{N}$, $k \geq 1$, $\frac{k+1}{2} \leq k$ holds, so (after deleting 0 capacity edges) $(c + \underline{1})/2 \leq c$, and therefore, dividing by 2 the primal optimum of (3), we immediately get a half-integer solution to (2).

This result extends a natural consequence for maximisation of the tight additive integrality gap known for plane demand flow problems with integer capacities and demands satisfying the cut condition: according to a result of Korach and Penn [12], if all demand edges lie in two faces of the supply graph, *there exists an integer multiflow satisfying all demands but at most 1.* This readily implies that increasing each capacity by 1, an integer flow of the same value as the maximum flow for the original capacities, can be reached.

From Frank and Szigeti [6] the same can be deduced for demand-edges lying on an arbitrary number k of faces. Indeed, increasing all capacities by 1, *the surplus of the cut condition will be at least k, which is the Frank-Szigeti condition for integer multiflows.* Theorem 3 states that this consequence is actually true in general, without requiring the integrality of the demands, and also for the maximisation problem; the same holds for maximum "packings of T-cuts".

5   From Half-Integer to Integer

In this section, we show how to round a half-integer flow to an integer one, losing at most one half of the flow value.

Theorem 4. *Let (G, H, c) be a plane multiflow problem, where $G = (V, E)$, $H = (V, F)$ and $c : E \to \mathbb{Z}_+$. Given a feasible half-integer multiflow f, there exists a feasible integer multiflow f', computable in polynomial time, and $|f'| \geq |f|/2$.*

Proof. Let (G, H, c) and f be as assumed in the condition, moreover that f is laminar (Lemma 1). We proceed by induction on the integer $2|f|$. We suppose that *all nonzero values $f(P) > 0$ ($P \in \mathcal{P}$) are actually $1/2$:* if $f(P) \geq 1$, we can decrease $f(P)$ by $\lfloor f(P) \rfloor$, as well as all capacities of edges of P, and the statement follows from the induction hypothesis. (Such a step can be repeated only a polynomial number of times, since $|\mathcal{P}| = O(|V|)$ by Lemma 2.)

To choose the values to round we replace $c(e)$, by $c(e)$ parallel edges of capacity 1 each.[4] Then take the plane dual of the resulting graph, which is $(G+H)^*$ with each edge $e^* \in E^*$ replaced by a path of size $c(e)$. We consider the laminar system $\mathcal{L}$ defining the paths $P \in \mathcal{P}$, $f(P) > 0$[5] in this subdivided graph so that every edge is contained in at most two sets $L \in \mathcal{L}$, and remains laminar (this is clearly possible, since all positive f_P values are $1/2$). For simplicity, we keep the notations of the original graph - as if what we get in this way were the given graph with all capacities equal to 1.

Let $I(\mathcal{L}) = (\mathcal{L}, M)$ be the intersection graph of the cuts defined by $\mathcal{L}$, that is, $M := \{L_1 L_2 : L_1, L_2 \in \mathcal{L}, \delta(L_1) \cap \delta(L_2) \neq \emptyset\}$. We have Claim a. and b. so far, and now we check Claim c.:

---

[4] This is not an allowed step for a polynomial algorithm, but it will not really be necessary to do it. The choice of the cuts to be rounded down or up will be clear from the proof without actually executing this subdivision. The choices for rounding concern a family of size $O(|V|)$.

[5] $\{\delta_{E^* \cup F^*}(L)^* : L \in \mathcal{L}\} = \{P \cup e_P : P \in \mathcal{P}, f(P) > 0\}$, as before.

Claim: a. Each $e \in E^*$ is contained in at most two sets in $\{\delta(L) : L \in \mathcal{L}\}$.

   b. $|f| = |\mathcal{L}|/2$.

   c. $I(\mathcal{L})$ is planar.

To check Claim c., note first its validity if $\mathcal{L}$ consists of disjoint sets. If this does not hold, not even by *complementing* some $L \in \mathcal{L}$ (i.e. replacing it by $V^* \backslash L$), then it is easy to find (possibly after coplemention) three sets $L_1 \subset L_0 \subset L_2$ in $\mathcal{L}$. We claim that L_0 is a cut-vertex in $I(\mathcal{L})$. By laminarity, every cut $\delta(L)$ $(L \in \mathcal{L})$ has either a shore A contained in L_0 (like L_1), or a shore B disjoint from L_0 (like $V \setminus L_2$). If there exists an $e \in \delta(A) \cap \delta(B)$, this would mean that e has an endpoint in $A \subseteq L_0$ and the other endpoint in $B \subseteq V \setminus L_0$, so $e \in \delta(L_0)$, contradicting Claim a. So L_0 *is a cut vertex, i.e.* $\mathcal{L} = \mathcal{L}_1 \cup \mathcal{L}_2$, $\mathcal{L}_1 \cap \mathcal{L}_2 = \{L_0\}$, *with no edge between* $\mathcal{L}_1$ *and* $\mathcal{L}_2$ *in* $I(\mathcal{L})$.

Since the graphs induced by $\mathcal{L}_i$ $(i = 1, 2)$ are both defined by flows of smaller value, we can apply the induction hypothesis to them: they are planar, so $I(\mathcal{L})$ is also planar, and the Claim follows.

To finish the proof of the theorem using Claim c., find a stable set of size $|\mathcal{L}|/4$ in $\mathcal{L}$, by Lemma 3, and increase the flow on the corresponding paths $\delta(L)^* \setminus F$ to 1, while decreasing the flow on the other paths to 0. This results in a feasible integer flow $|f'| \geq |\mathcal{L}|/4 = |f|/2$, finishing the proof of the bound.

For the computational complexity results, first recall that the support of the half integer laminar flow f obeys Lemma 2a. Then note, that among this linear number of sets, the proof finds in polynomial time at least one fourth of the cuts to round up, while the other cuts are rounded down, so that the capacity constraints are not violated. It is straightforward to mimic the subdivision of edges without doing it, and to compute an input to Lemma 3, in strongly polynomial time.   □

6   Lower Bounds on the Flow-Cut Gap

We show a class of plane multiflow instances G_k on which the half-integer flow-cut gap is tending to $1/2$ as $k \to \infty$. Cheriyan et.al. [3] used these instances to show integrality gap results for the Tree Augmentation Problem. Let $G_k = (V_k, E_k)$, $H_k = (V_k, F_k)$, $k \geq 3$ be an instance of the multiflow problem defined as follows: $V_k = \{a_1, a_2, \ldots, a_k\} \cup \{b_1, b_2, \ldots, b_k\}$, $E_k = \{(a_i, b_i) | i \in [1, k]\} \cup \{(a_i, a_{i+1}) | i \in [1, k-1]\}$ and $F_k = \{(b_i, b_{i+1}) | i \in [1, k-1]\} \cup \{(b_i, a_{i+2}) | i \in [1, k-2]\}$. The capacity of all edges in E_k is 1 (see Fig. 1).

Theorem 5. *The graph* $G_k + H_k$ *is planar for all* $k = 1, 2, \ldots$, *and the following hold:*

- *The minimum multicut capacity is* $k - 1$.
- *The maximum multiflow value is* $2(k-1)/3$.
- *The maximum half-integer multiflow value is* $k/2$.
- *The maximum integer multiflow value is* $\lfloor k/2 \rfloor$.

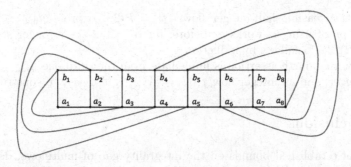

Fig. 1. G_8: supply edges are black, their capacity is 1; demand edges are red (thick). (Color figure online)

Proof. The minimum multicut capacity is clearly at most $k - 1$, since deleting each edge of the (a_1, a_{k-1})-path, the endpoints of all demand-edges are separated. We check now $|C| \geq k - 1$ for an arbitrary multicut C, by induction. For $k = 1, 2$ the statement is obvious. Deleting b_1 and a_1, the remaining (G', H') is (isomorphic to) G_{k-1}, H_{k-1} and $C' := C \setminus \{a_1 b_1, a_1 a_2\}$ is a multicut in it. By the induction hypothesis, the minimum multicut of (G', H') is of size $k - 2$.

Now either both $a_2 b_2 \in C'$ and $a_2 a_3 \in C'$ in which case C' is not an inclusion-wise minimal multicut for (G', H'), since deleting one of $a_1 b_1, a_1 a_2$ we already disconnect the same terminal pairs. So in this case $|C'| > k - 2$, and we are done; or one of $a_2 b_2, a_2 a_3$ is not in C', but then one of $a_1 b_1, a_1 a_2$ must be in it, since otherwise b_1 is not disconnected by C from b_2 or from a_3. So in this case $|C| \geq |C'| + 1 \geq k - 1$, finishing the proof of the first assertion.

For the maximum multiflow value μ, or maximum half-integer and integer multiflow values μ_{half}, μ_{int}, note that the supply edges form a tree, so the set $\mathcal{P}$ of paths between the endpoints of demand edges contains exactly one path P_e for each demand edge e, so $|\mathcal{P}| = 2k - 3$. Defining $f(P_{b_{k-1} b_k}) = 2/3$ and $f(P) = 1/3$ for each other path $P \in \mathcal{P}$, we have $|f| = 2(k-1)/3$. To prove that this is a maximum flow, note that $e_1 := a_1 b_1$ is contained in $P \in \mathcal{P}$ if and only if $e_2 := a_2 b_2$ is contained in it, so for any multiflow f,

$$\alpha := \alpha_f := \sum_{\{P \in \mathcal{P}: e_1 \in P\}} f(P) = \sum_{\{P \in \mathcal{P}: e_2 \in P\}} f(P)$$

Claim: $\mu \leq 2(k-1)/3$, and if $\alpha \leq 2/3$, then $\mu \leq 2(k-2)/3 + \alpha$.

Indeed, on the one hand, $\alpha = 2/3 + \varepsilon$ $(0 \leq \varepsilon \leq 1/3)$ causes $\alpha' \leq 2/3 - \varepsilon$ in (G', H') isomorphic to (G_{k-1}, H_{k-1}) with max flow value μ', after the deletion of $\{a_1, b_1\}$, and then by induction, $\mu \leq 2/3 + \varepsilon + \mu' \leq 2/3 + \varepsilon + 2(k-2)/3 - \varepsilon = 2(k-1)/3$. On the other hand, if $\alpha = 2/3 - \varepsilon$ we have by induction $\mu \leq 2/3 - \varepsilon + \mu' \leq 2/3 - \varepsilon + 2(k-2)/3 = 2(k-1)/3 - \varepsilon$, and the claim is proved, finishing the proof of the second statement.

The proof of the third assertion, concerning μ_{half} works similarly. Defining $f(P_{b_i b_{i+1}}) = 1/2$ $(i = 1, \ldots, k-1)$, and $f(P_{b_1 a_3}) = 1/2$, otherwise $f(P) = 0$, where f is a feasible half-integer flow, $|f| = k/2$. To prove $\mu_{\text{half}} \leq k/2$ an inductive proof using α works as before: for $\alpha \geq 1/2$ the stronger statement $\mu \leq (k-2)/2 + \alpha$ follows by induction.

Finally, the fourth assertion is immediate from the third one: $\mu_{\text{int}} \leq \mu_{\text{half}} \leq k/2$ is integer, and an integer flow f, $|f| = \lfloor k/2 \rfloor$ is also easy to construct.   □

7   Conclusions

This paper established bounds on the integrality gap of multiflows, developed approximation algorithms for them and bounded the flow-cut gap. Applying Theorem 1 first, then Theorem 2 to a maximum multiflow, and finally Theorem 4 to the result, we arrive at the following summary:

Theorem 6. *There exists a 1/4-approximation algorithm for integer plane multiflow maximisation, with an integer flow-cut gap of 8; there exists a 1/2-approximation algorithm for half-integer plane multiflows with a half-integer flow-cut gap of 4; the flow-cut gap is at most 2; the approximation algorithms provide lower bounds to the integrality or half-integrality gap equal to the approximation ratio.*

a. Multicuts and Flow-Cut Gap: The minimum multicut is NP-hard to find, but has a PTAS by Lemma 4 and [11]. The half-integer flow-cut gap is at least 2 by the example illustrated in Fig. 1 (and Theorem 5) and it is at most 4 by Theorem 6. The (fractional) flow-cut gap is in the interval $[3/2, 2]$, again by the same example and the theorem. For the integer flow-cut gap, a lower bound of 2 is shown by $G = K_4$ and H consisting of two demand edges forming a matching. The true value is thus wide open in the interval $[2, 8]$.

b. Half-Integrality Gap: We do not know the complexity of finding a half-integer multiflow of maximum value. The worst case ratio between the maximum value of a half-integer feasible flow and of a fractional flow is in the interval $[1/2, 3/4]$, again by Theorems 2 and 5. It remains an interesting open problem to pin down the exact half-integrality gap in this interval.

c. Integrality Gap: This lies in the interval $[1/4, 1/2]$, with lower bound given by Theorem 6, and upper bound by $G = K_4$ and H consisting of two demand edges forming a matching. Finally, the worst integer flow/half-integer flow ratio is closed: it is exactly $1/2$, as the same example of K_4 and Theorem 4 together show.

References

1. Appel, K., Haken, W.: A proof of the four color theorem. Discrete Math. **16**, 179–180 (1976)
2. Baker, B.S.: Approximation algorithms for np-complete problems on planar graphs. J. ACM **41**(1), 153–180 (1994)

3. Cheriyan, J., Karloff, H., Khandekar, R., Könemann, J.: On the integrality ratio for tree augmentation. Oper. Res. Lett. **36**(4), 399–401 (2008)
4. Edmonds, J., Rick, G.: A min-max relation for submodular functions in graphs. Ann. Discrete Math. **I**, 185–204 (1977)
5. Fiorini, S., Hardy, N., Reed, B., Vetta, A.: Approximate min-max relations for odd cycles in planar graphs. Math. Program. **110**(1), 71–91 (1995)
6. Frank, A., Szigeti, Z.: A note on packing paths in planar graphs. Math. Program. **70**, 201–209 (1995)
7. Garg, N., Vazirani, V.V., Yannakakis, M.: Approximate max-flow min-(multi) cut theorems and their applications. SIAM J. Comput. **25**(2), 235–251 (1996)
8. Garg, N., Vazirani, V.V., Yannakakis, M.: Primal-dual approximation algorithms for integral flow and multicut in trees. Algorithmica **18**(1), 3–20 (1997)
9. Hoffman, A.J., Kruskal, J.B.: Integral boundary points of convex polyhedra. In: Linear Inequalities and Related Systems, pp. 223–246 (1956)
10. Klein, P., Plotkin, S.A., Rao, S.: Excluded minors, network decomposition, and multicommodity flow. In: Proceedings of the Twenty-Fifth Annual ACM Symposium on Theory of Computing, pp. 682–690. ACM (1993)
11. Klein, P.N., Mathieu, C., Zhou, H.: Correlation clustering and two-edge-connected augmentation for planar graphs. In: 32nd International Symposium on Theoretical Aspects of Computer Science (STACS 2015). Schloss Dagstuhl-Leibniz-Zentrum fuer Informatik (2015)
12. Korach, E., Penn, M.: Tight integral duality gap in the chinese postman problem. Mateh. Program. **55**, 183–191 (1992)
13. Korte, B., Vygen, J.: Combinatorial Optimization: Theory and Algorithms, Collection Algorithms and Combinatorics, vol. 21, 5th edn. Springer, Heidelberg (2012). https://doi.org/10.1007/978-3-642-24488-9
14. Král, D., Voss, H.: Edge-disjoint odd cycles in planar graphs. J. Comb. Theor. Ser. B **90**, 107–120 (2004)
15. Lovász, L., Plummer, M.: Matching Theory. Akadémiai Kiadó and Springer (1986)
16. Middendorf, M., Pfeiffer, F.: On the complexity of the disjoint paths problem. Combinatorica **13**(1), 97–107 (1993)
17. Okamura, H., Seymour, P.D.: Multicommodity flows in planar graphs. J. Comb. Theor. Ser. B **31**(1), 75–81 (1981)
18. Robertson, N., Sanders, D.P., Seymour, P., Thomas, R.: Efficiently four-coloring planar graphs. In: Proceedings of the Twenty-Eighth Annual ACM Symposium on Theory of Computing, pp. 571–575 (1996)
19. Schrijver, A.: Theory of Linear and Integer Programming. Wiley and Chichester (1986)
20. Schrijver, A.: Combinatorial Optimization: Polyhedra and Efficiency, vol. 24. Springer, Heidelberg (2003)
21. Seguin-Charbonneau, L., Bruce Shepherd, F.: Maximum edge-disjoint paths in planar graphs with congestion 2. In: 2011 IEEE 52nd Annual Symposium on Foundations of Computer Science, pp. 200–209. IEEE (2011)
22. Seymour, P.D.: On odd cuts and plane multicommodity flows. Proc. London Math. Soc. **3**(1), 178–192 (1981)
23. Tardos, É., Vazirani, V.V.: Improved bounds for the max-flow min-multicut ratio for planar and k_r, r-free graphs. Inf. Process. Lett. **47**(2), 77–80 (1993)
24. Tutte, W.T.: Lectures on matroids. J. Res. Nat. Bur. Stan. B **69**, 1–47 (1965)
25. Williamson, D.P., Goemans, M.X., Mihail, M., Vazirani, V.V.: A primal-dual approximation algorithm for generalized steiner network problems. Combinatorica **15**(3), 435–454 (1995)

Stochastic Makespan Minimization in Structured Set Systems (Extended Abstract)

Anupam Gupta[1], Amit Kumar[2], Viswanath Nagarajan[3(✉)], and Xiangkun Shen[4]

[1] Carnegie Mellon University, Pittsburgh, USA
[2] Indian Institute of Technology, Delhi, Delhi, India
[3] University of Michigan, Ann Arbor, USA
viswa@umich.edu
[4] Yahoo! Research, New York City, USA

Abstract. We study stochastic combinatorial optimization problems where the objective is to minimize the expected maximum load (a.k.a. the makespan). In this framework, we have a set of n tasks and m resources, where each task j uses some subset of the resources. Tasks have random sizes X_j, and our goal is to non-adaptively select t tasks to minimize the expected maximum load over all resources, where the load on any resource i is the total size of all selected tasks that use i. For example, given a set of intervals in time, with each interval j having random load X_j, how do we choose t intervals to minimize the expected maximum load at any time? Our technique is also applicable to other problems with some geometric structure in the relation between tasks and resources; e.g., packing paths, rectangles, and "fat" objects. Specifically, we give an $O(\log \log m)$-approximation algorithm for all these problems.

Our approach uses a strong LP relaxation using the cumulant generating functions of the random variables. We also show an LP integrality gap of $\Omega(\log^* m)$ even for the problem of selecting intervals on a line.

1 Introduction

Consider the following task scheduling problem: an event center receives requests/tasks from its clients. Each task j specifies a start and end time (denoted (a_j, b_j)), and the amount x_j of some shared resource (e.g., staff support) that this task requires throughout its duration. The goal is to accept some target t number of tasks so that the maximum resource-utilization over time is as small as possible. Concretely, we want to choose a set S of tasks with $|S| = t$ to minimize

$$\underbrace{\max_{\text{times } \tau} \sum_{j \in S : \tau \in [a_j, b_j]} x_j}_{\text{usage at time } \tau}.$$

All missing proofs and details can be found in the full version [11].

© Springer Nature Switzerland AG 2020
D. Bienstock and G. Zambelli (Eds.): IPCO 2020, LNCS 12125, pp. 158–170, 2020.
https://doi.org/10.1007/978-3-030-45771-6_13

This can be modeled as an interval packing problem: if the sizes are identical, the natural LP is totally unimodular and we get an optimal algorithm. For general sizes, there is a constant-factor approximation algorithm [3].

However, in many settings, we may not know the resource consumption X_j precisely up-front, at the time we need to make a decision. Instead, we may be only given estimates. What if the requirement X_j is a random variable whose distribution is given to us? Again we want to choose S of size t, but this time we want to minimize the *expected* maximum usage:

$$\mathbb{E}\left[\max_{\text{times } \tau} \sum_{j \in S : \tau \in [a_j, b_j]} X_j\right].$$

Note that our decision to pick task j affects all times in $[a_j, b_j]$, and hence the loads on various places are no longer independent: how can we effectively reason about such a problem?

In this paper we consider general resource allocation problems of the following form. There are several tasks and resources, where each task j has some size X_j and uses some subset U_j of resources. That is, if task j is selected then it induces a load of X_j on every resource in U_j. Given a target t, we want to select a subset S of t tasks to minimize the *expected maximum load* over all resources. For the non-stochastic versions of these problems (when X_j is a single value and not a random variable), we can use the natural linear programming (LP) relaxation and randomized rounding to get an $O(\frac{\log m}{\log \log m})$-approximation algorithm; here m is the number of resources. However, much better results are known when the task-resource incidence matrix has some geometric structure. One such example appeared above: when the resources have some linear structure, and the tasks are intervals. Other examples include selecting rectangles in a plane (where tasks are rectangles and resources are points in the plane), and selecting paths in a tree (tasks are paths and resources are edges/vertices in the tree). This class of problems has received a lot of attention and has strong approximation guarantees, see e.g. [1–8].

However, the *stochastic* counterparts of these resource allocation problems remain wide open. Can we achieve good approximation algorithms when the task sizes X_j are random variables? We refer to this class of problems as stochastic makespan minimization (GENMAKESPAN). In the rest of this work, we assume that the distributions of all the random variables are known, and that the r.v.s X_js are independent.

1.1 Results and Techniques

We show that good approximation algorithms are indeed possible for GEN-MAKESPAN problems that have certain geometric structure. We consider the following two assumptions:

- *Deterministic problem assumption:* There is a linear-program based α approximation algorithm for a suitable deterministic variant of GENMAKESPAN.

– *Well-covered assumption:* for any subset $D \subseteq [m]$ of resources and tasks $L(D)$ incident to D, the tasks in $L(D)$ incident to any resource $i \in [m]$ are "covered" by at most λ resources in D.

These assumptions are formalized in Sect. 2. To give some intuition for these assumptions, consider intervals on the line. The first assumption holds by the results of [3]. The second assumption holds because each resource is some time τ, and the tasks using time τ can be covered by two resources in D, namely times $\tau_1, \tau_2 \in D$ such that $\tau_1 \leq \tau \leq \tau_2$.

Theorem 1 (Main (Informal)). *There is an $O(\alpha\lambda \log\log m)$-approximation algorithm for stochastic makespan minimization (GENMAKESPAN), with α and λ as in the above assumptions.*

We also show that both α and λ are constant in a number of geometric settings: for intervals on a line, for paths in a tree, for rectangles in a plane and for "fat" objects (such as disks) in a plane. Therefore, we obtain an $O(\log\log m)$-approximation algorithm in all these cases.

A first naive approach for GENMAKESPAN is (i) to write an LP relaxation with expected sizes $\mathbb{E}[X_j]$ as deterministic sizes and then (ii) to use any LP-based α-approximation algorithm for the deterministic problem. However, this approach only yields an $O(\alpha \frac{\log m}{\log\log m})$ approximation ratio, due to the use of union bounds in calculating the expected maximum. Our idea is to use the structure of the problem to improve the approximation ratio.

Our approach is as follows. First, we use the (scaled) logarithmic moment generating function (log-mgf) of the random variables X_j to define deterministic surrogates to the random sizes. Second, we formulate a strong LP relaxation with an exponential number of "volume" constraints that use the log-mgf values. These two ideas were used earlier for stochastic makespan minimization in settings where each task loads a single resource [10,14]. In the example above, where each task uses only a single time instant. However, we need a more sophisticated LP for GENMAKESPAN to be able to handle the combinatorial structure when tasks use many resources. Despite the large number of constraints, this LP can be solved approximately in polynomial time, using the ellipsoid method and using a maximum-coverage algorithm as the separation oracle. Third (and most important), we provide an iterative-rounding algorithm that partitions the tasks/resources into $O(\log\log m)$ many nearly-disjoint instances of the deterministic problem. The analysis of our rounding algorithm relies on both the assumptions above, and also on the volume constraints in our LP and on properties of the log-mgf.

We also show some limitations of our approach. For GENMAKESPAN involving intervals in a line (which is our simplest application), we prove that the integrality gap of our LP is $\Omega(\log^* m)$. This rules out a constant-factor approximation via this LP. For GENMAKESPAN on more general set-systems (without any structure), we prove that the integrality gap can be $\Omega(\frac{\log m}{(\log\log m)^2})$ even if all deterministic instances solved in our algorithm have an $\alpha = O(1)$ integrality gap. This suggests that we do need to exploit additional structure—such as the well-covered assumption above—in order to obtain significantly better approximation ratios via our LP.

1.2 Related Work

The deterministic counterparts of the problems studied here are well-understood. In particular, there are LP-based $O(1)$-approximation algorithms for intervals in a line [3], paths in a tree (with edge loads) [8] and rectangles in a plane (under some restrictions) [1].

Our techniques draw on prior work on stochastic makespan minimization for identical [14] and unrelated [10] resources; but there are also important new ideas. In particular, the use of log-mgf values as the deterministic proxy for random variables comes from [14] and the use of log-mgf values at multiple scales comes from [10]. The "volume" constraints in our LP also has some similarity to those in [10]: however, a key difference here is that the random variables loading different resources are correlated (whereas they were independent in [10]). Indeed, this is why our LP can only be solved approximately whereas the LP relaxation in [10] was optimally solvable. We emphasize that our main contribution is the rounding algorithm, which uses a new set of ideas; these lead to the $O(\log\log m)$ approximation bound, whereas the rounding in [10] obtained a constant-factor approximation. Note that we also prove a super-constant integrality gap in our setting, even for the case of intervals in a line.

The stochastic load balancing problem on unrelated resources has also been studied for general ℓ_p-norms (note that the makespan corresponds to the ℓ_∞-norm) and a constant-factor approximation is known [15]. We do not consider ℓ_p-norms in this paper.

2 Problem Definition and Preliminaries

We are given n tasks and m resources. Each task $j \in [n]$ uses some subset $U_j \subseteq [m]$ of resources. For each resource $i \in [m]$, define $L_i \subseteq [n]$ to be the tasks that utilize i. Each task $j \in [n]$ has a *random size* X_j. If a task j is selected into our set S, it adds a load of X_j to each resource in U_j: the load on resource $i \in [m]$ is $Z_i := \sum_{j \in S \cap L_i} X_j$. The makespan is the maximum load, i.e. $\max_{i=1}^m Z_i$. The goal is to select a subset $S \subseteq [n]$ with t tasks to minimize the expected *makespan*:

$$\min_{S \subseteq [n]:|S|=t} \quad \mathbb{E}\left[\max_{i=1}^m \sum_{j \in S \cap L_i} X_j\right]. \tag{1}$$

The distribution of each r.v. X_j is known (we use this knowledge only to compute some "effective" sizes below), and these distributions are independent.

For any subset $K \subseteq [m]$ of resources, let $L(K) := \cup_{i \in K} L_i$ be the set of tasks that utilize at least one resource in K.

2.1   Structure of Set Systems: The Two Assumptions

Our results hold when the following two properties are satisfied by the set system $([n], \mathcal{L})$, where $\mathcal{L}$ is the collection of sets L_i for each $i \in [m]$. Note that the set system has n elements (corresponding to tasks) and m sets (corresponding to resources).

A1 (α-packable): A set system $([n], \mathcal{L})$ is said to be α-*packable* if for any assignment of size $s_j \geq 0$ and reward $r_j \geq 0$ to each element $j \in [n]$, and any threshold parameter $\theta \geq \max_j s_j$, there is a polynomial-time algorithm that rounds a fractional solution y to the following LP relaxation into an integral solution $\widehat{y}$, losing a factor of at most $\alpha \geq 1$. (I.e., $\sum_j r_j \widehat{y}_j \geq \frac{1}{\alpha} \sum_j r_j y_j$.)

$$\max \left\{ \sum_{j \in [n]} r_j \cdot y_j : \sum_{j \in L} s_j \cdot y_j \leq \theta, \ \forall L \in \mathcal{L}, \ \text{and} \ 0 \leq y_j \leq 1, \ \forall j \in [n] \right\}. \quad (2)$$

We also assume that the support of $\widehat{y}$ is contained in the support of y.[1]

A2 (λ-safe): Let $[m]$ be the indices of the sets in $\mathcal{L}$; recall that these are the resources. The set system $([n], \mathcal{L})$ is λ-*safe* if for every subset $D \subseteq [m]$ of ("dangerous") resources, there exists a subset $M \supseteq D$ of ("safe") resources, such that (a) $|M|$ is polynomially bounded by $|D|$ and moreover, (b) for every $i \in [m]$, there is a subset $R_i \subseteq M$, $|R_i| \leq \lambda$, such that $L_i \cap L(D) \subseteq L(R_i)$. Recall that $L(D) = \cup_{h \in D} L_h$. We denote the set M as $\texttt{Extend}(D)$.

Let us give an example. Suppose P consists of m points on the real line, and consider n intervals $I_1, \ldots, I_n$ on the line. The set system is defined on n elements (one for each interval), with m sets with the set L_i for point $i \in P$ containing the indices of intervals that contain i. The λ-safe condition says that for any subset D of points in P, we can find a superset M which is not much larger such that for any point $i \in P$, there are λ points in M containing all the intervals that pass through both i and D. In other words, if these intervals contribute any load to i and D, they also contribute to one of these λ points. And indeed, choosing $M = D$ ensures that $\lambda = 2$: for any i we choose the nearest points in M on either side of i.

Other families that are α-packable and λ-safe include:

- Each element in $[n]$ corresponds to a path in a tree, with the set L_i being the subset of paths through node i.
- Elements in $[n]$ correspond to rectangles or disks in a plane, and each L_i consists of rectangles/disks containing a particular point i in the plane.

For a subset $X \subseteq [n]$, the projection of $([n], \mathcal{L})$ to X is the smaller set system $([n], \mathcal{L}|_X)$, where $\mathcal{L}|_X = \{L \cap X \mid L \in \mathcal{L}\}$. The following lemma formalizes that packability and safeness properties also hold for sub-families and disjoint unions.

---

[1] The support of vector $z \in \mathbb{R}_+^n$ is $\{j \in [n] : z_j > 0\}$ which corresponds to its positive entries.

Lemma 1. *Consider a set system* $([n], \mathcal{L})$ *that is α-packable and λ-safe. Then,*

(i) for all $X \subseteq [n]$, the set system $(X, \mathcal{L})$ is α-packable and λ-safe, and

(ii) given a partition $X_1, \ldots, X_s$ of $[n]$, and set systems $(X_1, \mathcal{L}_1), \ldots, (X_s, \mathcal{L}_s)$, where $\mathcal{L}_i = \mathcal{L}|_{X_i}$, the disjoint union of these systems is also α-packable.

We consider the GENMAKESPAN problem for settings where the set system $([n], \{L_i\}_{i \in [m]})$ is α-packable and λ-safe for small parameters α and λ.

Theorem 2 (Main Result). *For any instance of* GENMAKESPAN *where the corresponding set system* $([n], \{L_i\}_{i \in [m]})$ *is α-packable and λ-safe, there is an* $O(\alpha\lambda \cdot \log\log m)$-*approximation algorithm.*

2.2 Effective Size and Random Variables

In all the arguments that follow, imagine that we have scaled the instance so that the optimal expected makespan is between $\frac{1}{2}$ and 1. It is useful to split each random variable X_j into two parts:

- the *truncated* random variable $X'_j := X_j \cdot \mathbf{I}_{(X_j \leq 1)}$, and
- the *exceptional* random variable $X''_j := X_j \cdot \mathbf{I}_{(X_j > 1)}$.

These two kinds of random variables behave very differently with respect to the expected makespan. Indeed, the expectation is a good measure of the load due to exceptional r.v.s, whereas one needs a more nuanced notion for truncated r.v.s (as we discuss below). The following result was shown in [14]:

Lemma 2 (Exceptional Items Lower Bound). *Let* $\{X''_j\}$ *be non-negative discrete random variables each taking value zero or at least L. If $\sum_j \mathbb{E}[X''_j] \geq L$ then $\mathbb{E}[\max_j X''_j] \geq L/2$.*

We now consider the trickier case of truncated random variables X'_j. We want to find a deterministic quantity that is a good surrogate for each random variable, and then use this deterministic surrogate instead of the actual random variable. For stochastic load balancing, a useful surrogate is the *effective size*, which is based on the logarithm of the (exponential) moment generating function (also known as the cumulant generating function) [9,10,12,13].

Definition 1 (Effective Size). *For any r.v. X and integer $k \geq 2$, define*

$$\beta_k(X) := \frac{1}{\log k} \cdot \log \mathbb{E}\left[e^{(\log k) \cdot X}\right]. \tag{3}$$

Also define $\beta_1(X) := \mathbb{E}[X]$.

To see the intuition for the effective size, consider a set of independent r.v.s $Y_1, \ldots, Y_k$ all assigned to the same resource. The following lemma, whose proof is very reminiscent of the standard Chernoff bound (see [12]), says that the load is not much higher than the expectation.

Lemma 3 (Effective Size: Upper Bound). *For indep. r.v.s $Y_1, \ldots, Y_n$, if $\sum_i \beta_k(Y_i) \leq b$ then $\Pr[\sum_i Y_i \geq c] \leq \frac{1}{k^{c-b}}$.*

The usefulness of the effective size comes from a partial converse [14]:

Lemma 4 (Effective Size: Lower Bound). *Let $X_1, X_2, \cdots X_n$ be independent $[0,1]$ r.v.s, and $\{\widetilde{L_i}\}_{i=1}^m$ be a partition of $[n]$. If $\sum_{j=1}^n \beta_m(X_j) \geq 17m$ then*

$$\mathbb{E}\left[\max_{i=1}^m \sum_{j \in \widetilde{L_i}} X_j \right] = \Omega(1).$$

3   The General Framework

In this section we provide our main algorithm, which is used to prove Theorem 2. The idea is to write a suitable LP relaxation for the problem (using the effective sizes as deterministic surrogates for the stochastic jobs), to solve this exponentially-sized LP, and then to round the solution. The novelty of the solution is both in the LP itself, and in the rounding, which is based on a delicate decomposition of the instance into $O(\log \log m)$ many sub-instances and on showing that, loosely speaking, the load due to each sub-instance is at most $O(\alpha\lambda)$. By binary-searching on the value of the optimal makespan, and rescaling, we can assume that the optimal makespan is between $\frac{1}{2}$ and 1.

The LP Relaxation. Consider an instance $\mathcal{I}$ of GenMakespan given by a set of n tasks and m resources, with sets U_j and L_i as described in Sect. 2. We give an LP relaxation which is feasible if the optimal makespan is at most one.

Lemma 5. *Consider any feasible solution to $\mathcal{I}$ that selects a subset $S \subseteq [n]$ of tasks. If the expected maximum load $\mathbb{E}\left[\max_{i=1}^m \sum_{j \in L_i \cap S} X_j \right] \leq 1$, then*

$$\sum_{j \in S} \mathbb{E}[X_j''] \leq 2, \qquad and \tag{4}$$

$$\sum_{j \in L(K) \cap S} \beta_k(X_j') \leq b \cdot k, \text{ for all } K \subseteq [m], \quad where\ k = |K|, \tag{5}$$

for b being a large enough but fixed constant.

Lemma 5 allows us to write the following feasibility linear programming relaxation for GenMakespan (assuming the optimal value is 1). For every task j, we have a binary variable y_j, which is meant to be 1 if j is selected in the solution. Moreover, we can drop all tasks j with $c_j = \mathbb{E}[X_j''] > 2$ as such a task would never be part of an optimal solution- by (4). So in the rest of this paper we will assume that $\max_{j \in [n]} c_j \leq 2$. Further, note that we only use effective sizes β_k of *truncated* r.v.s, so we have $0 \leq \beta_k(X_j') \leq 1$ for all $k \in [m]$ and $j \in [n]$.

$$\sum_{j=1}^{n} y_j \geq t \tag{6}$$

$$\sum_{j=1}^{n} \mathbb{E}[X_j''] \cdot y_j \leq 2 \tag{7}$$

$$\sum_{j \in L(K)} \beta_k(X_j') \cdot y_j \leq b \cdot k \qquad \forall K \subseteq [m] \text{ with } |K| = k, \ \forall k = 1, 2, \cdots m, \tag{8}$$

$$0 \leq y_j \leq 1 \qquad \forall j \in [n]. \tag{9}$$

In the above LP, $b \geq 1$ denotes the universal constant multiplying k in the right-hand-side of (5). This can be solved approximately in polynomial time.

Theorem 3 (Solving the LP). *There is a polynomial time algorithm which given an instance $\mathcal{I}$ of* GENMAKESPAN *outputs one of the following:*

- *a solution $y \in \mathbb{R}^n$ to LP (6)–(9), except that the RHS of (8) is replaced by $\frac{e}{e-1}bk$, or*
- *a certificate that LP (6)–(9) is infeasible.*

The $\frac{e}{e-1}$ factor comes from checking feasibility via an approximation algorithm for the maximum coverage problem. In the rest of this section, we assume we have a feasible solution y to (6)–(9) since the approximate feasibility of (8) only affects the approximation ratio by a constant.

Rounding. We first give some intuition about the rounding algorithm. It involves formulating $O(\log \log m)$ many almost-disjoint instances of the deterministic reward-maximization problem (2) used in the definition of α-packability. The key aspect of each such deterministic instance is the definition of the sizes s_j: for the ℓth instance we use effective sizes $\beta_k(X_j')$ with parameter $k = 2^{2^\ell}$. We use the λ-safety property to construct these deterministic instances and the α-packable property to solve them. Finally, we show that the expected makespan induced by the selected tasks is at most $O(\alpha\beta)$ factor away from each such deterministic instance, which leads to an overall $O(\alpha\beta \log \log m)$ ratio.

Before delving into the details, let us formulate a generalization of the reward-maximization problem mentioned in (2), which we call the DETCOST problem. An instance $\mathcal{I}$ of the DETCOST problem consists of a set system $([n], \mathcal{S})$, with a size s_j and cost c_j for each element $j \in [n]$. It also has parameters $\theta \geq \max_j s_j$ and $\psi \geq \max_j c_j$. The goal is to find a maximum cardinality subset V of $[n]$ such that each set in $\mathcal{S}$ is "loaded" to at most θ, and the total cost of V is at most ψ. The DETCOST problem has the following LP relaxation:

$$\max\left\{ \sum_{j \in [n]} y_j : \sum_{j \in S} s_j \cdot y_j \leq \theta, \ \forall S \in \mathcal{S}; \ \sum_{j \in [n]} c_j \cdot y_j \leq \psi; \ 0 \leq y_j \leq 1 \ \forall j \in [n] \right\}. \tag{10}$$

Theorem 4. *If a set system satisfies the α-packable property, there is an $O(\alpha)$-approximation algorithm for* DETCOST *relative to the LP relaxation* (10).

We now give the rounding algorithm for the GENMAKESPAN problem. The procedure is described formally in Algorithm 1. The algorithm proceeds in $\log \log m$ iterations of the **for** loop in lines 3–7, since the parameter k is squared in line 3 for each iteration. In line 5, we identify resources i which are fractionally loaded to more than $2b$, where the load is measured in terms of $\beta_{k^2}(X'_j)$ values. The set of such resources is grouped in the set D_ℓ, and we define J_ℓ to be the tasks which can load these resources. Ideally, we would like to remove these resources and tasks, and iterate on the remaining tasks and resources. However, the problem is that tasks in J_ℓ also load resources other than D_ℓ, and so (D_ℓ, J_ℓ) is not independent of the rest of the instance. This is where we use the λ-safe property: we expand D_ℓ to a larger set of resources M_ℓ, which will be used to show that the effect of J_ℓ on resources outside D_ℓ will not be significant.

Algorithm 1: Rounding Algorithm

   Input : A fractional solution y to (6)–(9)
   Output: A subset of tasks.
1 Initialize remaining tasks $J \leftarrow [n]$;
2 **for** $\ell = 0, 1, \ldots, \log \log m$ **do**
3 │ Set $k \leftarrow 2^{2^\ell}$;
4 │ Initialize class-ℓ resources $D_\ell \leftarrow \emptyset$;
5 │ **while** *there is a resource* $i \in [m] : \sum_{j \in L_i \cap J} \beta_{k^2}(X'_j) \cdot y_j > 2b$ **do**
6 │ │ update $D_\ell \leftarrow D_\ell \cup \{i\}$;
7 │ └ Set $\widetilde{L_i} \leftarrow J \cap L_i$ and $J \leftarrow J \setminus \widetilde{L_i}$;
8 │ Define the class-ℓ tasks $J_\ell \leftarrow \bigcup_{i \in D_\ell} \widetilde{L_i}$;
9 │ Use λ-safety on the set system $(J_\ell, \{L_i \cap J_\ell\}_{i \in [m]})$ to get
   └ $M_\ell := \text{Extend}(D_\ell)$;
10 $\rho \leftarrow 1 + \log \log m$;
11 Define class-ρ tasks $J_\rho = J$ and class-ρ resources $M_\rho := D_\rho = [m] \setminus \left(\cup_{\ell=0}^{\rho-1} D_\ell\right)$;
12 Define an instance $\mathcal{C}$ of DETCOST as follows: the set system is the disjoint union of the set systems (J_ℓ, M_ℓ) for $\ell = 0, \ldots, \rho$. The other parameters are:

$$\text{Sizes } s_j = \beta_{2^{2^\ell}}(X'_j) \text{ for each } j \in J_\ell, \ \forall 0 \le \ell \le \rho, \text{ bound } \theta = 2\bar{\alpha}b,$$

$$\text{Costs } c_j = \mathbb{E}[X''_j] \text{ for each } j \in [n], \text{ bound } \psi = 2\bar{\alpha},$$

   where $\bar{\alpha}$ is the approximation ratio from Theorem 4 ;
13 Let $N_H = \{j \in [n] : y_j > 1/\bar{\alpha}\}$;
14 Let $\bar{y}_j = \bar{\alpha} \cdot y_j$ for $j \in [n] \setminus N_H$ and $\bar{y}_j = 0$ otherwise ;
15 Round $\bar{y}$ (as a feasible solution to (10)) using Theorem 4 to obtain N_L;
16 Output $N_H \cup N_L$.

Consider any iteration ℓ of the for-loop. We apply the λ-safety property to the set-system $(J_\ell, \{L_i \cap J_\ell\}_{i \in [m]})$ and set D_ℓ to get $M_\ell := \text{Extend}(D_\ell)$. We remove J_ℓ from the current set J of tasks, and continue to the next iteration. We abuse notation by referring to (J_ℓ, M_ℓ) as the following set system: each set is of the form $L_i \cap J_\ell$ for some $i \in M_\ell$. Having partitioned the tasks into classes $J_1, \ldots, J_\rho$, we consider the disjoint union $\mathcal{D}$ of the set systems (J_ℓ, M_ℓ), for $\ell = 1, \ldots, \rho$. While the sets D_ℓ are disjoint, the sets M_ℓ may not be disjoint. For each resource appearing in the sets M_ℓ of multiple classes, we make distinct copies in the combined set-system $\mathcal{D}$.

Finally, we set up an instance $\mathcal{C}$ of DETCOST (in line 11): the set system is the disjoint union of (J_ℓ, M_ℓ), for $\ell = 1, \ldots, \rho$. Every task $j \in J_\ell$ has size $\beta_{2^{2\ell}}(X'_j)$ and cost $\mathbb{E}[X''_j]$. The parameters θ and ψ are as mentioned in line 11. Our proofs show that the solution $\bar{y}$ defined in line 14 is a feasible solution to the LP relaxation (10) for $\mathcal{C}$. This allows us to use Theorem 4 to round $\bar{y}$ to an integral solution N_L. Finally, we output $N_H \cup N_L$, where N_H is defined in line 12.

The analysis is outlined in the appendix.

4 Applications

As discussed in Sect. 2, the problem of selecting intervals in a line satisfies the λ-safe property with $\lambda = 2$. Moreover, the α-packable property holds with $\alpha = O(1)$, which follows from [3]—indeed, the LP relaxation (2) corresponds to the unsplittable flow problem where all vertices have uniform capacity θ. So, Theorem 2 now implies the following.

Corollary 1. *There is an $O(\log \log m)$-approximation for GENMAKESPAN where resources are vertices on a line and tasks are intervals in this line.*

The full version [11] has a number of other applications:

Corollary 2. *There is an $O(\log \log m)$-approximation for GENMAKESPAN when*

- *resources are vertices in a tree and tasks are paths in this tree.*
- *resources are all points in the plane and tasks are rectangles, where the rectangles in a solution can be shrunk by a $(1-\delta)$-factor in either dimension; $\delta > 0$ is some constant.*
- *resources are all points in the plane and tasks are disks.*

5 Integrality Gap Lower Bounds

We consider two natural questions – (i) does one require any assumption on the underlying set system to obtain $O(1)$-approximation for GENMAKESPAN?, and (ii) what is the integrality gap of the LP relaxation given by the constraints (6)–(9) for settings where α and λ are constants? For the first question, we show that applying our LP based approach to general set systems only givesn an $\Omega\left(\frac{\log m}{(\log \log m)^2}\right)$ approximation ratio, and so we do require some conditions on the underlying set system. For the second question, we show that even for set systems given by intervals on a line (as in Sect. 4), the integrality gap of our LP relaxation is $\Omega(\log^* m)$. Hence this rules out getting a constant-factor approximation using our approach even when α and λ are constants.

A   Analysis Outline

We now show that the expected makespan for the solution produced by the rounding algorithm above is $O(\alpha\lambda\rho)$, where $\rho = \log\log m$ is the number of classes. In particular, we show that the expected makespan (taken over all resources) due to tasks of each class ℓ is $O(\alpha\lambda)$.

Using the terminating condition in line 5, we can show:

Lemma 6. *For any class ℓ, $0 \leq \ell \leq \rho$, and resource $i \in [m]$,*

$$\sum_{j \in J_\ell \cap L_i} \beta_r(X'_j) \cdot y_j \leq 2b, \qquad \text{where } r = 2^{2^\ell}.$$

Next, the sets D_ℓ cannot become too large (as a function of ℓ).

Lemma 7. *For any ℓ, $0 \leq \ell \leq \rho$, $|D_\ell| \leq k^2$, where $k = 2^{2^\ell}$. So $|M_\ell| \leq k^p$ for some constant p.*

Proof. The claim is trivial for the last class $\ell = \rho$ as $k \geq m$ in this case. Now consider any class $\ell < \rho$. For each $i \in D_\ell$, we know $\sum_{j \in \widetilde{L_i}} \beta_{k^2}(X'_j) \cdot y_j > 2b$, where $\widetilde{L_i}$ is as defined in line 7. Moreover, the subsets $\{\widetilde{L_i} : i \in D_\ell\}$ are disjoint as the set J gets updated (in line 7) after adding each $i \in D_\ell$. Suppose, for the sake of contradiction, that $|D_\ell| > k^2$. Then let $K \subseteq D_\ell$ be any set of size k^2. By the LP constraint (8) on this subset K,

$$2b \cdot k^2 < \sum_{i \in K} \sum_{j \in \widetilde{L_i}} \beta_{k^2}(X'_j) \cdot y_j \leq \sum_{j \in L(K)} \beta_{k^2}(X'_j) \cdot y_j \leq b|K| = b \cdot k^2,$$

which is a contradiction. This proves the first part of the claim. Finally, the λ-safe property implies that $|M_\ell|$ is polynomially bounded by $|D_\ell|$.   ∎

Using the definition of the DETCOST instance and Lemma 6, we can show:

Lemma 8. *The fractional solution $\bar{y}$ is feasible for the LP relaxation (10) corresponding to the DETCOST instance $\mathcal{C}$. Moreover, $\theta \geq \max_j s_j$ and $\psi \geq \max_j c_j$.*

The above lemmas show that the algorithm is well-defined, so we can indeed use Theorem 4 to round $\bar{y}$ into an integer solution. Recall that our final solution is $N = N_H \cup N_L$. The next two lemmas follow from this rounding step.

Lemma 9. $|N_H| + |N_L| \geq t$.

Lemma 10. *For any class $\ell \leq \rho$ and resource $i \in M_\ell$,*

$$\sum_{j \in N_\ell \cap L_i} \beta_k(X_j') \leq 4\bar{a}b, \quad where \ k = 2^{2^\ell}.$$

We now focus on a particular class $\ell \leq \rho$ and show that the expected makespan due to tasks in $N \cap J_\ell$ is small. Recall that $k = 2^{2^\ell}$. Let $N_\ell := N \cap J_\ell$ and let $\mathsf{Load}_i^{(\ell)} := \sum_{j \in N_\ell \cap L_i} X_j'$ be the load on any resource $i \in [m]$ due to class-ℓ tasks. We can now bound the makespan due to the truncated random variables.

Lemma 11. *For any class $\ell \leq \rho$, $\mathbb{E}\left[\max_{i \in M_\ell} \mathsf{Load}_i^{(\ell)}\right] \leq 4\bar{a}b + O(1)$, and*

$$\mathbb{E}\left[\max_{i=1}^{m} \mathsf{Load}_i^{(\ell)}\right] \leq 4\lambda\bar{a}b + O(\lambda) = O(\alpha\lambda).$$

Proof. Consider a resource $i \in M_\ell$. Lemmas 10 and 3 imply:

$$\Pr\left[\mathsf{Load}_i^{(\ell)} > 4\bar{a}b + \gamma\right] = \Pr\left[\sum_{j \in N_\ell \cap L_i} X_j' > 4\bar{a}b + \gamma\right] \leq k^{-\gamma}, \quad \forall \gamma \geq 0.$$

By a union bound, we get

$$\Pr\left[\max_{i \in M_\ell} \mathsf{Load}_i^{(\ell)} > 4\bar{a}b + \gamma\right] \leq |M_\ell| \cdot k^{-\gamma} \leq k^{p-\gamma}, \qquad \text{for all } \gamma \geq 0,$$

where p is the constant from Lemma 7. So the expectation

$$\mathbb{E}\left[\max_{i \in M_\ell} \mathsf{Load}_i^{(\ell)}\right] = \int_{\theta=0}^{\infty} \Pr\left[\max_{i \in M_\ell} \mathsf{Load}_i^{(\ell)} > \theta\right] d\theta$$

$$\leq 4\bar{a}b + p + 2 + \int_{\gamma=p+2}^{\infty} k^{-\gamma+p} d\gamma \leq 4\bar{a}b + p + 2 + \frac{1}{k(p+1)},$$

which completes the proof of the first statement.

We now prove the second statement. Consider any class $\ell < \rho$: by definition of J_ℓ, we know that $J_\ell \subseteq L(D_\ell)$. So the λ-safe property implies that for every resource i there is a subset $R_i \subseteq M_\ell$ of size at most λ such that $L_i \cap J_\ell \subseteq L(R_i) \cap J_\ell$. Because $N_\ell \subseteq J_\ell$, we also have $L_i \cap N_\ell \subseteq L(R_i) \cap N_\ell$. Therefore,

$$\mathsf{Load}_i^{(\ell)} \leq \sum_{z \in R_i} \mathsf{Load}_z^{(\ell)} \leq \lambda \max_{z \in M_\ell} \mathsf{Load}_z^{(\ell)}.$$

Taking expectation on both sides, we obtain the desired result.

Finally, for the last class $\ell = \rho$, note that any task in J_ρ loads the resources in $D_\rho = M_\rho$ only. Therefore, $\max_{i=1}^{m} \mathsf{Load}_i^{(\ell)} = \max_{z \in M_\ell} \mathsf{Load}_z^{(\ell)}$. The desired result now follows by taking expectation on both sides. ∎

Using Lemma 11 for all ρ classes, it follows that the expected makespan due to all truncated r.v.s is $O(\alpha\lambda\rho)$. For the exceptional random variables, we use:

Lemma 12. $\mathbb{E}\left[\sum_{j\in N} X_j''\right] = \sum_{j\in N} c_j \leq 4\bar{\alpha}$.

Adding the contributions from truncated and exceptional r.v.s, the overall expected makespan is $O(\alpha\lambda\rho)$, which completes the proof of Theorem 2.

References

1. Adamaszek, A., Chalermsook, P., Wiese, A.: How to tame rectangles: solving independent set and coloring of rectangles via shrinking. In: APPROX/RANDOM, pp. 43–60 (2015)
2. Agarwal, P.K., Mustafa, N.H.: Independent set of intersection graphs of convex objects in 2D. Comput. Geom. **34**(2), 83–95 (2006)
3. Chakrabarti, A., Chekuri, C., Gupta, A., Kumar, A.: Approximation algorithms for the unsplittable flow problem. Algorithmica **47**(1), 53–78 (2007)
4. Chalermsook, P.: Coloring and maximum independent set of rectangles. In: Goldberg, L.A., Jansen, K., Ravi, R., Rolim, J.D.P. (eds.) APPROX/RANDOM 2011. LNCS, vol. 6845, pp. 123–134. Springer, Heidelberg (2011). https://doi.org/10.1007/978-3-642-22935-0_11
5. Chalermsook, P., Chuzhoy, J.: Maximum independent set of rectangles. In: SODA, pp. 892–901 (2009)
6. Chan, T.M.: A note on maximum independent sets in rectangle intersection graphs. Inf. Process. Lett. **89**(1), 19–23 (2004)
7. Chan, T.M., Har-Peled, S.: Approximation algorithms for maximum independent set of pseudo-disks. Discret. Comput. Geom. **48**(2), 373–392 (2012)
8. Chekuri, C., Mydlarz, M., Shepherd, F.B.: Multicommodity demand flow in a tree and packing integer programs. ACM Trans. Algorithms **3**(3), 27 (2007)
9. Elwalid, A.I., Mitra, D.: Effective bandwidth of general markovian traffic sources and admission control of high speed networks. IEEE/ACM Trans. Netw. **1**(3), 329–343 (1993)
10. Gupta, A., Kumar, A., Nagarajan, V., Shen, X.: Stochastic load balancing on unrelated machines. In: SODA, pp. 1274–1285. Society for Industrial and Applied Mathematics (2018)
11. Gupta, A., Kumar, A., Nagarajan, V., Shen, X.: Stochastic makespan minimization in structured set systems. arXiv (2020). https://arxiv.org/abs/2002.11153
12. Hui, J.Y.: Resource allocation for broadband networks. IEEE J. Sel. Areas Commun. **6**(3), 1598–1608 (1988)
13. Kelly, F.P.: Notes on effective bandwidths. In: Stochastic Networks: Theory and Applications, pp. 141–168. Oxford University Press (1996)
14. Kleinberg, J., Rabani, Y., Tardos, E.: Allocating bandwidth for bursty connections. SIAM J. Comput. **30**(1), 191–217 (2000)
15. Molinaro, M.: Stochastic ℓ_p load balancing and moment problems via the l-function method. In: SODA, pp. 343–354 (2019)

Continuous Facility Location on Graphs

Tim A. Hartmann[1], Stefan Lendl[2], and Gerhard J. Woeginger[1]([⊠])

[1] Department of Computer Science, RWTH Aachen, Aachen, Germany
woeginger@algo.rwth-aachen.de
[2] Department of Mathematics, TU Graz, Graz, Austria

Abstract. We study a continuous facility location problem on undirected graphs where all edges have unit length and where the facilities may be positioned at the vertices as well as at interior points of the edges. The goal is to cover the entire graph with a minimum number of facilities with covering range $\delta > 0$. In other words, we want to position as few facilities as possible subject to the condition that every point on every edge is at distance at most δ from one of these facilities.

We investigate this covering problem in terms of the rational parameter δ. We prove that the problem is polynomially solvable whenever δ is a unit fraction, and that the problem is NP-hard for all non unit fractions δ. We also analyze the parametrized complexity with the solution size as parameter: The resulting problem is fixed parameter tractable for all $\delta < 3/2$, and it is $W[2]$-hard for all $\delta \geq 3/2$.

Keywords: Location theory · Graph theory · Parametrized complexity

1 Introduction

We investigate the algorithmic behavior of a continuous covering problem on graphs. Consider an undirected connected graph $G = (V, E)$, whose edges are rectifiable and have unit length. Denote by $P(G)$ the continuum set of points on all the edges and vertices. For two points $p, q \in P(G)$, denote by $d(p, q)$ the length of a shortest path connecting p and q in the underlying metric space. Point p is said to δ-cover point q for some positive real number δ, if $d(p, q) \leq \delta$ holds. A subset $S \subseteq P(G)$ is a δ-cover for G, if for every point $p \in P(G)$ there exists some $s \in S$ that δ-covers p. The objective is to compute for a given graph $G = (V, E)$ and for a given positive real number δ a minimum cardinality δ-cover $S \subseteq P(G)$. Such a minimizing set S is called an *optimal* δ-cover, and the cardinality $|S|$ is called the δ-*covering number* δ-COVER(G) of graph G.

Known and Related Results. The area of continuous facility location on graphs has been started by Megiddo and Tamir [8], who showed that computing an optimal δ-cover is NP-hard in case $\delta = 2$. Furthermore [8] contains a fast algorithm for the δ-covering number of trees. Kariv and Hakimi [7] establish many hardness results for discrete types of facility location problems on graphs, in

© Springer Nature Switzerland AG 2020
D. Bienstock and G. Zambelli (Eds.): IPCO 2020, LNCS 12125, pp. 171–181, 2020.
https://doi.org/10.1007/978-3-030-45771-6_14

which the facilities must be located on vertices. Fekete, Mitchell and Beurer [3] discuss a number of continuous facility location problems in a purely geometric setting in the Euclidean plane. We refer to the books [9] by Mirchandani and Francis and [1] by Drezner for comprehensive information on the area of facility location.

In a closely related line of research, Grigoriev et al. [6] study an obnoxious continuous facility location problem on graphs. The objective in [6] is not to cover, but to pack: Place as many facilities as possible on the graph, subject to the condition that any two facilities have at least distance δ from each other. This packing problem is polynomially solvable, if δ is a rational number with numerator 1 or 2, and it is NP-hard for all other rational values of δ.

Our Results. We provide a complete picture of the complexity of computing the δ-covering number for connected graphs $G = (V, E)$ and positive rational numbers δ. We trace out the boundary between polynomial time solvability and NP-hardness, as well as the boundary between parametrized tractability and parametrized intractability. With respect to polynomial time solvability, the picture is as follows:

- If δ is a unit fraction (that is, if $\delta = 1/c$ for some integer c), then the δ-covering number can be computed in polynomial time.
- If δ is not a unit fraction, then computing the δ-covering number is NP-hard.

The parametrized version of δ-covering takes the solution size of a δ-cover as parameter. The first intuition is that the problem should be easy for small values of δ and hard for large values of δ. Indeed, if δ is small (say $\delta \leq 1/4$), then δ-covering essentially boils down to covering all the edges of the input graph with the facilities; this has the flavor of the VERTEX COVER problem, which is known to be fixed parameter tractable. On the other hand, if δ is large (say $\delta \geq 4$), then the main goal should be to cover all the vertices of the input graph with the facilities, whereas the edges only play a minor role and will be covered without much additional effort. Hence these cases have the flavor of the DOMINATING SET problem, which belongs to the intractable problems in the parametrized world. This intuition turns out to be correct, and we will show that at the threshold $\delta = 3/2$ the parametrized complexity jumps from tractable to intractable:

- In the range $0 < \delta < 3/2$, the δ-covering number is fixed parameter tractable.
- In the range $\delta \geq 3/2$, the δ-covering number is $W[2]$-hard.

We stress that this transition occurs surprisingly sudden, and that there is no intermediate range of δ values for which this δ-covering problem is $W[1]$-complete (assuming FPT $\neq W[1]$ and $W[1] \neq W[2]$ as usual).

The paper is organized as follows. Section 2 states the basic notations and presents some technical observations. Section 3 gives the $W[2]$-hardness results, and Sect. 4 gives the NP-hardness results. The NP-hardness reductions are inductively structured, and break down exactly at the unit fractions. The polynomial time algorithm in Sect. 5 for the case $\delta = 1$ is mainly based on tools from matching theory. Section 6 contains one of our technical main results, an fpt-algorithm for the parametrized cases with $\delta < 3/2$.

2 Notation and Technical Preliminaries

All the graphs in this paper are undirected and connected, and all the edges have unit length. We use the word *vertex* in the graph-theoretic sense, and we use the word *point* to denote the elements of the geometric structure $P(G)$. For a graph $G = (V, E)$ and a vertex $v \in V$, we denote by $\Gamma(v)$ the set of neighbors of v and we denote $\Gamma^+(v) = \Gamma(v) \cup \{v\}$. For $V' \subseteq V$, we denote by $G[V']$ the subgraph induced by V'. A subset $C \subseteq V$ forms a *vertex cover* for the graph $G = (V, E)$, if every edge in E has at least one of its endpoints in C; the size of the smallest vertex cover in G is denoted by $\tau(G)$. A subset $M \subseteq E$ forms a *matching* for $G = (V, E)$, if no two edges in M share a common endpoint; the size of the largest matching in G is denoted by $\nu(G)$.

The *closed ball* $B^+(v, r)$ of radius r around point v contains all points $p \in P(G)$ with $d(v, p) \leq r$, and the corresponding *open ball* $B^-(v, r)$ contains all p with $d(v, p) < r$. For an edge $e = \{u, v\}$ and a real number λ with $0 \leq \lambda \leq 1$, we denote by $p(u, v, \lambda)$ the point on edge e that is at distance λ from vertex u. Note that $p(u, v, 0) = u$ and $p(u, v, 1) = v$, and note that $p(u, v, \lambda) = p(v, u, 1 - \lambda)$. For integers ℓ and k, the rational number k/ℓ is called ℓ-*simple*. A set $S \subseteq P(G)$ is ℓ-*simple*, if for every point $p(u, v, \lambda)$ in S the number λ is ℓ-simple.

Lemma 1. *Let $c \geq 1$ be an integer, let G be a graph, and let G' be the graph that results from G by subdividing every edge into c new edges. Then δ-COVER$(G) = (c \cdot \delta)$-COVER(G').*

Proof. As the subdivision stretches the metric space $P(G)$ by a factor c, the δ-covers in G are in one-to-one correspondence with the $(c \cdot \delta)$-covers in G'. $\quad \square$

Lemma 2. *Let $G = (V, E)$ be a graph and let $\delta = a/b$ with integers a and b. Then there exists an optimal δ-cover S^* that is $2b$-simple.* $\quad \square$

Lemma 3. *Let $G = (V, E)$ be a graph, and let a and b be positive integers. Then*

$$\frac{a}{2a + b}\text{-COVER}(G) = \frac{a}{b}\text{-COVER}(G) + |E|. \tag{1}$$

Proof. In this extended abstract, we only sketch some fragment of the proof for the cases $a/b < 1/2$. The remaining cases with $a/b \geq 1/2$ can be settled by similar arguments, but need a number of tedious case distinctions.

Hence let $\delta = a/b < 1/2$ and $\delta' = a/(2a + b)$. Since $\delta < 1/2$, every δ-cover S must contain at least one point from every edge in E. Now consider an edge $\{u, v\} \in E$, and let $s_i = p(u, v, \lambda_i)$ with $1 \leq i \leq m$ denote the points from S on this edge; we assume $0 \leq \lambda_1 < \cdots < \lambda_m < 1$. Since S is a δ-cover, we have $\lambda_{i+1} - \lambda_i \leq 2\delta$ for $1 \leq i \leq m - 1$.

We define a new set S' that contains $m + 1$ points $s_i' = p(u, v, \mu_i)$ with $0 \le i \le m$ as follows: We define the first point via $\mu_0 = \lambda_1 b/(2a+b)$ and the last point via $\mu_m = \lambda_m b/(2a+b)$. For $i = 1, \ldots, m-1$ we set $\mu_i = \mu_0 + 2ai/(2a+b)$. Note that $\mu_{i+1} - \mu_i = 2\delta'$ holds for $0 \le i \le m - 1$; this implies that the piece of the edge that lies between the two points s_0' and s_{m-1}' is δ'-covered by S'. Next we want to argue that also the piece between s_{m-1}' and s_m' is δ'-covered by S', which is equivalent to the inequality $\mu_m - \mu_{m-1} \le 2\delta'$. Since $\mu_{m-1} = \mu_0 + 2(m-1)a/(2a+b)$ and $\mu_m = \lambda_m b/(2a+b)$ and $\mu_0 = \lambda_1 b/(2a+b)$, the desired inequality can be rewritten into $\lambda_m - \lambda_1 \le 2(m-1)\delta$, which is easily seen to hold.

We repeat this translation process for every individual edge, and thereby translate the old δ-cover S into a new set S'. Since every edge receives one additional point, we have $|S'| = |S| + |E|$. We have argued above that the piece of the edge between s_0' and s_m' is δ'-covered by S'. A similar argument (that also takes points on other edges into account) shows that also the initial piece between point u and point s_0' and also the final piece between s_m' and point v are δ'-covered by S'. All in all this demonstrates that in the cases with $a/b < 1/2$ the left hand side in (1) is less or equal to the right hand side. The inequality in the other direction can be shown in a very similar way.   □

3   Parametrized Hardness Results

In this section we prove the following theorem.

Theorem 1. *For every fixed rational δ with $\delta \ge 3/2$, the δ-covering problem with the solution size k as parameter is $W[2]$-hard.*

The proof of Theorem 1 is done in three steps: Lemma 4 settles the cases with $3/2 \le \delta < 2$, Lemma 5 settles the cases with $\ell \le \delta < \ell+1/2$ for every $\ell \ge 2$, and Lemma 6 settles the remaining cases with $\ell + 1/2 \le \delta < \ell + 1$ for every $\ell \ge 2$. Some of our fpt-reductions are based on a $W[2]$-hard variant of the dominating set problem that we call COLORFUL DOMINATING SET:

> Instance: An undirected, connected graph $H = (V_H, E_H)$ whose vertex set V_H is partitioned into k color classes $V_1, V_2, \ldots, V_k$.
> Question: Do there exist vertices $u_i \in V_i$ for $1 \le i \le k$ that form a dominating set?

Now let us fix some rational number δ in the range $3/2 \le \delta < 2$. We fpt-reduce from an instance $H = (V_H, E_H)$ with color classes $V_1, V_2, \ldots, V_k$ of COLORFUL DOMINATING SET by constructing the following instance $G = (V_G, E_G)$ of δ-covering.

- The vertex set V_G contains every vertex $u \in V_H$ together with two copies u' and u''. Furthermore, there are $2k$ vertices $x_1, \ldots, x_k$ and $y_1, \ldots, y_k$.

- The edge set E_G contains for every i with $1 \le i \le k$ all the edges on $V_i \cup \{x_i, y_i\}$. Furthermore, E_G contains the triangle on u, u', u'' for every $u \in V_H$. Finally, for every edge $\{u, v\} \in E_H$ the set E_G contains $\{u, v\}$, the two cross-edges $\{u, v'\}$ and $\{u, v''\}$ and (by symmetry) the two cross-edges $\{v, u'\}$ and $\{v, u''\}$ (Fig. 1) .

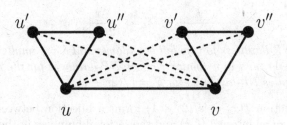

Fig. 1. Illustration for the fpt-reduction for $3/2 \le \delta < 2$. The cross-edges between the two triangles u, u', u'' and v, v', v'' are shown as dashed lines.

It can be shown (details omitted) that graph H possesses a colorful dominating set, if and only if δ-COVER$(G) \le k$. This yields the following.

Lemma 4. *For every fixed rational δ with $3/2 \le \delta < 2$, the δ-covering problem with the solution size k as parameter is $W[2]$-hard.* □

The following two Lemmas 5 and 6 settle the remaining cases for the proof of Theorem 1. The proofs are routine fpt-reductions from the classical $W[2]$-hard DOMINATING SET problem, and are omitted.

Lemma 5. *For every integer $\ell \ge 2$ and for every rational δ with $\ell \le \delta < \ell + 1/2$, the δ-covering problem with the solution size k as parameter is $W[2]$-hard.* □

Lemma 6. *For every integer $\ell \ge 2$ and for every rational δ with $\ell + 1/2 \le \delta < \ell + 1$, the δ-covering problem with solution size k as parameter is $W[2]$-hard.* □

4 NP-Hardness Results

Theorem 1 trivially implies the NP-hardness of computing the δ-covering number for all $\delta \ge 3/2$. The NP-hard cases in the range $\delta < 3/2$ will be identified and settled on the following pages, and thereby yield the following theorem.

Theorem 2. *For every fixed positive rational δ that is not a unit fraction, it is NP-hard to compute the δ-covering number of a graph.*

Our main tools are the following two Lemmas 7 and 8. The first lemma is an immediate consequence of Lemma 1.

Lemma 7. *Let $\delta > 0$ and let $c \geq 1$ be an integer. Suppose that the computation of the δ-covering number is NP-hard. Then also the computation of the $(c \cdot \delta)$-covering number is NP-hard.* □

Lemma 8. *Let ℓ and r be real numbers with $0 \leq \ell < r$. Suppose that for every rational δ with $\ell < \delta < r$, the computation of the δ-covering number is NP-hard. Then the computation of the δ'-covering number is also NP-hard for every rational δ' with*

$$\frac{\ell}{2\ell + 1} < \delta' < \frac{r}{2r + 1}. \tag{2}$$

Furthermore, NP-hardness for the left boundary point $\delta = \ell$ implies NP-hardness for $\delta' = \ell/(2\ell + 1)$, and in a symmetric way NP-hardness for the right boundary point $\delta = r$ implies NP-hardness for $\delta' = r/(2r + 1)$.

Proof. The function $f(x) = x/(2x + 1)$ forms a bijection between the rational numbers in the open interval (ℓ, r) and the rational numbers in the open interval $(\ell/(2\ell + 1), r/(2r + 1))$. Note that for $\delta = a/b$ we have $f(\delta) = a/(2a + b)$. Lemma 3 yields that δ-COVER$(G) = f(\delta)$-COVER$(G) - |E|$ for every graph $G = (V, E)$. Hence, computing the δ-covering number is polynomial time reducible to computing the $f(\delta)$-covering number.   □

Now we turn to the proof of Theorem 2, which is structured into three parts. The main trick is to alternately apply the tools developed in Lemmas 7 and 8.

In the **first part of the proof**, we define for every integer $j \geq 0$ a real number $\alpha_j = 3^j/(3^j - 1)$ and a half-open interval $A_j = [\alpha_{j+1}, \alpha_j)$. Note that with the help of the function $f(x) = x/(2x + 1)$ introduced in the proof of Lemma 8, we may equivalently write $\alpha_{j+1} = 3f(\alpha_j)$. We prove by induction on $j \geq 0$ that for every rational $\delta \in A_j$, the δ-covering number is NP-hard to compute. The anchor step with $j = 0$ states NP-hardness for $\delta \in A_0 = [3/2, \infty)$, and hence follows from Theorem 1. In the inductive step, we first apply Lemma 8 with $\ell = \alpha_j$ and $r = \alpha_{j-1}$ to interval A_{j-1} and thus deduce hardness for all rational δ in the range $f(\alpha_j) \leq \delta < f(\alpha_{j-1})$. Then we apply Lemma 7 with $c = 3$ to deduce NP-hardness for all δ in the range $3f(\alpha_j) \leq \delta < 3f(\alpha_{j-1})$ and hence for all $\delta \in A_j$. This completes the inductive step. Since the numbers α_j form a decreasing sequence that converges to the limit point 1, the union of these intervals A_j covers all numbers that are strictly larger than 1. Summarizing, in the first part we have established NP-hardness for all rational numbers $\delta > 1$.

In the **second part of the proof**, we consider the open intervals $A'_j = (1/(2j + 1), 1/(2j))$ for $j \geq 0$. We prove by induction on $j \geq 0$ that for every rational $\delta \in A'_j$, the δ-covering number is NP-hard to compute. The anchor step with $j = 0$ claims NP-hardness for $\delta \in A'_0 = (1, \infty)$, and hence follows from the above first part of our proof. In the inductive step, we apply Lemma 8 with $\ell = 1/(2j + 1)$ and $r = 1/(2j)$, which yields $f(\ell) = 1/(2j + 3)$ and $f(r) = 1/(2j + 2)$. Thereby we lift NP-hardness from interval A'_j to the next interval A'_{j+1}.

In the **third and last part of the proof**, we finally establish NP-hardness for all $\delta = a/b$ with $\gcd(a, b) = 1$ and $a \geq 2$. Note that this settles all the positive rational numbers except for the unit fractions, and hence completes the proof of

Theorem 2. The argument branches into two subcases. In the first subcase, we assume that the denominator b is odd. Since the integers $2a$ and b are relatively prime, there exists a positive integer c so that $bc \equiv 1 \bmod 2a$. In other words, there are positive integers c and j with $bc = 1 + 2aj$. Then $2aj < bc$ implies $a/b < c/(2j)$, and $bc < a + 2aj$ implies $a/b > c/(2j + 1)$. We summarize these inequalities as

$$\frac{c}{2j+1} < \delta < \frac{c}{2j}.$$

This means that $\delta/c \in A'_j$ (for the interval A'_j as introduced in the second part of the proof), and that consequently (δ/c)-covering is NP-hard. Then Lemma 7 implies that also δ-covering is NP-hard.

In the second subcase, we assume that the denominator b is even with $b = 2b'$; note that in this case the numerator satisfies $a \geq 3$. Since the integers a and b' are relatively prime, there exists a positive integer c so that $b'c \equiv 1 \bmod a$. In other words, there are positive integers c and j with $b'c = 1 + aj$. Then $aj < b'c$ implies $a/b < c/(2j)$, and $2b'c < a + 2aj$ implies $a/b > c/(2j + 1)$. Analogously to the first subcase, we conclude that $\delta/c \in A'_j$ and that δ-covering is NP-hard.

This finally completes the proof of Theorem 2. It is instructive to examine the places where the above argument breaks down for unit fractions $\delta = 1/b$.

5   The Polynomial Time Result for 1-Covering

In this section we derive a polynomial time algorithm for computing the 1-covering number of a graph $G = (V, E)$. A 1-cover is in *canonical* form, if it entirely consists of vertices and of midpoints of edges. By Lemma 2 every graph possesses an optimal 1-cover in canonical form. Our algorithm heavily relies on Edmonds-Gallai decompositions. The following theorem summarizes the result in a notation that is appropriate for our usage.

Theorem 3 *(Edmonds [2] and Gallai [4,5]). Let $G = (V, E)$ be a graph. The following decomposition of V into X, Y, Z can be computed in polynomial time.*

$X = \{v \in V \mid$ *there exists a maximum cardinality matching that misses $v\}$*
$Y = \{v \in V \mid v \notin X$ *and v is adjacent to some vertex in $X\}$*
$Z = V - (X \cup Y)$

Set X is the union of the odd-sized components of $G - Y$; every such odd-sized component is factor-critical. Set Z is the union of the even-sized components of $G - Y$. Every maximum cardinality matching in G induces perfect matchings on all components of Z, induces near-perfect matchings on all components of X, and matches the vertices in Y to vertices in different components of X. □

Based on such an Edmonds-Gallai decomposition of G, we define three subgraphs:

– $G_0 = (Z, E_0)$ is the subgraph of G induced by Z.

- Let X_1 contain the vertices that form components of size 1 in X, let $Y_1 = \Gamma(X_1)$, and let E_1 denote the set of edges between X_1 and Y_1. We define $G_1 = (X_1 \cup Y_1, E_1)$ as the corresponding (not necessarily induced) bipartite subgraph of G.
- Let $X_{\geq 3} = X - X_1$ denote the vertices of X that belong to (odd-sized) components of size at least 3. We define $G_{\geq 3} = G[X_{\geq 3}]$ as the subgraph of G induced by $X_{\geq 3}$, and we let $c_{\geq 3}$ denote the number of components in $G_{\geq 3}$.

Lemma 9. *Every 1-cover S satisfies $|S| \geq \nu(G_0) + \tau(G_1) + \nu(G_{\geq 3}) + c_{\geq 3}$.*

Proof. We separately derive three lower bounds for the three subgraphs G_0, G_1 and $G_{\geq 3}$. An appropriate combination of these bounds yields the inequality. The arguments are not difficult but lengthy, and hence are omitted in this extended abstract.   □

Our next goal is to construct in polynomial time a 1-cover S^* whose size matches the lower bound stated in Lemma 9, and which therefore is an optimal 1-cover. We start by computing a maximum cardinality matching $M \subseteq E$ for graph G and a minimum vertex cover $C \subseteq X_1 \cup Y_1$ for the bipartite subgraph G_1. The set S^* is defined as follows:

(i) S^* contains the midpoint of every edge in M that does not belong to the edges between X and Y.
(ii) S^* contains every vertex in $C \cap Y_1$, and furthermore contains every vertex in $C \cap X_1$ that is not saturated by M. If $x \in C \cap X_1$ is saturated by M, then S^* contains the midpoint of the matching edge incident to x.
(iii) S^* contains every vertex $x \in X_{\geq 3}$ that is not saturated by M and the midpoint of each edge $\{x, y\} \in M$ for which $x \in X_{\geq 3}$ and $y \in Y$.

Note that (i) contributes $\nu(G_0) + \nu(G_{\geq 3})$ points to S^*, (ii) contributes $\tau(G_1)$ points, and (iii) contributes $c_{\geq 3}$ points. Hence the size of S^* indeed matches the lower bound in Lemma 9. It remains to show that S^* is a 1-cover for G.

- As M induces a perfect matching on Z, the set S^* contains for every $z \in Z$ the midpoint of some incident edge. Hence every closed ball $B^+(z, 1/2)$ is 1-covered by S^*.
- Every vertex $y \in Y$ is saturated by the matching M. If $y \in C \cap Y_1$ then $y \in S^*$, and if $y \notin C \cap Y_1$ then S^* contains the midpoint of the matching edge incident to y. In either case the ball $B^+(y, 1/2)$ is 1-covered by S^*.
- Let $x \in X_1$. If $x \in C$, then S^* does either contain x itself or does contain the midpoint of an incident matching edge; in either case the ball $B^+(x, 1/2)$ is 1-covered by S^*. If $x \notin C$, then every vertex $y \in \Gamma(x)$ lies in $C \cap Y_1$; in this case S^* contains every neighbor $y \in \Gamma(x)$ so that the ball $B^+(x, 1)$ is 1-covered by S^*.
- If a vertex $x \in X_{\geq 3}$ is saturated by M, then S^* contains the midpoint of the incident matching edge. If $x \in X_{\geq 3}$ is not saturated by M, then by construction $x \in S^*$. In either case the ball $B^+(x, 1/2)$ is 1-covered by S^*.

All in all, we have shown that for every vertex $v \in V$ the entire ball $B^+(v, 1/2)$ is 1-covered by S^*. Hence all the edges in G are covered, and S^* indeed is a 1-cover for G. The above discussion yields a polynomial time algorithm for $\delta = 1$, and Lemma 1 extends this result to all unit fractions δ.

Theorem 4. *For every unit fraction δ, the δ-covering problem is solvable in polynomial time.*

6   The Fixed Parameter Tractable Cases

Throughout this section, we consider some fixed rational number $\delta < 3/2$ and some fixed integer k. We will develop an fpt-algorithm with parameter k for deciding whether an input graph G satisfies δ-COVER$(G) \leq k$, that is, a decision algorithm whose running time is bounded by some computable function $f(k)$ and by some polynomial function in the instance size $|G|$.

In a preliminary step, we consider a vertex v that is incident to $\ell \leq 4k$ edges. We denote by $g(\ell, \delta)$ the minimum size of a δ-cover for the open ball $B^-(v, 1-\delta)$. Note that $g(\ell, \delta) = 0$ whenever $\delta \geq 1$. Note furthermore that the value $g(\ell, \delta)$ can be computed in constant time, as the numbers $\ell \leq 4k$ and δ are fixed constants and do not depend on the input graph G.

Lemma 10. *For $\delta < 3/2$, every graph $G = (V, E)$ satisfies δ-COVER$(G) \geq \nu(G)/2$.*

Proof. Let $M \subseteq E$ be a maximum size matching in G, and let $S \subseteq P(G)$ be a minimum size δ-cover. Then any point $p \in S$ will δ-cover at most two of the midpoints of edges in M. (This is the only place in this section where we exploit the condition $\delta < 3/2$.) $\square$

Our fpt-algorithm first computes a maximum cardinality matching M for the input graph G. If $|M| > 2k$ then the algorithm outputs NO and stops; this step is justified by Lemma 10, as δ-COVER$(G) > k$ implies a negative answer to the decision problem. Hence from now on we will assume $|M| \leq 2k$. Then the end-vertices of the edges in M form a vertex cover C of size $|C| \leq 4k$ and the vertices in $I = V - C$ form an independent set. For $T \subseteq C$, we let I_T denote the set of all vertices $v \in I$ with $\Gamma(v) = T$.

Lemma 11. *Let S be an optimal δ-cover for the graph $G = (V, E)$ and let $T \subseteq C$. Then in the set I_T there are at most $|T|$ vertices v for which $S \cap B^-(v, 1)$ contains at least $g(|T|, \delta) + 1$ points.*

Proof. Suppose otherwise. Then there are at least $|T| + 1$ vertices v in I_T for which the ball $B^-(v, 1)$ contains at least $g(|T|, \delta) + 1$ points from S. For all other vertices $v \in I_T$, the ball $B^-(v, 1)$ contains at least $g(|T|, \delta)$ points from S. All in all, this yields that S contains at least $|I_T| g(|T|, \delta) + |T| + 1$ points from these open balls around vertices in I_T. We remove all these points from set S and replace them by the following points:

– We add all the points in T to S.
– For every vertex $v \in I_T$, we add $g(|T|, \delta)$ points to S that δ-cover $B^-(v, 1-\delta)$.

The resulting point set is again a δ-cover for G, but does contain strictly fewer points than the optimal δ-cover S. This yields the desired contradiction.   □

In the following paragraphs, we investigate a subset $T \subseteq C$ that satisfies $|I_T| \geq 2|T| + 1$. Recall that I_T is an independent set and that every vertex $v \in I_T$ satisfies $\Gamma(v) = T$. We denote by $\mathcal{B}_T$ the union of all the open balls $B^-(v, 1)$ with $v \in I_T$. Now consider some fixed optimal δ-cover S for graph G.

– For every vertex $t \in T$, consider a point $s_t \in S \cap \mathcal{B}_T$ that minimizes the distance $d(s_t, t)$. Whenever some point $p \notin \mathcal{B}_T$ is δ-covered by a point from $S \cap \mathcal{B}_T$, then this point p will also be covered by one of the points s_t with $t \in T$. If $s_t \in B^-(v, 1)$, then we say that the corresponding vertex v is *busy*. There are at most $|T|$ busy vertices in I_T.
– A vertex v in I_T is *heavy*, if $B^-(v, 1)$ contains at least $g(|T|, \delta) + 1$ points from S. By Lemma 11, there are at most $|T|$ heavy vertices in I_T.

Since $|I_T| \geq 2|T| + 1$, there exists a vertex $w \in I_T$ that is neither busy nor heavy. What can we say about the points in the set $S \cap B^-(w, 1)$? Since vertex w is not heavy, $S \cap B^-(w, 1)$ must contain exactly $g(|T|, \delta)$ points. Since vertex w is not busy, the points in $S \cap B^-(w, 1)$ are not required for δ-covering the points outside of $\mathcal{B}_T$. In other words, the only duty of these points in $S \cap B^-(w, 1)$ is to cover the open ball $B^-(w, 1)$. This observation allows us to eliminate vertex w, as it is dispensable and only imposes minor constraints on the structure of a δ-cover. We define a new instance by removing vertex w (together with its $|T|$ incident edges) from G and by simultaneously decreasing the parameter k by the value $g(|T|, \delta)$.

Lemma 12. *The new instance $G - w$ and $k - g(|T|, \delta)$ is a YES-instance of δ-covering, if and only if the old instance G and k is a YES-instance of δ-covering.*

Proof. For the (if) part, assume that G has a δ-cover of size at most k. Let S be an optimal δ-cover with $|S| \leq k$. We pick a vertex $w \in I_T$ that is neither busy nor heavy, and remove the $g(|T|, \delta)$ points in $S \cap B^-(w, 1)$ from S. This yields a δ-cover of size $k - g(|T|, \delta)$ for $G - w$.

For the (only if) part, assume that the graph $G - w$ has a δ-cover S' of size at most $k - g(|T|, \delta)$. We pick an arbitrary vertex $u \in I_T$ with $u \neq w$, and we use a clone of the set $S' \cap B^-(u, 1)$ to cover $B^-(w, 1)$. This extends S' to a δ-cover of G of size at most k.   □

The above discussion yields the following reduction rule: *"Whenever some subset $T \subseteq C$ satisfies $|I_T| \geq 2|T| + 1$, then we may remove a vertex $w \in I_T$ from G and decrease the parameter k by $g(|T|, \delta)$."* We apply this reduction rule over and over again, until no further reductions are possible. At termination, the vertex set of the residual graph will consist of a vertex cover C together with the shrunken independent sets I_T that now satisfy $|I_T| \leq 2|T| \leq 8k$. Hence the residual graph has at most $4k(1 + 2^{4k+1})$ vertices. As this size is bounded by a function in the parameter k and as the parameter is not part of the input, the resulting instance of δ-covering can be solved in constant time. Since the reduction rule can easily be implemented in polynomial time, we formulate the following summarizing theorem.

Theorem 5. *For every fixed rational δ with $\delta < 3/2$, the δ-covering problem with the solution size k as parameter allows an fpt-algorithm.*   □

Acknowledgement. Stefan Lendl acknowledges support by the Austrian Science Fund (FWF): W1230, Doctoral Program "Discrete Mathematics". Gerhard Woeginger acknowledges support by the DFG RTG 2236 "UnRAVeL".

References

1. Drezner, Z.: Facility Location: A Survey of Applications and Methods. Springer Series in Operations Research. Springer, New York (1995)
2. Edmonds, J.: Paths, trees, and flowers. Can. J. Math. **17**, 449–467 (1965)
3. Fekete, S.P., Mitchell, J.S.B., Beurer, K.: On the continuous Fermat-Weber problem. Oper. Res. **53**, 61–76 (2005)
4. Gallai, T.: Kritische graphen II. Magyar Tudományos Akadémia Matematikai Kutat'o Intézetének közleményei **8**, 373–395 (1963)
5. Gallai, T.: Maximale Systeme unabhñgiger Kanten. Magyar Tudományos Akadémia Matematikai Kutat'o Intézetének közleményei **9**, 401–413 (1964)
6. Grigoriev, A., Hartmann, T.A., Lendl, S., Woeginger, G.J.: Dispersing obnoxious facilities on a graph. In: Proceedings of the 36th Annual Symposium on Theoretical Aspects of Computer Science (STACS 2019), Leibniz International Proceedings in Informatics, LIPICS, vol. 126, pp. 33:1–33:11 (2019)
7. Kariv, O., Hakimi, S.L.: An algorithmic approach to network location problems, part I. The p-centers. SIAM J. Appl. Math. **37**, 513–538 (1979)
8. Megiddo, N., Tamir, A.: New results on the complexity of p-center problems. SIAM J. Comput. **12**, 751–758 (1983)
9. Mirchandani, P.B., Francis, R.L.: Discrete Location Theory. Wiley, New York (1990)

Recognizing Even-Cycle
and Even-Cut Matroids

Cheolwon Heo[(✉)] and Bertrand Guenin[(✉)]

Department of Combinatorics and Optimization, University of Waterloo,
200 University Avenue West, Waterloo, ON N2L 3G1, Canada
{cheo,bguenin}@uwaterloo.ca

Abstract. Even-cycle matroids are elementary lifts of graphic matroids.
Even-cut matroids are elementary lifts of cographic matroids. We give
a polynomial time algorithm to check if a binary matroid is an even-
cycle matroid. We also give a polynomial time algorithm to check if a
binary matroid is an even-cut matroid. These algorithms rely on structural properties of the class of pinch-graphic matroids.

Keywords: Binary matroids · Signed graphs · Complexity

1 Introduction

By a *cycle* in a graph we mean a subset of the edges with the property that
every vertex of the subgraph formed by these edges has even degree. A matroid
M is *graphic* if its circuits are precisely the inclusion-wise minimal non-empty
cycles of some graph G. The vertex-edge incidence matrix of G is a matrix representation of M over the two-element field. In particular, graphic matroids are
binary. Tutte [18] proved that one can recognize if a binary matroid is graphic in
polynomial-time. Seymour [15] extended this result and showed that there exists
a polynomial time algorithm to check if a matroid specified by an independence
oracle is graphic.

A *signed graph* is a pair (G, Σ) where G is a graph and $\Sigma \subseteq EG$. A cycle
$C \subseteq EG$ is *even* (resp. *odd*) if $|C \cap \Sigma|$ is even (resp. odd). M is an *even-cycle
matroid* if its circuits are precisely the inclusion-wise minimal non-empty even
cycles of some signed graph (G, Σ). The matrix obtained from the vertex-edge
incidence matrix of G by adding a row corresponding to the characteristic vector
of Σ is a matrix representation of M over the two-element field. In particular,
even-cycle matroids are binary and are elementary lifts of graphic matroids.
Even-cycle matroids are examples of *lift matroids* [21]. A *graft* is a pair (G, T)
where G is a graph and $T \subseteq VG$ where $|T|$ is even. Vertices in T are called
terminals. A cut $\delta(U) := \{uv \in EG : u \in U, v \notin U\} \neq \emptyset$ is *even* (resp. *odd*) if
$|T \cap U|$ is even (resp. odd). M is an *even-cut matroid* if its circuits are precisely

Supported by NSERC grant 238811 and ONR grant N00014-12-1-0049.

D. Bienstock and G. Zambelli (Eds.): IPCO 2020, LNCS 12125, pp. 182–195, 2020.
https://doi.org/10.1007/978-3-030-45771-6_15

the inclusion-wise minimal non-empty even cuts of some graft (G, T). Even-cut matroids are binary and are elementary lifts of cographic matroids [13].

We are ready to state the main results of our paper,

- Given a binary matroid M described by its $0, 1$ matrix representation A, we present an algorithm that will check if M is an even-cycle matroid, in time polynomial in the number of entries of A.
- Given a binary matroid M described by its $0, 1$ matrix representation A, we present an algorithm that will check if M is an even-cut matroid, in time polynomial in the number of entries of A.

We believe that these algorithms ought to be fast in practice but have not conducted numerical experiments. For both algorithms the bound on the running time depends on a constant c that arises from the Matroid Minors Project and that has no explicit bound [5]. However, the algorithm does not use the value c for its computation. In the interest of brevity we only discuss the first algorithm in this paper. The second algorithm is similar and is also based on the recognition of pinch-graphic matroids.

2   A Simple Algorithm for Recognizing Graphic Matroids

2.1   Reduction to the 3-Connected Case

Consider a matroid M with rank function r, a set $X \subseteq EM$ is k-separating if $r(X) + r(EM - X) - r(M) \leq k - 1$.[1] A set $X \subseteq EM$ is a k-separation if it is k-separating and $|X|, |EM - X| \geq k$. M is connected if it has no 1-separations, and M is 3-connected if it is connected and has no 2-separations. A matroid M has a 1-separation if and only if M can be expressed as a 1-sum, $M_1 \oplus_1 M_2$. A connected matroid has a 2-separation if and only if M can be expressed as a 2-sum, $M_1 \oplus_2 M_2$ [1,2,14]. Moreover, for $k \in \{1, 2\}$, $M = M_1 \oplus_k M_2$ is graphic if and only if both M_1 and M_2 are graphic [12], Corollary 7.1.26. Assume that we know how to check if a 3-connected binary matroid is graphic and suppose that we want to check if an arbitrary binary matroid M is graphic. If M is 3-connected use the algorithm for 3-connected matroids. Otherwise find a k-separation for $k \in \{1, 2\}$, express M as $M_1 \oplus_k M_2$ and recursively check if M_1 and M_2 are both graphic, if so then M is graphic otherwise M is not. We need to be able to check for the presence of 1- and 2-separations in a binary matroid in polynomial time. Cunningham and Edmonds [2] showed that the more general problem of checking if a matroid has a k-separating set with separators of size at least $\ell \geq k$ can be reduced to the matroid intersection problem [3,10] and be solved in polynomial-time for fixed values k and ℓ.

---

[1] For sets A, B we denote by $A - B$ the set $\{a \in A : a \notin B\}$.

2.2   Graph Representations

Given a graph G and $X \subseteq EG$, we write $G|X$ for the subgraph of G with edges X and vertices that correspond to endpoints of edges of X. We denote $\partial(X)$ the set of vertices common to $G|X$ and $G|(EG - X)$. Consider a graph G with a partition X, Y of its edge set where $G|X$ and $G|Y$ are connected and where $\partial(X)$ consists of two vertices v_1 and v_2. Let G' be obtained from G by identifying, for $i = 1, 2$, vertex v_i of $G|X$ with vertex v_{3-i} of $G|Y$. We say that G' is obtained from G by a 2-*flip* on the set X (resp. Y). We call a 1-*flip* the identification of two vertices in distinct components or splitting two blocks into different components. Two graphs are *equivalent* if they are related by a sequence of 1-flips and 2-flips. Given a graphic matroid M where the cycles of M correspond to the cycles of a graph G we say that G is a *graph representation* of M. Whitney [20] proved that any two graph representations of a graphic matroid are equivalent. This implies in particular,

Theorem 1. *A 3-connected graphic matroid has a unique graph representation.*

Given a graph G we denote by $G/I \setminus J$ the graph obtained from G by contracting edges I and deleting edges J. Given a matroid M we denote by $M/I \setminus J$ the matroid obtained from M by contracting elements I and deleting elements J. Consider a graphic matroid M with graph representation G. Then $H = G/I \setminus J$ is a graph representation of the minor $N = M/I \setminus J$. In particular, the class of graphic matroids is minor-closed. We say that the representation H of N *extends* to the representation G of M. Theorem 1 implies the following result:

Remark 1. Suppose N is a 3-connected graphic matroid with a graph representation H. If N is a minor of a 3-connected matroid M, then M is graphic if and only if the representation H of N extends to a representation of M.

2.3   The Algorithm

A *wheel* is the graph obtained by starting with a circuit with at least three edges, adding a new vertex (the hub) and connecting every vertex of the circuit to the hub. Consider a 3-connected binary matroid M and suppose that we wish to check if M is graphic. First we check if M is the graphic matroid of a wheel. Otherwise it follows from Tutte's Wheels-and-Whirls Theorem [19] that there exists an element e, and for either: $N = M/e$ or $N = M \setminus e$, N is 3-connected. Recursively, we check if N is graphic, if it is not, then neither is M. Otherwise, we check if the (unique) representation of N extends to M. If it does then M is graphic, otherwise it is not.

3   A First Attempt at Generalization

In this section we try to generalize the algorithm outlined in the previous section to recognize even-cycle matroids and identify some pitfalls.

3.1   Signed Graph Representations

We say that $\Gamma \subseteq EG$ is a *signature* of a signed graph (G, Σ) if (G, Σ) and (G, Γ) have the same even cycles. Equivalently, Γ is a signature of (G, Σ) if $\Gamma \triangle \Sigma := \Gamma \cup \Sigma - \Gamma \cap \Sigma$ is a cut of G [8]. The operation that consists of replacing a signature by another signature is called *resigning*. A pair of signed graphs are *equivalent* if they are related by a sequence of 1-flips, 2-flips, and resignings. Given an even-cycle matroid M where the cycles of M correspond to the even cycles of a signed graph (G, Σ) we say that (G, Σ) is a *signed graph representation* of M. We will see that in contrast to Theorem 1, 3-connected even-cycle matroids can have inequivalent representations. Consider a signed graph (G, Σ) and $I, J \subseteq EG$ where $I \cap J = \emptyset$. The minor $(G, \Sigma) \setminus I / J$ is the signed graph defined as follows: If there exists an odd circuit of (G, Σ) contained in I then $(G, \Sigma)/I \setminus J = (G/I \setminus J, \emptyset)$, otherwise there exists a signature Γ where $\Gamma \cap I = \emptyset$ and $(G, \Sigma)/I \setminus J = (G/I \setminus J, \Gamma - J)$. Note, minors are only defined up to resigning. Consider an even-cycle matroid M with a signed graph representation (G, Σ). Then $(H, \Gamma) = (G, \Sigma)/I \setminus J$ is a graph representation of the minor $N = M/I \setminus J$ [13], page 21. In particular, the class of even-cycle matroids is minor-closed. We say that the representation (H, Γ) of N *extends* to the representation (G, Σ) of M.

Remark 2. Suppose N is an even-cycle matroid that is a minor of a matroid M. Then M is an even-cycle matroid if and only if *some* signed graph representation of N extends to M.

3.2   A Bad Example

In light of the similarities between Remark 1 and Remark 2 it is natural to wonder if the strategy outlined in Sect. 2, can be used to check if a 3-connected binary matroid M is an even-cycle matroid. Namely, we would find e such that $N = M/e$ or $N = M \setminus e$ is 3-connected. We would then, recursively, find all signed graph representations of N (up to equivalence) and then check which of these representations extend to M. Alas this does not lead to a polynomial-time algorithm as we can have an exponential number of pairwise inequivalent representations as we illustrate next.

Consider a 2-connected graph H with subsets $X_1 \subset \ldots \subset X_k \subset EH$ $(k \geq 1)$ where for all $i \in \{1, \ldots, k\}$, $|\partial(X_i)| = 2$ and for all distinct $i, j \in \{1, \ldots, k\}$, $\partial(X_i) \cap \partial(X_j) = \emptyset$. Consider distinct vertices u_1, u_2, v_1, v_2 of H where $u_1, u_2 \in V(H|X_1) - \partial(X_1)$ and $v_1, v_2 \in V(H|(EH - X_k)) - \partial(X_k)$. Let G be obtained from H by identifying u_i and v_i for $i = 1, 2$. Let $\Sigma = \delta_H(u_1) \triangle \delta_H(u_2)$.[2] We call the signed graph (G, Σ) obtained from that construction a *donut*. This construction is illustrated in Fig. 1(i) for the case $k = 3$. In that example let $A = X_1$, $B = X_2 - X_1$, $C = X_3 - X_2$ and $D = EH - X_3$. The shaded region next to vertices $u_1 = v_1$ and $u_2 = v_2$ of G corresponds to edges in Σ. (G, Σ) is a representation of some matroid M. Let us now show how to construct other donuts that are also

---

[2] $\delta_H(u_i)$ denotes the set of non-loop edges of H incident to u_i.

representations of M. Let $S \subseteq \{1, \ldots, k\}$ and let H' be obtained from H by doing a 2-flip on the set X_i for each $i \in S$. Let G' be obtained from H by identifying for u_i and v_{3-i} for $i = 1, 2$. Then (G', Σ) is a donut that is also a representation of M, i.e. (G, Σ) and (G', Σ') have the same even cycles [13]. This construction is illustrated in Fig. 1(ii). In that example we pick $S = \{1, 2, 3\}$. There are 2^k donuts that we can obtain in that way and for suitable choice of graph H they will be pairwise inequivalent.

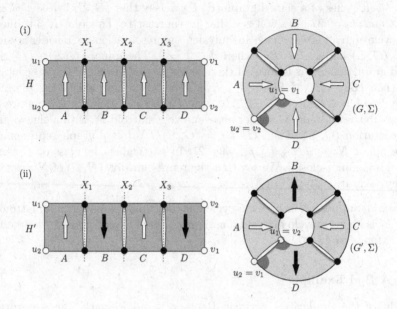

Fig. 1. Constructing donuts.

4   Pinch-Graphic Matroids

4.1   The Definition

Consider a signed graph (G, Σ). A pair of vertices a, b of G is a *blocking pair* if every odd circuit of (G, Σ) uses at least one of vertices a or b. Vertices a, b form a blocking pair if and only if there exists a signature $\Gamma \subseteq \delta(a) \cup \delta(b)$. Observe that the donuts defined in Sect. 3.2 all have a blocking pair. A matroid M is *pinch-graphic* if its circuits are precisely the non-empty inclusion-wise minimal even cycles of a signed graph (G, Σ) with a blocking pair. We say that (G, Σ) is a *blocking pair representation* of M.

Remark 3. Every graphic matroid is pinch-graphic and every pinch-graphic matroid is an even-cycle matroid.

The inclusions are strict in the previous remark, for instance, F_7^* is pinch-graphic but not graphic, and R_{10} is an even-cycle matroid that is not pinch-graphic. If a signed graph has a blocking pair then so does every minor. In particular, the class of pinch-graphic matroids is minor-closed. We saw that even-cycle matroids are elementary lifts of graphic matroids. Pinch-graphic matroids are also elementary projections of graphic matroids [13], page 30.

4.2 Even-Cycle Matroids that Are Not Pinch-Graphic

The matroids described in Sect. 3.2 are pinch-graphic. Hence, pinch-graphic matroids can have exponentially many pairwise inequivalent signed graph representations. This in stark contrast with even-cycle matroids that are not pinch graphic as the next result illustrates,

Theorem 2. *There exists a constant c such that every even-cycle matroid that is not pinch-graphic has fewer than c pairwise inequivalent signed graph representations.*

As an example an even-cycle matroid that contains R_{10} has at most 6 pairwise inequivalent representations [7]. Observe that there are no connectivity conditions in the previous result.

Let us prove Theorem 2 for the case of 3-connected matroids. A binary matroid is *minimally non-pinch-graphic* if it is not pinch-graphic but every proper minor is. A highlight of the Matroid Minor Project [5] is the fact that minor-closed classes of binary matroids are well-quasi ordered. Hence,

Theorem 3. *There exists a constant c, such that every minimally non-pinch-graphic matroid has at most c elements.*

For a matroid N, let $\lambda_1(N)$ denote the number of connected components of N. Now N can be constructed from a collection $\Lambda_2(N)$ of 3-connected matroids by 1-sum and 2-sum. Cunningham and Edmonds [2] showed that $\Lambda_2(N)$ is unique up to isomorphism. Let $\lambda_2(N)$ be the number of matroids in $\Lambda_2(N)$. Lemos and Oxley [11] proved the following result,

Theorem 4. *Let N be a non-empty matroid and M be a minor-minimal 3-connected matroid having N as a minor. Then*

$$|EM| - |EN| \leq 22\left(\lambda_1(N) - 1\right) + 5\left(\lambda_2(N) - 1\right).$$

The relation "(G, Σ) is equivalent to (G', Σ')" defines an equivalence relation. Hence, for an even-cycle matroid M, the set of signed graph representations can be partitioned into equivalence classes. Guenin, Pivotto, and Wollan proved [7],

Theorem 5. *Let M be a 3-connected matroid and let N be a 3-connected minor of M that is not pinch-graphic. For every equivalence class $\mathcal{F}$ of N, the set of extensions of $\mathcal{F}$ to M is the union of at most two equivalence classes.*

Let us now prove Theorem 2 for the 3-connected case, i.e. we have a 3-connected matroid M that is not pinch-graphic and we need to show that M has a constant number of pairwise inequivalent signed graph representations. Since M is not pinch-graphic, it has a minor N that is minimally non-pinch-graphic. By Theorem 3, $|E(N)| \leq c$ for some constant c. In particular, $\lambda_1(N), \lambda_2(N) \leq c$. Let N' be a minor-minimal matroid with the following properties: (a) N' is 3-connected, (b) N is a minor of N' and (c) N' is a minor of M. Since $M = N'$ satisfies (a)–(c), N' is well-defined. By Theorem 4, $|EN'| \leq c + 22(c-1) + 5(c-1) \leq 28c$. Thus N' has a constant number, say d, of equivalence classes. It follows by Theorem 5 that there are at most $2d$ equivalence classes for M.

4.3   Recognition: From Pinch-Graphic to Even-Cycle Matroids

Let us assume now that we have a polynomial time algorithm to check if a binary matroid is pinch-graphic. We can use it to check in polynomial-time if a matroid M is an even-cycle matroid. First we check if M is pinch-graphic, if it is, then M is an even-cycle matroid and we stop. Thus we may assume M is not pinch-graphic. If for every $e \in M$, M/e and $M \backslash e$ is pinch-graphic then M is minimally non-pinch graphic. It follows by Theorem 3 that M has constant size. Then we can find all representations of M, up to equivalence, in constant time (A finite algorithm for finding all representations, up to equivalence, of a an even-cycle matroid is given in [13], page 132.) Otherwise, there exists e such that $N = M/e$ or $N = M \backslash e$ is not pinch-graphic. Recursively, we construct every equivalence class (of signed graph representations) of N. Then we construct every equivalence class of M from the equivalence classes of N. If M has a signed graph representation, then it is an even-cycle matroid, otherwise it is not. Unlike the algorithm sketched in Sect. 2 we are keeping track of equivalence classes of signed graphs rather than individual signed graphs. It turns out that it suffices to keep track of one representative of each equivalence class. To construct the representative of equivalence classes of M from N we use the algorithm to check if a matroid is graphic as a subroutine. Note, by Theorem 2 the algorithm never has to keep track of more than a constant number of equivalence classes.

5   Internally 4-Connected Pinch-Graphic Matroids

In light of the result in Sect. 4.3 to obtain a polynomial time algorithm to recognize even-cycle matroids, it suffices to obtain a polynomial time algorithm to recognize pinch-graphic matroids. In this section we consider the case where we wish to check if an internally 4-connected matroid is pinch-graphic.

5.1   Connectivity Helps, up to a Point

A matroid M is *internally 4-connected* if it is 3-connected and for every $X \subseteq EM$ that is 3-separating, $\min\{|X|, |EM - X|\} \leq 3$. In Sect. 3.2 we saw that the pinch-graphic matroids M with a donut representation (G, Σ) can have an exponential number of pairwise inequivalent blocking pair representations. However, for large donuts ($k \geq 6$), M is not internally 4-connected. On the other hand, while Theorem 1 says that 3-connected graphic matroids have a unique representation, we give examples of internally 4-connected pinch-graphic matroids with a linear (in the rank) number of blocking pair representations. A *double wheel* is the signed graph obtained by starting with an odd circuit, adding a new vertex (the hub) and joining every vertex of the circuit with both an odd and an even edge. See Fig. 2(i) for an example of such a graph (G, Σ) with 6 vertices on the rim. Odd edges correspond to dashed lines. Every odd edge is incident to either a or b, in particular, a, b is a blocking pair. Get a new signed graph (H, Σ) by replacing odd edge 6 joining a and b by an odd loop, and moving the end of every other odd edge incident to a to b and vice-versa. This is known as a *Lovász-flip* and it preserves the even-cycles [6]. See Fig. 2(ii) for an example. We can repeat the same construction for every rim vertex (after resigning). For a general double wheel (G, Σ), this yields $|VG|$ pairwise inequivalent blocking pair representations of a pinch graphic matroid M and $r(M) = |VG|$.

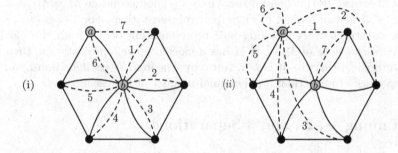

Fig. 2. (i) Double wheel (G, Σ). (ii) Double wheel after Lovász-flip (H, Σ).

5.2 Preserving Connectivity

A binary matroid is *minimally non-graphic* if it is non-graphic but every proper minor is graphic. Tutte [16,17] proved that the set of minimally non-graphic binary matroids is $\mathcal{S} = \{F_7, F_7^*, M(K_{3,3})^*, M(K_5)^*\}$ where F_7 is the Fano, $M(G)$ is the graphic matroid with graph representation G, and M^* is the dual of M. A matroid is *almost 4-connected* if it is 3-connected and $\min\{|X|, |EM - X|\} \leq 6$. Thus internally 4-connected matroids are almost 4-connected. We define a *good sequence* to be a sequence of almost 4-connected matroids $M_1, \ldots, M_k$ where $M_k \in \mathcal{S}$ and where for all $i \in \{1, \ldots, k-1\}$, M_{i+1} is a single element deletion or contraction of M_i. Combining Seymour's Splitter Theorem [14] with a Splitter Theorem by Geelen and Zhou for internally 4-connected binary matroids [4], we derive the following result,

Theorem 6. *If M is an internally 4-connected binary matroid that is not graphic then there exists a good sequence $M_1, \ldots, M_k$ where $M = M_1$.*

Next, we use connectivity to get a polynomial bound on the number of blocking pair representations.

Theorem 7. *Let $M_1, \ldots, M_k$ be a good sequence, then for all $i \in \{1, \ldots, k\}$, M_i has at most $\mathcal{O}(|EM_i|^4)$ distinct blocking pair representations.*

Note that this theorem only bounds the number of blocking pair representations. Pinch-graphic matroids may have signed-graph representations without blocking pairs that are very complicated. This is the main reason we use a two step approach: (1) recognize pinch-graphic matroid (and only consider blocking pair representations), then (2) recognize even-cycle matroids that are not pinch-graphic. We outline a proof of Theorem 7 in the Appendix A.

5.3   Recognition: Is an Internally 4-Connected Matroid pinch-Graphic?

Suppose we wish to check if an internally 4-connected binary matroid M is pinch-graphic. First we check if M is graphic, if it is, then M is pinch-graphic and we stop. Otherwise, by Theorem 6 there exists a good sequence $M = M_1, \ldots, M_k$. For $M_k \in \mathcal{S}$ we can find all blocking pair representations. For all $i \in \{1, \ldots, k-1\}$, we construct every blocking pair representation of M_i from the blocking pair representations of M_{i+1}. If M has a blocking pair representation, then it is pinch-graphic, otherwise it is not. Note, by Theorem 7 the algorithm never has to keep track of more than a polynomial number of representations.

6   Taming 1-, 2-, and 3-Separations

In Sect. 5.3 we saw an outline of a polynomial-time algorithm to recognize if an internally 4-connected binary matroid is pinch-graphic. In this section, we will sketch an algorithm that takes as input an arbitrary binary matroid M and in polynomial-time either: certifies that M is pinch-graphic, or not pinch-graphic, or finds a matroid N isomorphic to a minor of M where N is internally 4-connected and is pinch-graphic if and only if M is pinch-graphic. We can then combine the algorithm in Sect. 5.3 with this algorithm to recognize if an arbitrary binary matroid is pinch-graphic.

6.1   Reduction to the 3-Connected Case

A 3-connected binary matroid M has a 3-separation if and only if M can be expressed as a 3-*sum* $M_1 \oplus_3 M_2$ [14]. We first prove,

Proposition 1. *Let $M = M_1 \oplus_1 M_2$ be a binary matroid where M_1 is graphic. Then M is pinch-graphic if and only if M_2 is pinch-graphic.*

Proposition 2. *Let $M = M_1 \oplus_2 M_2$ be a connected binary matroid where M_1 is graphic. Then M is pinch-graphic if and only if M_2 is pinch-graphic.*

Proposition 3. *Let* $M = M_1 \oplus_3 M_2$ *be a 3-connected binary matroid where* M_1 *is graphic. Then* M *is pinch-graphic if and only if* M_2 *is pinch-graphic.*

Consider a binary matroid M with a k-separation X for some $k \in \{1, 2, 3\}$. We say that X is *reducible* if $M = M_1 \oplus_k M_2$ for some matroids M_1 and M_2 where at least one of M_1 or M_2 is graphic. We then prove that 1- and 2-separations in pinch-graphic matroids are reducible.

Proposition 4. *If* X *is a 1-separation of a pinch-graphic matroid* M *or* X *is a 2-separation of a connected pinch-graphic matroid, then* X *is reducible.*

Suppose you are now given a binary matroid M and would like to check if it is pinch-graphic. If M has a k-separation for $k \in \{1, 2\}$, then $M = M_1 \oplus_k M_2$. Then $X = EM_1 - EM_2$ is a k-separation. By Proposition 4, X is reducible, say M_1 is graphic. Then by Proposition 1 or 2, M_2 is pinch-graphic if and only if M is pinch-graphic. Then we apply the algorithm recursively to M_2. When we are done the resulting matroid is 3-connected. Thus it suffices to check if 3-connected binary matroids are pinch-graphic.

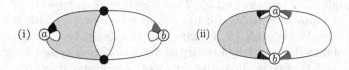

Fig. 3. Examples of 3-separations that are not reducible.

6.2   Structure of 3-Separations

There is no analogue of Proposition 4 for 3-separations. Indeed, 3-separations of pinch-graphic matroids need not be reducible as we illustrate next. Consider the signed graphs (G_1, Σ_1) and (G_2, Σ_2) illustrated in Fig. 3(i) and (ii) respectively. The shaded region corresponds to edges X. In black we indicate a signature with all edges incident to the blocking pair a, b. Then X is a 3-separation of the corresponding pinch-graphic matroid and in general X is not reducible. If a pinch-graphic matroid M has a blocking pair representation of the form given by (G_1, Σ_1) then X is a *compliant* separation. If M has a representation of the form given by (G_2, Σ_2) then X is a *recalcitrant* separation.

Given a 3-separation X, we can get a new 3-separation X' by moving elements that are in the co-closure of either side of the separation. We say that X and X' are *homologous*. There are at most 8 separations that are homologous to X. Next we characterize the structure of 3-separations in pinch-graphic matroids.

Theorem 8. *Let* M *be a 3-connected pinch-graphic matroid. For every proper 3-separation there exists a homologous proper 3-separation that is either reducible, compliant, or recalcitrant.*

Suppose you are given a 3-connected binary matroid M and would like to check if it is pinch-graphic. If it has a proper 3-separation X we consider all homologous proper 3-separations X'. If X' is reducible, then $M = M_1 \oplus_3 M_2$ for some M_1, M_2 where say M_1 is graphic. Then by Proposition 3, M_2 is pinch-graphic if and only if M is pinch-graphic and we apply the algorithm recursively to M_2. If X is compliant or recalcitrant then M is pinch-graphic. If none of the homologous separations are reducible, compliant, or recalcitrant, then by Theorem 8, M is not pinch-graphic. When we are done the matroid is internally 4-connected as required. We need to be able to check if a separation is in fact (a) reducible, (b) compliant, or (c) recalcitrant. Cases (a) and (b) can be reduced to checking if a binary matroid is graphic. Case (c) is more complex, there we either prove that the separation is recalcitrant or find another reducible separation.

A   Appendix: Outline of the Proof of Theorem 7

A.1   Pinch Cographic Matroids

Given an even-cut matroid M where the cycles of M correspond to the even cuts of a graft (G, T) we say that (G, T) is a *graft representation* of M. An even-cut matroid is *pinch-cographic* if it has a graft representation with at most four terminals. Consider a graft (H, T) with four terminals, i.e. $T = \{t_1, t_2, t_3, t_4\}$. Let G be obtained from H by identifying vertices t_1 and t_2 and by identifying vertices t_3 and t_4. Denote by a the vertex of G corresponding to $t_1 = t_2$ and by b the vertex of G corresponding to $t_3 = t_4$. Let $\Sigma = \delta_H(t_1) \triangle \delta_H(t_3)$. Then (G, Σ) is a signed graph with blocking pair a, b. We say that (G, Σ) is obtained from (H, T) by *folding* and that (H, T) is obtained from (G, Σ) by *unfolding*. Pinch-graphic and pinch-cographic matroids are duals [13], page 26.

Proposition 5. *Let (G, Σ) be a signed graph with a blocking pair and let (H, T) be obtained from (G, Σ) by unfolding. Let M be the pinch-graphic matroid with representation (G, Σ) and let N be the pinch-cographic matroid with representation (H, T). Then $M^* = N$.*

A.2   Sizes of Equivalence Classes

Recall that a pair of signed graphs are equivalent if they are related by 1-flips, 2-flips, and resigning.

Proposition 6. *There exists a constant c_1 such that for every non-graphic, pinch-graphic matroid that is almost 4-connected, the number of pairwise equivalent blocking pair representations is at most $c_1 r(M)^3$ where $r(M)$ denotes the rank of M.*

A pair of grafts (H, T) and (H', T') are *equivalent* if they have the same even-cuts and H and H' are related by 1-flips and 2-flips.

Proposition 7. *There exists a constant c_2 such that for every non-cographic, pinch-cographic matroid that is almost 4-connected, the number of pairwise equivalent graft representations with four terminals is at most c_2.*

We will require the following observations [7],

Remark 4. If a pair of signed graphs have the same even-cycles and a common odd cycle then they are equivalent. If a pair of grafts have the same even-cuts and a common odd cut then they are equivalent.

A.3 Counting Representations

Let $M_1, \ldots, M_k$ be a good sequence and let $i \in \{1, \ldots, k\}$. For $i \in \{1, \ldots, k\}$, let $f(i)$ denote the number of blocking pair representations of M_i. We will show,

$$f(i) \leq 8 + 6c_2 r(M_i) + c_1 r(M_i^*) r(M_i)^3 \in \mathcal{O}(|EM_i|^4). \tag{1}$$

Proceed by induction on $k - i$. If $k - i = 0$, i.e. $M_i = M_k$ then M_k is minimally non-graphic and $f(k) \leq 8$. Otherwise $M_{i+1} = M_i \backslash e$ or $M_{i+1} = M_i / e$.

Consider first the case where $M_{i+1} = M_i \backslash e$. By induction, (1) holds for $i+1$, i.e. $f(i+1) \leq 8 + 6c_2 r(M_{i+1}) + c_1 r(M_{i+1}^*) r(M_{i+1})^3$. Since $r(M_i^*) = r(M_{i+1}^*) + 1$ to prove that (1) holds for M_i we will show $f(i) \leq f(i+1) + c_1 r(M_i)^3$. Every blocking pair representation of M_i extends some blocking pair representation of M_{i+1}. If each of these representations of M_{i+1} extends to at most one representation of M_i then $f(i) \leq f(i+1)$. We prove that if a blocking pair representation of M_{i+1} extends to more than one blocking pair representation of M_i then in each of these representations, e is an odd loop. By Remark 4 each of these representations are pairwise equivalent. By Proposition 6 there are at most $c_1 r(M_i)^3$ of these representations. Thus $f(i) \leq f(i+1) + c_1 r(M_i)^3$ as required.

Consider now the case where $M_{i+1} = M_i / e$. By induction, (1) holds for $i + 1$ and since $r(M_i) = r(M_{i+1}) + 1$ to prove that (1) holds for M_i it suffices to show $f(i) \leq f(i + 1) + 6c_2$. We only consider an example here where we have two distinct blocking pair representations (G_1, Σ) and (G_2, Σ) of M_i where $(G, \Sigma) := (G_1, \Sigma)/e = (G_2, \Sigma)/e$ is a representation of M_{i+1} with $\Sigma \subseteq \delta_G(a) \cup \delta_G(b)$ and $a, b \in VG$, and where G_1 is obtained from G by splitting vertex a into two vertices, say a', a'' so that all edges in $\delta_G(a) \cap \Sigma$ are incident to a' and no edge of $\delta_G(a) \cap \Sigma$ is incident to a'' and by joining a', a'' by e and where G_2 is obtained from G by applying the same construction but now to vertex b. (G, Σ) is illustrated in Fig. 4(i) and (G_1, Σ) and (G_2, Σ) in Fig. 4(ii). We say that (G_i, Σ) is obtained from (G, Σ) by *splitting a signature*. How many representations of M_i are obtained in that way? For each representation obtained by splitting a signature, let (H, T) be the graft obtained by unfolding that representation. See Fig. 4(iii) for an illustration where square vertices correspond to terminals. Observe that e is an odd cut of (H, T). By Remark 4, each of these grafts are equivalent. By Propositions 5 and 7 there are at most c_2 such grafts. Moreover, every representation of M_i obtained by splitting a signature is obtained by folding such a graft. There are $\binom{4}{2}$ ways of folding a graft, hence, at most $6c_2$ representations obtained by splitting a signature. Hence, $f(i) \leq f(i+1) + 6c_2$ in this case. The analysis for the other cases is similar.

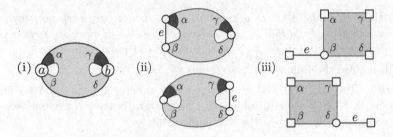

Fig. 4. Non unique extension and unfolding.

References

1. Bixby, R.E.: Composition and decomposition of matroids and related topics. Ph.D. thesis, Cornell University (1972)
2. Cunningham, W.H.: A combinatorial decomposition theory. Ph.D. thesis, University of Waterloo (1973)
3. Edmonds, J.: Matroid intersection. In: Hammer, P.L., Johnson, E.L., Korte, B.H. (eds.) Discrete Optimization I. Annals of Discrete Mathematics, vol. 4, pp. 39–49, North-Holland, Amsterdam (1979)
4. Geelen, J.F., Zhou, X.: A splitter theorem for internally 4-connected binary matroids. SIAM J. Discret. Math. **20**, 578–587 (2016)
5. Geelen, J.F., Gerards, A.M.H., Whittle, G.: Towards a matroid-minor structure theory. In: Grimmett, G., Mcdiarmid, C. (eds.) Combinatorics, Complexity, and Chance. A tribute to Dominic Welsh, Oxford Lecture Series in Mathematics and its Applications, vol. 34, pp. 72–82. Oxford University Press, Oxford (2007)
6. Gerards, A.M.H.: Personal communication
7. Guenin, B., Pivotto, I., Wollan, P.: Stabilizer theorems for even cycle matroids. J. Comb. Theory Ser. B **118**, 44–75 (2016)
8. Harary, F.: On the notion of balance of a signed graph. Michigan Math. J. **2**, 143–146 (1953)
9. Heo, C.: Recognizing even-cycle and even-cut matroids. Master's thesis, University of Waterloo (2016)
10. Lawler, E.: Combinatorial Optimization: Networks and Matroids. Holt, Rinehart and Winston, New York (2001). Reprinted 2001, Dover, Mineola
11. Lemos, M., Oxley, J.: On the minor-minimal 3-connected matroids having a fixed minor. Eur. J. Comb. **24**, 1097–1123 (2003)
12. Oxley, J.: Matroid Theory, 2nd edn. Oxford University Press Inc., New York (2011)
13. Pivotto, I.: Even cycle and even cut matroids. Ph.D. thesis, University of Waterloo (2011)
14. Seymour, P.D.: Decomposition of regular matroids. J. Comb. Theory Ser. B **28**, 305–359 (1980)
15. Seymour, P.D.: Recognizing graphic matroids. Combinatorica **1**, 75–78 (1981)
16. Tutte, W.T.: Matroids and graphs. Trans. Am. Math. Soc. **90**, 527–552 (1959)
17. Tutte, W.T.: Lectures on matroids. J. Res. Natl. Bur. Stand. Sect B **69B**, 1–47 (1965)
18. Tutte, W.T.: An algorithm for determining whether a given binary matroid is graphic. Proc. Am. Math. Soc. **11**, 905–917 (1960)

19. Tutte, W.T.: Connectivity in matroids. Canad. J. Math. **18**, 1301–1324 (1966)
20. Whitney, H.: 2-isomorphic graphs. Am. J. Math. **55**, 245–254 (1933)
21. Zaslavsky, T.: Biased graphs. II. The three matroids. J. Comb. Theory Ser. B **51**, 46–72 (1991)

A Combinatorial Algorithm for Computing the Rank of a Generic Partitioned Matrix with 2×2 Submatrices

Hiroshi Hirai[1] and Yuni Iwamasa[2]

[1] The University of Tokyo, Tokyo 113-8656, Japan
hirai@mist.i.u-tokyo.ac.jp
[2] Kyoto University, Kyoto 606-8501, Japan
iwamasa@i.kyoto-u.ac.jp

Abstract. In this paper, we consider the problem of computing the rank of a block-structured symbolic matrix (a generic partitioned matrix) $A = (A_{\alpha\beta}x_{\alpha\beta})$, where $A_{\alpha\beta}$ is a 2×2 matrix over a field $\mathbf{F}$ and $x_{\alpha\beta}$ is an indeterminate for $\alpha = 1, 2, \ldots, \mu$ and $\beta = 1, 2, \ldots, \nu$. This problem can be viewed as an algebraic generalization of the bipartite matching problem, and was considered by Iwata and Murota (1995). One of recent interests on this problem lies in the connection with non-commutative Edmonds' problem by Ivanyos, Qiao and Subrahamanyam (2018), and Garg, Gurvits, Oliveiva and Wigderson (2019), where a result by Iwata and Murota implicitly says that the rank and the non-commutative rank (nc-rank) are the same for this class of symbolic matrices.

The main result of this paper is a combinatorial $O((\mu\nu)^2 \min\{\mu, \nu\})$-time algorithm for computing the symbolic rank of a (2×2)-type generic partitioned matrix of size $2\mu \times 2\nu$. Our algorithm is based on the Wong sequence algorithm by Ivanyos, Qiao, and Subrahamanyam for the nc-rank of a general symbolic matrix, but is simpler. Our proposed algorithm requires no blow-up operation, no field extension, and no additional care for bounding the bit-size. Moreover it naturally provides a maximum rank completion of A for an arbitrary field $\mathbf{F}$.

Keywords: Generic partitioned matrix · Edmonds' problem · Non-commutative Edmonds' problem

1 Introduction

The maximum matching problem in a bipartite graph G has a natural algebraic interpretation: It amounts to the symbolic rank computation of the matrix A

Hiroshi Hirai was supported by JSPS KAKENHI Grant Number JP17K00029 and JST PRESTO Grant Number JPMJPR192A, Japan. Yuni Iwamasa was supported by JSPS KAKENHI Grant Number JP19J01302, Japan.

D. Bienstock and G. Zambelli (Eds.): IPCO 2020, LNCS 12125, pp. 196–208, 2020.
https://doi.org/10.1007/978-3-030-45771-6_16

defined by $(A)_{ij} := x_{ij}$ if $ij \in E(G)$ and zero otherwise, where x_{ij} is a variable for each edge ij and the row and column sets of A are identified with the color classes of G. Such an algebraic interpretation is known for other matching-type combinatorial optimization problems. Examples include linear matroid intersection, nonbipartite matching, and their generalizations (e.g., linear matroid matching); see [16]. *Edmonds' problem* [2] is a general algebraic formulation, which asks to compute the rank of a symbolic matrix A represented by

$$A = A_1 x_1 + A_2 x_2 + \cdots + A_m x_m. \tag{1.1}$$

Here A_i is a matrix over a field $\mathbf{F}$ and x_i is a variable for each i. Although a randomized polynomial-time algorithm for Edmonds' problem is known [15,17] (if $|\mathbf{F}|$ is large), a deterministic polynomial-time algorithm is not known, which is one of the prominent open problems in theoretical computer science; see e.g., [13]. Known polynomial-time algorithms for the above-mentioned matching-type problems can be viewed as solutions for special Edmonds' problems.

The present article addresses the rank computation (Edmonds' problem) of a matrix of the following block-matrix structure

$$A = \begin{pmatrix} A_{11}x_{11} & A_{12}x_{12} & \cdots & A_{1\nu}x_{1\nu} \\ A_{21}x_{21} & A_{22}x_{22} & \cdots & A_{2\nu}x_{2\nu} \\ \vdots & \vdots & \ddots & \vdots \\ A_{\mu 1}x_{\mu 1} & A_{\mu 2}x_{\mu 2} & \cdots & A_{\mu\nu}x_{\mu\nu} \end{pmatrix}, \tag{1.2}$$

where $A_{\alpha\beta}$ is a 2×2 matrix over a field $\mathbf{F}$ and $x_{\alpha\beta}$ is a variable for $\alpha = 1, 2, \ldots, \mu$ and $\beta = 1, 2, \ldots, \nu$. Recall that bipartite matching is precisely the case where each $A_{\alpha\beta}$ is a 1×1 matrix. This type of matrices, which we call (2×2)-*type generic partitioned matrices*, was considered in detail by Iwata and Murota [12], subsequent to the study on partitioned matrices of general type [8]. They established a min-max formula (i.e., good characterization) for the rank of this class of matrices, which involves the minimization of a submodular function on the lattice of vector subspaces. A combinatorial polynomial-time rank-computation has not been known and has been desired. Our main result solves this issue.

Theorem 1. *There exists a combinatorial $O((\mu\nu)^2 \min\{\mu, \nu\})$-time algorithm for Edmonds' problem for a (2×2)-type generic partitioned matrix of the form* (1.2).

This result links the recent development on Edmonds' problem in a noncommutative setting. *Noncommutative Edmonds' problem* [10] asks to compute the rank of matrix of the form (1.1), where x_i and x_j are supposed to be noncommutative, i.e., $x_i x_j \neq x_j x_i$. In this setting, the rank concept can be defined (via the free skew field or the inner rank of a matrix over a ring), and is called *noncommutative rank* or *nc-rank*. Nc-rank is an upper bound of (usual) rank. Surprisingly, the nc-rank can be computed in deterministic polynomial time. The algorithms were given by Garg, Gurvits, Oliveira, and Wigderson [5] for the case of $\mathbf{F} = \mathbf{Q}$, and by Ivanyos, Qiao and Subrahmanyam [11] for an arbitrary field.

The former algorithm (*operator scaling*) is an analytical algorithm motivated from quantum information theory. The latter one, which we call *IQS-algorithm*, is an augmenting-path type algorithm. It utilizes *Wong sequence* [9]—a vector-space analogue of an alternating space-walk—and the formula of nc-rank earlier proved by Fortin and Reutenauer [3]. In fact, for (2×2)-type generic partitioned matrices, the rank formula proved by Iwata and Murota is the same as the nc-rank formula by Fortin and Reutenauer. This means that the rank and nc-rank are the same for this class of matrices and that the polynomial solvability follows from these results. Thus the real contribution of this paper is in the term "combinatorial" in the theorem, which is explained as follows.

Our proposed algorithm is viewed as a combinatorial enhancement of IQS-algorithm for (2×2)-type generic partitioned matrices. As mentioned, IQS-algorithm is an augmenting-path type algorithm: Given a *substitution* $\tilde{A}$ obtained from A by substituting a value in $\mathbf{F}$ to each x_i, construct the Wong sequence for $(A, \tilde{A})$, which is an analogue of augmenting path search in the auxiliary graph. If an augmenting path exists, then one can/try to find other substitution $\tilde{A}'$ with rank $\tilde{A}' >$ rank $\tilde{A}$ and repeat it with updating $\tilde{A} \leftarrow \tilde{A}'$. If an augmenting path does not exist, then one obtains a certificate of the optimality in the nc-rank formula. Here, for reaching rank $\tilde{A} =$ nc-rank A, the algorithm conducts the *blow-up* operation, which replaces A by a larger matrix $A^{(d)} := \sum_{i=1}^{m} A_i \otimes X_i$ for $d \times d$ matrices $X_1, X_2, \ldots, X_m$ of variable entries. It is known that nc-rank A is equal to $1/d$ times the rank of a substitution $\tilde{A}^{(d)}$ for some d (if $|\mathbf{F}|$ is large). The blow-up steps (and field extensions of $\mathbf{F}$) make the algorithm considerably difficult and slow. For our special case, the rank and nc-rank are equal, and therefore a *blow-up free* algorithm is expected and desirable; a naive application of IQS-algorithm cannot avoid the blow-up.

Our algorithm is the first blow-up free one for Edmonds' problem for the class of (2×2)-type generic partitioned matrices. The key concept that we introduce in this paper is a *matching*. It is actually a 2-matching in the graph consisting of edges $\alpha\beta$ with nonzero $A_{\alpha\beta}$, which inherits the 2×2 structure of our matrix A and gives a canonical substitution of A. Incorporating the idea of Wong sequence, we introduce the auxiliary digraph and an augmenting path for a matching. Then our algorithm goes on analogous to the augmenting path algorithm, as one expects. It requires no blow-up operation, no field extension, and no additional care for bounding the bit size. Moreover it naturally provides a maximum rank completion (substitution) $\tilde{A}$ of A for an arbitrary field $\mathbf{F}$. This reveals that the maximum rank completion problem for a (2×2)-type generic partitioned matrix is polynomially solvable for arbitrary $\mathbf{F}$, while this problem is known to be NP-hard in general [1].

All proofs are omitted and will be given in the full version of this paper.

Related Works. It is an interesting research direction to construct a polynomial-time blow-up free algorithm for general matrices A with rank $A =$ nc-rank A. Such an algorithm can decide whether rank and nc-rank are equal, and then leads to a solution of (commutative) Edmonds' problem. A representative example of a matrix A with rank $A =$ nc-rank A is a matrix such that each A_i in (1.1) is a rank-1 matrix. In this case, the rank computation is equivalent to linear matroid intersection problem [16]. Edmonds' matroid intersection algorithm becomes obviously blow-up free. In fact, it can naturally be interpreted from the Wong sequence [7]. Ivanyos, Karpinski, Qiao and Santha [9] gave a Wong-sequence-based blow-up free algorithm for matrices A having an "implicit" rank-1 expression, that is, A becomes to consist of rank-1 summands by some (unknown) linear transformation of variables.

Computation of nc-rank is formulated as submodular function minimization on the modular lattice of vector subspaces. Based on this fact, Hamada and Hirai [6] developed a conceptually different algorithm from [5] and [11]. Via an analogue of the Lovász extension, they solved the problem as a geodesically-convex optimization on a CAT(0)-space. For the case of a (2×2)-type generic partitioned matrix, the submodular function is defined on the direct product of modular lattices of rank-2 with infinite size. Following a pioneering work by Kuivinen [14], Fujishige, Király, Makino, Takazawa and Tanigawa [4] showed the oracle tractability of submodular function minimization on *diamond*, which is a direct product of modular lattices of rank-2 with "finite" size.

Notations. For a positive integer k, we denote $\{1, 2, \ldots, k\}$ by $[k]$. A $p \times q$ matrix B over a field $\mathbf{K}$ is regarded as the bilinear map defined by $B(u, v) := u^\top B v$ for $u \in \mathbf{K}^p$ and $v \in \mathbf{K}^q$. For a vector space $U \subseteq \mathbf{K}^p$, let $U^{\perp_B} \subseteq \mathbf{K}^q$ denote the orthogonal vector space with respect to B:

$$U^{\perp_B} := \{v \in \mathbf{K}^q \mid B(u, v) = 0 \text{ for all } u \in U\}.$$

For a vector space $V \subseteq \mathbf{K}^q$, $V^{\perp_B}$ is defined analogously. We denote by $\ker_{\mathrm{L}}(B)$ and $\ker_{\mathrm{R}}(B)$ the left kernel and the right kernel of B, respectively. That is, $\ker_{\mathrm{L}}(B) := (\mathbf{K}^q)^{\perp_B}$ and $\ker_{\mathrm{R}}(B) := (\mathbf{K}^p)^{\perp_B}$.

A (2×2)-type generic partitioned matrix A of the form (1.2) is regarded as a matrix over the field $\mathbf{F}(x)$ of rational functions with variables $x_{\alpha\beta}$ for $\alpha \in [\mu]$ and $\beta \in [\nu]$. Symbols α, β, and γ are used to represent elements of $[\mu]$, $[\nu]$, and $[\mu] \sqcup [\nu]$, respectively, where $\sqcup$ denotes the direct sum. We often drop "$\in [\mu]$" from "$\alpha \in [\mu]$" if it is clear in the context. For each α and β, consider 2-dimensional $\mathbf{F}$-vector spaces $U_\alpha = \mathbf{F}^2$ and $V_\beta = \mathbf{F}^2$. Each submatrix $A_{\alpha\beta}$ is considered as a bilinear form $U_\alpha \times V_\beta \to \mathbf{F}$. The vector spaces U_α and V_β are also regarded as ones over $\mathbf{F}(x)$ by the scalar extensions $\mathbf{F}(x) \otimes U_\alpha$ and $\mathbf{F}(x) \otimes V_\beta$, respectively. Letting $U := \mathbf{F}(x)^{2\mu} = \bigoplus_\alpha U_\alpha$ and $V := \mathbf{F}(x)^{2\nu} = \bigoplus_\beta V_\beta$, the matrix A is viewed as a bilinear form $U \times V \to \mathbf{F}(x)$.

2   Matching

In this section, we introduce the concept of *matching* for a (2×2)-type generic partitioned matrix A of the form (1.2). First define the (undirected) bipartite graph $G := ([\mu], [\nu]; E)$ by $E := \{\alpha\beta \mid A_{\alpha\beta} \neq O\}$. An edge $\alpha\beta \in E$ is said to be *rank-k* $(k = 1, 2)$ if $\operatorname{rank} A_{\alpha\beta} = k$. For notational simplicity, the subgraph $([\mu], [\nu]; I)$ for $I \subseteq E$ is also denoted by I. For a node γ, let $\deg_I(\gamma)$ denote the degree of γ in I, i.e., the number of edges in I incident to γ. An edge $\alpha\beta \in I$ is said to be *isolated* if $\deg_I(\alpha) = \deg_I(\beta) = 1$. A connected component of I is said to be *rank-1* if it contains a rank-1 edge.

An edge subset $I \subseteq E$ is called a *matching* if it satisfies the following combinatorial and algebraic conditions (M1)–(M4):

(M1) $\deg_I(\gamma) \leq 2$ for each node γ.

Suppose that I satisfies (M1). Then each connected component of I forms a path or a cycle. Thus I is 2-edge-colorable. Namely, there are two edge classes such that any of two incident edges are in the different classes. An edge in one color class is called a $+$-*edge* and an edge in the other color class is called a $-$-*edge*.

A *valid labeling* for I is a node-labeling that assigns two distinct 1-dimensional subspaces for each node, $U_\alpha^+, U_\alpha^- \subseteq U_\alpha$ for α and $V_\beta^+, V_\beta^- \subseteq V_\beta$ for β, such that for each edge $\alpha\beta$ in I it hold

$$A_{\alpha\beta}(U_\alpha^+, V_\beta^-) = A_{\alpha\beta}(U_\alpha^-, V_\beta^+) = \{0\}, \tag{2.1}$$

$$(\ker_{\mathrm{L}}(A_{\alpha\beta}), \ker_{\mathrm{R}}(A_{\alpha\beta})) = \begin{cases} (U_\alpha^+, V_\beta^+) & \text{if } \alpha\beta \text{ is a rank-1 } +\text{-edge,} \\ (U_\alpha^-, V_\beta^-) & \text{if } \alpha\beta \text{ is a rank-1 } -\text{-edge,} \end{cases} \tag{2.2}$$

(M2) I has a valid labeling.

For a path component C of I, an *end edge* is an edge $\alpha\beta \in C$ with $\deg_I(\alpha) = 1$ or $\deg_I(\beta) = 1$.

(M3) For each path component of I with length at least two, both end edges are rank-1.

(M4) Every cycle component of I contains at least one rank-1 edge.

For $I \subseteq E$, let A_I denote the matrix obtained from A by replacing each submatrix $A_{\alpha\beta}$ with $\alpha\beta \notin I$ by the 2×2 zero matrix. If I is a matching, then rank A_I has a simple combinatorial formula. Let us define

$$r(I) := |I| + \text{the number of isolated rank-2 edges in } I.$$

Then the following holds.

Theorem 2. *For a matching I, it holds* rank $A_I = r(I)$.

Our algorithm gives a constructive proof of the existence of a matching I with rank $A = $ rank A_I. By combining this with Iwata–Murota's minimax formula [12] (or Fortin–Reutenauer's one [3] with rank $A = $ nc-rank A), we obtain the following combinatorial and algebraic minimax theorem between matchings and vector spaces.

Theorem 3. *The maximum of $r(I)$ over matchings I is equal to the minimum of $2\mu + 2\nu - \sum_\alpha \dim X_\alpha - \sum_\beta \dim Y_\beta$ over all vector spaces $X_\alpha \subseteq U_\alpha$ and $Y_\beta \subseteq V_\beta$ such that $A_{\alpha\beta}(X_\alpha, Y_\beta) = \{0\}$ for α and β.*

From Theorem 2, we obtain an explicit expression of the left/right kernel of A_I. For a matching I, define

$$\ker_\alpha(I) := \begin{cases} U_\alpha & \text{if } \deg_I(\alpha) = 0, \\ \ker_L(A_{\alpha\beta}) & \text{if } \alpha \text{ is incident only to a rank-1 edge } \alpha\beta \text{ in } I, \quad (2.3) \\ \{0\} & \text{otherwise.} \end{cases}$$

Also let $\ker_\beta(I)$ be the vector subspace of V_β such that "L", α, and U in (2.3) are replaced by "R", β, and V, respectively.

Corollary 1. *If I is a matching, then $\ker_L(A_I) = \bigoplus_\alpha \ker_\alpha(I)$ and $\ker_R(A_I) = \bigoplus_\beta \ker_\beta(I)$.*

3 Algorithm

Our proposed algorithm is an augmenting-path type algorithm. The following lemma will be used for verifying the optimality of a matching I.

Lemma 1 ([10]). *If there exist vector spaces $U^* \subseteq \mathbf{F}(x)^{2\mu}$ and $V^* \subseteq \mathbf{F}(x)^{2\nu}$ satisfying $(V^*)^{\perp_A} = U^*$, $(U^*)^{\perp_{A_I}} = V^*$, and $U^* \supseteq \ker_L(A_I)$, then $\operatorname{rank} A = \operatorname{rank} A_I$.*

A pair (U^*, V^*) of vector spaces satisfying the conditions in Lemma 1 is called an *optimality witness* of I.

The outline of our algorithm is the following. Let I be a matching of A. We first search an optimality witness of I or an *augmenting trail* for I; this procedure can be interpreted as a combinatorial/unsynchronized version of the computation of the Wong sequence for (A, A_I). The former case establishes $\operatorname{rank} A = \operatorname{rank} A_I$ by Lemma 1, and hence output $r(I)$ by Theorem 2. In the latter case, we have $\operatorname{rank} A > \operatorname{rank} A_I$ and update a matching I to another matching I^* with $\operatorname{rank} A_{I^*} > \operatorname{rank} A_I$ via the augmenting trail.

3.1 Augmenting Trail

For a matching I, the *auxiliary graph* $\vec{G}_I := ([\mu], [\nu]; \vec{E}_I)$ is defined as follows: For each edge $\alpha\beta \notin I$, consider a directed edge $\beta\alpha$. In addition, if $\alpha\beta$ is rank-2, then one more edge $\beta\alpha$ is added. For each isolated rank-2 edge $\alpha\beta$ in I, consider two directed edges $\alpha\beta$ (of the same direction). For the other rank-2 edges $\alpha\beta$ in I (i.e., $\alpha\beta$ belongs to a rank-1 component of I), consider two directed edges $\alpha\beta$ and $\beta\alpha$. For each rank-1 edge $\alpha\beta$ in I, consider one directed edge $\alpha\beta$.

A walk $(\gamma_1\gamma_2, \gamma_2\gamma_3, \ldots, \gamma_{k-1}\gamma_k)$ in $\vec{G}_I$ is called a *path* if all vertices $\gamma_1, \gamma_2, \ldots, \gamma_k$ are distinct, and is called a *trail* if all edges $\gamma_1\gamma_2, \gamma_2\gamma_3, \ldots, \gamma_{k-1}\gamma_k$ are distinct. An ordered tuple $P := (Z_1, \gamma_1\gamma_2, Z_2, \gamma_2\gamma_3, \ldots, \gamma_{k-1}\gamma_k, Z_k)$ is called a *space-trail* for I if $(\gamma_1\gamma_2, \gamma_2\gamma_3, \ldots, \gamma_{k-1}\gamma_k)$ forms a trail in $\vec{G}_I$ and $Z_1, Z_2, \ldots, Z_k$ are vector subspaces of $\mathbf{F}^2$ satisfying $Z_{i+1} = (Z_i)^{\perp_{\gamma_i\gamma_{i+1}}}$ for $i = 1, 2, \ldots, k-1$. Here we abbreviate $\perp_{A_{\alpha\beta}}$ as $\perp_{\alpha\beta}$ or $\perp_{\beta\alpha}$ for $\alpha\beta \in E$. We call γ_1, $\gamma_1\gamma_2$, and

X_1 the *initial vertex*, the *initial edge*, and the *initial space*, respectively. Also we call γ_k, $\gamma_{k-1}\gamma_k$, and X_k the *last vertex*, the *last edge*, and the *last space*, respectively. The associated trail $(\gamma_1\gamma_2, \gamma_2\gamma_3, \ldots, \gamma_{k-1}\gamma_k)$ is also denoted by P, and the associated undirected edge set $\{\gamma_1\gamma_2, \gamma_2\gamma_3, \ldots, \gamma_{k-1}\gamma_k\}$ is denoted by $E(P)$.

An *outer alternating trail* P for I is a space-trail $(Y_1, \beta_1\alpha_1, X_1, \alpha_1\beta_2, \ldots, \beta_k\alpha_k, X_k)$ satisfying the following (O1)–(O3):

(O1) $\dim Y_1 \geq 1$.
(O2) $\beta_i\alpha_i \notin I$ and $Y_i \neq \ker_{\mathrm{R}}(A_{\alpha_i\beta_i})$ for $i = 1, 2, \ldots, k$.
(O3) $\alpha_i\beta_{i+1} \in I$ is an isolated rank-2 edge for $i = 1, 2, \ldots, k-1$.

Also an *inner alternating path* Q for I is a space-trail $(X_1, \alpha_1\beta_1, Y_1, \beta_1\alpha_2, \ldots, \alpha_k\beta_k, Y_k)$ satisfying the following (I1) and (I2).

(I1) Q forms a path in $\vec{G}_I$ that belongs to a rank-1 component of I.
(I2) $\alpha_1\beta_1, \alpha_2\beta_2, \ldots, \alpha_k\beta_k$ satisfy
  - $\alpha_i\beta_i$ is a +-edge and $(X_i, Y_i) = (U_{\alpha_i}^+, V_{\beta_i}^-)$ for $i = 1, 2, \ldots, k-1$, or
  - $\alpha_i\beta_i$ is a --edge and $(X_i, Y_i) = (U_{\alpha_i}^-, V_{\beta_i}^+)$ for $i = 1, 2, \ldots, k-1$.

By the construction of $\vec{G}_I$ and (I1), $\beta_i\alpha_{i+1}$ is rank-2 for $i = 1, 2, \ldots, k-1$

Let P be an outer alternating trail, and Q an inner alternating path. An ordered pair (Q, P) is said to be *compatible* if the last vertex and the last space of Q coincide with the initial vertex and the initial space of P, respectively. Also (P, Q) is said to be *compatible* if the last vertex α of P and the initial vertex α' of Q are the same, and the last space X of P and the initial space X' of Q satisfy

$$X \neq \begin{cases} U_{\alpha'}^- & \text{if } X' = U_{\alpha'}^+, \\ U_{\alpha'}^+ & \text{if } X' = U_{\alpha'}^-. \end{cases} \tag{3.1}$$

An ordered tuple T of space-trails is called an *augmenting trail* for I if the following (A1)–(A5) hold:

(A1) $T = (P_0, Q_1, P_1, \ldots, Q_m, P_m)$ such that
  - P_i is an outer alternating trail for $i = 0, 1, 2, \ldots, m$,
  - Q_j is an inner alternating path for $j = 1, 2, \ldots, m$, and
  - (P_i, Q_{i+1}) and (Q_{i+1}, P_{i+1}) are compatible for $i = 0, 1, 2, \ldots, m-1$.
(A2) The initial space of P_0 coincides with $\ker_\beta(I)$, where β is the initial node of P_0.
(A3) The last space of P_m does not include $\ker_\alpha(I)$, where α is the last node of P_m.
(A4) The union of the trails $P_0, Q_1, P_1, \ldots, Q_m, P_m$ also forms a trail in $\vec{G}_I$.
(A5) For each α and β, no subspaces of U_α and of V_β appear twice in T.

Theorem 4. *For a matching I and an augmenting trail for I, we can construct a matching I^* with $\operatorname{rank} A_{I^*} > \operatorname{rank} A_I$ in $O(|E|^2)$ time.*

An overview of the augmentation procedure is given in Sect. 3.3.

3.2 Finding an Augmenting Trail

In this subsection, we present an algorithm for finding either an optimality witness or an augmenting trail. During the procedure, each vertex γ has (at most two) vector subspaces of $\mathbf{F}^2$, which are used as *labels*. Each pair (γ, Z) of a vertex γ and a label Z at γ has a back pointer to another pair (γ', Z') which represents that "Z is added to the label set $\mathtt{label}(\gamma)$ through Z' and $\gamma'\gamma$." When I is not maximum, i.e., $\mathrm{rank}\, A_I < \mathrm{rank}\, A$, we can construct an augmenting trail by tracking back pointers and composing them.

The formal description of our procedure is given as follows. For each vertex γ, let $\mathtt{label}(\gamma)$ be a label set (or a set of vector spaces) of γ. While updating, set $U_\alpha^* := U_\alpha \cap \bigcap_{X \in \mathtt{label}(\alpha)} X$ for each α and $V_\beta^* := \{0\} + \sum_{Y \in \mathtt{label}(\beta)} Y$ for each β, and $U^* := \bigoplus_\alpha U_\alpha^*$ and $V^* := \bigoplus_\beta V_\beta^*$. As initialization, let $\mathtt{label}(\beta) := \{\ker_\beta(I)\}$ for β with $\ker_\beta(I) \neq \{0\}$, and $\mathtt{label}(\gamma) := \emptyset$ for other γ. Hence, in the initial phase, it holds $U_\alpha^* = U_\alpha$ and $V_\beta^* = \ker_\beta(I)$, implying $V^* = (U^*)^{\perp_{A_I}}$ by Corollary 1. This property $V^* = (U^*)^{\perp_{A_I}}$ is kept during the update.

The procedure is the following. Check first whether there is a triple (α, β, Y) such that $\alpha\beta \in E \setminus I$, $Y \in \mathtt{label}(\beta)$, and $Y^{\perp_{\alpha\beta}} \not\supseteq U_\alpha^*$.

If there is no such a triple, namely, if all triples (α, β, Y) with $\alpha\beta \in E \setminus I$ and $Y \in \mathtt{label}(\beta)$ satisfy $Y^{\perp_{\alpha\beta}} \supseteq U_\alpha^*$, then output the pair (U^*, V^*), which is an optimality witness of I. Stop the procedure.

If there is such a triple (α, β, Y), then choose any (α, β, Y). There are two cases: (i) α does not belong to a rank-1 component in I, and (ii) α belongs to a rank-1 component C in I.

(i). This case corresponds to an expansion of an outer alternating trail. Add $Y^{\perp_{\alpha\beta}}$ to $\mathtt{label}(\alpha)$ and define the back pointer from $(\alpha, Y^{\perp_{\alpha\beta}})$ to (β, Y). Recall that $\ker_\alpha(I) = U_\alpha$ if $\deg_I(\alpha) = 0$, and $\ker_\alpha(I) = \{0\}$ if α is incident to an isolated rank-2 edge in I.

- In the former case, output a tuple of space-trails by tracking back pointers from $(\alpha, Y^{\perp_{\alpha\beta}})$, which is an augmenting trail.
- In the latter case, let $\alpha\beta'$ be the isolated rank-2 edge incident to α. Add $(Y^{\perp_{\alpha\beta}})^{\perp_{\alpha\beta'}}$ to $\mathtt{label}(\beta')$ and define the back pointer from $(\beta', (Y^{\perp_{\alpha\beta}})^{\perp_{\alpha\beta'}})$ to $(\alpha, Y^{\perp_{\alpha\beta}})$. Return to the check phase.

(ii). This case corresponds to an addition of an inner alternating trail. Update

$$\mathtt{label}(\alpha) \leftarrow \begin{cases} \mathtt{label}(\alpha) \cup \{U_\alpha^+\} & \text{if } Y^{\perp_{\alpha\beta}} = U_\alpha^+, \\ \mathtt{label}(\alpha) \cup \{U_\alpha^-\} & \text{if } Y^{\perp_{\alpha\beta}} = U_\alpha^-, \\ \mathtt{label}(\alpha) \cup \{U_\alpha^+, U_\alpha^-\} & \text{if } U_\alpha^+ \neq Y^{\perp_{\alpha\beta}} \neq U_\alpha^-; \end{cases}$$

see (3.1). Define the back pointer from (α, X) to (β, Y) for each label X newly added to $\mathtt{label}(\alpha)$. Since α belongs to a rank-1 component C, if $\deg_I(\alpha) = 1$ then α is incident to a rank-1 edge in I; see (M3). Hence $\ker_\alpha(I) \neq \{0\}$ if $\deg_I(\alpha) = 1$, and $\ker_\alpha(I) = \{0\}$ if $\deg_I(\alpha) = 2$.

- If there is a newly added X with $\ker_\alpha(I) \not\subseteq X$, i.e., α is incident only to one rank-1 edge in I and $\{0\} \neq \ker_\alpha(I) \neq X$, then output a tuple of space-trails by tracking the back pointers from (α, X), which is an augmenting trail.
- Consider that $\ker_\alpha(I) \subseteq X$ for all new $X \in \mathtt{label}(\alpha)$. Then, for each such X, do the following; here we only consider the case of $X = U_\alpha^+$ (for $X = U_\alpha^-$ we change all signs in the below argument). Let $\alpha_1 := \alpha$ and $\alpha_1\beta_1$ is a $+$-edge in I. Define $(\alpha_1\beta_1, \beta_1\alpha_2, \ldots, \alpha_k\beta_k)$ by the longest path in $\vec{G}_I$ satisfying that $\alpha_1\beta_1, \beta_1\alpha_2, \ldots, \alpha_k\beta_k$ belong to C (and hence I) and $U_{\alpha_i}^+ \notin \mathtt{label}(\alpha_i)$ for all $i \geq 2$. Note that $\alpha_1\beta_1, \alpha_2\beta_2, \ldots, \alpha_k\beta_k$ are $+$-edges and $\beta_1\alpha_2, \beta_2\alpha_3, \ldots, \beta_{k-1}\alpha_k$ are rank-2 $-$-edges. For $i = 1, 2, \ldots, k$, add $U_{\alpha_i}^+$ and $V_{\beta_i}^-$ to $\mathtt{label}(\alpha_i)$ and $\mathtt{label}(\beta_i)$, respectively, and define the back pointers from $(\alpha_i, U_{\alpha_i}^+)$ to $(\beta_{i-1}, V_{\beta_{i-1}}^-)$ (except for $i = 1$) and from $(\beta_i, V_{\beta_i}^-)$ to $(\alpha_i, U_{\alpha_i}^+)$. Then return to the check phase.

Theorem 5. *The following hold:*

(1) *If the above algorithm outputs the pair (U^*, V^*), then it is an optimality witness of I.*
(2) *If the above algorithm outputs the ordered tuple of space-walks, then it is an augmenting trail for I.*
(3) *The running-time of the above algorithm is $O(|E|)$.*

3.3 Augmentation

In this subsection, we present an overview of the augmentation procedure for a given matching I and an augmenting trail T with respect to I. The procedure is an inductive construction that repeats to replace (I, T) by (I', T') until reducing to the base case, where I' is a matching with $r(I) = r(I')$ and T' is an augmenting trail for I' "shorter" than T. In the base case (I, T), we can construct a matching I^* with $r(I^*) > r(I)$ from T.

We first consider the base case where T consists only of a single outer alternating trail $P = (Y_1, \beta_1\alpha_1, X_1, \alpha_1\beta_2, \ldots, \beta_k\alpha_k, X_k)$ that is a path with $k \geq 1$. Let $I^* := I \cup E(P)$, which satisfies $r(I^*) > r(I)$ and (M1); (M1) follows from $\deg_I(\beta_1) \leq 1$, $\deg_I(\alpha_k) \leq 1$, and (O3). We show that I^* satisfies (M2), i.e., I^* has a valid labeling. We can assume that if $\deg_I(\alpha_k) = 1$, then α_k is incident to a $+$-edge in I. Define $U_{\alpha_k}^+$ as $\ker_I(\alpha_k)$ if $\deg_I(\alpha_k) = 1$, and $U_{\alpha_k}^+$ as any 1-dimensional subspace of U_{α_k} different from X_k if $\deg_I(\alpha_k) = 0$. Accordingly, define $V_{\beta_i}^- := (U_{\alpha_i}^+)^{\perp_{\alpha_i\beta_i}}$ and $U_{\alpha_{i-1}}^+ := (V_{\beta_i}^-)^{\perp_{\alpha_{i-1}\beta_i}}$ for $i = k, k-1, \ldots, 1$. They are well-defined 1-dimensional subspaces (since $U_{\alpha_i}^+ \neq \ker_L A_{\alpha_i\beta_i}$ and $V_{\beta_i}^- = \ker_R A_{\alpha_i\beta_i}$ if $\alpha_i\beta_i$ is rank-1). The other vector subspaces $(U_{\alpha_i}^-, V_{\beta_i}^+)$ are defined as (X_i, Y_i) if both X_i and Y_i are 1-dimensional. If some Y_i is 2-dimensional (equivalently X_{i-1} is zero-dimensional), then $\deg_I(\beta_1) = 0$ and $\beta_j\alpha_j$ is rank-2 for $j = 1, 2, \cdots, i-1$. In this case, consider the maximum index i with this property. Define $V_{\beta_i}^+$ as any 1-dimensional subspace different from $\ker_R A_{\alpha_i\beta_i} = V_{\beta_i}^-$. Accordingly, define $U_{\alpha_{j-1}}^- := (V_{\beta_j}^+)^{\perp_{\alpha_{j-1}\beta_j}}$ and

$V_{\beta_{j-1}}^+ := (U_{\alpha_{j-1}}^-)^{\perp \alpha_{j-1} \beta_{j-1}}$. Then the resulting labeling is valid for I^*. Indeed, the orthogonal property (2.1) is satisfied by the construction. Observe that $U_{\alpha_i}^+ \neq U_{\alpha_i}^-$ implies $V_{\beta_i}^- \neq V_{\beta_i}^+$ and $V_{\beta_i}^- \neq V_{\beta_i}^+$ implies $U_{\alpha_{i-1}}^+ \neq U_{\alpha_{i-1}}^-$. Therefore, from $U_{\alpha_k}^+ \neq U_{\alpha_k}^-$ we have $U_{\alpha_i}^+ \neq U_{\alpha_i}^-$ and $V_{\beta_i}^+ \neq V_{\beta_i}^-$ for all i. Suppose that β_1 is incident to $\beta_1 \alpha'$ that is an end edge of a path component C of I. If C contains α_k, then $\beta_1 \alpha'$ is a $+$-edge. Otherwise we can assume by re-coloring that $\beta_1 \alpha'$ is a $+$-edge. Thus we obtain a valid labeling for I^*. In addition, if I^* satisfies (M3) and (M4), then I^* is a desired augmentation. Hence assume that I^* violates (M3) or (M4). Then (i) $\deg_I(\beta_1) = 0$ and $\beta_1 \alpha_1$ is rank-2, or (ii) $\deg_I(\alpha_k) = 0$ and $\beta_k \alpha_k$ is rank-2, or both. Suppose that (i) occurs. Consider a path component of a path $(\beta_1 \alpha_1, \alpha_1 \beta_2, \ldots, \beta_k \alpha_k, \ldots)$ in I^*. To make I^* satisfy (M3) and (M4), remove edges $\alpha_1 \beta_2, \alpha_2 \beta_3, \ldots \alpha_\ell \beta_{\ell+1}$ for the minimum $\ell \geq 1$ such that $\alpha_{\ell+1} \beta_{\ell+1}$ is rank-1. If such an ℓ does not exist, then all $+$-edges in the path are removed. This modification keeps r. The resulting I^* is a desired one. For the case of (ii), do the same procedure from α_k in a reverse way.

Next we consider the general case of $T = (P_0, Q_1, P_1, \ldots, Q_m, P_m)$ that is composed of several outer and inner alternating trails. For the space-limitation, we only explain the procedure for the case where P_m is a path, Q_m belongs to a cycle component C, and P_k, Q_k for $k < m$ do not meet nodes in Q_m and in P_m. For explanation, we further assume that the last node of P_m is incident to a rank-1 edge of I. Suppose that $Q_m = (X_1, \alpha_1 \beta_1, Y_1, \beta_1 \alpha_2, \ldots, \alpha_k \beta_k, Y_k)$ and $\alpha_1 \beta_1$ is a $+$-edge, and suppose that the cycle component C is formed by edges of Q_m and edges $\alpha_1 \beta_1', \beta_1' \alpha_2', \ldots, \alpha_p' \beta_k$. Also suppose (for explanation) that there is at least one rank-1 $-$-edge in C. Define $Q := (\beta_k \alpha_k, \ldots, \beta_1 \alpha_1, \alpha_1 \beta_1', \ldots, \beta_\ell' \alpha_{\ell+1}')$ by the longest path in $\vec{G}_I$ satisfying that all edges belong to C and all $-$-edges in Q are rank-2. Let $I' := I \cup E(P_m) \setminus \{$all $+$-edges in $E(Q)\}$. Then I' satisfies (M1), (M3), and (M4), and $r(I') = r(I)$. We show (M2). As above, assign two vector subspaces on the nodes in P_m from $(U_{\tilde{\alpha}}^+, U_{\tilde{\alpha}}^-) := (X_{\tilde{\alpha}}, \ker_I(\tilde{\alpha}))$ for the last node $\tilde{\alpha}$ and the last space $X_{\tilde{\alpha}}$ of P_m, where we can assume that $\tilde{\alpha}$ is incident to a $-$-edge of I. Then $V_{\beta_k}^+$ is different from $V_{\beta_k}^- = Y_k$. Subsequently, propagate $V_{\beta_k}^-$ on the path $P = (\beta_k \alpha_p', \ldots, \beta_{\ell+1}' \alpha_{\ell+1}')$ to assign $U_{\alpha_j'}^+, V_{\beta_j'}^-$. Then this labeling is valid for I'.

Here the labeling may change on the path P. If some previous inner alternating path $Q_{m'}$ $(m' < m)$ uses the path, then update the spaces in $Q_{m'}$ accordingly. However the last space of $Q_{m'}$ may be replaced by a different space. Then the compatibility condition for $Q_{m'}$ and $P_{m'+1}$ is violated. In this case, we back to consider I and T. We can extend $Q_{m'}$ to β_k, and join P_m to it with the start space $V_{\beta_k}^-$. Then the last space of P_m is different from $\ker_I(\tilde{\alpha})$. Thus we obtain an augmenting trail $T' := (P_0, \ldots, Q_{m'}, P_m)$ for I of a shorter length.

Therefore we assume that such a case does not occur. We modify T so that T is an augmenting trail for I'. First remove Q_m and P_m from T. Suppose that the last space of the outer path P_{m-1} is X, which is different from $U_{\alpha_1}^-$ by the compatibility of (P_{m-1}, Q_m). Join the space-walk $(X, \alpha_1 \beta_1', Y_1', \beta_1' \alpha_2', \ldots, \beta_\ell' \alpha_{\ell+1}', X_{\ell+1}')$ to P_{m-1}, where $Y_i' := (X_i')^{\perp \alpha_i' \beta_i'}$ and $X_{i+1}' := (Y_i')^{\perp \alpha_{i+1}' \beta_i'}$ with $\alpha_1' := \alpha$,

$X_1' := X$, and $\alpha_{\ell+1}' := \alpha_{\ell+1}$. Then the resulting P_{m-1} is an outer alternating trail for I' such that the last space $X_{\ell+1}'$ is different from $U_{\ell+1}^- = \ker_{I'}(\alpha_{\ell+1})$. Thus $T' := (P_0, Q_1, \ldots, P_{m-1})$ is a shorter augmenting trail for I'

In this way, with modifying a matching, we can shorten an augmenting trail to reach the base case. There remain several other cases to be dealt with, e.g., P_m has repeated edges, or meets a previous outer alternating trail $P_{m'}$, etc, which are not explained here by space-limitation and will be given in the full version of this paper.

A   Bit Complexity

In the case of $\mathbf{F} = \mathbf{Q}$, the required bit-size during the algorithm is bounded by a polynomial in the input bit-size as follows. Without loss of generality, we assume that each entry of $A_{\alpha\beta}$ is integer.

Consider the algorithm for finding an augmenting trail. During the algorithm, a 1-dimensional vector subspace $Z \subseteq \mathbf{F}^2$ is represented as a nonzero vector $z \in Z$. In the initial phase, for each $\alpha\beta \in I$ such that $\alpha\beta$ is rank-1 and $\deg_I(\beta) = 1$, we can take an integer nonzero vector $y_\beta \in \ker_R(A_{\alpha\beta})$ with the bit-length bounded in a polynomial of the bit-size of $A_{\alpha\beta}$. In the update phase, we compute $X^{\perp\alpha\beta}$ and $Y^{\perp\alpha\beta}$ for $X \subseteq U_\alpha$ and $Y \subseteq V_\beta$, respectively. This can be simulated as follows. Here we only consider the case of computing $X^{\perp\alpha\beta}$. Suppose $\operatorname{rank} A_{\alpha\beta} = 1$ and that we have an integer nonzero vector $x \in X$ at hand. Then $X^{\perp\alpha\beta} = V_\beta$ if $x \in \ker_L(A_{\alpha\beta})$, and $X^{\perp\alpha\beta} = \ker_R(A_{\alpha\beta})$ if $x \notin \ker_L(A_{\alpha\beta})$. Suppose $\operatorname{rank} A_{\alpha\beta} = 2$ and $A_{\alpha\beta} = \begin{bmatrix} a & b \\ c & d \end{bmatrix}$ and $x = \begin{bmatrix} s \\ t \end{bmatrix}$. Then a nonzero vector $y = \begin{bmatrix} -(cs + dt) \\ as + bt \end{bmatrix}$ belongs to $X^{\perp\alpha\beta}$. By $\log(|cs + dt|) = \log|c| + \log|s|$ if $dt = 0$, $\log(|cs + dt|) = \log|d| + \log|t|$ if $cs = 0$, and $\log(|cs + dt|) \leq \log(|csdt|(1/|cs| + 1/|dt|)) \leq \log 2 + \log|c| + \log|d| + \log|s| + \log|t|$, we have $\texttt{bit}(y) = \texttt{bit}(A) + \texttt{bit}(x) + O(1)$, where $\texttt{bit}(\cdot)$ is the bit-length of the argument. Hence the bit-length is polynomially bounded in finding an augmenting space-walk.

The case of the augmentation procedure is similar. Thus, in the whole step, the bit-size is polynomially bounded.

B   Constructing an Augmenting Space-Walk and Computing the Wong Sequence

We formally introduce Wong sequence. Let B be a $p \times q$ symbolic matrix of the form (1.1), and $\tilde{B}$ a substitution of B. The (orthogonal version of the) *Wong sequence* [9] for $(B, \tilde{B})$ is a sequence $X_0, Y_1, X_1, \ldots$ of vector spaces determined by $X_0 := \ker_L(\tilde{B})$, $Y_i := (X_{i-1})^{\perp B}$, and $X_i := (Y_i)^{\perp B}$ for $i = 1, 2, \ldots$. Here, for a vector space $Y \subseteq \mathbf{F}^q$, define $Y^{\perp B}$ by the set of vectors $x \in \mathbf{F}^p$ such that x is orthogonal to any $y \in Y$ with respect to any substitution $\tilde{B}$ of B. That is, $Y^{\perp B} := \bigcap_{i=1}^m Y^{\perp B_i}$. By easy observations, the limits X_∞, Y_∞ of the Wong sequence for $(B, \tilde{B})$ can be obtained by $O(\min\{p, q\})$ computations, and satisfy

$Y_\infty = (X_\infty)^{\perp_{\tilde{B}}}$ and $X_\infty = (Y_\infty)^{\perp_B}$. In [9], they showed $\operatorname{rank} B = \operatorname{rank} \tilde{B}$ if $\ker_L(\tilde{B}) \subseteq X_\infty$, which is used as an "optimality witness" of $\tilde{B}$ in a polynomial-time algorithm for noncommutative Edmonds' problem [11].

Our labeling procedure can be viewed as a nontrivial specialization of the computation of the Wong sequence for $(A, \tilde{A}_I)$, where A is a (2×2)-type generic partitioned matrix of the form (1.2) and I is a matching of A. Recall $U :=$ $\mathbf{F}^{2\mu} = \bigoplus_\alpha U_\alpha$ and $V := \mathbf{F}^{2\nu} = \bigoplus_\beta V_\beta$. For a vector subspace $Y \subseteq V$, we denote by $\operatorname{Pr}_\beta(Y)$ the projection of Y to the β-th coordinate, namely, $\operatorname{Pr}_\beta(Y) :=$ $\{y_\beta \in V_\beta \mid y = (y_1, y_2, \ldots, y_\nu) \in Y\}$. By the partitioned structure of A, we can see $(Y)^{\perp_A} = \left(\bigoplus_\beta \operatorname{Pr}_\beta(Y)\right)^{\perp_A}$. Hence it suffices to obtain $\operatorname{Pr}_\beta(Y)$ for β for computing the Wong sequence for $(A, \tilde{A})$. The vector space Y_β^* in the labeling procedure plays a role of the above $\operatorname{Pr}_\beta(Y)$. Furthermore, while the computation of the Wong sequence corresponds to the breadth-first search of an auxiliary graph, our algorithm does not.

C   Blow-Up Free Algorithm for Edmonds' Problem

We see that Edmonds' problem can be solved in polynomial time if there exists a polynomial-time blow-up free algorithm for general symbolic matrices A of the form (1.1) with $\operatorname{rank} A = \text{nc-rank } A$. A framework of a blow-up free algorithm is the following, where we are given a symbolic matrix A and an initial substitution $\tilde{A}$ of A:

1. Compute the Wong sequence for $(A, \tilde{A})$ and obtain the limits X_∞ and Y_∞ such that $(Y_\infty)^{\perp_A} = X_\infty$ and $(X_\infty)^{\perp_{\tilde{A}}} = Y_\infty$.
2. If X_∞ includes $\ker_L(\tilde{A})$, then output $\operatorname{rank} A = \operatorname{rank} \tilde{A}$ (see [10]). Otherwise apply an augmenting step: Obtain another substitution $\tilde{A}'$ of A with $\operatorname{rank} \tilde{A}' > \operatorname{rank} \tilde{A}$ and go to 1 with $\tilde{A} \leftarrow \tilde{A}'$, or output "we cannot augment $\tilde{A}$ without a blow-up procedure." Here the latter implies $\operatorname{rank} A < \text{nc-rank } A$.

By using the above blow-up free algorithm, we can check if A is full rank or not. Indeed, if $\operatorname{rank} A = \text{nc-rank } A$, then we can compute $\operatorname{rank} A$ by the algorithm. On the one hand, $\operatorname{rank} A < \text{nc-rank } A$ implies that A is not full. Hence a polynomial-time blow-up free algorithm naturally provides the polynomial-time solvability of the full-rank decision version of Edmonds' problem.

The tractability of Edmonds' problem follows from that of the full-rank decision version of Edmonds' problem. Note that the family of linear independent column sets of A forms a matroid, and the full-rank decision version of Edmonds' problem can play a role as an independence oracle of the matroid. Thus, by a greedy algorithm for the matroid, we can obtain a basis in polynomial time; the size of the basis is equal to $\operatorname{rank} A$.

References

1. Buss, J.F., Frandsen, G.S., Shallit, J.O.: The computational complexity of some problems of linear algebra. J. Comput. Syst. Sci. **58**, 572–596 (1999)

2. Edmonds, J.: Systems of distinct representatives and linear algebra. J. Res. Nat. Bur. Stand. **71B**(4), 241–245 (1967)
3. Fortin, M., Reutenauer, C.: Commutative/noncommutative rank of linear matrices and subspaces of matrices of low rank. Séminaire Lotharingien de Combinatoire **52**, B52f (2004)
4. Fujishige, S., Király, T., Makino, K., Takazawa, K., Tanigawa, S.: Minimizing submodular functions on diamonds via generalized fractional matroid matchings. EGRES Technical reports, TR-2014-14 (2014)
5. Garg, A., Gurvits, L., Oliveira, R., Wigderson, A.: Operator scaling: theory and applications. Found. Comput. Math. (2019)
6. Hamada, M., Hirai, H.: Maximum vanishing subspace problem, CAT(0)-space relaxation, and block-triangularization of partitioned matrix (2017). arXiv:1705.02060
7. Ishikawa, T.: Max-rank matrix completion via Wong sequence. Bachelor thesis, The University of Tokyo (2018). (in Japanese)
8. Ito, H., Iwata, S., Murota, K.: Block-triangularizations of partitioned matrices under similarity/equivalence transformations. SIAM J. Matrix Anal. Appl. **15**(4), 1226–1255 (1994)
9. Ivanyos, G., Karpinski, M., Qiao, Y., Santha, M.: Generalized Wong sequences and their applications to Edmonds' problems. J. Comput. Syst. Sci. **81**, 1373–1386 (2015)
10. Ivanyos, G., Qiao, Y., Subrahmanyam, K.V.: Non-commutative Edmonds' problem and matrix semi-invariants. Comput. Complex. **26**, 717–763 (2017). https://doi.org/10.1007/s00037-016-0143-x
11. Ivanyos, G., Qiao, Y., Subrahmanyam, K.V.: Constructive non-commutative rank computation is in deterministic polynomial time. Comput. Complex. **27**, 561–593 (2018). https://doi.org/10.1007/s00037-018-0165-7
12. Iwata, S., Murota, K.: A minimax theorem and a Dulmage-Mendelsohn type decomposition for a class of generic partitioned matrices. SIAM J. Matrix Anal. Appl. **16**(3), 719–734 (1995)
13. Kabanets, V., Impagliazzo, R.: Derandomizing polynomial identity tests means proving circuit lower bounds. Comput. Complex. **13**, 1–46 (2004)
14. Kuivinen, F.: On the complexity of submodular function minimisation on diamonds. Discret. Optim. **8**, 459–477 (2011)
15. Lovász, L.: On determinants, matchings, and random algorithms. In: International Symposium on Fundamentals of Computation Theory (FCT 1979) (1979)
16. Lovász, L.: Singular spaces of matrices and their application in combinatorics. Boletim da Sociedade Brasileira de Matemática **20**(1), 87–99 (1989). https://doi.org/10.1007/BF02585470
17. Schwartz, J.T.: Fast probabilistic algorithms for verification of polynomial identities. J. ACM **27**(4), 701–717 (1980)

Fair Colorful k-Center Clustering

Xinrui Jia[✉], Kshiteej Sheth, and Ola Svensson

School of Computer and Communication Sciences, EPFL, Lausanne, Switzerland
{xinrui.jia,kshiteej.sheth,ola.svensson}@epfl.ch

Abstract. An instance of *colorful k-center* consists of points in a metric space that are colored red or blue, along with an integer k and a coverage requirement for each color. The goal is to find the smallest radius ρ such that there exist balls of radius ρ around k of the points that meet the coverage requirements.

The motivation behind this problem is twofold. First, from fairness considerations: each color/group should receive a similar service guarantee, and second, from the algorithmic challenges it poses: this problem combines the difficulties of clustering along with the subset-sum problem. In particular, we show that this combination results in strong integrality gap lower bounds for several natural linear programming relaxations.

Our main result is an efficient approximation algorithm that overcomes these difficulties to achieve an approximation guarantee of 3, nearly matching the tight approximation guarantee of 2 for the classical k-center problem which this problem generalizes.

Keywords: Approximation algorithms · k-Center · Clustering and facility location · Fairness

1 Introduction

In the *colorful k-center* problem introduced in [4], we are given a set of n points P in a metric space partitioned into a set R of red points and a set B of blue points, along with parameters k, r, and b.[1] The goal is to find a set of k centers $C \subseteq P$ that minimizes ρ so that balls of radius ρ around each point in C cover at least r red points and at least b blue points.

This generalization of the classic k-center problem has applications in situations where fairness is a concern. For example, if a telecommunications company is required to provide service to at least 90% of the people in a country, it would be cost effective to only provide service in densely populated areas. This is at

---

[1] Our results (in particular the 3-approximation algorithm) rather immediately generalize to any constant number of colors. However, to keep the exposition of our ideas as clean as possible, we have restricted this abstract to the version with two colors.

Supported by the Swiss National Science Foundation project 200021-184656 "Randomness in Problem Instances and Randomized Algorithms".

D. Bienstock and G. Zambelli (Eds.): IPCO 2020, LNCS 12125, pp. 209–222, 2020.
https://doi.org/10.1007/978-3-030-45771-6_17

odds with the ideal that at least some people in every community should receive service. In the absence of color classes, an approximation algorithm could be "unfair" to some groups by completely considering them as outliers. The inception of fairness in clustering can be found in the recent paper [7] (see also [1,3]), which uses a related but incomparable notion of fairness. Their notion of fairness requires *each individual cluster* to have a balanced number of points from each color class, which leads to very different algorithmic considerations and is motivated by other applications, such as "feature engineering".

The other motive for studying the colorful k-center problem derives from the algorithmic challenges it poses. One can observe that it generalizes the *k-center problem with outliers*, which is equivalent to only having red points and needing to cover at least r of them. This outlier version is already more challenging than the classic k-center problem: only recent results give tight 2-approximation algorithms [5,11], improving upon the 3-approximation guarantee of [6]. In contrast, such algorithms for the classic k-center problem have been known since the '80s [9,12]. That the approximation guarantee of 2 is tight, even for classic k-center, was proved in [13].

At the same time, a subset-sum problem with polynomial-sized numbers is embedded within the colorful k-center problem. To see this, consider n numbers $a_1, \ldots, a_n$ and let $A = \sum_{i=1}^{n} a_i$. Construct an instance of the colorful k-center problem with $r = k \cdot A + A/2$, $b = k \cdot A - A/2$, and for every $i \in \{1, \ldots, n\}$, a ball of radius one containing $A + a_i$ red points and $A - a_i$ blue points. These balls are assumed to be far apart so that any single ball that covers two of these balls must have a very large radius. It is easy to see that the constructed colorful k-center instance has a solution of radius one if and only if there is a size k subset of the n numbers whose sum equals $A/2$.

We use this connection to subset-sum to show that the standard linear programming (LP) relaxation of the colorful k-center problem has an unbounded integrality gap even after a linear number of rounds of the powerful Lasserre/Sum-of-Squares hierarchy (see Sect. 3). We remark that the standard linear programming relaxation gives a 2-approximation algorithm for the outliers version even without applying lift-and-project methods. Another natural approach for strengthening the standard linear programming relaxation is to add flow-based inequalities specially designed to solve subset-sum problems. However, in Appendix B, we prove that they do not improve the integrality gap due to the clustering feature of the problem. This shows that clustering and the subset-sum problem are intricately related in colorful k-center. This interplay makes the problem more complex and prior to our work only a randomized constant-factor approximation algorithm was known when the points are in $\mathbb{R}^2$ with an approximation guarantee greater than 6 [4].

Our main result overcomes these difficulties and we give a nearly tight approximation guarantee:

Theorem 1. *There is a 3-approximation algorithm for the colorful k-center problem.*

As aforementioned, our techniques can be easily extended to a constant number of color classes but we restrict the discussion here to two colors. On a very high level, our algorithm manages to decouple the clustering and the subset-sum aspects. First, our algorithm guesses certain centers of the optimal solution that it then uses to partition the point set into a "dense" part P_d and a "sparse" part P_s. The dense part is clustered using a subset-sum instance while the sparse set is clustered using the techniques of Bandyapadhyay, Inamdar, Pai, and Varadarajan [4] (see Sect. 2.1). Specifically, we use the pseudo-approximation of [4] that satisfies the coverage requirements using $k+1$ balls of at most twice the optimal radius.

While our approximation guarantee is nearly tight, it remains an interesting open problem to give a 2-approximation algorithm or to show that the ratio 3 is tight. One possible direction is to understand the strength of the relaxation obtained by combining the Lasserre/Sum-of-Squares hierarchy with the flow constraints. While we show that individually they do not improve the integrality gap, we believe that their combination can lead to a strong relaxation.

Organization. We begin by giving some notation and definitions and describing the pseudo-approximation algorithm in [4]. In fact, we then describe a 2-approximation algorithm on a certain class of instances that are *well-separated*, and the 3-approximation follows almost immediately. This 2-approximation proceeds in two phases: the first is dedicated to the guessing of certain centers, while the second processes the dense and sparse sets. In Sect. 3 we present our integrality gap under the Sum-of-Squares hierarchy, and Appendix B contains an integrality gap example for the flow constraints.

2   A 3-Approximation Algorithm

In this section we present our 3-approximation algorithm. We briefly describe the pseudo-approximation algorithm of Bandhyapadhyay et al. [4] since we use it as a subroutine in our algorithm.

Notation: We assume that our problem instance is normalized to have an optimal radius of one and we refer to the set of centers in an optimal solution as OPT. The set of all points at distance at most ρ from a point j is denoted by $\mathcal{B}(j, \rho)$ and we refer to this set as a *ball of radius ρ at j*. We write $\mathcal{B}(p)$ for $\mathcal{B}(p, 1)$. By a *ball of OPT* we mean $\mathcal{B}(p)$ for some $p \in OPT$.

2.1   The Pseudo-Approximation Algorithm

The algorithm of Bandhyapadhyay et al. [4] first guesses the optimal radius for the instance (there are at most $O(n^2)$ distinct values the optimal radius can take), which we assume to be one after normalization, and considers the natural LP relaxation LP1 depicted on the left in Fig. 1. The variable x_i indicates how much point i is fractionally opened as a center and z_i indicates the amount that i is covered by centers.

Given a fractional solution to LP1, the algorithm of [4] finds a clustering of the points. The clusters that are produced are of radius two, and with a simple modification (details and proof can be found in the full version), can be made to have a special structure that we call a flower:

<table>
<tr><td>

LP1

$$\sum_{i \in \mathcal{B}(j)} x_i \geq z_j, \quad \forall j \in P$$

$$\sum_{i \in P} x_i \leq k$$

$$\sum_{j \in R} z_j \geq r,$$

$$\sum_{j \in B} z_j \geq b,$$

$$z_j, x_i \in [0,1], \quad \forall i, j \in P.$$

</td><td>

LP2

maximize $\quad \sum_{j \in S} r_j y_j$

subject to $\quad \sum_{j \in S} b_j y_j \geq b,$

$$\sum_{j \in S} y_j \leq k,$$

$$y_j \in [0,1] \quad \forall j \in S.$$

</td></tr>
</table>

Fig. 1. The linear programs used in the pseudo-approximation algorithm.

Definition 1. *For $j \in P$, a **flower** centered at j is the set $\mathcal{F}(j) = \cup_{i \in \mathcal{B}(j)} \mathcal{B}(i)$.*

More specifically, given a fractional solution (x, z) to LP1, Algorithm 1 in [4] produces a set of points $S \subseteq P$ and a cluster $C_j \subseteq P$ for every $j \in S$ such that:

1. The set S is a subset of the points $\{j \in P : z_j > 0\}$ with positive z-values.
2. For each $j \in S$, we have $C_j \subseteq \mathcal{F}(j)$ and the clusters $\{C_j\}_{j \in S}$ are pairwise disjoint.
3. If we let $r_j = |C_j \cap R|$ and $b_j = |C_j \cap B|$ for $j \in S$, then the linear program LP2 (depicted on the right in Fig. 1) has a feasible solution y of value at least r.

As LP2 has only two non-trivial constraints, any extreme point will have at most two variables attaining strictly fractional values. So at most $k + 1$ variables of y are non-zero. The pseudo-approximation of [4] now simply takes those non-zero points as centers. Since each flower is of radius two, this gives a 2-approximation algorithm that opens at most $k + 1$ centers. (Note that, as the clusters $\{C_j\}_{j \in S}$ are pairwise disjoint, at least b blue points are covered, and at least r red points are covered since the value of the solution is at least r.)

Obtaining a constant-factor approximation algorithm that only opens k centers turns out to be significantly more challenging. Nevertheless, the above techniques form an important subroutine in our algorithm. Given a fractional solution (x, z) to LP1, we proceed as above to find S and an extreme point to LP2 of value at least r. However, instead of selecting all points with positive y-value, we, in the case of two fractional values, only select the one whose cluster covers more blue points. This gives us a solution of at most k centers whose clusters cover at least b blue points. Furthermore, the number of red points that are covered is at least $r - \max_{j \in S} r_j$ since we disregarded at most one center. As $S \subseteq \{j : z_j > 0\}$ (see first property above) and $C_j \subseteq \mathcal{F}(j)$ (see second property above), we have $\max_{j \in S} r_j \leq \max_{j:z_j>0} |\mathcal{F}(j) \cap R|$. We summarize the obtained properties in the following lemma.

Lemma 1. *Given a fractional solution (x, z) to LP1, there is a polynomial time algorithm that outputs at most k clusters of radius two that cover at least b blue points and at least $r - \max_{j:z_j>0} |\mathcal{F}(j) \cap R|$ red points.*

We can thus find a 2-approximate solution that covers sufficiently many blue points but may cover fewer red points than necessary. The idea now is that, if the number of red points in any cluster is not too large, i.e., $\max_{j:z_j>0} |\mathcal{F}(j) \cap R|$ is "small", then we can hope to meet the coverage requirements for the red points by increasing the radius around some opened centers. Our algorithm builds on this intuition to get a 2-approximation algorithm using at most k centers for *well-separated* instances as defined below.

Definition 2. *An instance of colorful k-center is **well-separated** if there does not exist a ball of radius three that covers at least two balls of OPT.*

Our main result of this section can now be stated as follows:

Theorem 2. *There is a 2-approximation algorithm for well-separated instances.*

The above theorem immediately implies Theorem 1, i.e., the 3-approximation algorithm for general instances. Indeed, if the instance is not well-separated, we can find a ball of radius three that covers at least two balls of OPT by trying all n points and running the pseudo-approximation of [4] on the remaining uncovered points with $k - 2$ centers. In the correct iteration, this gives us at most $k - 1$ centers of radius two, which when combined with the ball of radius three that covers two balls of OPT, is a 3-approximation.

Our algorithm for well-separated instances now proceeds in two phases with the objective of finding a subset of P on which the pseudo-approximation algorithm produces subsets of flowers containing not too many red points. In addition, we maintain a partial solution set of centers (some guessed in the first phase), so that we can expand the radius around these centers to recover the deficit of red points from closing one of the fractional centers.

2.2   Phase I

In this phase we will guess some balls of OPT that can be used to construct a bound on $\max_{j:z_j>0} |R \cap \mathcal{F}(j)|$. To achieve this, define the notion of **Gain**(p, q) for any point $p \in P$ and $q \in \mathcal{B}(p)$.

Definition 3. *For any $p \in P$ and $q \in \mathcal{B}(p)$, let*

$$\boldsymbol{Gain}(p, q) := R \cap (\mathcal{F}(q) \setminus \mathcal{B}(p))$$

be the set of red points added to $\mathcal{B}(p)$ by forming a flower centered at q.

Our algorithm in this phase proceeds by guessing three centers c_1, c_2, c_3 of the optimal solution OPT:

> For $i = 1, 2, 3$, guess the center c_i in OPT and calculate the point $q_i \in \mathcal{B}(c_i)$ such that the number of red points in $\boldsymbol{Gain}(c_i, q_i) \cap P_i$ is maximized over all possible c_i, where
>
> $$P_1 = P$$
> $$P_i = P_{i-1} \setminus \mathcal{F}(q_{i-1}) \text{ for } 2 \leq i \leq 4.$$

The time it takes to guess c_1, c_2, and c_3 is $O(n^3)$ and for each c_i we find the $q_i \in \mathcal{B}(c_i)$ such that $|\boldsymbol{Gain}(c_i, q_i) \cap P_i|$ is maximized by trying all points in $\mathcal{B}(c_i)$ (at most n many).

For notation, define $\boldsymbol{Guess} := \cup_{i=1}^{3} \mathcal{B}(c_i)$ and let

$$\tau = |\boldsymbol{Gain}(c_3, q_3) \cap P_3|.$$

The important properties guaranteed by the first phase is summarized in the following lemma.

Lemma 2. *Assuming that c_1, c_2, and c_3 are guessed correctly, we have that*

1. *the $k - 3$ balls of radius one in $OPT \setminus \{c_i\}_{i=1}^{3}$ are contained in P_4 and cover $b - |B \cap \boldsymbol{Guess}|$ blue points and $r - |R \cap \boldsymbol{Guess}|$ red points; and*
2. *the three clusters $\mathcal{F}(q_1), \mathcal{F}(q_2)$, and $\mathcal{F}(q_3)$ are contained in $P \setminus P_4$ and cover at least $|B \cap \boldsymbol{Guess}|$ blue points and at least $|R \cap \boldsymbol{Guess}| + 3 \cdot \tau$ red points.*

Proof. (1) We claim that the intersection of any ball of $OPT \setminus \{c_i\}_{i=1}^{3}$ with $\mathcal{F}(q_i)$ in P is empty, for all $1 \leq i \leq 3$. Then the $k - 3$ balls in $OPT \setminus \{c_i\}_{i=1}^{3}$ satisfy the statement of (1). To prove the claim, suppose that there is $p \in OPT \setminus \{c_i\}_{i=1}^{3}$ such that $\mathcal{B}(p) \cap \mathcal{F}(q_i) \neq \emptyset$ for some $1 \leq i \leq 3$. Note that $\mathcal{F}(q_i) = \cup_{i \in \mathcal{B}(q_i)} \mathcal{B}(i)$, so this implies that $\mathcal{B}(p) \cap \mathcal{B}(q') \neq \emptyset$, for some $q' \in \mathcal{B}(q_i)$. Hence, a ball of radius three around q' covers both $\mathcal{B}(p)$ and $\mathcal{B}(c_i)$ as $c_i \in \mathcal{B}(q_i)$, which contradicts that the instance is well-separated.

(2) Note that for $1 \leq i \leq 3$, $\mathcal{B}(c_i) \cup \boldsymbol{Gain}(c_i, q_i) \subseteq \mathcal{F}(q_i)$, and that $\mathcal{B}(c_i)$ and $\boldsymbol{Gain}(c_i, q_i)$ are disjoint. The balls $\mathcal{B}(c_i)$ cover at least $|B \cap \boldsymbol{Guess}|$ blue points and $|R \cap \boldsymbol{Guess}|$ red points, while $\sum_{i=1}^{3} |\boldsymbol{Gain}(c_i, q_i) \cap P_i| \geq 3\tau$. $\qquad \square$

2.3   Phase II

Throughout this section we assume c_1, c_2, and c_3 have been guessed correctly in Phase I so that the properties of Lemma 2 hold. Furthermore, by the selection and the definition of τ, we also have

$$|\mathbf{Gain}(p,q) \cap P_4| \leq \tau \qquad \text{for any } p \in P_4 \cap OPT \text{ and } q \in \mathcal{B}(p) \cap P_4. \quad (1)$$

This implies that $\mathcal{F}(p) \setminus \mathcal{B}(p)$ contains at most τ red points of P_4. However, to apply Lemma 1 we need that the number of red points of P_4 in the whole flower $\mathcal{F}(p)$ is bounded. To deal with balls with many more than τ red points, we will iteratively remove *dense* sets from P_4 to obtain a subset P_s of *sparse* points.

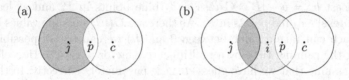

Fig. 2. The shaded regions are subsets of $\mathbf{Gain}$(c,p), which contain the darkly shaded regions that have $> \tau$ red points. (Color figure online)

Definition 4. *When considering a subset of the points $P_s \subseteq P$, we say that a point $j \in P_s$ is **dense** if the ball $\mathcal{B}(j)$ contains strictly more than $2 \cdot \tau$ red points of P_s. For a dense point j, we also let $I_j \subseteq P_s$ contain those points $i \in P_s$ whose intersection $\mathcal{B}(i) \cap \mathcal{B}(j)$ contains strictly more than τ red points of P_s.*

We remark that in the above definition, we have in particular that $j \in I_j$ for a dense point $j \in P_s$. Our iterative procedure now works as follows:

Initially, let $I = \emptyset$ and $P_s = P_4$. While there is a dense point $j \in P_s$:
  – Add I_j to I and update P_s by removing the points $D_j = \cup_{i \in I_j} \mathcal{B}(i) \cap P_s$.

Let $P_d = P_4 \setminus P_s$ denote those points that were removed from P_4. We will cluster the two sets P_s and P_d of points separately. Indeed, the following lemma says that a center in $OPT \setminus \{c_i\}_{i=1}^3$ either covers points in P_s or P_d but not points from both sets. Recall that D_j denotes the set of points that are removed from P_s in the iteration when j was selected and so $P_d = \cup_j D_j$.

Lemma 3. *For any $c \in OPT \setminus \{c_i\}_{i=1}^3$ and any $I_j \in I$, either $c \in I_j$ or $\mathcal{B}(c) \cap D_j = \emptyset$.*

Proof. Let $c \in OPT \setminus \{c_i\}_{i=1}^3$, $I_j \in I$, and suppose $c \notin I_j$. If $\mathcal{B}(c) \cap D_j \neq \emptyset$, there is a point p in the intersection $\mathcal{B}(c) \cap \mathcal{B}(i)$ for some $i \in I_j$. Suppose first that $\mathcal{B}(c) \cap \mathcal{B}(j) \neq \emptyset$. Then, since $c \notin I_j$, the intersection $\mathcal{B}(c) \cap \mathcal{B}(j)$ contains fewer than τ red points from D_j (recall that D_j contains the points of $\mathcal{B}(j)$ in P_s at the time j was selected). But by the definition of dense clients, $\mathcal{B}(j) \cap D_j$

has more than $2 \cdot \tau$ red points, so $(\mathcal{B}(j) \setminus \mathcal{B}(c)) \cap D_j$ has more than τ red points. This region is a subset of $\mathbf{Gain}(c, p) \cap P_4$, which contradicts (1). This is shown in Fig. 2(a). Now consider the second case when $\mathcal{B}(c) \cap \mathcal{B}(j) = \emptyset$ and there is a point p in the intersection $\mathcal{B}(c) \cap \mathcal{B}(i)$ for some $i \in I_j$ and $i \neq j$. Then, by the definition of I_j, $\mathcal{B}(i) \cap \mathcal{B}(j)$ has more than τ red points of D_j. However, this is also a subset of $\mathbf{Gain}(c, p) \cap P_4$ so we reach the same contradiction. See Fig. 2(b).

□

Our algorithm now proceeds by guessing the number k_d of balls of $OPT \setminus \{c_i\}_{i=1}^3$ contained in P_d. We also guess the numbers r_d and b_d of red and blue points, respectively, that these balls cover in P_d. Note that after guessing k_d, we know that the number of balls in $OPT \setminus \{c_i\}_{i=1}^3$ contained in P_s equals $k_s = k - 3 - k_d$. Furthermore, by the first property of Lemma 2, these balls cover at least $b_s = b - |B \cap \mathbf{Guess}| - b_d$ blue points in P_s and at least $r_s = r - |R \cap \mathbf{Guess}| - r_d$ red points in P_s. As there are $O(n^3)$ possible values of k_d, b_d, and r_d (each can take a value between 0 and n) we can try all possibilities by increasing the running time by a multiplicative factor of $O(n^3)$. Henceforth, we therefore assume that we have guessed those parameters correctly. In that case, we show that we can recover an equally good solution for P_d and a solution for P_s that covers b_s blue points and almost r_s red points:

Lemma 4. *There exist two polynomial-time algorithms $\mathcal{A}_d$ and $\mathcal{A}_s$ such that if k_d, r_d, and b_d are guessed correctly then*

- *$\mathcal{A}_d$ returns k_d balls of radius one that cover b_d blue points of P_d and r_d red points of P_d;*
- *$\mathcal{A}_s$ returns k_s balls of radius two that cover at least b_s blue points of P_s and at least $r_s - 3 \cdot \tau$ red points of P_s.*

Proof. We first describe and analyze the algorithm $\mathcal{A}_d$ followed by $\mathcal{A}_s$.

The algorithm $\mathcal{A}_d$ for the dense point set P_d. By Lemma 3, we have that all k_d balls in $OPT \setminus \{c_i\}_{i=1}^3$ that cover points in P_d are centered at points in $\cup_j I_j$. Furthermore, we have that each I_j contains at most one center of OPT. This is because every $i \in I_j$ is such that $\mathcal{B}(i) \cap \mathcal{B}(j) \neq \emptyset$ and so, by the triangle inequality, $\mathcal{B}(j, 3)$ contains all balls $\{\mathcal{B}(i)\}_{i \in I_j}$. Hence, by the assumption that the instance is well-separated, the set I_j contains at most one center of OPT.

We now reduce our problem to a 3-dimensional subset-sum problem. For each $I_j \in I$, form a group consisting of an item for each $p \in I_j$. The item corresponding to $p \in I_j$ has the 3-dimensional value vector $(1, |\mathcal{B}(p) \cap D_j \cap B|, |\mathcal{B}(p) \cap D_j \cap R|)$. Our goal is to find k_d items such that at most one item per group is selected and their 3-dimensional vectors sum up to (k_d, b_d, r_d). Such a solution, if it exists, can be found by dynamic programming that has a table of size $O(n^4)$. The recurrence and precise details of this are given in Appendix A. Furthermore, since the D_j's are disjoint by definition, this gives k_d centers that cover b_d blue points and r_d red points in P_d, as required in the statement of the lemma.

It remains to show that such a solution exists. Let $o_1, o_2, \ldots, o_{k_d}$ denote the centers of the balls in $OPT \setminus \{c_i\}_{i=1}^{3}$ that cover points in P_d. Furthermore, let $I_{j_1}, \ldots, I_{j_{k_d}}$ be the sets in I such that $o_i \in I_{j_i}$ for $i \in \{1, \ldots, k_d\}$. Notice that by Lemma 3 we have that $\mathcal{B}(o_i) \cap P_d$ is disjoint from $P_d \setminus D_{j_i}$ and contained in D_{j_i}. It follows that the 3-dimensional vector corresponding to an OPT center o_i equals $(1, |\mathcal{B}(p) \cap P_d \cap B|, |\mathcal{B}(p) \cap P_d \cap R|)$. Therefore, the sum of these vectors corresponding to $o_1, \ldots, o_{k_d}$ results in the vector (k_d, b_d, r_d), where we used that our guesses of k_d, b_d, and r_d were correct.

The algorithm $\mathcal{A}_s$ for the sparse point set P_s. Assuming that the guesses are correct we have that $OPT \setminus \{c_i\}_{i=1}^{3}$ contains k_s balls that cover b_s blue points of P_s and r_s red points of P_s. Hence, LP1 has a feasible solution (x, z) to the instance defined by the point set P_s, the number of balls k_s, and the constraints b_s and r_s on the number of blue and red points to be covered, respectively. Lemma 1 then says that we can in polynomial-time find k_s balls of radius two such that at least b_s blue balls of P_s are covered and at least

$$r_s - \max_{j:z_j>0} |\mathcal{F}(j) \cap R|$$

red points of P_s are covered. Here, $\mathcal{F}(j)$ refers to the flower restricted to the point set P_s.

To prove the second part of Lemma 4, it is thus sufficient to show that LP1 has a feasible solution where $z_j = 0$ for all $j \in P_s$ such that $|\mathcal{F}(j) \cap R| > 3 \cdot \tau$. In turn, this follows by showing that, for any such $j \in P_s$ with $|\mathcal{F}(j) \cap R| > 3 \cdot \tau$, no point in $\mathcal{B}(j)$ is in OPT (since then $z_j = 0$ in the integral solution corresponding to OPT).

To see why this holds, suppose towards a contradiction that there is a $c \in OPT$ such that $c \in \mathcal{B}(j)$. First, since there are no dense points in P_s, we have that the number of red points in $\mathcal{B}(c) \cap P_s$ is at most $2 \cdot \tau$. Therefore the number of red points of P_s in $\mathcal{F}(j) \setminus \mathcal{B}(c)$ is strictly more than τ. In other words, we have $\tau < |\mathbf{Gain}(c, j) \cap P_s| \leq |\mathbf{Gain}(c, j) \cap P_4|$ which contradicts (1). $\qquad\square$

Equipped with the above lemma we are now ready to finalize the proof of Theorem 2.

Proof of Theorem 2. Our algorithm guesses the optimal radius, the centers c_1, c_2, c_3 in Phase I, and k_d, r_d, b_d in Phase II. There are at most $\binom{n}{2}$ choices of the optimal radius, n choices for each c_i, and $n + 1$ choices of k_d, r_d, b_d (ranging from 0 to n). We can thus try all these possibilities in polynomial time and, since all other steps in our algorithm run in polynomial time, the total running time will be polynomial. The algorithm tries all these guesses and outputs the best solution found over all choices. For the correct guesses, we output a solution with $3 + k_d + k_s = k$ balls of radius at most two. Furthermore, by the second property of Lemma 2 and the two properties of Lemma 4, we have that

- the number of blue points covered is at least $|B \cap \mathbf{Guess}| + b_d + b_s = b$; and
- the number of red points covered is at least $|R \cap \mathbf{Guess}| + 3\tau + r_d + r_s - 3\tau = r$.

We have thus given a polynomial-time algorithm that returns a solution where the balls are of radius at most twice the optimal radius. $\qquad\square$

3   Sum-of-Squares Integrality Gap

By "integrality gap" of the feasibility relaxation LP1 we mean the largest ratio between ρ such that LP1 with balls of radius ρ is feasible, and the radius of the optimal integral solution, for any instance of colorful k-center. In essence, the pseudo-approximation algorithm finds too many centers because the clustering S comes from LP1 which has unbounded integrality gap and there is no feasible subset-sum solution with the clusters in S as the items.

The Sum-of-Squares (abbreviated SoS, equivalently Lasserre [15,16]) hierarchy is a method of strengthening linear programs that has been used with varying degrees of success in constraint satisfaction problems, set-cover, and graph coloring, to just name a few examples [2,8,17]. For a succinct definition see [14]. A known weakness of SoS is in recognizing the infeasibility of knapsack constraints. Likewise, we show that SoS does not improve the integrality gap of LP1 in a small number of rounds.

We use a reduction from Grigoriev's SoS lower bound for knapsack [10].

Theorem 3 (Grigoriev). *At least* $\min\{2\lfloor \min\{k/2, n - k/2\}\rfloor + 3, n\}$ *rounds of SoS are required to recognize that the following is infeasible for* $k \in \mathbb{Z}$ *odd.*

$$\sum_{i=1}^{n} 2x_i = k, \quad x_i \in \{0,1\} \ \forall i. \tag{2}$$

Consider an instance of colorful k-center with $8n$ points, $k = n$, $r = b = 2n$, and n odd. The points are in pairs of clusters of radius one. There are three blue points and one red point in one cluster and one blue point and three red points in the other, as shown in Fig. 3. In an optimal integer solution, one center needs to cover two of these clusters while a fractional solution satisfying LP1 can open $1/2$ of a center around each cluster. Hence, LP1 has an unbounded integrality gap since the clusters can be arbitrarily far apart. This instance takes an odd number of copies of the integrality gap example given in [4].

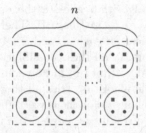

Fig. 3. Integrality gap example for linear rounds of SoS (Color figure online)

We can do a simple mapping from a feasible solution for the tth round of SoS on (2) to our variables in the tth round of SoS on LP1 for this instance to demonstrate that the infeasibility of balls of radius one is not recognized. Intuitively, we can assign a variable x_i to each pair of clusters of radius one as shown in Fig. 3, corresponding to opening each cluster in the pair by x_i amount. Then a feasible opening of centers reduces to a feasible assignment of variables in Theorem 3. Formal details can be found in the full version.

Theorem 4. *The integrality gap of LP1 with $8n$ points persists up to n rounds of Sum-of-Squares.* □

Appendix A   Dynamic Programming for Dense Points

In this section we describe the dynamic programming algorithm discussed in Lemma 4. As stated in the proof of Lemma 4, given $I = \cup_j I_j$ and correct guesses for k_d, b_d, r_d, we need to find k_d balls of radius one centered at points in I covering b_d blue and r_d red points with at most one point from each $I_j \in I$ picked as a center. To do this, we first order the sets in I arbitrarily as $I = \{I_{j_1}, \ldots, I_{j_m}\}, m = |I|$. We create a 4-dimensional table T of dimension (m, b_d, r_d, k_d). $T[m', b', r', k']$ stores whether there is a set of k' balls in the first m' sets of I covering b' blue and r' red points. The recurrence relation for T is

$$T[0, 0, 0, 0] = \text{True}$$
$$T[0, b', r', k'] = \text{False}, \quad \text{for any } b', r', k' \neq 0$$

$$T[m', b', r', k'] = \begin{cases} \text{True} & \text{if } T[m'-1, b', r', k'] = \text{True} \\ \text{True} & \text{if } \exists c \in I_{j_{m'}}, \text{ s.t. } T[m'-1, b'', r'', k'-1] = \text{True, for} \\ & b'' = b' - |\mathcal{B}(c) \cap B|, r'' = r' - |\mathcal{B}(c) \cap R| \\ \text{False} & \text{otherwise} \end{cases}.$$

The table T has size $O((m+1) \cdot (n+1) \cdot (n+1) \cdot (n+1)) = O(n^4)$ since the first parameter has range from 0 to m, and the other parameters can have value 0 up to at most n. Moreover, since $|I_j| \leq n$ for all $I_j \in I$, we can compute the whole table in time $O(n^5)$ using for e.g. the bottom-up approach. We can also remember the choices in a separate table and so we can find a solution in time $O(n^5)$ if it exists.

Appendix B   Flow Constraints

In this section we add additional constraints to LP1 that incorporate natural subset-sum requirements for the fractional centers produced by LP1. The objective is to obtain a better clustering, but we show that this fails to reduce the integrality gap.

We define an instance of a multi-dimensional subset-sum problem. Each point $p \in P$ corresponds to an item with three dimensions: a dimension of size one, $|B \cap \mathcal{B}(p)|$, and $|R \cap \mathcal{B}(p)|$. We set up a flow network with an $(n+1) \times n \times n \times k$ grid of nodes and we name the nodes with the coordinate (w, x, y, z) of its position.

Flow 1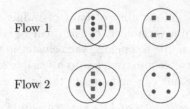

Flow 2

Fig. 4. $k = 3$, $r = b = 8$ (Color figure online)

The source is located at $(0,0,0,0)$ and we add an extra node t for the sink. Assign an arbitrary order to the points in P. For the item corresponding to $i \in P$, for each $x \in [n]$, $y \in [n]$, $z \in [k]$:

1. Add an edge from (i,x,y,z) to $(i+1,x,y,z)$ with flow variable $e_{i,x,y,z}$.
2. With $b_i := |B \cap \mathcal{B}(i)|$ and $r_i := |R \cap \mathcal{B}(i)|$, if $z < k$ add an edge from (i,x,y,z) to $(i+1,\min\{x+b_i,n\},\min\{y+b_i,n\},z+1)$ with flow variable $f_{i,x,y,z}$.

For each $x \in [b,n]$, $y \in [r,n]$:

3. Add an edge from $(n+1,x,y,k)$ to t with flow variable $g_{x,y}$.

Set the capacities of all edges to one. In addition to the constraint that there should be one unit of flow from s to t and the usual flow conservation constraints, we add to LP1 the constraints

$$x_i = \sum_{x,y \in [n], z \in [k]} f_{i,x,y,z} \quad \text{for all } i \in P \tag{3}$$

$$1 - x_i = \sum_{x,y \in [n], z \in [k]} e_{i,x,y,z} \quad \text{for all } i \in P. \tag{4}$$

We refer to the resulting linear program as LP3. Notice that by definition, any path P from s to t defines a set C_P of at most k centers by taking those points c for which $f_{c,x,y,z} \in P$ for some x, y, and z. Moreover, as t can only be reached from a coordinate with $x \geq b$ and $y \geq r$ we have that $\sum_{c \in C_P} |\mathcal{B}(c) \cap B| \geq b$ and $\sum_{c \in C_P} |\mathcal{B}(c) \cap R| \geq r$. It follows that C_P forms a solution to the problem of radius one *if the balls are disjoint*. In particular, our integrality gap instances for the Sum-of-Squares hierarchy do not fool LP3.

However, the example in Fig. 4 shows that in an instance where balls overlap, the integrality gap remains large. Here, the fractional assignment of open centers is $1/2$ for each of the six balls and this gives a fractional covering of 8 red and 8 blue points as required. This assignment also satisfies the flow constraints because the three balls at the top of the diagram define a path disjoint from the three at the bottom. By double counting the five points in the intersection of two balls we cover 8 red and 8 blue points with each set of three balls. Hence, we can send flow along each path. However, this does not give a feasible integral solution with three centers as any set of three clusters does not contain enough points. In fact, the four clusters can be placed arbitrarily far from each other and in this way we have an unbounded integrality gap since one ball needs to cover two clusters.

References

1. Anagnostopoulos, A., et al.: Principal fairness: removing bias via projections. CoRR abs/1905.13651 (2019)
2. Arora, S., Ge, R.: New tools for graph coloring. In: Goldberg, L.A., Jansen, K., Ravi, R., Rolim, J.D.P. (eds.) APPROX/RANDOM -2011. LNCS, vol. 6845, pp. 1–12. Springer, Heidelberg (2011). https://doi.org/10.1007/978-3-642-22935-0_1
3. Backurs, A., Indyk, P., Onak, K., Schieber, B., Vakilian, A., Wagner, T.: Scalable fair clustering. In: Proceedings of the 36th International Conference on Machine Learning, ICML, pp. 405–413 (2019)
4. Bandyapadhyay, S., Inamdar, T., Pai, S., Varadarajan, K.R.: A constant approximation for colorful k-center. In: 27th Annual European Symposium on Algorithms, ESA, pp. 12:1–12:14 (2019)
5. Chakrabarty, D., Goyal, P., Krishnaswamy, R.: The non-uniform k-center problem. In: 43rd International Colloquium on Automata, Languages, and Programming, ICALP, pp. 67:1–67:15 (2016)
6. Charikar, M., Khuller, S., Mount, D.M., Narasimhan, G.: Algorithms for facility location problems with outliers. In: Proceedings of the 12th Annual ACM-SIAM Symposium on Discrete Algorithms (SODA), pp. 642–651 (2001)
7. Chierichetti, F., Kumar, R., Lattanzi, S., Vassilvitskii, S.: Fair clustering through fairlets. In: Advances in Neural Information Processing Systems (NIPS), pp. 5029–5037 (2017)
8. Chlamtac, E., Friggstad, Z., Georgiou, K.: Understanding set cover: sub-exponential time approximations and lift-and-project methods. CoRR abs/1204.5489 (2012)
9. Gonzalez, T.F.: Clustering to minimize the maximum intercluster distance. Theoret. Comput. Sci. **38**, 293–306 (1985)
10. Grigoriev, D.: Complexity of positivstellensatz proofs for the knapsack. Comput. Complex. **10**(2), 139–154 (2001)
11. Harris, D.G., Pensyl, T., Srinivasan, A., Trinh, K.: A lottery model for center-type problems with outliers. ACM Trans. Algorithms **15**(3), 36:1–36:25 (2019)
12. Hochbaum, D.S., Shmoys, D.B.: A best possible heuristic for the k-center problem. Math. Oper. Res. **10**(2), 180–184 (1985)
13. Hsu, W.L., Nemhauser, G.L.: Easy and hard bottleneck location problems. Discrete Appl. Math. **1**(3), 209–215 (1979)
14. Karlin, A.R., Mathieu, C., Nguyen, C.T.: Integrality gaps of linear and semi-definite programming relaxations for knapsack. In: Günlük, O., Woeginger, G.J. (eds.) IPCO 2011. LNCS, vol. 6655, pp. 301–314. Springer, Heidelberg (2011). https://doi.org/10.1007/978-3-642-20807-2_24
15. Lasserre, J.B.: An explicit exact SDP relaxation for nonlinear 0–1 programs. In: Aardal, K., Gerards, B. (eds.) IPCO 2001. LNCS, vol. 2081, pp. 293–303. Springer, Heidelberg (2001). https://doi.org/10.1007/3-540-45535-3_23

16. Lasserre, J.B.: Global optimization with polynomials and the problem of moments. SIAM J. Optim. **11**(3), 796–817 (2001)
17. Tulsiani, M.: CSP gaps and reductions in the Lasserre hierarchy. In: Proceedings of the 41st Annual ACM Symposium on Theory of Computing, STOC, pp. 303–312 (2009)

Popular Branchings and Their Dual Certificates

Telikepalli Kavitha[1], Tamás Király[2], Jannik Matuschke[3], Ildikó Schlotter[4], and Ulrike Schmidt-Kraepelin[5($\boxtimes$)]

[1] TIFR, Mumbai, India
kavitha@tifr.res.in
[2] Eötvös University, Budapest, Hungary
tkiraly@cs.elte.hu
[3] KU Leuven, Leuven, Belgium
jannik.matuschke@kuleuven.be
[4] Budapest University of Technology and Economics, and KRTK KTI,
Budapest, Hungary
ildi@cs.bme.hu
[5] Technische Universität Berlin, Berlin, Germany
u.schmidt-kraepelin@tu-berlin.de

Abstract. Let G be a digraph where every node has preferences over its incoming edges. The preferences of a node extend naturally to preferences over *branchings*, i.e., directed forests; a branching B is *popular* if B does not lose a head-to-head election (where nodes cast votes) against any branching. Such popular branchings have a natural application in liquid democracy. The popular branching problem is to decide if G admits a popular branching or not. We give a characterization of popular branchings in terms of *dual certificates* and use this characterization to design an efficient combinatorial algorithm for the popular branching problem. When preferences are weak rankings, we use our characterization to formulate the *popular branching polytope* in the original space and also show that our algorithm can be modified to compute a branching with *least unpopularity margin*. When preferences are strict rankings, we show that "approximately popular" branchings always exist.

1 Introduction

Let G be a directed graph where every node has preferences (in partial order) over its incoming edges. When G is simple, the preferences can equivalently be defined on in-neighbors. We define a *branching* as a subgraph of G that is a directed forest where any node has in-degree at most 1; a node with in-degree 0 is a *root*. The problem we consider here is to find a branching that is *popular*.

Given any pair of branchings, we say a node u prefers the branching where it has a more preferred incoming edge (being a root is u's worst choice). If neither incoming edge is preferred to the other, then u is indifferent between the two

Part of this work was done at the 9th Emléktábla workshop in Gárdony, Hungary.

branchings. So any pair of branchings, say B and B', can be compared by asking for the majority opinion, i.e., every node opts for the branching that it prefers, and it abstains if it is indifferent between them. Let $\phi(B, B')$ (resp., $\phi(B', B)$) be the number of nodes that prefer B (resp., B') in the B-vs-B' comparison. If $\phi(B', B) > \phi(B, B')$, then we say B' is *more popular* than B.

Definition 1. *A branching B is popular in G if there is no branching that is more popular than B. That is, $\phi(B, B') \geq \phi(B', B)$ for all branchings B' in G.*

An Application in Computational Social Choice. We see the main application of popular branchings within *liquid democracy*. Suppose there is an election where a specific issue should be decided upon, and there are several proposed alternatives. Every individual voter has an opinion on these alternatives, but might also consider certain other voters as being better informed than her. Liquid democracy is a novel voting scheme that provides a middle ground between the feasibility of representative democracy and the idealistic appeal of direct democracy [4]: Voters can choose whether they delegate their vote to another, well-informed voter or cast their vote themselves. As the name suggests, voting power flows through the underlying network, or in other words, delegations are transitive. During the last decade, this idea has been implemented within several online decision platforms such as *Sovereign* and *LiquidFeedback*[1] and was used for internal decision making at Google [22] and political parties, such as the German *Pirate Party* or the Swedish party *Demoex*.

In order to circumvent *delegation cycles*, e.g., a situation in which voter x delegates to voter y and vice versa, and to enhance the expressiveness of delegation preferences, several authors proposed to let voters declare a set of acceptable representatives [20] together with a preference relation among them [5,22,30]. Then, a mechanism selects one of the approved representatives for each voter, avoiding delegation cycles. Similarly as suggested in [6], we additionally assume that voters accept themselves as their least preferred approved representative.

This reveals the connection to branchings in simple graphs where nodes correspond to voters and the edge (x, y) indicates that voter x is an approved delegate of voter y.[2] Every root in the branching casts a weighted vote on behalf of all her descendants. We assume that voters rate branchings only based on their predecessors. This is justified when approved delegates are considered to be more competent on the issue as well as in assessing the competence of others. What is a good mechanism to select representatives for voters? A crucial aspect in liquid democracy is the *stability* of the delegation process [3,14]. For the model described above, we propose popular branchings as a new concept of stability.

Not every directed graph admits a popular branching. Consider the following simple graph on four nodes a, b, c, d where a, b (similarly, c, d) are each other's top choices, while a, c (similarly, b, d) are each other's second choices. There is no edge

---

[1] See www.democracy.earth and www.interaktive-demokratie.org, respectively.

[2] Typically, such a delegation is represented by an edge (y, x); for the sake of consistency with downward edges in a branching, we use (x, y).

between a, d (similarly, b, c). Consider the branching $B = \{(a, b), (a, c), (c, d)\}$. A more popular branching is $B' = \{(d, c), (c, a), (a, b)\}$. Observe that a and c prefer B' to B, while d prefers B to B' and b is indifferent between B and B'. We can similarly obtain a branching $B'' = \{(b, a), (b, d), (d, c)\}$ that is more popular than B'. It is easy to check that this instance has no popular branching.

1.1   Our Problem and Results

The popular branching problem is to decide if a given digraph G admits a popular branching or not, and if so, to find one. We show that determining whether a given branching B is popular is equivalent to solving a min-cost arborescence problem in an extension of G with appropriately defined edge costs (these edge costs are a *function* of the branching). The dual LP to this arborescence problem gives rise to a laminar set system that serves as a certificate for the popularity of B if it is popular. This dual certificate proves crucial in devising an algorithm for efficiently solving the popular branching problem.

Theorem 2. *Given a directed graph G where every node has preferences in arbitrary partial order over its incoming edges, there is a polynomial-time algorithm to decide if G admits a popular branching or not, and if so, to find one.*

The proof of Theorem 2 is presented in Sect. 3; it is based on a characterization of popular branchings that we develop in Sect. 2. In applications like liquid democracy, it is natural to assume that the preference order of every node is a *weak ranking*, i.e., a ranking of its incoming edges with possible ties. In this case, the proof of correctness of our popular branching algorithm leads to a formulation of the *popular branching polytope* $\mathcal{B}_G$, i.e., the convex hull of incidence vectors of popular branchings in G.

Theorem 3. *Let G be a digraph on n nodes and m edges where every node has a weak ranking over its incoming edges. The popular branching polytope of G admits a formulation of size $O(2^n)$ in $\mathbb{R}^m$. Moreover, this polytope has $\Omega(2^n)$ facets.*

We also show an extended formulation of $\mathcal{B}_G$ in $\mathbb{R}^{m+mn}$ with $O(mn)$ constraints. When G has edge costs and node preferences are weak rankings, the min-cost popular branching problem can be efficiently solved. So we can efficiently solve extensions of the popular branching problem, such as finding one that minimizes the largest rank used or one with given forced/forbidden edges.

Relaxing Popularity. Since popular branchings need not always exist in G, this motivates relaxing popularity to *approximate popularity*—do approximately popular branchings always exist in any instance G? An approximately popular branching B may lose an election against another branching, however the extent of this defeat will be bounded. There are two measures of unpopularity: *unpopularity factor $u(\cdot)$* and *unpopularity margin $\mu(\cdot)$*. These are defined as follows:

$$u(B) = \max_{\phi(B',B)>0} \frac{\phi(B', B)}{\phi(B, B')} \quad \text{and} \quad \mu(B) = \max_{B'} \phi(B', B) - \phi(B, B').$$

A branching B is popular if and only if $u(B) \leq 1$ or $\mu(B) = 0$. We show the following results.

Theorem 4 ($\star^3$). *A branching with minimum unpopularity margin in a digraph where every node has a weak ranking over its incoming edges can be efficiently computed. In contrast, when node preferences are in arbitrary partial order, the minimum unpopularity margin problem is* NP-*hard.*

Theorem 5 ($\star$). *Let G be a digraph where every node has a strict ranking over its incoming edges. Then there always exists a branching B in G with $u(B) \leq \lfloor \log n \rfloor$. Moreover, for every n, we can show an instance G_n on n nodes with strict rankings such that $u(B) \geq \lfloor \log n \rfloor$ for every branching B in G_n.*

Hardness Results for Restricted Popular Branching Problems. A natural optimization problem here is to compute a popular branching where no tree is large. In liquid democracy, a large-sized tree shows a high concentration of power in the hands of a single voter, and this is harmful for social welfare [20]. When there is a fixed subset of root nodes in a directed graph, it was shown in [20] that it is NP-hard to find a branching that minimizes the size of the largest tree. To translate this result to popular branchings, we need to allow ties, whereas Theorem 6 below holds even for strict rankings. Another natural restriction is to limit the out-degree of nodes; Theorem 6 also shows that this variant is computationally hard.

Theorem 6 ($\star$). *Given a digraph G where each node has a strict ranking over its incoming edges, it is* NP-*hard to decide if there exists*

(a) a popular branching in G where each node has at most 9 descendants;
(b) a popular branching in G with maximum out-degree at most 2.

1.2 Background and Related Work

The notion of popularity was introduced by Gärdenfors [19] in 1975 in the domain of bipartite matchings. Algorithmic questions in popular matchings have been well-studied for the last 10–15 years [1,2,8,9,15,16,21,23–26,28,31].

Algorithms for popular matchings were first studied in the *one-sided* preferences model where vertices on only one side of the bipartite graph have preferences over their neighbors. Popular matchings need not always exist here and there is an efficient algorithm to solve the popular matching problem [1]. The functions unpopularity factor/margin were introduced in [31] to measure the *unpopularity* of a matching; it was shown in [31] that it is NP-hard to compute a matching that minimizes either of these quantities. In the domain of bipartite matchings with *two-sided* strict preferences, popular matchings always exist since stable matchings always exist [18] and every stable matching is popular [19].

---

[3] Theorems marked by an asterisk ($\star$) are proved in the full version [27] of our paper.

The concept of popularity has previously been applied to (undirected) spanning trees [10–12]. In contrast to our setting, voters have rankings over the entire edge set. This allows for a number of different ways to derive preferences over trees, most of which lead to hardness results.

Techniques. We characterize popular branchings in terms of *dual certificates*. This is analogous to characterizing popular matchings in terms of *witnesses* (see [15,24,26]). However, witnesses of popular matchings are in $\mathbb{R}^n$ and these are far simpler than dual certificates. A dual certificate is an appropriate family of subsets of the node set V. A certificate of size k implies that the unpopularity margin of the branching is at most $n - k$. Our algorithm constructs a partition $\mathcal{X}'$ of V such that if G admits popular branchings, then there has to be *some* popular branching in G with a dual certificate of size n supported by $\mathcal{X}'$. Moreover, when nodes have weak rankings, $\mathcal{X}'$ supports some dual certificate of size n to *every* popular branching in G: this leads to the formulation of $\mathcal{B}_G$ (see Sect. 4). Our positive results on low unpopularity branchings are extensions of our algorithm.

Notation. The preferences of node v on its incoming edges are given by a strict partial order $\prec_v$, so $e \prec_v f$ means that v prefers edge f to edge e. We use $e \sim_v f$ to denote that v is indifferent between e and f, that is, neither $e \prec_v f$ nor $e \succ_v f$ holds. The relation $\prec_v$ is a *weak ranking* if $\sim_v$ is transitive. In this case, $\sim_v$ is an equivalence relation and there is a strict order on the equivalence classes. When each equivalence class has size 1, we call it a *strict ranking*.

2   Dual Certificates

We add a dummy node r to $G = (V_G, E_G)$ as the root and make (r, v) the least preferred incoming edge of any node v in G. Let $D = (V \cup \{r\}, E)$ be the resulting graph where $V = V_G$ and $E = E_G \cup \{(r, u) : u \in V\}$. An *r-arborescence* in D is an out-tree with root r (throughout the paper, all arborescences are assumed to be rooted at r and to span V, unless otherwise stated).

Note that there is a one-to-one correspondence between branchings in G and arborescences in D (simply make r the parent of all roots of the branching). A branching is popular in G if and only if the corresponding arborescence is popular among all arborescences in D.[4] We will therefore prove our results for arborescences in D. The corresponding results for branchings in G follow immediately by projection, i.e., removing node r and its incident edges.

Let A be an arborescence in D. There is a simple way to check if A is popular in D. Let $A(v)$ be the incoming edge of v in A. For $e = (u, v)$ in D, define:

$$c_A(e) := \begin{cases} 0, & \text{if } e \succ_v A(v), \text{ i.e., } v \text{ prefers } e \text{ to } A(v); \\ 1, & \text{if } e \sim_v A(v), \text{ i.e., } v \text{ is indifferent between } e \text{ and } A(v); \\ 2, & \text{if } e \prec_v A(v), \text{ i.e., } v \text{ prefers } A(v) \text{ to } e. \end{cases}$$

---

[4] Note that, by the special structure of D, an arborescence is popular among all arborescences in D if and only if it is a popular branching in D.

Observe that $c_A(A) = |V| = n$ since $c_A(e) = 1$ for every $e \in A$. Let A' be any arborescence in D and let $\Delta(A, A') = \phi(A, A') - \phi(A', A)$ be the difference in the number of votes for A and the number of votes for A' in the A-vs-A' comparison. Observe that $c_A(A') = \Delta(A, A') + n$. Thus, $c_A(A') \geq n = c_A(A)$ if and only if $\Delta(A, A') \geq 0$. So we can conclude the following.

Proposition 7. *Let A' be a min-cost arborescence in D with respect to c_A. Then $\mu(A) = n - c_A(A')$. In particular, A is popular in D if and only if it is a min-cost arborescence in D with edge costs given by c_A.*

Consider the following linear program LP1, which computes a min-cost arborescence in D, and its dual LP2. For any non-empty $X \subseteq V$, let $\delta^-(X)$ be the set of edges entering the set X in the graph D.

$$\text{minimize} \sum_{e \in E} c_A(e) \cdot x_e \tag{LP1}$$

$$\text{subject to} \quad \sum_{e \in \delta^-(X)} x_e \geq 1 \quad \forall X \subseteq V, \ X \neq \emptyset$$

$$x_e \geq 0 \quad \forall e \in E.$$

$$\text{maximize} \sum_{X \subseteq V, X \neq \emptyset} y_X \tag{LP2}$$

$$\text{subject to} \quad \sum_{X : \delta^-(X) \ni e} y_X \leq c_A(e) \quad \forall e \in E$$

$$y_X \geq 0 \quad \forall X \subseteq V, \ X \neq \emptyset.$$

For any feasible solution y to LP2, let $\mathcal{F}_y := \{X \subseteq V : y_X > 0\}$ be the support of y. Inspired by Edmonds' branching algorithm [13], Fulkerson [17] gave an algorithm to find an optimal solution y to LP2 such that y is integral. From an alternative proof in [29], we obtain the following useful lemma.

Lemma 8. *There exists an optimal, integral solution y^* to LP2 such that $\mathcal{F}_{y^*}$ is laminar.*

Let y be an optimal, integral solution to LP2 such that $\mathcal{F}_y$ is laminar. Note that for any nonempty $X \subseteq V$, there is an $e \in A \cap \delta^-(X)$ and thus $y_X \leq c_A(e) = 1$. This implies that $y_X \in \{0, 1\}$ for all X. We conclude that $\mathcal{F}_y$ is a dual certificate for A in the sense of the following definition.

Definition 9. *A dual certificate for A is a laminar family $\mathcal{Y} \subseteq 2^V$ such that $|\{X \in \mathcal{Y} : e \in \delta^-(X)\}| \leq c_A(e)$ for all $e \in E$.*

For the remainder of this section, let $\mathcal{Y}$ be a dual certificate maximizing $|\mathcal{Y}|$.

Lemma 10. *Arborescence A has unpopularity margin $\mu(A) = n - |\mathcal{Y}|$. Furthermore, the following three statements are equivalent:*

(1) A is popular.

(2) $|\mathcal{Y}| = n$.

(3) $|A \cap \delta^-(X)| = 1$ for all $X \in \mathcal{Y}$ and $|\{X \in \mathcal{Y} : e \in \delta^-(X)\}| = 1$ for all $e \in A$.

Proof. Let x and y be the characteristic vectors of A and $\mathcal{Y}$, respectively. By Proposition 7, A is popular if and only if x is an optimal solution to LP1. This is equivalent to (2) because $c_A(A) = n$. Note also that (3) is equivalent to x and y fulfilling complementary slackness, which is equivalent to x being optimal. □

Lemma 10 establishes the following one-to-one correspondence between the nodes in V and the sets of $\mathcal{Y}$: For every set $X \in \mathcal{Y}$, there is a unique edge $(u, v) \in A$ that enters X. We call v the *entry-point* for X. Conversely, we let Y_v be the unique set in $\mathcal{Y}$ for which v is the entry-point; thus $\mathcal{Y} = \{Y_v : v \in V\}$.

Observation 11. *For every $v \in V$ we have $|\{X \in \mathcal{Y} : v \in X\}| \leq 2$.*

Observation 11 is implied by the fact that $e = (r, v)$ is an edge in D for every $v \in V$ and $c_A(e) \leq 2$. Laminarity of $\mathcal{Y}$ yields the following corollary:

Corollary 12. *If $|\mathcal{Y}| = n$, then $w \in Y_v \setminus \{v\}$ for some $v \in V$ implies $Y_w = \{w\}$.*

The following definition of the set of *safe* edges $S(X)$ with respect to a subset $X \subseteq V$ will be useful. $S(X)$ is the set of edges (u, v) in $E[X] := E \cap (X \times X)$ such that properties 1 and 2 hold:

1. (u, v) is *undominated* in $E[X]$, i.e., $(u, v) \not\prec_v (u', v) \; \forall (u', v) \in E[X]$.
2. (u, v) *dominates* $(w, v) \; \forall w \notin X$, i.e., $(u, v) \succ_v (w, v) \; \forall (w, v) \in \delta^-(X)$.

The interpretation of $S(X)$ is the following. Suppose that the dual certificate $\mathcal{Y}$ proves the popularity of A. Let $X \in \mathcal{Y}$ with $|X| > 1$. By Corollary 12, for every node $v \in X$ other than the entry-point in X we have $\{v\} = Y_v \in \mathcal{Y}$. So edges in $\delta^-(v)$ within $E[X]$ enter exactly one dual set, i.e., $\{v\}$, while any edge (w, v) where $w \notin X$ enters two of the dual sets: X and $\{v\}$. This induces exactly the constraints (1) and (2) given above on $(u, v) \in A$ (see LP2), showing that the edge $A(v)$ must be safe, as stated in Observation 13.

Observation 13. *If A is popular, then $A \cap E[X] \subseteq S(X)$ for all $X \in \mathcal{Y}$.*

3   Popular Branching Algorithm

We are now ready to present our algorithm for deciding if D admits a popular arborescence or not. For each $v \in V$, step 1 builds the largest set X_v such that v can reach all nodes in X_v using edges in $S(X_v)$. The collection $\mathcal{X} = \{X_v : v \in V\}$ will be laminar (see Lemma 14). To construct the sets X_v we make use of the *monotonicity* of S: $X \subseteq X'$ implies $S(X) \subseteq S(X')$.

In steps 2–3, the algorithm contracts each maximal set in $\mathcal{X}$ into a single node and builds a graph D' on these nodes and r. For each set X here that has been contracted into a node, edges incident to X in D' are undominated edges from other nodes in D' to the *candidate entry-points* of X, which are nodes $v \in X$ such that $X = X_v$. Our proof of correctness (see Theorems 15–16) shows that D admits a popular arborescence if and only if D' admits an arborescence.

Our algorithm for computing a popular arborescence in D is given below.

1. For each $v \in V$ do:
   - let $X_v^0 = V$ and $i = 0$;
   - while v does not reach all nodes in the graph $D_v^i = (X_v^i, S(X_v^i))$ do:
     X_v^{i+1} = the set of nodes reachable from v in D_v^i; let $i = i + 1$.
   - let $X_v = X_v^i$.
2. Let $\mathcal{X} = \{X_v : v \in V\}$, $\mathcal{X}' = \{X_v \in \mathcal{X} : X_v \text{ is } \subseteq\text{-maximal in } \mathcal{X}\}$, $E' = \emptyset$.
3. For every edge $e = (u, v)$ in D such that $X_v \in \mathcal{X}'$ and $u \notin X_v$ do:
   - if e is undominated (i.e., $e \not\prec_v e'$) among all edges $e' \in \delta^-(X_v)$, then

$$f(e) = \begin{cases} (U, X_v) & \text{where } u \in U \text{ and } U \in \mathcal{X}', \\ (r, X_v) & \text{if } u = r; \end{cases}$$

   - let $E' := E' \cup \{f(e)\}$.
4. If $D' = (\mathcal{X}' \cup \{r\}, E')$ contains an arborescence $\tilde{A}$, then
   - let $A' = \{e : f(e) \in \tilde{A}\}$;
   - let $R = \{v \in V : |X_v| \geq 2 \text{ and } v \text{ has an incoming edge in } A'\}$;
   - for each $v \in R$: let A_v be an arborescence in $(X_v, S(X_v))$;
   - return $A^* = A' \cup_{v \in R} A_v$.
5. Else return "*No popular arborescence in D*".

Correctness of the Above Algorithm. We will first show the easy direction, that is, if the algorithm returns an edge set A^*, then A^* is a popular arborescence in D. The following lemma will be key to this. Note that the set X_u, for each $u \in V$, is defined in step 1. Lemmas marked by ($\circ$) are proved in the appendix.

Lemma 14 ($\circ$). $\mathcal{X} = \{X_v : v \in V\}$ *is laminar. If* $u \in X_v$, *then* $X_u \subseteq X_v$.

Theorem 15 ($\star$). *If the above algorithm returns an edge set* A^*, *then* A^* *is a popular arborescence in* D.

Proof (Sketch). It is straightforward to verify that A^* is an arborescence in D. To prove the popularity of A^*, we construct a dual certificate $\mathcal{Y}$ of size n for A^*, by setting $\mathcal{Y} := \{X_v : v \in R\} \cup \{\{v\} : v \in V \setminus R\}$.

Note that $|\mathcal{Y}| = |R| + |V \setminus R| = n$. It remains to show that any edge $(w, v) \in E$ satisfies the constraints in LP2; let (u, v) be the incoming edge of v in A^*.

Suppose $v \in R$; then $(u, v) \in A'$ and $u \notin X_v$. Consider any edge (w, v): this enters one set of $\mathcal{Y}$ iff $w \notin X_v$ and no set iff $w \in X_v$. Hence, it suffices to show that $c_{A^*}((w, v)) \in \{1, 2\}$ for $w \notin X_v$. By construction of E', (w, v) does not dominate (u, v) and therefore $c_{A^*}((w, v)) \in \{1, 2\}$.

Suppose $v \in V \setminus R$. Let s be v's local root, i.e., the unique $s \in R$ with $v \in X_s$. Then $(u, v) \in A_s \subseteq S(X_s)$ by construction of A_s. Any edge $(w, v) \in \delta^-(v)$ enters at most two sets of $\mathcal{Y}$: $\{v\}$ and possibly X_s. If, on the one hand,

$(w, v) \in \delta^-(X_s)$, then $(u, v) \in S(X_s)$ dominates (w, v) by property 2 of $S(X_s)$, and hence $c_{A^*}((w, v)) = 2$. If, on the other hand, $w \in X_s$, then $(u, v) \in S(X_s)$ is not dominated by (w, v) by property 1 of $S(X_s)$, and hence $c_{A^*}((w, v)) \geq 1$. Thus, any edge satisfies the constraints in LP2, proving the theorem. □

Theorem 16. *If D admits a popular arborescence, then our algorithm finds one.*

Before we prove Theorem 16, we need Lemma 17 and Lemma 18.

Lemma 17 (○). *Let A be a popular arborescence and $\mathcal{Y}$ a dual certificate for A of size n. Then $Y_v \subseteq X_v$ for any $v \in V$.*

Lemma 18 (○). *Let A be a popular arborescence in D and let $X \in \mathcal{X}'$. Then A enters X exactly once, and it enters X at some node v such that $X = X_v$.*

Proof of Theorem 16. Assume there exists a popular arborescence A in D; then there exists a dual certificate $\mathcal{Y}$ of size n for A. We will show there exists an arborescence in D'. By Lemma 18, for each $X \in \mathcal{X}'$ there exists exactly one edge $e_X = (u, v)$ of A that enters X and v is a candidate entry-point of X.

We claim that (u, v) is not dominated by any $(u', v) \in \delta^-(X)$. Recall that by Lemma 17, we know $Y_v \subseteq X_v = X$. If some $(u', v) \in \delta^-(X)$ dominates $(u, v) \in A$, its cost must be $c_A((u', v)) = 0$. However, (u', v) clearly enters $Y_v \subseteq X$, and thus violates LP2, contradicting our assumption that $\mathcal{Y}$ is a dual solution. Hence, e_X is undominated among the edges of $\delta^-(X) \cap \delta^-(v)$ and therefore our algorithm creates an edge $f(e_X)$ in E' pointing to X. Using the fact that A is an arborescence in D, it is straightforward to verify that the edges $\{f(e_X) : X \in \mathcal{X}'\}$ form an arborescence $\tilde{A}$ in D'. Thus our algorithm returns an edge set A^*, which by Theorem 15 must be a popular arborescence in D. □

It is easy to see that step 1 (the bottleneck step) takes $O(mn)$ time per node. Hence the running time of the algorithm is $O(mn^2)$; thus Theorem 2 follows.

3.1 A Simple Extension of Our Algorithm: Algorithm MINMARGIN

Our algorithm can be extended to compute an arborescence with minimum *unpopularity margin* when nodes have weak rankings. When D' does not admit an arborescence, algorithm MINMARGIN below computes a max-size branching $\tilde{B}$ in D' and adds edges from r to all root nodes in $\tilde{B}$ to construct an arborescence. This arborescence in D' is then transformed into an arborescence in D.

1. Let D' be the graph constructed in our algorithm for Theorem 2, and let $\tilde{B}$ be a branching of maximum cardinality in D'.
2. Let $B' = \{e \mid f(e) \in \tilde{B}\}$, $R_1 = \{v \in V \mid \delta^-(v) \cap B' \neq \emptyset\}$, $R_2 = \emptyset$.
3. For each $X \in \mathcal{X}'$ which is a root in the branching $\tilde{B}$, select one arbitrary $v \in V$ such that $X_v = X$, add v to R_2 and (r, v) to B'.
4. For each $v \in R_1 \cup R_2$: let A_v be an arborescence in $(X_v, S(X_v))$.
5. Return $A^* := B' \bigcup_{v \in R_1 \cup R_2} A_v$.

Theorem 19 (⋆). *When nodes have weak rankings, Algorithm MINMARGIN returns an arborescence with minimum unpopularity margin in $D = (V \cup \{r\}, E)$.*

4   The Popular Arborescence Polytope of D

We now describe the popular arborescence polytope of $D = (V \cup \{r\}, E)$ in $\mathbb{R}^m$. Throughout this section we assume that every node has a weak ranking over its incoming edges. The arborescence polytope $\mathcal{A}$ of D is described below [29].

$$\sum_{e \in E[X]} x_e \leq |X| - 1 \qquad \forall X \subseteq V, \, |X| \geq 2. \tag{1}$$

$$\sum_{e \in \delta^-(v)} x_e = 1 \quad \forall v \in V \quad \text{and} \quad x_e \geq 0 \; \forall e \in E. \tag{2}$$

We will define a subgraph $D^* = (V \cup \{r\}, E_{D^*})$ of D: this is essentially the *expanded* version of the graph D' from our algorithm. The edge set of D^* is:

$$E_{D^*} = \bigcup_{X \in \mathcal{X}'} S(X) \cup \{(u, v) \in E : X_v \in \mathcal{X}', u \notin X_v, \text{ and } (u, v) \text{ is}$$

$$\text{undominated in } \delta^-(X_v)\}.$$

Thus each set $X \in \mathcal{X}'$, which is a node in D', is replaced in D^* by the nodes in X and with edges in $S(X)$ between nodes in X. We also replace edges in D' between sets in $\mathcal{X}'$ by the original edges in E.

Lemma 20. *If every node has a weak ranking over its incoming edges, then every popular arborescence in D is an arborescence in D^* that includes exactly $|X| - 1$ edges from $S(X)$ for each $X \in \mathcal{X}'$.*

Proof. Let A be a popular arborescence in D and let $X \in \mathcal{X}'$. By Lemma 18 we know $|A \cap \delta^-(X)| = 1$; moreover, the proof of Theorem 16 tells us that the unique edge in $A \cap \delta^-(X)$ is contained in D^*. So A contains $|X| - 1$ edges from $E[X]$ for each $X \in \mathcal{X}'$. It remains to show that these $|X| - 1$ edges are in $S(X)$.

Let $u \in X$. Suppose $A(u) \in E[X] \setminus S(X)$. This means that either (i) $A(u)$ is dominated by some edge in $E[X] \cup \delta^-(X)$ or (ii) u is indifferent between $A(u)$ and some edge in $\delta^-(X)$. Let $\mathcal{Y}$ be a dual certificate of A. We know that $Y_u \subseteq X_u \subseteq X$ (by Lemma 17). Since the entry point of A into X is not in Y_u, there is an edge $e \in S(X) \cap \delta^-(Y_u)$.

Let e enter $w \in Y_u$. Since $e \in S(X)$, we have $e \succ_w A(w)$ or $e \sim_w A(w)$, hence $c_A(e) \in \{0, 1\}$. If $w \neq u$, then e enters two sets Y_u and $\{w\}$—thus the constraint in LP2 corresponding to edge e is violated. If $w = u$ then $e \succ_u A(u)$ (since $A(u) \in E[X] \setminus S(X)$, $e \in S(X)$, and u has a weak ranking over its incoming edges): so $c_A(e) = 0$. Since e enters one set Y_u, the constraint corresponding to e in LP2 is again violated. So $A(u) \in S(X)$, i.e., $A \cap E[X] \subseteq S(X)$. □

Hence, every popular arborescence in D satisfies constraints (1)–(2) along with constraints (3) given below, where E_{D^*} is the edge set of D^*.

$$\sum_{e \in E[X]} x_e = |X| - 1 \; \forall X \in \mathcal{X}', \, |X| \geq 2 \quad \text{and} \quad x_e = 0 \; \forall e \in E \setminus E_{D^*} \tag{3}$$

Note that constraints (3) define a face $\mathcal{F}$ of the arborescence polytope $\mathcal{A}$ of D. Thus every popular arborescence in D belongs to face $\mathcal{F}$.

Consider a vertex in face $\mathcal{F}$: this is an arborescence A in D of the form $A' \cup_{X \in \mathcal{X}'} A_X$ where (i) A_X is an arborescence in $(X, S(X))$ whose root is an entry-point of X and (ii) $A' = \{e_X : X \in \mathcal{X}'\}$ where e_X is an edge in D^* entering the root of A_X. Theorem 15 proved that such an arborescence A is popular in D. Thus we can conclude Theorem 21 which proves the upper bound in Theorem 3. The lower bound in Theorem 3 is given in the appendix.

Theorem 21. *If every node has a weak ranking over its incoming edges, then face $\mathcal{F}$ (defined by constraints (1)–(3)) is the popular arborescence polytope of D.*

A compact extended formulation of this polytope and all missing proofs are in the full version of our paper [27]. We also discuss popular *mixed branchings* (probability distributions over branchings) in the full version [27].

Acknowledgments. Thanks to Markus Brill for helpful discussions on liquid democracy, and to Nika Salia for our conversations in Gárdony. Telikepalli Kavitha is supported by the DAE, Government of India, under project no. 12-R&D-TFR-5.01-0500, Tamás Király is supported by NKFIH grant no. K120254 and by the HAS, grant no. KEP-6/2017, Ildikó Schlotter was supported by the Hungarian Academy of Sciences under its Momentum Programme (LP2016-3/2018) and Cooperation of Excellences Grant (KEP-6/2018), and the Hungarian Scientific Research Fund, NFKIH, grants no. K128611 and K124171, and Ulrike Schmidt-Kraepelin by the Deutsche Forschungsgemeinschaft (DFG) under grant BR 4744/2-1.

Appendix: Missing Proofs from Sects. 3 and 4

Lemma 14 (∘). $\mathcal{X} = \{X_v : v \in V\}$ *is laminar. If* $u \in X_v$, *then* $X_u \subseteq X_v$.

Proof. We first show that $X_u^i \subseteq X_v^i$ for any i, where we set $X_v^i := X_v$ whenever X_v^i is not defined by the above algorithm. The claim clearly holds for $i = 0$. Let i be the smallest index such that $x \in X_u^i \setminus X_v^i$ for some node x; we must have $x \in X_u^{i-1} \cap X_v^{i-1}$. By the definition of X_u^i, x is reachable from u in $S(X_u^{i-1})$. Note that $X_u^{i-1} \subseteq X_v^{i-1}$ implies $S(X_u^{i-1}) \subseteq S(X_v^{i-1})$, which yields that x is reachable from u in $S(X_v^{i-1})$ as well. Moreover, u is reachable from v in $S(X_v^{i-1}) \supseteq S(X_v)$ because $u \in X_v$ and $S(\cdot)$ is monotone. Hence it follows that x is reachable from v in $S(X_v^{i-1})$ via u, contradicting the assumption that $x \notin X_v^i$. This proves the second statement of the lemma.

Now we will show the laminarity of $\mathcal{X}$. For contradiction, assume there exist $s, t \in V$ such that X_s and X_t cross, i.e., their intersection is non-empty, and neither contains the other. Then, by the second statement of the lemma, neither $s \in X_t$ nor $t \in X_s$ can hold. So we have that $s \notin X_t$ and $t \notin X_s$.

Let (x, y) be an edge in $S(X_t)$ such that $y \in X_s \cap X_t$ but $x \in X_t \setminus X_s$; since each node in X_t is reachable from t in $S(X_t)$, such an edge exists. Since $y \in X_s \setminus \{s\}$, there also exists an edge (u, y) in $S(X_s)$. As $x \notin X_s$ but $(u, y) \in S(X_s)$, we know that $(u, y) \succ_y (x, y)$ which contradicts $(x, y) \in S(X_t)$. $\square$

Lemma 17 (◦). *Let A be a popular arborescence and $\mathcal{Y}$ a dual certificate for A of size n. Then $Y_v \subseteq X_v$ for any $v \in V$.*

Proof. If $Y_v = \{v\}$, then $Y_v \subseteq X_v$ is trivial, so suppose that Y_v is not a singleton. We know from Corollary 12 that Y_w is a singleton set for each $w \in Y_v \setminus \{v\}$. Moreover, for every $(u, w) \in A$ with $w \in Y_v \setminus \{v\}$ it holds that $u \in Y_v$ since this edge would otherwise enter two sets; however, $c_A((u, w)) = 1$ as $(u, w) \in A$.

Assume for contradiction that $Y_v \setminus X_v \neq \emptyset$. Let i be the last iteration when $Y_v \subseteq X_v^i$. Then there exists a subset of Y_v which is not reachable by edges in $S(X_v^i)$, i.e., $\delta^-(Y_v \setminus X_v^{i+1}) \cap S(X_v^i) = \emptyset$. On the other hand, we know that the arborescence A can only enter nodes in $Y_v \setminus \{v\}$ by edges from $E[Y_v]$, and therefore, it needs to contain at least one edge from $\delta^-(Y_v \setminus X_v^{i+1}) \cap \delta^+(X_v^{(i+1)})$. Let (u, w) be this edge (see Fig. 1). By construction of X_v^i and X_v^{i+1}, we know that one of the following cases has to be true.

Case 1. There exists an edge $(x, w) \in E[X_v^i]$ which dominates (u, w). Note that we do not know if $(x, w) \in E[Y_v]$ or not. However, $c_A((x, w)) = 0$ in either case, but by Corollary 12, (x, w) enters at least one set in $\mathcal{Y}$, namely $\{w\}$. This is a violation of LP2 and it contradicts $\mathcal{Y}$ being a dual certificate for A.

Case 2. There exists an edge $(x, w) \in \delta^-(X_v^i)$ which is not dominated by (u, w). Note that $c_A((x, w)) \in \{0, 1\}$, but $(x, w) \in \delta^-(Y_v)$ and so the edge (x, w) enters two dual sets: Y_v and $\{w\}$. This contradicts $\mathcal{Y}$ being a dual solution. □

Lemma 18 (◦). *Let A be a popular arborescence in D and let $X \in \mathcal{X}'$. Then A enters X exactly once, and it enters X at some node v such that $X = X_v$.*

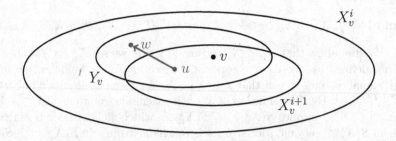

Fig. 1. Illustration of the situation in the proof of Lemma 17.

Proof. Let $X \in \mathcal{X}'$ and let A be a popular arborescence which enters X at some node $v \in V$ through an edge $(u, v) \in A \cap \delta^-(X)$. Moreover, let $\mathcal{Y}$ be a dual certificate for A, and let Y_v be the set whose entry-point is v.

Let $\mathsf{entry}(X) := \{w \in V : X_w = X\}$. We first show that $\mathsf{entry}(X) \subseteq Y_v$. Assume for contradiction that there exists $w \in \mathsf{entry}(X)$ such that $w \notin Y_v$. Since $X_w = X$ we know that there exists a w-v path P in $(X, S(X))$. Hence, there exists an edge $e \in P$ which enters Y_v. If the head of e is v, we know that e

dominates $(u, v) \in \delta^-(X)$ and hence $c_A(e) = 0$, a contradiction to the feasibility of $\mathcal{Y}$. If v is not the head of e, then e not only enters Y_v, but also the singleton set corresponding to its head. However, $c_A(e) \leq 1$ since e is an undominated edge by $e \in S(X)$, a contradiction to the feasibility of $\mathcal{Y}$.

To prove that $v \in \text{entry}(X)$, let us choose some $s \in \text{entry}(X)$. By the previous paragraph and Lemma 17, we get $s \in Y_v \subseteq X_v$, from which Lemma 14 implies $X_s \subseteq X_v$. Because $s \in \text{entry}(X)$, we have $X = X_s \subseteq X_v$. Because $X \in \mathcal{X}'$ is inclusionwise maximal in $\mathcal{X}$, we get $X = X_v$, proving $v \in \text{entry}(X)$.

It remains to prove that A enters X only once. Suppose for contradiction that there exist two nodes $v, v' \in \text{entry}(X)$ such that $(u, v), (u', v') \in A \cap \delta^-(X)$. By $\emptyset \neq \text{entry}(X) \subseteq Y_v \cap Y_{v'}$ and the laminarity of $\mathcal{Y}$, we can assume w.l.o.g. that $Y_v \subseteq Y_{v'}$. Moreover, since $Y_{v'} \subseteq X$, the arborescence edge (u, v) enters both Y_v and $Y_{v'}$, a contradiction to the feasibility of the dual solution $\mathcal{Y}$.   □

Lower Bound for the Popular Arborescence Polytope of D. Let $D = (V \cup \{r\}, E)$ be the complete graph where every node $v \in V$ regards all other nodes $u \in V$ as top-choice in-neighbors and r as its second-choice in-neighbor. Here $\mathcal{X}' = \{V\}$ and D^* is the complete bidirected graph on V along with edges (r, v) for all $v \in V$. We claim that in any minimal system contained in (1)–(3), the constraint $\sum_{e \in E[X]} x_e \leq |X| - 1$ for every $X \subset V$ with $|X| \geq 2$ has to be present. This is because a cycle on the nodes in X along with any rooted arborescence A on $V \setminus X$ plus (r, v), where v is the root of A, satisfies all the remaining constraints. Thus any minimal system of inequalities from (1)–(3) has to contain $2^n - n - 2$ inequalities from (1): one for every $X \subset V$ with $|X| \geq 2$. Since inequalities in a minimal system are in one-to-one correspondence with the facets of the polyhedron they describe [7, Theorem 3.30], the lower bound given in Theorem 3 follows.

References

1. Abraham, D.J., Irving, R.W., Kavitha, T., Mehlhorn, K.: Popular matchings. SIAM J. Comput. **37**(4), 1030–1034 (2007)
2. Biró, P., Irving, R.W., Manlove, D.F.: Popular matchings in the marriage and roommates problems. In: Calamoneri, T., Diaz, J. (eds.) CIAC 2010. LNCS, vol. 6078, pp. 97–108. Springer, Heidelberg (2010). https://doi.org/10.1007/978-3-642-13073-1_10
3. Bloembergen, D., Grossi, D., Lackner, M.: On rational delegations in liquid democracy. In: Proceedings of the 33rd AAAI Conference on Artificial Intelligence (AAAI) (2019)
4. Blum, C., Zuber, C.I.: Liquid democracy: Potentials, problems, and perspectives. J. Polit. Philos. **24**(2), 162–182 (2016)
5. Brill, M.: Interactive democracy. In: Proceedings of the 17th International Conference on Autonomous Agents and Multiagent Systems (AAMAS) Blue Sky Ideas Track, pp. 1183–1187 (2018)
6. Christoff, Z., Grossi, D.: Binary voting with delegable proxy: An analysis of liquid democracy. In: Proceedings of the 16th Conference on Theoretical Aspects of Rationality and Knowledge (TARK), pp. 134–150 (2017)

7. Conforti, M., Cornuéjols, G., Zambelli, G.: Integer Programming. GTM, vol. 271. Springer, Cham (2014). https://doi.org/10.1007/978-3-319-11008-0
8. Cseh, Á., Huang, C.C., Kavitha, T.: Popular matchings with two-sided preferences and one-sided ties. SIAM J. Disc. Math. **31**(4), 2348–2377 (2017)
9. Cseh, Á., Kavitha, T.: Popular edges and dominant matchings. Math. Program. **172**(1), 209–229 (2017)
10. Darmann, A.: Popular spanning trees. Int. J. Found. Comput. Sci. **24**(5), 655–677 (2013)
11. Darmann, A.: It is difficult to tell if there is a Condorcet spanning tree. Math. Methods Oper. Res. **84**(1), 93–104 (2016). https://doi.org/10.1007/s00186-016-0535-3
12. Darmann, A., Klamler, C., Pferschy, U.: Finding socially best spanning trees. Theory Decis. **70**(4), 511–527 (2011)
13. Edmonds, J.: Optimum branchings. J. Res. Nat. Bureau Stan. **71B**(4), 233–240 (1967)
14. Escoffier, B., Gilbert, H., Pass-Lanneau, A.: The convergence of iterative delegations in liquid democracy in a social network. In: Fotakis, D., Markakis, E. (eds.) SAGT 2019. LNCS, vol. 11801, pp. 284–297. Springer, Cham (2019). https://doi.org/10.1007/978-3-030-30473-7_19
15. Faenza, Y., Kavitha, T.: Quasi-popular matchings, optimality, and extended formulations. In: Proceedings of the 31st Annual ACM-SIAM Symposium on Discrete Algorithms (SODA), pp. 325–344 (2020)
16. Faenza, Y., Kavitha, T., Powers, V., Zhang, X.: Popular matchings and limits to tractability. In: Proceedings of the 30th Annual ACM-SIAM Symposium on Discrete Algorithms (SODA), pp. 2790–2809 (2019)
17. Fulkerson, D.R.: Packing rooted directed cuts in a weighted directed graph. Math. Program. **6**(1), 1–13 (1974)
18. Gale, D., Shapley, L.S.: College admissions and the stability of marriage. Am. Math. Mon. **69**(1), 9–15 (1962)
19. Gärdenfors, P.: Match making: Assignments based on bilateral preferences. Behav. Sci. **20**(3), 166–173 (1975)
20. Gölz, P., Kahng, A., Mackenzie, S., Procaccia, A.D.: The fluid mechanics of liquid democracy. In: Christodoulou, G., Harks, T. (eds.) WINE 2018. LNCS, vol. 11316, pp. 188–202. Springer, Cham (2018). https://doi.org/10.1007/978-3-030-04612-5_13
21. Gupta, S., Misra, P., Saurabh, S., Zehavi, M.: Popular matching in roommates setting is NP-hard. In: Proceedings of the 30th Annual ACM-SIAM Symposium on Discrete Algorithms (SODA), pp. 2810–2822 (2019)
22. Hardt, S., Lopes, L.: Google votes: A liquid democracy experiment on a corporate social network. Technical report, Technical Disclosure Commons (2015)
23. Huang, C.C., Kavitha, T.: Popular matchings in the stable marriage problem. Inf. Comput. **222**, 180–194 (2013)
24. Huang, C.C., Kavitha, T.: Popularity, mixed matchings, and self-duality. In: Proceedings of the 28th Annual ACM-SIAM Symposium on Discrete Algorithms (SODA), pp. 2294–2310 (2017)
25. Kavitha, T.: A size-popularity tradeoff in the stable marriage problem. SIAM J. Comput. **43**(1), 52–71 (2014)
26. Kavitha, T.: Popular half-integral matchings. In: Proceedings of the 43rd International Colloquium on Automata, Languages, and Programming (ICALP), Leibniz International Proceedings in Informatics (LIPIcs), vol. 55, pp. 22:1–22:13. Schloss Dagstuhl-Leibniz-Zentrum fuer Informatik (2016)

27. Kavitha, T., Király, T., Matuschke, J., Schlotter, I., Schmidt-Kraepelin, U.: Popular branchings and their dual certificates. Technical report 1912.01854 (2019). https://arxiv.org/abs/1912.01854
28. Kavitha, T., Mestre, J., Nasre, M.: Popular mixed matchings. Theor. Comput. Sci. **412**(24), 2679–2690 (2011)
29. Korte, B., Vygen, J.: Combinatorial Optimization. Springer, Heidelberg (2012). https://doi.org/10.1007/978-3-642-24488-9
30. Kotsialou, G., Riley, L.: Incentivising participation in liquid democracy with breadth first delegation. Technical report 1811.03710 (2018). https://arxiv.org/abs/1811.03710
31. McCutchen, R.M.: The least-unpopularity-factor and least-unpopularity-margin criteria for matching problems with one-sided preferences. In: Laber, E.S., Bornstein, C., Nogueira, L.T., Faria, L. (eds.) LATIN 2008. LNCS, vol. 4957, pp. 593–604. Springer, Heidelberg (2008). https://doi.org/10.1007/978-3-540-78773-0_51

Sparse Graphs and an Augmentation Problem

Csaba Király[1,2]([✉]) [iD] and András Mihálykó[2] [iD]

[1] MTA-ELTE Egerváry Research Group, Budapest, Hungary
[2] Department of Operations Research, ELTE Eötvös Loránd University,
Budapest, Hungary
{cskiraly,mihalyko}@cs.elte.hu

Abstract. For two integers $k > 0$ and ℓ, a graph $G = (V, E)$ is called (k, ℓ)-tight if $|E| = k|V| - \ell$ and $|E(X)| \leq k|X| - \ell$ for all $X \subseteq V$ for which $k|X| - \ell \geq 0$. G is called (k, ℓ)-redundant if $G - e$ has a spanning (k, ℓ)-tight subgraph for all $e \in E$. We consider the following augmentation problem. Given a graph $G = (V, E)$ that has a (k, ℓ)-tight spanning subgraph, find a graph $H = (V, F)$ with minimum number of edges, such that $G + H$ is (k, ℓ)-redundant.

In this paper, we give a polynomial algorithm and a min-max theorem for this augmentation problem when the input is (k, ℓ)-tight. For general inputs, we give a polynomial algorithm when $k \geq \ell$ and show the NP-hardness of the problem when $k < \ell$. Since (k, ℓ)-tight graphs play an important role in rigidity theory, these algorithms can be used to make several types of rigid frameworks redundantly rigid by adding a smallest set of new bars.

1 Introduction

Let k be a positive integer and ℓ be an integer such that $\ell < 2k$. A (multi)graph $G = (V, E)$ is called **(k, ℓ)-sparse** if $i_G(X) \leq k|X| - \ell$ for all $X \subseteq V$ for which $k|X| - \ell \geq 0$, where $i_G(X)$ denotes the number of edges of G induced by $X \subseteq V$. A (k, ℓ)-sparse graph is called **(k, ℓ)-tight** if $|E| = k|V| - \ell$. A graph G is called **(k, ℓ)-rigid** if G has a (k, ℓ)-tight spanning subgraph. We will call an edge e of G a **(k, ℓ)-redundant edge** if $G - e$ is (k, ℓ)-rigid. A graph G is called a **(k, ℓ)-redundant graph** if each edge of G is (k, ℓ)-redundant. We consider the following augmentation problem that we call the **general (augmentation) problem**.

Problem. Let k and ℓ be integers with $k \geq 0$ and $\ell < 2k$ and let $G = (V, E)$ be a (k, ℓ)-rigid graph. Find a graph $H = (V, F)$ on the same vertex set with minimum number of edges, such that $G + H = (V, E \cup F)$ is (k, ℓ)-redundant.

We call the special case of this problem, where the input graph G is (k, ℓ)-tight, the **reduced (augmentation) problem**. In this extended abstract, we give a min-max theorem and an $O(n^2)$ running time algorithm on a graph with n vertices for fixed k and ℓ for the reduced problem. We also show how this

© Springer Nature Switzerland AG 2020
D. Bienstock and G. Zambelli (Eds.): IPCO 2020, LNCS 12125, pp. 238–251, 2020.
https://doi.org/10.1007/978-3-030-45771-6_19

algorithm can be extended to solve the general problem in the same running time when $\ell \leq k$. In contrast, we note that the general problem is NP-hard whenever $\ell > k$. We leave all the detailed proofs to the full version [16].

Sparsity properties are important in rigidity theory as they can be used in the characterization of many rigidity classes. For example, the generically rigid graphs in $\mathbb{R}^2$ are exactly the $(2, 3)$-rigid graphs by the fundamental theorem of Pollaczek-Geiringer [23] and Laman [17]. The 'body-bar graph' induced by a given graph G is generically rigid in $\mathbb{R}^d$ if and only if G is $\left(\binom{d+1}{2}, \binom{d+1}{2} \right)$-rigid by Tay's theorem [25]. Note that G is (k, k)-rigid if and only if G contains k edge-disjoint spanning trees by the fundamental result of Nash-Williams [21].

Besides the effect of redundancy, redundant rigidity is an important concept in rigidity theory since variants of Hendrickson's result [8] show that redundant rigidity is often a necessary condition of 'global rigidity' which plays a crucial role in many applications [1, 28, 29]. Furthermore, in some cases, for example for 'body-bar graphs' (see [3]), redundant rigidity is also a sufficient condition of global rigidity. It is thus natural to ask how many new edges are needed to make a rigid graph redundantly rigid.

Known Special Cases. The general problem for $(1, 1)$-rigid graphs is the well-studied 2-edge-connectivity augmentation problem solved by Eswaran and Tarjan [4]. The general problem for $k = \ell$ was solved by Frank and T. Király [6] who gave a polynomial algorithm to augment a graph to an h-times (k, k)-redundant graph using polyhedral techniques. The algorithm, that will be presented here, is a more efficient solution for this problem, however, it does not deal with the case of $h \geq 2$ and also the (k, k)-rigidity of the input is needed. García and Tejel [7] showed that the general problem is NP-hard for $(2, 3)$-rigid graphs but can be solved in polynomial time for $(2, 3)$-tight graphs. We use similar techniques to [7], however, our method is based on a new min-max theorem for the reduced problem.

To obtain the solution for the general problem by using the reduced problem when $k \geq \ell$, we need more general concepts. The idea of our method comes from Jackson and Jordán [9] who proved that the (k, k)-redundant edges of a (k, k)-rigid graph $\bar{G}$ form induced subgraphs of $\bar{G}$ with disjoint vertex sets. If we contract these subgraphs into single vertices one can show that the resulting graph is (k, k)-tight for which we can use the algorithm for the reduced problem. In order to extend this idea for $k > \ell$, we need the following generalization of sparsity. Suppose throughout this paper that

(A0) $\ell \in \mathbb{Z}$ and $m : V \rightarrow \mathbb{Z}_+$ such that $m(u) + m(v) \geq \ell$ holds for each $u, v \in V$ and equality is only allowed when $m(u) = m(v) = \ell = 0$.

A graph $G = (V, E)$ is called (m, ℓ)**-sparse** if $i_G(X) \leq m(X) - \ell$ holds for every $X \subseteq V$ for which $m(X) - \ell \geq 0$, where $m(X) := \sum_{v \in X} m(v)$. An (m, ℓ)-sparse graph is called (m, ℓ)**-tight** if $|E| = m(V) - \ell$. Observe that each subgraph of an (m, ℓ)-sparse graph is (m, ℓ)-sparse. For simplicity, we will call a set $X \subseteq V$ (m, ℓ)**-tight in** G if the **induced subgraph** $G[X]$ of X in G is (m, ℓ)-tight.

A graph G is called (m, ℓ)-**rigid** if G has an (m, ℓ)-tight spanning subgraph. We call an edge e of G an (m, ℓ)-**redundant edge** if $G - e$ is (m, ℓ)-rigid. A graph G is called an (m, ℓ)-**redundant graph** if each edge of G is (m, ℓ)-redundant. Note that when $m \equiv k$, an (m, ℓ)-sparse/tight/rigid/redundant graph is (k, ℓ)-sparse/tight/rigid/redundant, respectively. For our solution of the general problem in Sect. 6, we need an algorithm for the reduced problem on (m, ℓ)-tight graphs. Hence we shall solve the reduced problem also for (m, ℓ)-tight graphs.

2   Preliminaries

In this section we list some basic properties of (m, ℓ)-sparse graphs. It follows from the definition that an (m, ℓ)-tight subgraph of an (m, ℓ)-sparse graph is always an induced subgraph. The following statement can be proved by using standard submodular techniques.

Lemma 1. *Let $G = (V, E)$ be an (m, ℓ)-sparse graph, and let $T_1 = (V_1, E_1)$ and $T_2 = (V_2, E_2)$ be (m, ℓ)-tight subgraphs of G. If $m(V_1 \cap V_2) \geq \ell$, then $T_1 \cup T_2$ is an (m, ℓ)-tight graph and there are no edges between $V_1 - V_2$ and $V_2 - V_1$. If $|V_1 \cap V_2| \geq 1$, then $T_1 \cap T_2$ is (m, ℓ)-tight as well.*

We note that, by (A0), the assumption on $m(V_1 \cap V_2)$ always holds when $|V_1 \cap V_2| \geq 2$, and also when $\ell \leq 0$ (since $m \geq 0$).

It is known that the edge sets of the (m, ℓ)-sparse subgraphs of a given graph form a matroid which is a generalization of the well-known 2-dimensional rigidity matroid (see Frank [5, Sect. 13.5], Lorea [19], and Whiteley [27, Appendix A]). A circuit of this matroid is called an (m, ℓ)-**circuit**. It follows from matroid theory (see details in [5, Chap. 5]) that for an (m, ℓ)-sparse graph $G = (V, E)$ and $i, j \in V$ for which $G + ij$ is not (m, ℓ)-sparse, $G + ij$ contains a unique (m, ℓ)-circuit $C_{(m,\ell)}^G(ij)$. In this case $T_{(m,\ell)}^G(ij) := C_{(m,\ell)}^G(ij) - ij$ is (m, ℓ)-tight. Let $V_{(m,\ell)}^G(ij)$ denote the vertex set of $T_{(m,\ell)}^G(ij)$. The main property of $T_{(m,\ell)}^G(ij)$ is the following.

Lemma 2. *Let $G = (V, E)$ be an (m, ℓ)-sparse graph and $i, j \in V$. Assume that $G + ij$ is not (m, ℓ)-sparse. If $G' = (V', E')$ is an (m, ℓ)-tight subgraph of G with $i, j \in V'$, then $T_{(m,\ell)}^G(ij) \subseteq G'$. Hence $T_{(m,\ell)}^G(ij) = \bigcap \{T_h : T_h$ is an (m, ℓ)-tight subgraph of G inducing i and $j\}$.*

Let $R_{(m,\ell)}^G(i_1 j_1, \ldots, i_r j_r)$ denote the subgraph induced by the (m, ℓ)-redundant edges of $G = (V, E)$ in $G \cup \{i_1 j_1, \ldots, i_r j_r\}$ for $i_1, \ldots, i_r, j_1, \ldots, j_r \in V$. For the sake of simplicity, when the graph G or (m, ℓ) is clear from the context, we will omit the superscript G or subscript (m, ℓ), respectively, from all of the above notation. Note that $R(ij) = T(ij)$ for any $i, j \in V$. The following lemma extends this simple fact by generalizing [7, Lemma 4] .

Lemma 3. *If G is an (m, ℓ)-tight graph, then*

$$R(i_1 j_1, \ldots, i_r j_r) = T(i_1 j_1) \cup \cdots \cup T(i_r j_r).$$

2.1   Algorithmic Preliminaries

To give a polynomial algorithm for our (general or reduced) augmentation problem, one first needs an algorithm for testing the (m, ℓ)-sparsity of a graph. Such a polynomial algorithm already exists for each pair of m and ℓ (see the works of Hendrickson and Jacobs [11,12] and Berg and Jordán [2] for the case where $k = 2$ and $\ell = 3$, the paper by Lee and Streinu [18] for general k and $\ell \geq 0$, the book of Frank [5, Sect. 13.5.4] for the (m, ℓ) case, and the note of the first author [15] for the $\ell < 0$ case). We note that in the main applications of (k, ℓ)-sparse graphs k and ℓ are fixed constants which translates to the general (m, ℓ)-sparse case that

($*$) there exists a constant $c > 0$ such that $m(v) \leq c$ for all $v \in V$ and $|\ell| \leq c$.

Thus we give the running time of our algorithms by assuming this condition. Observe that ($*$) implies that an (m, ℓ)-sparse graph on V has $O(|V|)$ edges.

We shall use the algorithm provided by the following theorem (which can be constructed based on the algorithms in [5,15,18]) as a subroutine in our algorithms.

Theorem 1 (Based on [5,15,18]). *There exists an algorithm which decides in $O(|V|^2)$ time whether its input graph $G = (V, E)$ is (m, ℓ)-sparse. It has the following outputs:*

If G is (m, ℓ)-sparse, then it outputs this fact along with an orientation D of all the edges in G minus a set F' of at most $-\ell$ edges when $\ell < 0$. If G is also (m, ℓ)-tight, then it also outputs this fact.

If G is not (m, ℓ)-sparse, then it outputs a maximal (m, ℓ)-sparse subgraph $H = (V, F)$ of G along with an orientation D of the edges in H minus a set F' of $-\ell$ edges when $\ell < 0$. It also outputs the set R of edges in H which are (m, ℓ)-redundant in G.

Furthermore, if it returns that G is (m, ℓ)-sparse (including the case when G is (m, ℓ)-tight), then by only using the extra data in the output one can decide in $O(|V|)$ extra time whether $G + e$ is (m, ℓ)-sparse for any new edge e, and if the answer is no, then output the (m, ℓ)-tight subgraph $T(e)$ of G.

3   Preprocessing

It is easier to formulate our results by assuming some extra conditions for our (m, ℓ)-tight graph. In most of our results we make the following assumption.

(A) Assuming (A0) for m and ℓ, $G = (V, E)$ is an (m, ℓ)-tight graph on at least 4 vertices such that either $\ell \leq 0$ or all of the following three conditions hold.

    (A1) There exists no $v \in V$ such that $m(v) = 0$ and v is an isolated vertex.

    (A2) There exist $u, v \in V$ such that $V(uv) \neq \{u, v\}$.

    (A3) There exists no $v \in V$ such that $V(uv) = \{u, v\}$ for all $u \in V - v$ and $V - v$ induces an (m, ℓ)-tight graph.

We note that conditions (A1)–(A3) (and thus (A)) automatically hold for (k, ℓ)-tight graphs with sufficiently many vertices and also for (m, ℓ)-tight graphs that arise from (k, ℓ)-rigid graphs in the algorithm of Sect. 6. However, it is also true that these conditions do not affect the reduced augmentation problem in any significant way, as the following lemma states.

Lemma 4. *Assume (A0), $\ell > 0$, and (*). Let $G = (V, E)$ be an (m, ℓ)-tight graph with $|V| \geq 4$. We can determine whether G violates (A1), (A2) or (A3) and we can construct a graph G' such that either G' does not violate any of (A1), (A2) and (A3) or G' has less than four vertices, and from the optimal solution of the reduced augmentation problem on G' we can compute the optimal solution of the reduced augmentation problem on G. All these computations can be done in $O(|V|^2)$ running time.*

4   The Min-Max Theorem for the Reduced Problem

Throughout this whole section, we shall assume (A). We call a set of vertices $\emptyset \neq C \subsetneq V$ **(m, ℓ)-co-tight** in G if $i_G(V - C) = m(V - C) - \ell$, that is, if the complement of C is an (m, ℓ)-tight set in G. Equivalently, C is (m, ℓ)-co-tight if $|\widehat{E}(C)| = m(C)$ where $\widehat{E}(C)$ denotes the set of edges that are incident with at least one vertex in C. Note that for every $X \subset V$ for which $m(V - X) \geq \ell$, $|\widehat{E}(X)| \geq m(X)$ holds by the sparsity of $V - X$. For the sake of brevity let us denote the inclusion-wise minimal (m, ℓ)-co-tight sets as **(m, ℓ)-MCT** sets. We omit the (m, ℓ) prefix of the notions (m, ℓ)-tight (m, ℓ)-co-tight, and (m, ℓ)-MCT when it is clear from the context.

Lemmas 2 and 3 imply that if $G + H$ is (m, ℓ)-redundant for a graph $H = (V, F)$ and C is a co-tight set in G then there exists an edge $uv \in F$ such that $u \in C$ or $v \in C$. This observation immediately gives a lower bound for the cardinality of the optimal solution of the reduced problem.

Lemma 5. *The minimum number of edges that augment G to an (m, ℓ)-redundant graph is at least the half of the maximal number of pairwise disjoint (m, ℓ)-co-tight sets.*

We say that a set U **covers** a set family $\mathcal{C}$, if $|U \cap C| \geq 1$ for every $C \in \mathcal{C}$. The definition of MCT sets implies the following.

Lemma 6. *Let $\mathcal{C}$ be the family of all (m, ℓ)-MCT sets of G. Suppose that $U \subseteq V$ is a set that covers $\mathcal{C}$. If $V' \subseteq V$ such that $U \subseteq V'$ and V' induces an (m, ℓ)-tight subgraph in G, then $V' = V$. In particular, for two vertices $u, v \in V$, the set $\{u, v\}$ covers $\mathcal{C}$ if and only if $G + uv$ is (m, ℓ)-redundant.*

Note that it is possible that there are no co-tight sets in a graph G. For example, $K_6 - e$ is $(3, 4)$-tight and there are no $(3, 4)$-co-tight sets in it. By Lemma 6, if there are no MCT sets in G then $G + uv$ is (m, ℓ)-redundant for any pair $u, v \in V$. Thus we may suppose that G contains at least one MCT set. We shall show that in this case the lower bound of Lemma 5 is sharp. From now on, let $\mathcal{C} = \{C_1, \ldots, C_t\}$ be the family of all (m, ℓ)-MCT sets in G. The following statement on the structure of MCT sets is the key of our solution to the reduced problem. It is an extension of a result of Jordán [13, Theorem 3.9.13] that states the same for $(2, 3)$-MCT sets.

Lemma 7. *The members of $\mathcal{C}$ are either pairwise disjoint or there exists a pair $u, v \in V$ such that $T(uv) = G$.*

The proof of Lemma 7 is quite complex hence we sketch some of its details in Appendix A. When there exists a pair $u, v \in V$ such that $T(uv) = G$, Lemma 6 implies that the lower bound of Lemma 5 is sharp. A set $X = \{x_1, \ldots, x_t\} \subseteq V$ is called a **transversal** of a family $\{S_1, \ldots, S_t\}$ if $x_i \in S_i$ for each $i \in \{1, \ldots, t\}$. When there is no pair $u, v \in V$ such that $T(uv) = G$, Lemma 7 claims that the members of $\mathcal{C}$ are disjoint and $|\mathcal{C}| = t \geq 3$. In this case we shall show that any graph isomorphic to a star on a transversal of $\mathcal{C}$ augments G to an (m, ℓ)-redundant graph. Next, by extending the ideas of [7, Lemma 15] we reduce the number of the edges of this star to the optimum.

Let $N(X)$ denote the neighbor set of $X \subseteq V$ in a graph $G = (V, E)$, that is, $N(X) := \{v \in V - X : v \text{ has at least one neighbor in } X\}$. One can show that the members of $\mathcal{C}$ are not only disjoint but there is no edge between any two of them. This observation and Lemma 1 lead to the following statement.

Lemma 8. *Suppose that the members of $\mathcal{C}$ are pairwise disjoint and $t \geq 3$. Let $\{x_1, \ldots, x_t\}$ be a transversal of $\mathcal{C}$ where $x_i \in C_i$ for $i \in \{1, \ldots, t\}$. Then $V(x_i x_j) \supseteq C_i \cup N(C_i) \cup C_j \cup N(C_j)$ for each $i, j \in \{1, \ldots, t\}$ with $i \neq j$.*

Based on Lemmas 1 and 8 we can construct a (not yet optimal) edge set, that augments G to an (m, ℓ)-redundant graph, as follows.

Lemma 9. *Suppose that the members of $\mathcal{C}$ are pairwise disjoint and $t \geq 3$. Let $\{x_1, \ldots, x_t\}$ be a transversal of $\mathcal{C}$. If H is a connected graph on the vertices $\{x_1, \ldots, x_t\}$, (in particular, if H is isomorphic to $K_{1,t-1}$) then $G + H$ is (m, ℓ)-redundant.*

The cardinality of the edge set provided by Lemma 9 can be decreased by iteratively using the following statement.

Lemma 10. *Let $\mathcal{C} = \{C_1, \ldots, C_t\}$ be the family of all (m, ℓ)-MCT sets in G. Suppose that the members of $\mathcal{C}$ are pairwise disjoint and $t \geq 4$. Let $\{x_1, \ldots, x_t\}$ be a transversal of $\mathcal{C}$ and let x_i, x_j, x_k, y be distinct vertices from this transversal set. Let $T^* = T(yx_i) \cup T(yx_j) \cup T(yx_k)$. Then $T^* = T(yx_i) \cup T(x_j x_k)$ or $T^* = T(yx_j) \cup T(x_i x_k)$ holds.*

This allows us to state that the lower bound in Lemma 5 can be reached.

Theorem 2. *Assume (A). If there exists any (m, ℓ)-co-tight set in G, then*

$$\min\{|F| : H = (V, F) \text{ is a graph for which } G + H \text{ is } (m, \ell)\text{-redundant}\}$$
$$= \max\left\{\left\lceil \frac{|\mathcal{C}|}{2} \right\rceil : \mathcal{C} \text{ is a family of disjoint } (m, \ell)\text{-co-tight sets}\right\}.$$

Otherwise, $G + uv$ is (m, ℓ)-redundant for every pair $u, v \in V$.

5   The Algorithm for the Reduced Problem

Our goal is to show that the reduced augmentation problem can be solved in $O(|V|^2)$ time (under the assumption (*)). According to Lemma 4 the preprocessing can be done in this time complexity.

By Sect. 4, the MCT sets of G play an important role in giving a solution to the reduced problem. Let G/X denote the graph arising from G by shrinking X into a single vertex. To give an efficient algorithm for finding MCT sets, we need the following statement.

Lemma 11. *Assume (A). Suppose that $T \subsetneq V$ is an (m, ℓ)-tight set in G. Let t' be the new vertex that arises after shrinking T in G. Let $\ell' := \max(\ell, 0)$ and let $m' : V(G/T) \to \mathbb{Z}_+$ be a map such that $m'(v) = m(v)$ when $v \in V(G/T) \cap V$ and $m'(t') = \ell'$. Then S is an (m', ℓ')-tight set in G/T such that $t' \in S$ if and only if $S - t' \cup T$ is (m, ℓ)-tight in G. Furthermore, $S \subseteq V(G/T) - t'$ is an (m', ℓ')-tight set in G/T if and only if S is an (m, ℓ)-tight set in G and $\ell \geq 0$, or $S \cup T$ is an (m, ℓ)-tight set in G such that no edge of G connects S and T and $\ell < 0$.*

By Lemma 11, the following algorithm finds an MCT set in $O(|V|^2)$ time.

Algorithm 1. INPUT: *A graph $G = (V, E)$ along with $m : V \to \mathbb{Z}_+$ and $\ell \in \mathbb{Z}$ such that (A) and (*) hold, and two vertices $u, v \in V$.*
OUTPUT: *An (m, ℓ)-MCT set in G that does not contain u, v or an edge e such that $T(e) = G$.*

0. Run the algorithm of Theorem 1 on G, let D be the output orientation.
1. Using the output of STEP *0, calculate $T := V_{(m,\ell)}^G(uv)$.*
*2. **If** $T = V$ **then Output:** the edge uv, **STOP**.*
3. Shrink T to t' according to Lemma 11, that is, $G' := G/T$, $D' := D/T$,
   $\ell' := \max(\ell, 0)$, $m'(u) := m(u)$ for each $u \in V(G') \cap V$, $m'(t') := \ell'$.
4. $v := t'$ (hence $m'(v) = \ell'$), $V^* := V(G') - v$.
*5. **While** $V^* \neq \emptyset$, **do:***
6.   Calculate $T' := V_{(m',\ell')}^{G'}(uv)$ using D'.
*7.   **If** $V' = V(G')$, **then** $V^* := V^* - u$.*
*8.   **Else,***
          Shrink T' to t', so $G' := G'/T', D' := D'/T'$.
9. $v := t', V^* := V^* \cap V(G') - v$.
*10. **Output:** $V(G') - v$.*

It can be checked in polynomial time whether $G + uv$ is (m, ℓ)-redundant for some pair $u, v \in V$. The naïve algorithm has $O(|V|^3)$ running time, however, as the following lemma shows, this can be improved to $O(|V|^2)$.

Lemma 12. *Assume (A) and (*). Then there exists an algorithm which outputs in $O(|V|^2)$ time an edge e for which $T(e) = G$ or, if there is no such edge, a vertex v of an (m, ℓ)-MCT set in G.*

Such an algorithm can be given by using some details of the proof of Lemma 7 and Algorithm 1. We provide some details in Appendix B. Now, we focus our attention to the case where there is no edge e that augments G to a redundant graph. We aim to find a minimal transversal set of the MCT sets of G starting from a vertex v that is in a MCT set. This is solved by the following algorithm in $O(|V|^2)$ time. From this point on, our algorithm's idea is a generalization of that of García and Tejel [7] (see Sect. 7 for some explanation).

Algorithm 2. INPUT: *A graph $G = (V, E)$ along with $m : V \to \mathbb{Z}_+$ and $\ell \in \mathbb{Z}$ such that (A) and (*) hold and there exists no edge e that augments G to an (m, ℓ)-redundant graph, and a vertex $v \in V$ from an (m, ℓ)-MCT set.*
OUTPUT: *A transversal of the MCT sets of G: $X = \{v, x_2, \ldots, x_t\}$.*

0. Run the algorithm of Theorem 1 on G.
1. Initialize $X = \emptyset$. All vertices are unmarked. Mark v.
2. Explore all vertices $j \in V$:
    If** j is unmarked, **then
        Calculate $T(vj)$ by using the output of STEP *0;*
        Mark all unmarked vertices in $V(vj)$;
        $X := (X - V(vj)) + j$.
*3. **Output**: $X + v$.*

Finally, by using Lemma 10 and the above algorithms, we can give the following algorithm to find an optimal solution of the reduced problem in $O(|V|^2)$ time.

Algorithm 3. INPUT: *A graph $G = (V, E)$ along with $m : V \to \mathbb{Z}_+$ and $\ell \in \mathbb{Z}$ such that (A) and (*) hold.*
OUTPUT: *A minimum cardinality edge set F that augments G to an (m, ℓ)-redundant graph.*

1. Run the algorithm of Lemma 12 (Algorithm 4 in Appendix B). Result: edge e or vertex y.
    ***If** the result is an edge e, **then Output** $\{e\}$, **STOP**.*
2. Generate a transversal system X of the (m, ℓ)-MCT stets of G by using Algorithm 2 with y.
3. $F := \{vy | v \in X - y\}$.
4. Calculate $T(f)$ for all $f \in F$ by using the output of STEP *0 of Algorithm 2.*
*5. **While** $d_F(y) \geq 3$, **do***
    Choose three neighbours of y in F, say $x_i, \dot{x}_j, x_k$.
    Calculate $T(x_j x_k)$ by using the output of STEP *0 of Algorithm 2.*
    If** $T(yx_i) \cup T(x_j x_k) = T(yx_i) \cup T(yx_j) \cup T(yx_k)$, **then
        $F := F - \{yx_j, yx_k\} + \{x_j x_k\}$.
    ***Else**, $F := F - \{yx_i, yx_k\} + \{x_i x_k\}$.*
*6. **Output**: F.*

It follows by Lemma 4 that we do not need to assume that (A1), (A2) and (A3) hold for G. This implies the following.

Theorem 3. *Assume (A0) and (*). Let $G = (V, E)$ be an (m, ℓ)-tight graph. There exists an algorithm that gives an optimal solution for the reduced augmentation problem in $O(|V|^2)$ time.*

6   The General Augmentation Problem

García and Tejel [7] showed that the general augmentation problem is NP-hard for $(2,3)$-rigid graphs by reducing it to the set cover problem. Based on their method and the inapproximability of the set cover problem [20], we can show the following.

Theorem 4. *Let k and ℓ be two integers such that $1 < k < \ell < 2k$. Then the general problem for (k, ℓ)-rigid graphs is NP-hard, moreover, there is no polynomial algorithm that gives a constant factor approximation to this problem unless $P = NP$.*

In this section we will show that, in any other case (that is, if $\ell \le k$), there exists an $O(|V|^2)$ time algorithm that gives an optimal solution for the general problem. Moreover, we give our solution for all (m, ℓ)-rigid graphs for which $m \ge \ell$. $\bar{G} = (V, \bar{E})$ will denote an (m, ℓ)-rigid graph and $G = (V, E)$ will denote an (m, ℓ)-tight spanning subgraph of $\bar{G}$. Obviously, every edge in $\bar{E} - E$ is (m, ℓ)-redundant in $\bar{G}$. By Lemma 3, the (m, ℓ)-redundant edges of G in $\bar{G}$ are the edges of $R^G(\bar{E} - E) = \bigcup_{uv \in \bar{E} - E} T^G(uv)$. As we already have solved the augmentation problem for (m, ℓ)-tight graphs, we assume that $\bar{E} - E \ne \emptyset$.

As mentioned in the Introduction, the idea of our method comes from Jackson and Jordán [9].

Since $m \le \ell$, the (m, ℓ)-redundant edges of G in $\bar{G}$ form some vertex disjoint (m, ℓ)-redundant induced subgraphs of G by Lemmas 1 and 3. (Note that we have only one such subgraph when $\ell < 0$.) By shrinking each of these subgraphs to a single vertex and by defining $\ell' := \max(\ell, 0)$ and m' to be ℓ' on each of the shrunken vertices and to be $m(v)$ on each non-shrunken vertex v, we get the shrunken graph $G' = (V', E')$ along with the map $m' : V' \to \mathbb{Z}_+$. The following statement follows by Lemma 11.

Proposition 1. *Let G be an (m, ℓ)-tight graph and let $G' = (V', E')$ and $m' : V' \to \mathbb{Z}_+$ arise from G as we defined above. Then G' is (m', ℓ')-tight. Moreover, the shrunken image of an (m, ℓ)-tight subgraph of G is (m', ℓ')-tight, which contains the only shrunken vertex when $\ell < 0$. Furhermore, the pre-image of any (m', ℓ')-tight subgraph of G' is either (m, ℓ)-tight or, when $\ell < 0$, it gets (m, ℓ)-tight if we union it to the sole shrunken (m, ℓ)-tight subgraph of G.*

By Proposition 1, a covering of the edges of G which are not (m, ℓ)-redundant in $\bar{G}$ with (m, ℓ)-tight subgraphs gives a covering of G' with (m', ℓ')-tight subgraphs. Hence the minimum number of edges that we need to make G (m, ℓ)-redundant is at least the minimum number of edges that we need to make G' (m', ℓ')-redundant. The following statement shows that these two values are equal.

Proposition 2. *Let F' denote an edge set of minimum cardinality on V' for which $G' \cup F'$ is (m', ℓ')-redundant. Let F be an arbitrary pre-image of F', that is, we get F' from F by our shrinking procedure. Then $\bar{G} \cup F$ is (m, ℓ)-redundant.*

With Proposition 2, we have reduced the problem of augmenting an (m, ℓ)-rigid graph to an (m, ℓ)-redundant graph to the problem of augmenting an (m', ℓ')-tight graph to an (m', ℓ')-redundant graph that we can solve in $O(|V|^2)$ time by Theorem 3 as (A0) holds obviously. Note that the contraction that we used can be done in $O(|V|^2)$ time by Theorem 1. This implies the following.

Theorem 5. *There exists an $O(|V|^2)$ time algorithm to obtain a set of edges F of minimum cardinality for any input of $m : V \to \mathbb{Z}_+$, $\ell \in \mathbb{Z}$ for which $m \geq \ell$ and (*) hold, and of an (m, ℓ)-rigid graph $\bar{G} = (V, \bar{E})$, such that $\bar{G} \cup F$ is (m, ℓ)-redundant.*

7 Concluding Remarks

We conclude this extended abstract by mentioning three facts which we shall only explicate precisely in the full version of this paper [16].

As we observed before Algorithm 2, the second part of our algorithm generalizes the steps of the algorithm of García and Tejel [7] for all (m, ℓ)-tight graphs. This is the consequence of the fact that the 'classes of extreme vertices' defined in [7] are exactly the $(2, 3)$-MCT sets when there exist at least three disjoint such sets. The main difference between the two algorithms is their first part where we check this latter condition and output an element of an MCT set. To do this, [7] only needs to run an algorithm like Algorithm 2 by starting it with a vertex of minimum degree. We note that this idea only works when $\ell \leq \frac{3}{2}k$.

In several applications of sparse graphs in rigidity theory, parallel edges are meaningless, that is, we need to assume that our graph is simple (see e.g. [10, 22]). We shall show that our concept is capable of optimally augmenting G to a simple (m, ℓ)-redundant graph (if there exists any non-graph edge of the input). Indeed, when our algorithm outputs at least two edges, those are not in G by Lemma 8. On the other hand, if there exists no non-graph edge which augments G to an (m, ℓ)-redundant graph, then there exists no sole edge augmenting G to (m, ℓ)-redundant. Hence, if the output of our algorithm is a sole edge, one of the non-graph edges augments G to an (m, ℓ)-redundant graph. We note that looking for such a single edge may lead to an $O(|V|^3)$ time algorithm.

In some other applications (e.g. in [14, 24, 26]), instead of a sparsity condition on G, we have a sparsity condition on cG (that is, the graph that arises from G by replacing each edge e of G by c parallel copies of e). By Lemma 3, $cG + H$ is (m, ℓ)-redundant if and only if $c(G + H)$ is (m, ℓ)-redundant. Hence the augmentation problem is also solvable in this case.

Acknowledgements. Project no. NKFI-128673 has been implemented with the support provided from the National Research, Development and Innovation Fund of Hungary, financed under the FK_18 funding scheme. The first author was supported by the János Bolyai Research Scholarship of the Hungarian Academy of Sciences and by the ÚNKP-19-4 New National Excellence Program of the Ministry for Innovation and Technology. The authors are grateful to Tibor Jordán for the inspiring discussions and his comments.

Appendix A: Sketch of the Proof of Lemma 7

Here we give the main steps of the proof of Lemma 7. Recall that we assume (A) and $\mathcal{C}$ is the family of all (m, ℓ)-MCT sets of G. When $\ell \leq 0$, the proof is straightforward from Lemmas 1 and 6. In the case of $\ell > 0$, we need to use our assumptions (A1), (A2), and (A3) at some points. Our first statement follows by Lemma 1.

Lemma 13. *If X and Y are two (m, ℓ)-MCT sets in G, such that $X \cap Y \neq \emptyset$, then $m(V - (X \cup Y)) < \ell$. In particular, $|X \cup Y| \geq |V| - 1$.*

If $X \cup Y = V$ holds whenever X and Y are intersecting MCT sets, then the proof is straightforward. Hence we may assume that $|X \cup Y| = |V| - 1$ for some $X, Y \subset V$. For a vertex $v \in V$, let $\mathcal{C}(v) := \{C \in \mathcal{C} : v \notin C\}$. The first key of the proof is the following lemma which can be proved by Lemmas 6 and 13. Assumption (A1) is used in its proof.

Lemma 14. *Suppose that $\ell > 0$. Assume that there exists two (m, ℓ)-MCT sets $X, Y \in \mathcal{C}$ such that $X \cap Y \neq \emptyset$ and $X \cup Y = V - v$ for some $v \in V$. Then $\mathcal{C}(v)$ is a co-partition of $V - v$ with $|\mathcal{C}(v)| \geq 3$ or there exists a vertex $u \in V - v$ such that $T(uv) = G$.*

For a vertex $v \in V$ and a set $W \subseteq V - v$, let $\widetilde{W}^v := V - v - W$. Lemma 6 implies that, if we take two members W_1 and W_2 of the co-partition $\mathcal{C}(v)$ and take $w_1 \in \widetilde{W}_1^v$ and $w_2 \in \widetilde{W}_2^v$, then V is the only (m, ℓ)-tight set in G which contains w_1, w_2 and v. Using this observation we can prove a much stronger statement which claims that in many cases $V(w_1 w_2)$ is also V.

Lemma 15. *Suppose that $\ell > 0$. Let $v \in V$ be a vertex for which the family $\mathcal{C}(v)$ is a co-partition of $V - v$ with $|\mathcal{C}(v)| \geq 3$. Suppose that there exists a vertex $u \in V - v$ with $m(u) \leq m(v)$. Let $W_1, W_2 \in \mathcal{C}(v)$ and let $w_1 \in \widetilde{W}_1^v$ and $w_2 \in \widetilde{W}_2^v$. Suppose that V' is an (m, ℓ)-tight set in G with $w_1, w_2 \in V'$. Then either $V' = V$ or $V' = \{w_1, w_2\}$. In particular, either $V(w_1 w_2) = V$ (and $T(w_1 w_2) = G$) or $V(w_1 w_2) = \{w_1, w_2\}$.*

Based on Lemma 15, using assumption (A2) one can prove the following.

Lemma 16. *Suppose that $\ell > 0$. Let $v \in V$ be a vertex for which the family $\mathcal{C}(v)$ is a co-partition of $V - v$ with $|\mathcal{C}(v)| \geq 3$. Then $m(v) < m(u)$ holds for every $u \in V - v$ or there exist two vertices $x, y \in V - v$ such that $T(xy) = G$.*

Finally, to finish the proof of Lemma 7, we can assume by Lemma 16 that, whenever X and Y are intersecting MCT sets with $X \cup Y = V - v$ for a $v \in V$, then $m(v) < m(u)$ holds for every $u \in V - v$. In this case, one can prove that $\{v\}$ is an MCT set and hence $d(v) = m(v)$. Now the proof follows from (A0), (A3) and the fact that the degree of v in any (m, ℓ)-tight subgraph on more than 3 vertices is at least $m(v)$.

Appendix B: The algorithm of Lemma 12

In this section we give the algorithm of Lemma 12. First we solve the case when we have an MCT set consisting of a single vertex.

Lemma 17. *Assume (A). If we are given an (m, ℓ)-MCT singleton set $C = \{v\}$, then we can check if there exists an edge e such that $T(e) = G$ in $O(|V|^2)$ time.*

Based on the steps of the proof of Lemma 7 in Appendix A we can provide the following algorithm for Lemma 12.

Algorithm 4. INPUT: *A graph $G = (V, E)$ along with $m : V \to \mathbb{Z}_+$ and $\ell \in \mathbb{Z}$ such that (A) and (*) hold.*
OUTPUT: *If there exists an edge e such that $T(e) = G$, then e, otherwise a vertex v of an (m, ℓ)-MCT set.*

1. *Choose two vertices $u, v \in V$, such that $|V(uv)| > 2$. Also suppose that $m(v) \geq m(u)$.*
2. *Run Algorithm 1 with u and v. Result: edge e or MCT set Z.*
   If the result is an edge e, then Output e, STOP.
3. *If $|Z| = 1$, then check every edge from it by the algorithm of Lemma 17.*
   If this outputs an edge e, then Output e, STOP.
   Else, Output the single element z of Z, STOP.
4. *Let $z \in Z$ be such that $m(z)$ is not the unique minimum of m.*
5. *Run Algorithm 1 with z, and v. Result: edge e or MCT set C.*
   If the result is an edge e, then Output e, STOP.
7. *If $|C| = 1$, then check every edge from it by the algorithm of Lemma 17.*
   If this outputs an edge e, then Output e, STOP.
   Else, Output z, STOP.
8. *Let $c \in C - Z$.*
9. *Run Algorithm 1 with z and c. Result: edge e or MCT set S.*
   If the result is an edge e, then Output e, STOP.
10. *If $Z \cap C = \emptyset$, $C \cap S = \emptyset$, and $S \cap Z = \emptyset$, then Output: z.*
11. *Else, check each possible edge from v, z and c, it gives a suitable edge e.*
    Output e.

Proof Sketch of the Correctness of Algorithm 4: By Lemma 17 and by the correctness of Algorithm 1, we can see that any output from STEPS $2 - 9$ is correct. It is also clear by Algorithm 1 that the sets Z, C and S are MCT sets, thus the output in STEP 10 is also correct by Lemma 7. Hence we only need to see that STEP 11 gives a suitable edge. By Lemma 14, if no suitable edge is given, then $\mathcal{C}(v)$, $\mathcal{C}(z)$ and $\mathcal{C}(c)$ are co-partitions. However, this contradicts Lemma 15.   □

References

1. Aspnes, J., et al.: A theory of network localization. IEEE Trans. Mob. Comput. **5**(12), 1663–1678 (2006)
2. Berg, A.R., Jordán, T.: Algorithms for graph rigidity and scene analysis. In: Di Battista, G., Zwick, U. (eds.) ESA 2003. LNCS, vol. 2832, pp. 78–89. Springer, Heidelberg (2003). https://doi.org/10.1007/978-3-540-39658-1_10
3. Connelly, R., Jordán, T., Whiteley, W.: Generic global rigidity of body-bar frameworks. J. Comb. Theory Ser. B **103**(6), 689–705 (2013)
4. Eswaran, K.P., Tarjan, R.E.: Augmentation problems. SIAM J. Comput. **5**(4), 653–665 (1976)
5. Frank, A.: Connections in Combinatorial Optimization. Oxford University Press, Oxford (2011)
6. Frank, A., Király, T.: Combined connectivity augmentation and orientation problems. Discrete Appl. Math. **131**(2), 401–419 (2003)
7. García, A., Tejel, J.: Augmenting the rigidity of a graph in $\mathbb{R}^2$. Algorithmica **59**(2), 145–168 (2011)
8. Hendrickson, B.: Conditions for unique graph realizations. SIAM J. Comput. **21**(1), 65–84 (1992)
9. Jackson, B., Jordán, T.: Brick partitions of graphs. Discrete Math. **310**(2), 270–275 (2010)
10. Jackson, B., Nixon, A.: Global rigidity of generic frameworks on the cylinder. J. Comb. Theory Ser. B **139**, 193–229 (2019)
11. Jacobs, D.J., Hendrickson, B.: An algorithm for two dimensional rigidity percolation: the pebble game. J. Comput. Phys. **137**, 346–365 (1997)
12. Jacobs, D.J., Thorpe, M.F.: Generic rigidity percolation: the pebble game. Phys. Rev. Lett. **75**, 4051–4054 (1995)
13. Jordán, T.: Combinatorial rigidity: graphs and matroids in the theory of rigid frameworks. Discrete Geometric Analysis. vol. 34 of MSJ Memoirs, pp. 33–112. Mathematical Society of Japan, Japan (2016)
14. Jordán, T., Király, Cs., Tanigawa, S.: Generic global rigidity of body-hinge frameworks. J. Comb. Theory Ser. B **117**, 59–76 (2016)
15. Király, Cs.: An efficient algorithm for testing (k, ℓ)-sparsity when $\ell < 0$. Technical Report (Quick Proof) QP-2019-04, Egerváry Research Group, Budapest (2019). www.cs.elte.hu/egres
16. Király, Cs., Mihálykó, A.: Sparse graphs and an augmentation problem. Technical Report TR-2019-14, Egerváry Research Group, Budapest (2019). www.cs.elte.hu/egres
17. Laman, G.: On graphs and rigidity of plane skeletal structures. J. Eng. Math. **4**, 331–340 (1970)
18. Lee, A., Streinu, I.: Pebble game algorithms and sparse graphs. Discrete Math. **308**(8), 1425–37 (2008)
19. Lorea, M.: On matroidal families. Discrete Math. **28**(1), 103–106 (1979)
20. Lund, C., Yannakakis, M.: On the hardness of approximating minimization problems. J. ACM **41**(5), 960–981 (1994)
21. Nash-Williams, C.S.J.A.: Decomposition of finite graphs into forests. J. Lond. Math. Soc. **39**, 12 (1961)
22. Nixon, A., Owen, J.C., Power, S.C.: Rigidity of frameworks supported on surfaces. SIAM J. Discrete Math. **26**(4), 1733–1757 (2012)

23. Pollaczek-Geiringer, H.: Über die Gliederung ebener Fachwerke. ZAMM - J. Appl. Math. Mech. / Zeitschrift für Angewandte Mathematik und Mechanik **7**(1), 58–72 (1927)

24. Tay, T.-S.: Linking $(n-2)$-dimensional panels in n-space II: $(n-2,2)$-frameworks and body and hinge structures. Graphs Comb. **5**(1), 245–73 (1989)

25. Tay, T.-S.: Henneberg's method for bar and body frameworks. Struct. Topol. **17**, 53–8 (1991)

26. Whiteley, W.: The union of matroids and the rigidity of frameworks. SIAM J. Discrete Math. **1**(2), 237–55 (1988)

27. Whiteley, W.: Some matroids from discrete applied geometry. In: Bonin, J.E., Oxley, J.G., Servatius, B. (eds.) Matroid Theory of Contemporary Mathematics, vol. 197, pp. 171–311. AMS (1996)

28. Whiteley, W.: Rigidity of molecular structures: generic and geometric analysis. In: Thorpe, M.F., Duxbury, P.M. (eds.) Rigidity Theory and Applications, pp. 21–46. Springer, Boston (2002). https://doi.org/10.1007/0-306-47089-6_2

29. Yu, C., Anderson, B.D.O.: Development of redundant rigidity theory for formation control. Int. J. Robust Nonlinear Control **19**(13), 1427–1446 (2009)

About the Complexity of Two-Stage Stochastic IPs

Kim-Manuel Klein[✉][iD]

University of Kiel, Christian-Albrechts-Platz 4, 24118 Kiel, Germany
kmk@informatik.uni-kiel.de

Abstract. We consider so called 2-stage stochastic integer programs (IPs) and their generalized form of multi-stage stochastic IPs. A 2-stage stochastic IP is an integer program of the form $\max\{c^T x \mid \mathcal{A}x = b, l \leq x \leq u, x \in \mathbb{Z}^{s+nt}\}$ where the constraint matrix $\mathcal{A} \in \mathbb{Z}^{rn \times s + nt}$ consists roughly of n repetitions of a block matrix $A \in \mathbb{Z}^{r \times s}$ on the vertical line and n repetitions of a matrix $B \in \mathbb{Z}^{r \times t}$ on the diagonal.

In this paper we improve upon an algorithmic result by Hemmecke and Schultz from 2003 [16] to solve 2-stage stochastic IPs. The algorithm is based on the Graver augmentation framework where our main contribution is to give an explicit doubly exponential bound on the size of the augmenting steps. The previous bound for the size of the augmenting steps relied on non-constructive finiteness arguments from commutative algebra and therefore only an implicit bound was known that depends on parameters r, s, t and Δ, where Δ is the largest entry of the constraint matrix. Our new improved bound however is obtained by a novel theorem which argues about the intersection of paths in a vector space. As a result of our new bound we obtain an algorithm to solve 2-stage stochastic IPs in time $f(r, s, \Delta) \cdot poly(n, t)$, where f is a doubly exponential function.

Keywords: Integer programming · 2-stage stochastic programming · Graver complexity

1 Introduction

Integer programming is one of the most fundamental problems in algorithm theory. Many problems in combinatorial optimization and other areas can be modeled by integer programs. An *integer program* (IP) is thereby of the form

$$\max\{c^T x \mid Ax = b, l \leq x \leq u, x \in \mathbb{Z}^n\}$$

This work was mostly done during the authors time at EPFL. The project was supported by the Swiss National Science Foundation (SNSF) within the project Convexity, geometry of numbers, and the complexity of integer programming (Nr.163071). A full version of the paper is available at https://arxiv.org/abs/1901.01135.

D. Bienstock and G. Zambelli (Eds.): IPCO 2020, LNCS 12125, pp. 252–265, 2020.
https://doi.org/10.1007/978-3-030-45771-6_20

for some matrix $A \in \mathbb{Z}^{m \times n}$, a right hand side $b \in \mathbb{Z}^m$, a cost vector $c \in \mathbb{Z}^n$ and lower and upper bounds $l, u \in \mathbb{Z}^n$. The famous algorithm of Kannan [20] computes an optimal solution of the IP in time of roughly $n^{O(n)} \cdot poly(m, \log \Delta)$, where Δ is the largest entry of A and b.

In recent years there was significant progress in the development of algorithms for IPs when the constraint matrix A has a specific structure. Consider for example the class of integer programs with a constraint matrix $\mathcal{N}$ of the form

$$\mathcal{N} = \begin{pmatrix} A & A & \cdots & A \\ B & 0 & \cdots & 0 \\ 0 & B & \ddots & \vdots \\ \vdots & \ddots & \ddots & 0 \\ 0 & \cdots & 0 & B \end{pmatrix}$$

for some block matrices $A \in \mathbb{Z}^{r \times s}$ and $B \in \mathbb{Z}^{r \times t}$. An IP of this specific structure is called an n-*fold* IP. This class of IPs has found numerous applications in the area of string algorithms [22], social choice games [23] and scheduling [17,21]. State-of-the-art algorithms compute a solution of an n-fold IP in time $\Delta^{O(r^2 s)} \cdot poly(n, t)$ [9,18,25], where Δ is the largest entry in the matrices A and B.

1.1 Two-Stage Stochastic Integer Programming

Stochastic programming deals with uncertainty of decision making over time [19]. One of the basic models in stochastic programming is 2-stage stochastic programming. In this model one has to decide on a solution at the first stage and in the second stage there is an uncertainty where n possible scenarios can happen. Each of n possible scenarios might have a different optimal solution and the goal is to minimize the costs of the solution of the first stage in addition to the expected costs of the solution of the second stage. In the case that said scenarios can be modeled by an (integer) linear program, we are talking about 2-*stage stochastic (integer) linear programs*. 2-stage stochastic linear programs that do not contain any integer variable are well understood (we refer to standard text books [3,19]). In contrast, 2-stage stochastic programs that contain integer variables are hard to solve and are the topic of ongoing research. Typically, those IPs are investigated in the context of decomposition based methods (we refer to a tutorial [26] or a survey [29] on the topic). For recent progress on 2-stage stochastic programs we refer to [1,4,29]. The interest in solving 2-stage stochastic (I)LPs efficiently stems from their wide range of applications for example in modeling manufacturing processes [8] or energy planing [15].

In this paper we consider 2-stage stochastic IPs with only integral variables. These so called pure integral 2-stage stochastic IPs have also been considered in the literature from a practical perspective (see [12,31]). The considered IP is then of the form

$$\max c^T x \tag{1}$$
$$\mathcal{A} x = b$$
$$l \leq x \leq u$$
$$x \in \mathbb{Z}^{s+nt}$$

for given objective vector $c \in \mathbb{Z}_{\geq 0}^{s+nt}$, upper and lower bound $\ell, u \in \mathbb{Z}_{\geq 0}^{s+nt}$. The constraint matrix $\mathcal{A}$ has the shape

$$\mathcal{A} = \begin{pmatrix} A^{(1)} & B^{(1)} & 0 & \cdots & 0 \\ A^{(2)} & 0 & B^{(2)} & \ddots & \vdots \\ \vdots & \vdots & & \ddots & 0 \\ A^{(n)} & 0 & \cdots & 0 & B^{(n)} \end{pmatrix}$$

for given block matrices $A^{(1)}, \ldots, A^{(n)} \in \mathbb{Z}^{r \times s}$ and $B^{(1)}, \ldots, B^{(n)} \in \mathbb{Z}^{r \times t}$. Typically, 2-stage stochastic IPs are written in a slightly different (equivalent) form that explicitly involves the scenarios and the probability distribution of the scenarios of the second stage. In this presented form, roughly speaking, the solution for the first stage scenario is encoded in the variables corresponding to vertical block matrices. A solution for each of the second stage scenarios is encoded in the variables corresponding to one of the diagonal block matrices and the expectation for the second stage scenarios can be encoded in a linear objective function. Since we do not rely on known techniques of stochastic programming in this paper, we omit the technicalities surrounding 2-stage stochastic IPs and simply refer to a survey for further details [29].

Despite their similarity, it seems that 2-stage IPs are significantly harder to solve than n-fold IPs. While it is known that the 2-stage stochastic IP with constraint matrix $\mathcal{S}$ can be solved in running time of the form $f(r, s, t, \Delta) \cdot poly(n)$ for some computable function f, which was developed by Hemmecke and Schultz [16], the actual dependence on the parameters r, s, t, Δ was unknown (we elaborate on this further in the coming section). Their algorithm is based on the augmentation framework which we also discuss in the following section.

1.2   The Augmentation Framework

Suppose we have an initial feasible solution z_0 of an IP $\max\{c^T x \mid Ax = b, l \leq x \leq u, x \in \mathbb{Z}^n\}$ and our goal is to find an optimal solution. The idea behind the augmentation framework (see [7]) is to compute an augmenting (integral) vector y in the kernel, i.e. $y \in ker(A)$ with $c^T y > 0$. A new solution z' with improved objective can then be defined by $z' = z_0 + \lambda y$ for a suitable $\lambda \in \mathbb{Z}_{\geq 0}$. This procedure can be iterated until a solution with optimal objective is obtained eventually.

We call an integer vector $y \in ker(A)$ a *cycle*. A cycle can be decomposed if there exist integral vectors $u, v \in ker(A) \setminus \{0\}$ with $y = u + v$ and $u_i \cdot v_i \geq 0$ for all i (i.e. the vectors are sign-compatible with y). An integral vector $y \in ker(A)$

that can not be decomposed is called a *Graver element* [13] or we simply say that it is *indecomposable*. The set of all indecomposable elements is called the *Graver basis*.

The power of the augmentation framework is based on the observation that the size of Graver elements can be bounded. With the help of these bounds, good augmenting steps can be computed by a dynamic program and finally the corresponding IP can be solved efficiently.

In the case that the constraint matrix has a very specific structure, one can sometimes show improved bounds. Specifically, if the constraint matrix A has a 2-stage stochastic shape with identical block matrices in the vertical and diagonal line, then Hemmecke and Schultz [16] were able to prove a bound for the size of Graver elements that only depends on the parameters r, s, t and Δ. The presented bound is an existential result and uses so called saturation results from commutative algebra. As MacLagan's theorem is used in the proof of the bound, no explicit function can be derived. It is only known that the dependence on the parameters is lower bounded by Ackerman's function [27]. This implies that the parameter dependence of r, s, t and Δ for the bound of the the running time of the algorithm by Hemmecke and Schultz is at least ackermanian. In a very recent paper it was even conjectured that an algorithm with an explicit bound on parameters r, s, t and Δ in the running time to solve IPs of the form (1) does not exist [24].

Very recently, improved bounds for Graver elements of general matrices and matrices with specific structure like n-fold [9] or 4-block structure [5] were developed. They are based on the Steinitz Lemma, which was previously also used by Eisenbrand and Weismantel [11] in the context of integer programming.

Lemma 1 (Steinitz [14,30]). *Let $v_1, \ldots, v_n \in \mathbb{R}^m$ be vectors with $\|v_i\|_\infty \leq \Delta$ for $1 \leq i \leq n$. Assuming that $\sum_{i=1}^n v_i = 0$ then there is a permutation Π such that for each $k \in \{1, \ldots, n\}$ the norm of the partial sum $\left\|\sum_{i=1}^k v_{\Pi(i)}\right\|_\infty$ is bounded by $m\Delta$.*

The Steinitz Lemma was used by Eisenbrand, Hunkenschröder and Klein [9] to bound the size of Graver elements for a given matrix A.

Theorem 1 (Eisenbrand, Hunkenschröder, Klein [9]). *Let $A \in \mathbb{Z}^{m \times n}$ be an integer matrix where every entry of A is bounded by Δ in absolute value. Let $g \in \mathbb{Z}^n$ be an element of the Graver Basis of A then $\|g\|_1 \leq (2m\Delta + 1)^m$.*

1.3 Our Results

The main result of this paper is to prove a new structural lemma that enhances the toolset of the augmentation framework. We show that this Lemma can be directly used to obtain an explicit bound for Graver elements of the constraint matrix of 2-stage stochastic IPs. But we think that it might also be of independent interest as it provides interesting structural insights in vector sets.

Lemma 2. *Given multisets $T_1, \ldots, T_n \subset \mathbb{Z}_{\geq 0}^d$ where all elements $t \in T_i$ have bounded size $\|t\|_\infty \leq \Delta$. Assuming that the total sum of all elements in each set is equal, i.e.*

$$\sum_{t \in T_1} t = \ldots = \sum_{t \in T_n} t$$

then there exist nonempty submultisets $S_1 \subseteq T_1, \ldots, S_n \subseteq T_n$ of bounded size $|S_i| \leq (d\Delta)^{O(d(\Delta^{d^2}))}$ such that

$$\sum_{s \in S_1} s = \ldots = \sum_{s \in S_n} s.$$

Note that Lemma 2 only makes sense when we consider the T_i to be multisets as the number of different sets without allowing multiplicity of vectors would be bounded by $2^{(\Delta+1)^d}$.

A geometric interpretation of Lemma 2 is given in the following figure. On the left side we have n-paths consisting of sets of vectors and all path end at the same point b.

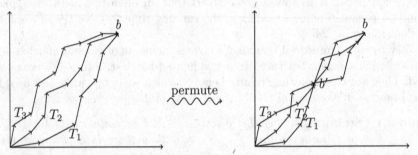

Then Lemma 2 shows, that there always exist permutations of the vectors of each path such that all paths meet at a point b' of bounded size. The bound does only depend on Δ and the dimension d and is thus independent of the number of paths n and the size of b. For the proof of Lemma 2 we need basic properties for the intersection of integer cones. We show that those properties can be obtained by using the Steinitz Lemma.

We show that Lemma 2 has strong implications in the context of integer programming. Using Lemma 2, we can show that the size of Graver elements of matrix $\mathcal{A}$ is bounded by $(rs\Delta)^{O(rs((2r\Delta+1)^{rs^2}))}$. Within the framework of Graver augmenting steps the bound implies that 2-stage stochastic IPs can be solved in time $(rs\Delta)^{O(rs^2((2r\Delta+1)^{rs^2}))} \cdot n^2 t^2 \varphi \log^2(nt)$, where φ is the encoding length of the instance. With this we improve upon an implicit bound for the size of the Graver elements matrix 2-stage stochastic constraint matrices due to Hemmecke and Schultz [16].

Furthermore, we show that our Lemma can also be applied to bound the size of Graver elements of constraint matrices that have a multi-stage stochastic structure. Multi-stage stochastic IPs are a well known generalization of 2-stage stochastic IPs. By this, we improve upon a result of Aschenbrenner and Hemmecke [2]. Due to space constraints however, this section is omitted here and we refer to the full version of the paper. Very recently, our Lemma 2 has been used by Eisenbrand et al. [10] to solve multi-stage stochastic IPs with a nearly linear dependence on n.

To complement our results for the upper bound, we also present a lower bound for the size of Graver elements of matrices that have a 2-stage stochastic IP structure. The given lower bound is for the case of $r = 1$. In this case we present a matrix where the Graver elements have a size of $2^{\Omega(\Delta^s)}$. The proof of the lower bound can be found in the full version of the paper.

2   The Complexity of Two-Stage Stochastic IPs

First, we argue about the application of our main Lemma 2. In the following we show that the infinity-norm of Graver elements of matrices with a 2-stage stochastic structure can be bounded using Lemma 2.

Given the block structure of the IP 1, we define for a vector $y \in \mathbb{Z}^{s+nt}$ with $\mathcal{A}y = 0$ the vector $y^{(0)} \in \mathbb{Z}_{\geq 0}^s$ which consists of the entries of y that belong to the vertical block matrices $A^{(i)}$ and we define $y^{(i)} \in \mathbb{Z}_{\geq 0}^t$ to be the entries of y that belong to the diagonal block matrix $B^{(i)}$.

Theorem 2. *Let y be a Graver element of the constraint matrix $\mathcal{A}$ of IP (1). Then $\|y\|_\infty$ is bounded by $(rs\Delta)^{O(rs((2r\Delta+1)^{rs^2}))}$. More precisely,*

$$\left\| y^{(i)} \right\|_1 \leq (rs\Delta)^{O(rs((2r\Delta+1)^{rs^2}))}$$

for every $0 \leq i \leq n$.

Proof. Let $y \in \mathbb{Z}^{s+nt}$ be a cycle of IP (1), i.e. $\mathcal{A}y = 0$. Consider a submatrix of the matrix $\mathcal{A}$ denoted by $(A^{(i)}B^{(i)}) \in \mathbb{Z}^{r\times(s+t)}$ consisting of the block matrix $A^{(i)}$ of the vertical line and the block matrix $B^{(i)}$ of the diagonal line. Consider further the corresponding variables $v^{(i)} = \begin{pmatrix} y^{(0)} \\ y^{(i)} \end{pmatrix} \in \mathbb{Z}^{s+t}$ of the respective matrix $A^{(i)}$ and $B^{(i)}$. Since $\mathcal{A}y = 0$, we also have that $(A^{(i)}B^{(i)})v^{(i)} = 0$. Hence, we can decompose $v^{(i)}$ into a multiset C_i of indecomposable elements, i.e. $v^{(i)} = \sum_{z\in C_i} z$. By Theorem 1 we obtain the bound $\|z\|_1 \leq (2r\Delta+1)^r$ for each $z \in C_i$.

Since all matrices $(A^{(i)}B^{(i)})$ share the same set of variables in the overlapping block matrices $A^{(i)}$, we can not choose indecomposable elements independently in each block to obtain a cycle of smaller size for the entire matrix $\mathcal{A}$. Let $p : \mathbb{Z}^{s+t} \to \mathbb{Z}^s$ be the projection that maps a cycle z of a block matrix $(A^{(i)}B^{(i)})$ to the variables in the overlapping part, i.e. $p(z) = p(\begin{pmatrix} z^{(0)} \\ z^{(i)} \end{pmatrix}) = z^{(0)}$. In the case that $\|y\|_\infty$ is large we will show that we can find a cycle $\bar{y}$ of smaller length with $|\bar{y}_i| \leq |y_i|$ and therefore show that y can be decomposed. In order to obtain this cycle $\bar{y}$ for the entire matrix $\mathcal{A}$, we have to find a multiset of cycles

$\bar{C}_i \subset C_i$ in each block matrix $(A^{(i)} B^{(i)})$ such that the sum of the projected parts is identical, i.e. $\sum_{z \in \bar{C}_1} p(z) = \ldots = \sum_{z \in \bar{C}_n} p(z)$. We apply Lemma 2 to the multisets $p(C_1), \ldots, p(C_n)$, where $p(C_i) = \{p(z) \mid z \in C_i\}$ is the multiset of projected elements in C_i with $\|p(z)\|_1 \le (2r\Delta + 1)^r$. Note that $\sum_{x \in p(C_1)} x = \ldots = \sum_{x \in p(C_n)} x = y^{(0)}$ and hence the conditions to apply Lemma 2 are fulfilled. Since every $v^{(i)}$ is decomposed in a sign compatible way, every entry of the vector in $p(C_i)$ has the same sign. Hence we can flip the negative signs in order to apply Lemma 2.

By Lemma 2, there exist submultisets $S_1 \subseteq p(C_1), \ldots, S_n \subseteq p(C_n)$ such that $\sum_{x \in S_1} x = \ldots = \sum_{x \in S_n} x$ and $|S_i| \le (s \|z\|_1)^{O(s(\|z\|_1^s))} = (rs\Delta)^{O(rs((2r\Delta+1)^{rs^2}))}$. As there exist submultisets $\bar{C}_1 \subseteq C_1, \ldots \bar{C}_n \subseteq C_n$ with $p(\bar{C}_1) = S_1, \ldots p(\bar{C}_n) = S_n$, we can use those submultisets $\bar{C}_i$ to define a solution $\bar{y}$ with $|\bar{y}_i| \le |y_i|$. For $i > 0$ let $\bar{y}^{(i)} = \sum_{z \in \bar{C}_i} \bar{p}(z)$, where $\bar{p}(z)$ is the projection that maps a cycle $z \in \bar{C}_i$ to the part that belongs to matrix $B^{(i)}$, i.e. $\bar{p}\left(\binom{z^{(0)}}{z^{(i)}}\right) = z^{(i)}$. Let $\bar{y}^{(0)} = \sum_{z \in \bar{C}_i} p(z)$ for an arbitrary $i > 0$, which is well defined as the sum is identical for all multisets $\bar{C}_i$. As the cardinality of the multisets $\bar{C}_i$ is bounded, we know by construction of $\bar{y}$ that the one-norm of every $y^{(i)}$ is bounded by

$$\left\| y^{(i)} \right\|_1 \le (2r\Delta + 1)^r \cdot (rs\Delta)^{O(rs((2r\Delta+1)^{rs^2}))} = (rs\Delta)^{O(rs((2r\Delta+1)^{rs^2}))}.$$

This directly implies the infinity-norm bound for y as well.

Computing the Augmenting Step. As a direct consequence of the bound for the size of the Graver elements, we obtain by the framework of augmenting steps an efficient algorithm to compute an optimal solution of a 2-stage stochastic IP. For this we can use the algorithm by Hemmecke and Schultz [16] or a more recent result by Koutecky, Levin and Onn [25] which gives a strongly polynomial algorithm. Using these algorithms directly would result in an algorithm with a running time of the form $f(r, s, t, \Delta) \cdot poly(n)$ for some doubly exponential function involving parameters r, s, t and Δ. However, in the following we explain briefly how the augmenting step can be computed in order to obtain an algorithm with running time that is polynomial in t.

Let $z \in \mathbb{Z}_{\ge 0}^{s+nt}$ be a feasible solution of IP (1) and let $\lambda \in \mathbb{Z}_{\ge 0}$ be a multiplicity (which can be guessed). A core ingredient in the augmenting framework is to find an augmenting step. Therefore, we have to compute a Graver element $y \in ker(\mathcal{A})$ such that $z + \lambda y$ is a feasible solution of IP (1) and the objective $\lambda c^T y$ is maximal over such all Graver elements.

Let $L = (rs\Delta)^{O(rs((2r\Delta+1)^{rs^2}))}$ be the bound for $\left\| y^{(i)} \right\|_1$ that we obtain from the previous Lemma. To find the optimal augmenting step, it is sufficient to solve the IP $\max\{c^T x \mid \mathcal{A}x = 0, \bar{\ell} \le x \le \bar{u}, \|x\|_\infty \le L\}$ for modified upper and lower bounds $\bar{\ell}, \bar{u}$ according to the multiple λ and the feasible solution z. Having the best augmenting step at hand, one can show that the objective value improves by a factor of $1 - \frac{1}{2n}$. This is due to the fact (see [6]) that the difference $z - z^*$ between z and an optimal solution z^* can be represented by

$$z - z^* = \sum_{i=1}^{2n} \lambda_i g_i$$

for Graver elements $g_1, \ldots g_{2n} \in \mathbb{Z}_{\geq 0}^d$ and multiplicities $\lambda_1, \ldots, \lambda_{2n} \in \mathbb{Z}_{\geq 0}$.

In the following we briefly show how to solve the IP $\max\{c^T x \mid \mathcal{A}x = 0, \bar{\ell} \leq x \leq \bar{u}, \|x\|_\infty \leq L\}$ in order to compute the augmenting step. The algorithm works as follows:

- Compute for every $y^{(0)}$ with $\|y^{(0)}\|_1 \leq L$ the objective value of the cycle y consisting of $y^{(0)}, \bar{y}^{(1)}, \ldots, \bar{y}^{(n)}$, where $\bar{y}^{(i)}$ for $i > 0$ are the optimal solutions of the IP

$$\max (c^{(i)})^T \bar{y}^{(i)}$$
$$B^{(i)} \bar{y}^{(i)} = -A^{(i)} y^{(0)}$$
$$\bar{\ell}^{(i)} \leq \bar{y}^{(i)} \leq \bar{u}^{(i)}$$

  where $\bar{\ell}^{(i)}, \bar{u}^{(i)}$ are the upper and lower bounds for the variables $\bar{y}^{(i)}$ and $c^{(i)}$ their corresponding objective vector. Note that the first set of constraints of the IP ensure that $\mathcal{A}y = 0$. The IPs can be solved with the algorithm of Eisenbrand and Weismantel [11] in time $O(\Delta^{O(r^2)})$ each.
- Return the cycle with maximum objective.

As the number of different vectors $y^{(0)}$ with 1-norm $\leq L$ is bounded by $(L+1)^s = (rs\Delta)^{O(rs^2((2r\Delta+1)^{rs^2}))}$ step 1 of the algorithm takes time $(rs\Delta)^{O(rs^2((2r\Delta+1)^{rs^2}))} \cdot n^2 t^2 \varphi \log^2(nt)$.

Putting all things together, we obtain the following theorem regarding the worst-case complexity for solving 2-stage stochastic IPs. For details regarding the remaining parts of the augmenting framework like finding an initial feasible solution or a bound on the required augmenting steps we refer to [9] and [25].

Theorem 3. *A 2-stage stochastic IP of the form (1) can be solved in time*

$$(rs\Delta)^{O(rs^2((2r\Delta+1)^{rs^2}))} \cdot n^2 t^2 \varphi \log^2(nt),$$

where φ is the encoding length of the IP.

3   About the Intersection of Integer Cones

In order to prove our main Lemma 2, we need two observations about the intersection of integer cones. An integer cone is defined for a given (finite) generating set $B \subset \mathbb{Z}_{\geq 0}^d$ of elements by

$$int.cone(B) = \{\sum_{b \in B} \lambda_b b \mid \lambda \in \mathbb{Z}_{\geq 0}^B\}.$$

Note that the intersection of two integer cones is again an integer cone, as the intersection is closed under addition and scalar multiplication of positive integers.

We say that an element b of an integer cone $int.cone(B)$ is *indecomposable* if there do not exist elements $b_1, b_2 \in int.cone(B) \setminus \{0\}$ such that $b = b_1 + b_2$. We can assume that the generating set B of an integer cone consists just of the set of indecomposable elements as any decomposable element can be removed from the generating set.

In the following we allow to use a vector set B as a matrix and vice versa where the elements of the set B are the columns of the matrix B. This way we can multiply B with a vector, i.e. $B\lambda = \sum_{b \in B} \lambda_b b$ for some $\lambda \in \mathbb{Z}^B$. For the proof of the lemma, we refer to the full version of the paper.

Lemma 3. *Consider integer cones $int.cone(B^{(1)})$ and $int.cone(B^{(2)})$ for some generating sets $B^{(1)}, B^{(2)} \subset \mathbb{Z}^d$ where each element $x \in B^{(1)} \cup B^{(2)}$ has bounded norm $\|x\|_\infty \leq \Delta$. Consider the integer cone of the intersection*

$$int.cone(\hat{B}) = int.cone(B^{(1)}) \cap int.cone(B^{(2)})$$

for some generating set of elements $\hat{B}$. Then for each indecomposable element $b \in \hat{B}$ of the intersection cone with $b = B^{(1)}\lambda = B^{(2)}\gamma$ for some $\lambda \in \mathbb{Z}_{\geq 0}^{B^{(1)}}$ and $\gamma \in \mathbb{Z}_{\geq 0}^{B^{(2)}}$, we have that $\|\lambda\|_1, \|\gamma\|_1 \leq (2d\Delta + 1)^d$. Furthermore, the norm of b is bounded by $\|b\|_\infty \leq \Delta(2d\Delta + 1)^d$.

Using a similar argumentation as in the previous lemma, we can consider the intersection of several integer cones. Note that we can not simply use the above Lemma inductively as this would lead to worse bounds. Due to space constraints we omit the proof and refer to the full version of the paper.

Lemma 4. *Consider integer cones $int.cone(B^{(1)}), \ldots, int.cone(B^{(\ell)})$ for some generating sets $B^{(1)}, \ldots, B^{(\ell)} \subset \mathbb{Z}_{\geq 0}^d$ with $\|x\|_\infty \leq \Delta$ for each $x \in B^{(i)}$. Consider the integer cone of the intersection*

$$int.cone(\hat{B}) = \bigcap_{i=1}^{\ell} int.cone(B^{(i)})$$

for some generating set of elements $\hat{B}$.

Then for each indecomposable element $b \in \hat{B}$ with $B^{(i)}\lambda^{(i)} = b$ for some $\lambda^{(i)} \in \mathbb{Z}_{\geq 0}^{B^{(i)}}$ in the intersection cone, we have that $\|\lambda^{(i)}\|_1 \leq O((d\Delta)^{d(\ell-1)})$ for all $1 \leq i \leq \ell$.

4   Proof of Lemma 2

Using the results from the previous section, we are now finally able to prove our main Lemma 2.

To get an intuition for the problem however, we start by giving a sketch of the proof for the 1-dimensional case. In this case, the multisets T_i consist solely of natural numbers, i.e $T_1, \ldots, T_n \subset \mathbb{Z}_{\geq 0}$. Suppose that each set T_i consists only of many copies of a single integral number $x_i \in \{1, \ldots, \Delta\}$. Then it is easy to find a common multiple as $\frac{\Delta!}{1} \cdot 1 = \frac{\Delta!}{2} \cdot 2 = \ldots = \frac{\Delta!}{\Delta} \cdot \Delta$. Hence one can choose the subsets consisting of $\frac{\Delta!}{x_i}$ copies of x_i. Now suppose that the multisets T_i can be arbitrary. If $|T_i| \leq \Delta \cdot \Delta! = \Delta^{O(\Delta)}$ we are done. But on the other hand, if $|T_i| \geq \Delta \cdot \Delta!$, by the pigeonhole principle there exists a single element $x_i \in \{1, \ldots, \Delta\}$ for every T_i that appears at least $\Delta!$ times. Then we can argue as in the previous case where we needed at most $\Delta!$ copies of a number $x_i \in \{1, \ldots, \Delta\}$. Note that the cardinality of the sets T_i has to be of similar size. As the elements of each set sums up to the same value, the cardinality of two sets T_i, T_j can only differ by a factor of Δ. This proves the lemma in the case $d = 1$.

In the case of higher dimensions, the lemma seems much harder to prove. But in principle we use generalizations of the above techniques. Instead of single natural numbers however, we have to work with bases of corresponding basic feasible LP solutions and the intersection of the integer cone generated by those bases.

Proof. First, we describe the multisets $T_1, \ldots, T_n \subset \mathbb{Z}_{\geq 0}^d$ by multiplicity vectors $\lambda^{(1)}, \ldots, \lambda^{(n)} \in \mathbb{Z}_{\geq 0}^P$, where $P \subset \mathbb{Z}^d$ is the set of non-negative integer points p with $\|p\|_\infty \leq \Delta$. Each $\lambda_p^{(i)}$ thereby states the multiplicity of vector p in T_i. Hence $\sum_{t \in T_i} t = \sum_{p \in P} \lambda_p^{(i)} p$ and our objective is to find vectors $y^{(1)}, \ldots, y^{(n)} \in \mathbb{Z}_{\geq 0}^P$ with $y^{(i)} \leq \lambda^{(i)}$ such that $\sum_{p \in P} y_p^{(1)} p = \ldots = \sum_{p \in P} y_p^{(n)} p$.

Consider the linear program

$$\sum_{p \in P} x_p p = b \qquad (2)$$

$$x \in \mathbb{R}_{\geq 0}^P$$

Note that each $\lambda^{(i)}$ is a feasible solution of the LP. First, we are interested in the set of all possible basic feasible solutions $x^{(1)}, \ldots, x^{(\ell)} \in \mathbb{R}_{\geq 0}^d$ of the LP corresponding to bases $B^{(1)}, \ldots, B^{(\ell)} \in \mathbb{Z}_{\geq 0}^{d \times d}$ with $B^{(i)} x^{(i)} = b$. In the proof we consider the special case first, that each multiset T_i corresponds to one of those basic feasible solution $x^{(j)}$. In the 1-dimensional case this would mean that each set consists only of a single number. In this case, we argued above that we can simply pick the least common multiple. What is the correspondence of a least common multiple in a d-dimensional space? This is where the intersection of integer cones come into play. Note that the intersection of integer cones in dimension 1 is just the least common multiple, i.e. $int.cone(z_1) \cap int.cone(z_2) = int.cone(lcm(z_1, z_2))$ for some $z_1, z_2 \in \mathbb{Z}_{\geq 0}$.

For the remaining part of the proof, we refer the reader to the appendix or the full version of the paper. There, we prove two claims that correspond to the two previously described cases of the one dimensional case.

Continuation of the Proof of Lemma 2

In the proof we need the notion of a cone which is simply the relaxation of an integer cone. For a generating set $B \subset \mathbb{Z}_{\geq 0}^d$, a cone is defined by

$$cone(B) = \{\sum_{b \in B} \lambda_b b \mid \lambda \in \mathbb{R}_{\geq 0}^B\}.$$

Proof. **Claim 1:** If for all i we have $\left\|x^{(i)}\right\|_1 > d \cdot O((d\Delta)^{d(\ell-1)})$ then there exist non-zero vectors $y^{(1)}, \ldots, y^{(\ell)} \in \mathbb{Z}_{\geq 0}^d$ with $y^{(1)} \leq x^{(1)}, \ldots, y^{(\ell)} \leq x^{(\ell)}$ and $\left\|y^{(i)}\right\|_1 \leq d \cdot O((d\Delta)^{d(\ell-1)})$ such that $B^{(1)}y^{(1)} = \ldots = B^{(\ell)}y^{(\ell)}$.

Note that all basic feasible solutions $x^{(i)} \in \mathbb{R}_{\geq 0}^d$ have to be of similar size. Since $Bx^{(i)} = b$ holds for all $1 \leq i \leq \ell$ we know that $\left\|x^{(i)}\right\|_1$ and $\left\|x^{(j)}\right\|_1$ can only differ by a factor of $d\Delta$ for all $1 \leq i, j \leq \ell$. Hence all basic feasible solutions $x^{(i)}$ have to be either small or all have to be large. This claim considers the case that the size of all $x^{(i)}$ is large.

Proof of the Claim: Note that $B^{(i)}x^{(i)} = b$ and hence $b \in cone(B^{(i)})$. In the following, our goal is to find a non-zero point $q \in \mathbb{Z}_{\geq 0}^d$ such that $q = B^{(1)}y^{(1)} = \ldots = B^{(\ell)}y^{(\ell)}$ for some vectors $y^{(1)}, \ldots, y^{(\ell)} \in \mathbb{Z}_{\geq 0}^d$. However, this means that q has to be in the integer cone $int.cone(B^{(i)})$ for every $1 \leq i \leq \ell$ and therefore in the intersection of all the integer cones, i.e. $q \in \bigcap_{i=1}^n int.cone(B^{(i)})$. By Lemma 4 there exists a set of generating elements $\hat{B}$ such that

- $int.cone(\hat{B}) = \bigcap_{i=1}^n int.cone(B^{(i)})$ and $int.cone(\hat{B}) \neq \{0\}$ as $b \in cone(\hat{B})$ and
- each generating vector $p \in \hat{B}$ can be represented by $p = B^{(i)}\lambda$ for some $\lambda \in \mathbb{Z}_{\geq 0}^d$ with $\|\lambda\|_1 \leq O((d\Delta)^{d(\ell-1)})$ for each basis $B^{(i)}$.

As $b \in cone(\hat{B})$ there exists a vector $\hat{x} \in \mathbb{R}_{\geq 0}^{\hat{B}}$ with $\hat{B}\hat{x} = b$. Our goal is to show that there exists a non-zero vector $q \in \hat{B}$ with $\hat{x}_q \geq 1$. In this case b can be simply written by $b = q + q'$ for some $q' \in cone(\hat{B})$. As q and q' are contained in the intersection of all cones, there exists for each generating set $B^{(j)}$ a vectors $y^{(j)} \in \mathbb{Z}_{\geq 0}^{B^{(j)}}$ and $z^{(j)} \in \mathbb{R}_{\geq 0}^{B^{(j)}}$ such that $B^{(j)}y^{(j)} = q$ and $B^{(j)}z^{(j)} = q'$. Hence $x^{(j)} = y^{(j)} + z^{(j)}$ and we finally obtain that $x^{(j)} \geq y^{(j)}$ for $y^{(j)} \in \mathbb{Z}_{\geq 0}^{B^{(j)}}$ which shows the claim.

Therefore it only remains to prove the existence of the point q with $\hat{x}_q \geq 1$. By Lemma 4, each vector $p \in \hat{B}$ can be represented, by $p = B^{(i)}x^{(p)}$ for some $x^{(p)} \in \mathbb{Z}_{\geq 0}^{B^{(i)}}$ with $\left\|x^{(p)}\right\|_1 \leq O((d\Delta)^{d(\ell-1)})$ for every basis $B^{(i)}$.

As $B^{(i)}x^{(i)} = b = \sum_{p \in \hat{B}} \hat{x}_p p = \sum_{p \in \hat{B}} \hat{x}_p(B^{(i)}x^{(p)})$, every $x^{(i)}$ can be written by $x^{(i)} = \sum_{p \in \hat{B}} x^{(p)}\hat{x}_p$ and we obtain a bound on $\left\|x^{(i)}\right\|_1$ assuming that every for every $p \in \hat{B}$ we have $\hat{x}_p < 1$.

$$\left\|x^{(i)}\right\|_1 \leq \sum_{p \in \hat{B}} \left\|x^{(p)}\hat{x}_p\right\|_1 \overset{\hat{x}_p < 1}{<} \sum_{p \in \hat{B}} \left\|x^{(p)}\right\|_1 \leq d \cdot O((d\Delta)^{d(\ell-1)}).$$

The last inequality follows as we can assume by Caratheodory's theorem [28] that the number of non-zero components of $\hat{x}$ is less or equal than d. Hence if $\left\|x^{(i)}\right\|_1 \geq d \cdot O((d\Delta)^{d(\ell-1)})$ then there has to exist a vector $q \in \hat{B}$ with $x_q \geq 1$ which proves the claim.

Claim 2: For every vector $\lambda^{(i)} \in \mathbb{Z}_{\geq 0}^P$ with $\sum_{p \in P} \lambda_p p = b$ there exists a basic feasible solution $x^{(k)}$ of LP (2) with basis $B^{(k)}$ such that $\frac{1}{\ell} x^{(k)} \leq \lambda^{(i)}$ in the sense that $\frac{1}{\ell} x_p^{(k)} \leq \lambda_p^{(i)}$ for every $p \in B^{(k)}$.

Proof of the Claim: The proof of the claim can be easily seen as each multiplicity vector $\lambda^{(i)}$ is also a solution of the linear program (2). By standard LP theory, we know that each solution of the LP is a convex combination of the basic feasible solutions $x^{(1)}, \ldots, x^{(\ell)}$. Hence, each multiplicity vector $\lambda^{(i)}$ can be written as a convex combination of $x^{(1)}, \ldots, x^{(\ell)}$, i.e. for each $\lambda^{(i)}$, there exists a $t \in \mathbb{R}_{\geq 0}^\ell$ with $\|t\|_1 = 1$ such that $\lambda^{(i)} = \sum_{j=1}^\ell t_j \bar{x}^{(j)}$, where $\bar{x}_p^{(j)} = \begin{cases} x_p^{(j)} & \text{if } p \in B^{(j)} \\ 0 & \text{otherwise} \end{cases}$.

By the pigeonhole principle, there exists for each multiplicity vector $\lambda^{(i)}$ an index k with $t_k \geq \frac{1}{\ell}$ which proves the claim.

Using the above two claims, we can now prove the claim of the lemma by showing that for each $\lambda^{(i)}$, there exist a vector $y^{(i)} \leq \lambda^{(i)}$ with bounded 1-norm such that $\sum_{p \in P} y_p^{(1)} p = \ldots = \sum_{p \in P} y_p^{(n)} p$.

First, consider the case that there exists a basic feasible solution $x^{(j)}$ of LP 2 with $\left\|x^{(j)}\right\|_1 \leq \ell d \cdot O((d\Delta)^{d(\ell-1)})$. In this case we have for all $1 \leq i \leq n$ that $\left\|\lambda^{(i)}\right\|_1 \leq \ell d^2 \Delta \cdot O((d\Delta)^{d(\ell-1)})$ as the size of solutions of LP (2) can not differ by a factor of more than $d\Delta$ (this is because for every $p, p' \in P$ the sizes $\|p\|_1, \|p'\|_1$ can not differ by a factor of more than $d\Delta$).

Now, assume that for all basic feasible solutions $x^{(i)}$ we have $\left\|x^{(i)}\right\|_1 > \ell d \cdot O((d\Delta)^{d(\ell-1)})$. We can argue by Claim 2 that for each $\lambda^{(i)}$ (with $1 \leq i \leq n$) we find one of the basic feasible solutions $x^{(k)}$ ($1 \leq k \leq \ell$) with $\frac{1}{\ell} x^{(k)} \leq \lambda^{(i)}$. As $\frac{1}{\ell} x^{(i)} \geq d \cdot O((d\Delta)^{d(\ell-1)})$ for all $1 \leq i \leq \ell$, we can apply the first claim to vectors $\frac{1}{\ell} x^{(1)}, \ldots, \frac{1}{\ell} x^{(\ell)}$ with $\frac{1}{\ell} b = \frac{1}{\ell} B x^{(1)} = \ldots = \frac{1}{\ell} B x^{(\ell)}$, we obtain vectors $y^{(1)} \leq \frac{1}{\ell} x^{(1)}, \ldots, y^{(\ell)} \leq \frac{1}{\ell} x^{(\ell)}$ with $By^{(1)} = \ldots = By^{(\ell)}$. Hence, we find for each $\lambda^{(i)}$ a vector $y^{(k)} \in \mathbb{Z}_{\geq 0}^{B^{(k)}}$ with $y^{(k)} \leq \lambda^{(i)}$.

Finally we obtain that

$$\left\|y^{(j)}\right\|_1 \leq d^2 \Delta \ell \cdot O((d\Delta)^{d(\ell-1)}) = (d\Delta)^{O(d(\Delta^{d^2}))}$$

using that ℓ is bounded by $\binom{|P|}{d} \leq |P|^d$ and $|P| \leq \Delta^d$.

References

1. Ahmed, S., Tawarmalani, M., Sahinidis, N.V.: A finite branch-and-bound algorithm for two-stage stochastic integer programs. Math. Program. **100**(2), 355–377 (2004)

2. Aschenbrenner, M., Hemmecke, R.: Finiteness theorems in stochastic integer programming. Found. Comput. Math. **7**(2), 183–227 (2007)
3. Birge, J.R., Louveaux, F.: Introduction to Stochastic Programming. Springer, Heidelberg (2011). https://doi.org/10.1007/978-1-4614-0237-4
4. Carøe, C.C., Tind, J.: L-shaped decomposition of two-stage stochastic programs with integer recourse. Math. Program. **83**(1–3), 451–464 (1998)
5. Chen, L., Xu, L., Shi, W.: On the Graver basis of block-structured integer programming. arXiv preprint arXiv:1805.03741 (2018)
6. Cook, W., Fonlupt, J., Schrijver, A.: An integer analogue of Caratheodory's theorem. J. Comb. Theory Ser. B **40**(1), 63–70 (1986)
7. De Loera, J.A., Hemmecke, R., Köppe, M.: Algebraic and Geometric Ideas in the Theory of Discrete Optimization. SIAM, New York (2012)
8. Dempster, M.A.H., Fisher, M., Jansen, L., Lageweg, B., Lenstra, J.K., Rinnooy Kan, A.: Analytical evaluation of hierarchical planning systems. Oper. Res. **29**(4), 707–716 (1981)
9. Eisenbrand, F., Hunkenschröder, C., Klein, K.: Faster algorithms for integer programs with block structure. In: 45th International Colloquium on Automata, Languages, and Programming, ICALP 2018, Prague, Czech Republic, 9–13 July 2018, pp. 49:1–49:13 (2018)
10. Eisenbrand, F., Hunkenschröder, C., Klein, K.M., Koutecký, M., Levin, A., Onn, S.: An algorithmic theory of integer programming (2019)
11. Eisenbrand, F., Weismantel, R.: Proximity results and faster algorithms for integer programming using the Steinitz lemma. In: Proceedings of the Twenty-Ninth Annual ACM-SIAM Symposium on Discrete Algorithms, pp. 808–816. SIAM (2018)
12. Gade, D., Küçükyavuz, S., Sen, S.: Decomposition algorithms with parametric gomory cuts for two-stage stochastic integer programs. Math. Program. **144**(1–2), 39–64 (2014)
13. Graver, J.E.: On the foundations of linear and integer linear programming I. Math. Program. **9**(1), 207–226 (1975)
14. Grinberg, V.S., Sevast'yanov, S.V.: Value of the Steinitz constant. Funct. Anal. Appl. **14**(2), 125–126 (1980)
15. Haneveld, W.K.K., van der Vlerk, M.H.: Optimizing electricity distribution using two-stage integer recourse models. In: Uryasev, S., Pardalos, P.M. (eds.) Stochastic Optimization: Algorithms and Applications, vol. 54, pp. 137–154. Springer, Heidelberg (2001). https://doi.org/10.1007/978-1-4757-6594-6_7
16. Hemmecke, R., Schultz, R.: Decomposition of test sets in stochastic integer programming. Math. Program. **94**(2–3), 323–341 (2003)
17. Jansen, K., Klein, K., Maack, M., Rau, M.: Empowering the configuration-IP - new PTAS results for scheduling with setups times. CoRR abs/1801.06460 (2018). http://arxiv.org/abs/1801.06460
18. Jansen, K., Lassota, A., Rohwedder, L.: Near-linear time algorithm for n-fold ILPs via color coding. arXiv preprint arXiv:1811.00950 (2018)
19. Kall, P., Wallace, S.W.: Stochastic Programming. Springer, Heidelberg (1994)
20. Kannan, R.: Minkowski's convex body theorem and integer programming. Math. Oper. Res. **12**(3), 415–440 (1987)
21. Knop, D., Koutecký, M.: Scheduling meets n-fold integer programming. J. Sched. **21**(5), 493–503 (2018)

22. Knop, D., Koutecký, M., Mnich, M.: Combinatorial n-fold integer programming and applications. In: Pruhs, K., Sohler, C. (eds.) 25th Annual European Symposium on Algorithms (ESA 2017). Leibniz International Proceedings in Informatics (LIPIcs), vol. 87, pp. 54:1–54:14. Schloss Dagstuhl-Leibniz-Zentrum fuer Informatik, Dagstuhl (2017)

23. Knop, D., Koutecký, M., Mnich, M.: Voting and bribing in single-exponential time. In: 34th Symposium on Theoretical Aspects of Computer Science, STACS 2017, Hannover, Germany, 8–11 March 2017, pp. 46:1–46:14 (2017)

24. Knop, D., Pilipczuk, M., Wrochna, M.: Tight complexity lower bounds for integer linear programming with few constraints. arXiv preprint arXiv:1811.01296 (2018)

25. Koutecký, M., Levin, A., Onn, S.: A parameterized strongly polynomial algorithm for block structured integer programs. In: 45th International Colloquium on Automata, Languages, and Programming, ICALP 2018, Prague, Czech Republic, 9–13 July 2018, pp. 85:1–85:14 (2018)

26. Küçükyavuz, S., Sen, S.: An introduction to two-stage stochastic mixed-integer programming. In: Leading Developments from INFORMS Communities, pp. 1–27. INFORMS (2017)

27. Pelupessy, F., Weiermann, A.: Ackermannian lower bounds for lengths of bad sequences of monomial ideals over polynomial rings in two variables. Math. Theory Comput. Practice 276 (2009)

28. Schrijver, A.: Theory of Linear and Integer Programming. Wiley, Hoboken (1998)

29. Schultz, R., Stougie, L., Van Der Vlerk, M.H.: Two-stage stochastic integer programming: a survey. Statistica Neerlandica 50(3), 404–416 (1996)

30. Steinitz, E.: Bedingt konvergente reihen und konvexe systeme. Journal für die reine und angewandte Mathematik 143, 128–176 (1913)

31. Zhang, M., Küçükyavuzvuz, S.: Finitely convergent decomposition algorithms for two-stage stochastic pure integer programs. SIAM J. Optim. 24(4), 1933–1951 (2014)

Packing Under Convex Quadratic Constraints

Max Klimm[1]([⊠]), Marc E. Pfetsch[2]([⊠]), Rico Raber[3]([⊠]),
and Martin Skutella[3]([⊠])

[1] School of Business and Economics, HU Berlin,
Spandauer Str. 1, 10178 Berlin, Germany
max.klimm@hu-berlin.de
[2] Department of Mathematics, TU Darmstadt,
Dolivostr. 15, 64293 Darmstadt, Germany
pfetsch@mathematik.tu-darmstadt.de
[3] Institute of Mathematics, TU Berlin,
Straße des 17. Juni 136, 10623 Berlin, Germany
{raber,skutella}@math.tu-berlin.de

Abstract. We consider a general class of binary packing problems with a convex quadratic knapsack constraint. We prove that these problems are APX-hard to approximate and present constant-factor approximation algorithms based upon three different algorithmic techniques: (1) a rounding technique tailored to a convex relaxation in conjunction with a non-convex relaxation whose approximation ratio equals the golden ratio; (2) a greedy strategy; (3) a randomized rounding method leading to an approximation algorithm for the more general case with multiple convex quadratic constraints. We further show that a combination of the first two strategies can be used to yield a monotone algorithm leading to a strategyproof mechanism for a game-theoretic variant of the problem. Finally, we present a computational study of the empirical approximation of the three algorithms for problem instances arising in the context of real-world gas transport networks.

1 Introduction

We consider packing problems with a convex quadratic knapsack constraint of the form

$$\max \left\{ p^\top x \ : \ x^\top W x \le c, \ x \in \{0,1\}^n \right\}, \qquad (P)$$

where $W \in \mathbb{Q}_{\ge 0}^{n \times n}$ is a symmetric positive semi-definite (psd) matrix with non-negative entries, $p \in \mathbb{Q}_{\ge 0}^n$ is a non-negative profit vector, and $c \in \mathbb{Q}_{\ge 0}$ is a non-negative budget. Such convex and quadratically constrained packing problems are clearly NP-complete since they contain the classical (linearly constrained)

We acknowledge funding through the DFG CRC/TRR 154, Subproject A007.

D. Bienstock and G. Zambelli (Eds.): IPCO 2020, LNCS 12125, pp. 266–279, 2020.
https://doi.org/10.1007/978-3-030-45771-6_21

NP-complete knapsack problem [12] as a special case when W is a diagonal matrix. In this paper, we therefore focus on the development of approximation algorithms. For some $\rho \in [0, 1]$, an algorithm is a ρ-approximation algorithm if its runtime is polynomial in the input size and for every instance, it computes a solution with objective value at least ρ times that of an optimum solution. The value ρ is then called the *approximation ratio* of the algorithm. We note that the assumption on W being psd is necessary in order to allow for sensible approximation. To see this, observe that when W is the adjacency matrix of an undirected graph and $c = 0$, (P) encodes the problem of finding an independent set of maximal weight, which is NP-hard to approximate within a factor better than $n^{-(1-\epsilon)}$ for any $\epsilon > 0$, even in the unweighted case [9].

The packing problems that we consider also have a natural interpretation in terms of mechanism design. Consider a situation where a set of n selfish agents demands a service, and the subsets of agents that can be served simultaneously are modeled by a convex quadratic packing constraint. Each agent j has private information p_j about its willingness to pay for receiving the service. In this context, a (direct revelation) mechanism takes as input the matrix W and the budget c. It then elicits the private value p_j from agent j. Each agent j may misreport a value p'_j instead of their true value p_j if this is to their benefit. The mechanism then computes a solution $x \in \{0,1\}^n$ to (P) as well as a payment vector $g \in \mathbb{Q}_{\geq 0}^n$. A mechanism is strategyproof if no agent has an interest in misreporting p_j, no matter what the other agents report.

Before we present our results on approximation ratios and mechanisms for non-negative, convex, and quadratically constrained packing problems, we give two real-world examples that fall into this category.

Example 1. (Welfare maximization in gas supply networks). Consider a gas pipeline modeled by a directed graph $G = (V, E)$ with different entry and exit nodes. There is a set of n transportation requests (s_j, t_j, q_j, p_j), $j \in [n] :=$ $\{1, \dots, n\}$, each specifying an entry node $s_j \in V$, an exit node $t_j \in V$, the amount of gas to be transported $q_j \in \mathbb{Q}_{\geq 0}$, and an economic value $p_j \in \mathbb{Q}_{\geq 0}$. One model for gas flows in pipe networks is given by the Weymouth equations [27] of the form $\beta_e\, q_e\, |q_e| = \pi_u - \pi_v$ for all $e = (u, v) \in E$. Here, the parameter $\beta_e \in \mathbb{Q}_{>0}$ is a pipe specific value that depends on physical properties of the pipe segment modeled by the edge, such as length, diameter, and roughness. Positive flow values $q_e > 0$ denote flow from u to v, while a negative q_e indicates flow in the opposite direction. The value π_u denotes the square of the pressure at node $u \in V$. In real-life gas networks, there is typically a bound $c \in \mathbb{Q}_{\geq 0}$ on the maximal difference of the squared pressures in the network. For the operation of gas networks, it is a natural problem to find the welfare-maximal subset of transportation requests that can be satisfied simultaneously while satisfying the pressure constraint.

To illustrate this problem, we consider the particular case in which the network has a path topology similar to the one depicted in Fig. 1. We assume that for each request the entry node is left of the exit node. Thus, the pressure in the pipe is decreasing from left to right. For $j \in [n]$, let $E_j \subseteq E$ denote the set of

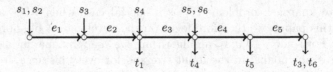

Fig. 1. Gas network with feed-in and feed-out nodes.

edges on the unique (s_j, t_j)-path in G. Indexing the vertices $v_0, \ldots, v_k$ and edges $e_1, \ldots, e_k$ from left to right, the maximal squared pressure difference in the pipe is given by

$$\pi_0 - \pi_k = \sum_{i=1}^{k} (\pi_{i-1} - \pi_i) = \sum_{i=1}^{k} \beta_{e_i} \Big(\sum_{j \in [n]: e_i \in E_j} q_j x_j \Big)^2,$$

where $x_j \in \{0, 1\}$ indicates whether transportation request $j \in [n]$ is being served. For the matrix $W = (w_{ij})_{i,j \in [n]}$ defined by $w_{ij} = \sum_{e \in E_i \cap E_j} \beta_e q_i q_j$, the pressure constraint can be formulated as $x^\top W x \leq c$. To see that the matrix W is positive semi-definite, we write $W = \sum_{e \in E} \beta_e q^e (q^e)^\top$, where $q^e \in \mathbb{Q}_{\geq 0}^n$ is defined as $q_i^e = q_i$ if $e \in E_i$, and $q_i^e = 0$, otherwise.

Gas networks are particularly interesting from a mechanism design perspective, since several countries employ or plan to employ auctions to allocate gas network capacities [20], but theoretical and experimental work uses only linear flow models [16,23], thus ignoring the physics of the gas flow.

Example 2 (Processor speed scaling). Consider a mobile device with battery capacity c and k compute cores. Further, there is a set of n tasks (q^j, p_j), each specifying a load $q^j \in \mathbb{Q}_{\geq 0}^k$ for the k cores and a profit p_j. The computations start at time 0 and all computations have to be finished at time 1. In order to adapt to varying workloads, the compute cores can run at different speeds. In the speed scaling literature, it is a common assumption that energy consumption of core i when running at speed s is equal to $\beta_i s^2$, where $\beta_i \in \mathbb{Q}_{>0}$ is a core-specific parameter [2,11,28].[1] The goal is to select a profit-maximal subset of tasks that can be scheduled in the available time with the available battery capacity. Given a subset of tasks, it is without loss of generality to assume that each core runs at the minimal speed such that the core finishes at time 1, i.e., every core i runs at speed $\sum_{j \in [n]} x_j q_i^j$ so that the total energy consumption is $\sum_{i=1}^{k} \beta_i (\sum_{j \in [n]} x_j q_i^j)^2$. The energy constraint can thus be formulated as a convex quadratic constraint.

Mechanism design problems for processor speed scaling are interesting when the tasks are controlled by selfish agents and access to computation on the energy-constrained device is determined via an auction.

---

[1] Other works assume that the relationship is cubic, but experiments conducted by Wierman et al. [28] suggest that the relationship is closer to quadratic than cubic.

Our Results and Paper Outline. In Sect. 3 we derive a ϕ-approximation algorithm for packing problems with convex quadratic constraints where $\phi = (\sqrt{5} - 1)/2 \approx 0.618$ is the inverse golden ratio. The algorithm first solves a convex relaxation and scales the solution by ϕ, which turns it into a feasible solution to a second non-convex relaxation. The latter relaxation has the property that any solution can be transformed into a solution with at most one fractional component without decreasing the objective value. In the end, the algorithm returns the integral part of the transformed solution. Combining this procedure with a partial enumeration scheme yields a ϕ-approximation; see Theorem 1. In Sect. 4 we prove that the greedy algorithm, when combined with partial enumeration, is a constant-factor approximation algorithm with an approximation ratio of at least $(1 - \sqrt{3}/e) \approx 0.363$. In Sect. 5, we show that a combination of the results from the previous section allows to derive a strategyproof mechanism with constant approximation ratio. In Sect. 6 we derive a randomized constant-factor approximation algorithm for the more general problem with a constant number of r convex quadratic packing constraints. The algorithm solves a convex relaxation, scales the solution, and performs randomized rounding based on that scaled solution. Combining this algorithm with partial enumeration yields a constant-factor approximation; see Theorem 4. In Sect. 7 we show that packing problems with convex quadratic constraints of type (P) are APX-hard; see Theorem 5. Finally, in the appendix, we apply the three algorithms to several instances of the problem type described in Example 1 based on real-world data from the GasLib library [25].

Due to space constraints, the proofs are deferred to the full version of the paper [13].

Related Work. When W is a non-negative diagonal matrix, the quadratic constraint in (P) becomes linear and the problem is then equivalent to the 0-1-knapsack problem which admits a *fully polynomial-time approximation scheme* (FPTAS) [10]. Another interesting special case is when W is completely-positive, i.e., it can then be written as $W = UU^\top$ for some matrix $U \in \mathbb{Q}_{\geq 0}^{n \times k}$ with non-negative entries. The minimal k for which W can be expressed in this way is called the *cp-rank* of W, see [3] for an overview on completely positive matrices. The quadratic constraint in (P) can then be expressed as $\|U^\top x\|_2 \leq \sqrt{c}$. For the case that $U \in \mathbb{Q}_{\geq 0}^{n \times 2}$, this problem is known as the 2-weighted knapsack problem for which Woeginger [29] showed that it does not admit an FPTAS, unless $\mathsf{P} = \mathsf{NP}$. Chau et al. [5] settled the complexity of this problem showing that it admits a *polynomial-time approximation scheme* (PTAS). Elbassioni et al. [6] generalized this result to matrices with constant cp-rank.

Exchanging constraints and objective in (P) leads to knapsack problems with quadratic objective function and a linear constraint first studied by Gallo [7]. These problems have a natural graph-theoretic interpretation where nodes and edges have profits, the nodes have weights, and the task is to choose a subset of nodes so as to maximize the total profit of the induced subgraph. Rader and Woeginger [22] give an FPTAS when the graph is edge series-parallel. Pferschy and Schauer [21] generalize this result to graphs of bounded treewidth. They also give a PTAS for graphs not including a forbidden minor which includes planar graphs.

Mechanism design problems with a knapsack constraint are contained as a special case when W is a diagonal matrix. For this special case, Mu'alem and Nisan [17] give a mechanism that is strategyproof and yields a 1/2-approximation. Briest et al. [4] give a general framework that allows to construct a mechanism that is an FPTAS for the objective function. Aggarwal and Hartline [1] study knapsack auctions with the objective to maximize the sum of the payments to the mechanism.

2   Preliminaries

For ease of exposition, we assume that all matrices and vectors are integer. Let $[n] := \{1, \ldots, n\}$ and $W = (w_{ij})_{i,j \in [n]} \in \mathbb{N}^{n \times n}$ be a symmetric psd matrix. Furthermore, let $p \in \mathbb{N}^n$ be a profit vector and let $c \in \mathbb{N}$ be a budget. We consider problems of the form (P), i.e., $\max\{p^\top x : x^\top W x \leq c, x \in \{0,1\}^n\}$. Throughout the paper, we denote the characteristic vector of a subset $S \subseteq [n]$ by $\chi_S \in \{0,1\}^n$, i.e., $\chi_i = 1$ if $i \in S$ and $\chi_i = 0$, otherwise.

We first state the intuitive result that after fixing $x_i = 1$ for $i \in N_1 \subseteq [n]$ and fixing $x_i = 0$ for $i \in N_0$ (with $N_0 \cap N_1 = \emptyset$), we again obtain a packing problem with a convex and quadratic packing constraint.

Lemma 1. *Let $W \in \mathbb{N}^{n \times n}$ be symmetric psd, $p \in \mathbb{N}^n$, and $c \in \mathbb{N}$. Further, let N_0, $N_1 \in 2^{[n]}$ with $N_0 \cap N_1 = \emptyset$ and $N_0 \cup N_1 \subsetneq [n]$ be arbitrary. Then, there exist $\tilde{n} \in \mathbb{N}$, $\tilde{W} \in \mathbb{N}^{\tilde{n} \times \tilde{n}}$ symmetric psd, $\tilde{p} \in \mathbb{N}^{\tilde{n}}$, and $\tilde{c} \in \mathbb{N}$ such that*

$$\max\{p^\top x : x^\top W x \leq c, \ x \in \{0,1\}^n, \ x_i = 0 \ \forall i \in N_0, \ x_i = 1 \ \forall i \in N_1\}$$
$$= p^\top \chi_{N_1} + \max\{\tilde{p}^\top \tilde{x} : \tilde{x}^\top \tilde{W} \tilde{x} \leq \tilde{c}, \ \tilde{x} \in \{0,1\}^{\tilde{n}}\}.$$

By Lemma 1, the following assumptions are without loss of generality.

Lemma 2. *It is without loss of generality to assume that $0 < w_{ii} \leq c$ and $p_i > 0$ for all $i \in [n]$.*

3   A Golden Ratio Approximation Algorithm

In this section, we derive a ϕ-approximation algorithm for packing problems with convex quadratic constraints of type (P) where $\phi = (\sqrt{5} - 1)/2 \approx 0.618$ is the inverse golden ratio. To this end, we first solve a convex relaxation of the problem. We then use the resulting solution to compute a feasible solution to another non-convex relaxation of the problem. The second relaxation has the property that any solution can be transformed so that it has at most one fractional value, and the transformation does not decrease the objective value. Together with a partial enumeration scheme in the spirit of Sahni [24], this yields a ϕ-approximation.

Denote by $d \in \mathbb{N}^n$ the diagonal of $W \in \mathbb{N}^{n \times n}$ and let $D := \operatorname{diag}(d) \in \mathbb{N}^{n \times n}$ be the corresponding diagonal matrix. For a vector $x \in \{0, 1\}^n$ we have $x_i^2 = x_i$ for all $i \in [n]$ and, thus, we obtain $x^\top W x \geq x^\top D x = d^\top x$ for all $x \in \{0, 1\}^n$. We arrive at the following relaxation of (P):

$$\max \left\{ \lfloor p^\top x \rfloor \ : \ x^\top W x \leq c, \ d^\top x \leq c, \ x \in [0, 1]^n \right\}. \tag{R_1}$$

Algorithm 1: Golden ratio algorithm

---

1 **foreach** $H \subseteq [n]$ with $|H| \leq 3$ **do**
2     $y^H \leftarrow$ sol. of (R_1) with $x_i = 1 \, \forall i \in H$,
       $x_i = 0 \, \forall i \in \{j \in [n] \setminus H : p_j > \min_{h \in H} p_h\}$;
3     $z^H \leftarrow$ transf. of ϕy^H containing at most one fractional variable;
4     $\bar{z}^H \leftarrow \lfloor z^H \rfloor$;
5 $H^* \leftarrow \operatorname{argmax} \{ p^\top \bar{z}^H : H \subseteq [n] \text{ with } |H| \leq 3 \}$;
6 **return** $\bar{z}^{H^*}$;

The following lemma shows that we can compute an exact optimal solution to (R_1) in polynomial time. For the proof, we use binary search for the optimal objective value in order to bring the quadratic constraint into the objective. The resulting quadratic program with linear constraints can then be solved to optimality by the ellipsoid method [15].

Lemma 3. *The relaxation (R_1) can be solved exactly in polynomial time.*

We proceed to propose a second relaxation of (P). To this end, note that for every $x \in \{0, 1\}^n$ we have $x^\top W x = x^\top (W - D) x + x^\top D x = x^\top (W - D) x + d^\top x$. Relaxing the integrality condition yields the following relaxation of (P):

$$\max \left\{ p^\top x \ : \ x^\top (W - D) x + d^\top x \leq c, \ x \in [0, 1]^n \right\}. \tag{R_2}$$

Note that since the trace of $W - D$ is zero, $W - D$ has a negative eigenvalue unless all eigenvalues are zero. Hence, $W - D$ is not positive semi-definite, unless W is a diagonal matrix. Therefore, the relaxation (R_2) is in general not convex.

We proceed to show that (R_2) always has an optimal solution for which at most one variable is fractional. For $x \in \mathbb{R}^n$, let $N_0(x) := \{i \in [n] : x_i = 0\}$, $N_1(x) := \{i \in [n] : x_i = 1\}$, and $N_f(x) := [n] \setminus (N_1(x) \cup N_0(x))$.

Lemma 4. *For any feasible solution x of (R_2), one can construct a feasible solution $\bar{x}$ with $|N_f(\bar{x})| \leq 1$ and $p^\top \bar{x} \geq p^\top x$ in linear time.*

We proceed to devise a ϕ-approximation algorithm. The algorithm iterates over all sets $H \subseteq [n]$ with $|H| \leq 3$. For each set H it computes an optimal solution y^H to the convex relaxation (R_1) with the additional constraints that

$x_i = 1$ for all $i \in H$, and $x_i = 0$ for all $i \in \{j \in [n] \setminus H : p_j > \min_{h \in H} p_h\}$. Then, we scale down y^H by a factor of ϕ and show that ϕy^H is a feasible solution to the non-convex relaxation (R_2). By Lemma 4, we can transform this solution into another solution z^H with at most one fractional variable. The integral part of z^H is our candidate solution for the starting set H. In the end, we return the best thus computed candidate over all possible sets H; see Algorithm 1.

Theorem 1. *Algorithm 1 computes a ϕ-approximation for (P).*

As a result of Theorem 1, we can derive an upper bound on the optimal value of (R_1). This will turn out to be useful when constructing a monotone greedy algorithm in the next section.

Corollary 1. *Let x^* and y^* be optimal solutions to (P) and (R_1), respectively. Then $p^\top y^* \leq \frac{2}{\phi} p^\top x^*$.*

4   The Greedy Algorithm

In this section we analyze the greedy algorithm and show that, when combined with a partial enumeration scheme in the spirit of Sahni [24], it is at least a $(1 - \sqrt{3}/e)$-approximation for packing problems with quadratic constraints of type (P). Even though this approximation ratio is thus not better than the one guaranteed by the golden ratio algorithm (Theorem 1), it is worth analyzing it for several reasons. Firstly, it is simple to understand as well as to implement and turns out to have a much better running time in practice than the golden ratio algorithm; see the computational results in the appendix. And, secondly, the greedy algorithm serves as a main building block to devise a strategyproof mechanism with constant welfare guarantee; see Sect. 5.

For a set $S \subseteq [n]$, we write $w(S) := \chi_S^\top W \chi_S$. The core idea of the greedy algorithm is as follows. Assume that we have an initial solution $S \subset [n]$. Amongst all remaining items in $[n] \setminus S$, we pick an item i that maximizes the ratio between profit gain and weight gain, i.e.,

$$i \in \operatorname*{argmax}_{j \in [n] \setminus S} \frac{p_j}{w(S \cup \{j\}) - w(S)}.$$

If adding i to the solution set would make it infeasible, i.e., $w(S \cup \{i\}) > c$, then we delete i from $[n]$. Otherwise, we add i to S. We repeat this process until $[n] \setminus S$ is empty.

It is known from the knapsack problem that, when starting the greedy algorithm as described above with the empty set as initial set, then the produced solution can be arbitrarily bad compared to an optimal solution. However, the greedy algorithm can be turned into a constant-factor approximation by using partial enumeration: For all feasible subsets $U \subseteq [n]$ with $|U| \leq 2$, we run the greedy algorithm starting with U as initial set. In the end we return the best solution set found in this process.

The analysis of the algorithm follows a similar approach as the analysis of Sviridenko [26] for the greedy algorithm for maximizing a submodular function under a linear knapsack constraint. The non-linearity of the constraint in our case makes the analysis more complicated, though. We first need the following technical lemma.

Lemma 5. *Let* $w_0, \ldots, w_m \in \mathbb{N}$ *with* $0 = w_0 < w_1 < \cdots < w_m$, *and let* $\theta_i \geq 0$, $i \in [m]$. *Then,* $\sum_{i=1}^{m} \theta_i(w_i - w_{i-1}) \geq (1 - \frac{\sqrt{3}}{e}) \min_{t=0,\ldots,m-1} \sum_{i=1}^{t} \theta_i(w_i - w_{i-1}) + \theta_{t+1}(w_m + 2\sqrt{w_t w_m})$.

We can now prove the approximation ratio of the greedy algorithm.

Theorem 2. *The Greedy algorithm with partial enumeration is an approximation algorithm with approximation ratio* $1 - \frac{\sqrt{3}}{e}$ *for* (P).

For an upper bound of ϕ on the approximation ratio of the greedy algorithm, we refer to the full version of the paper [13].

5   Monotone Algorithms

To illustrate the need for monotone algorithms, reconsider the situation described in Example 1 with a set of n selfish agents requesting permission to send gas through a pipeline. Each agent j has a private value p_j expressing the monetary gain from being allowed to send the gas. A natural objective of a system provider is to maximize social welfare, i.e., to solve (P). Since the true value p_j is the private information of agent j, the system designer has to employ a mechanism that incentivizes the agents to report their true values p_j.[2] It is without loss of generality [8,18] to assume the following form of a direct revelation mechanism. The mechanism elicits a (potentially misrepresented) bid p_j' from each agent j and computes a solution $x(p') \in \{0,1\}^n$ to (P) based on these values. Further, the mechanism computes a payment g_j for each agent j. The utility of agent j, when their true valuation is p_j and the agents report p', is then $p_j x_j(p') - g_j(p')$. The mechanism is strategyproof if truthtelling is a dominant strategy of each agent j in the underlying game where each agent chooses a value to report.

Myerson [18] shows that an algorithm $\mathcal{A}$ can be turned into a strategyproof mechanism if and only if it is monotone in the following sense. Let $x(p')$ denote the feasible solution to (P) computed by $\mathcal{A}$ as a function of the reported valuations. Then $\mathcal{A}$ is monotone if for all agents j the function $x_j(p')$ is nondecreasing in p_j' for all fixed values p_i' with $i \neq j$. For a monotone algorithm, charging every

---

[2] We here make the standard assumption that the true values of the source vertex s_j, the target vertex t_j, and the quantity of gas q_j are public knowledge. This is reasonable since these values are physically measurable by the system provider so that misreporting them would be pointless for the agent. This assumption is also frequently made in the knapsack auction literature [1,4,17].

agent j with $x_j(p') = 1$ the critical bid $\inf \{z \in \mathbb{R}_{\geq 0} : x_j(z, p'_{-j}) = 1\}$ and charging all other agents nothing yields a strategyproof mechanism. Here, $x_j(z, p'_{-j})$ denotes the binary variable x_j as a function of the bid of agent j, when the bids p'_{-j} of the other agents are fixed.

We note that the algorithms designed in Sects. 3 and 4 are unlikely to be monotone, since the partial enumeration schemes in both of them are not monotone. On the other hand, without the enumeration scheme, they do not provide a constant approximation, even when W is a diagonal matrix. However, by combining ideas from both algorithm, we derive a monotone algorithm with constant approximation guarantee.

Theorem 3. *Algorithm 2 is a monotone α-approximation algorithm for (P), where $\alpha = \left(1 - \frac{\sqrt{3}}{e}\right) / \left(1 + \frac{4}{\sqrt{5}-1}\right) \approx 0.086$. The corresponding critical payments can be computed in polynomial time.*

Algorithm 2: Monotone greedy algorithm

1   $y^* \leftarrow$ solution of (R_1);
2   **if** $\max\limits_{i \in [n]} p_i \geq \left(1 - \frac{\sqrt{3}}{e}\right) / \left(1 + \frac{4}{\sqrt{5}-1}\right) p^\top y^*$ **then**
3   **return** χ_{i^*} for $i^* \in \underset{i \in [n]}{\operatorname{argmax}} \, p_i$;
4   **else**
5   **return** solution of Greedy algorithm without partial enumeration.

Algorithm 3: Randomized rounding

1   $y \leftarrow \varepsilon$-optimal solution of (R^k);
2   **repeat**
3       $x \leftarrow$ realization of $\mathrm{Ber}(\alpha y)$
4   **until** x is feasible for (P^k);
5   **return** x;

---

6   Constantly Many Packing Constraints

In this section we generalize Problem (P) by allowing a constant number of convex quadratic constraints and derive a constant-factor approximation algorithm using randomized rounding combined with partial enumeration. To this end, let $r \in \mathbb{N}$ be a constant natural number, and for every $k \in [r]$ let $W^k = (w^k_{ij})_{i,j \in [n]} \in \mathbb{N}^{n \times n}$ be a symmetric psd matrix with non-negative entries. Furthermore, let $p \in \mathbb{N}^n$ and $c^k \in \mathbb{N}$, $k \in [r]$. We consider packing problems with r convex quadratic knapsack constraints of the form

$$\max \left\{ p^\top x \ : \ x^\top W^k x \leq c^k \text{ for all } k \in [r], \ x \in \{0,1\}^n \right\}. \qquad (P^k)$$

Denote by d^k the vector consisting of the diagonal elements of W^k. We obtain the following convex relaxation of (P^k),

$$\max \left\{ p^\top x \ : \ x^\top W^k x \le c^k, \ (d^k)^\top x \le c^k \text{ for all } k \in [r], \ x \in [0,1]^n \right\}. \quad (R^k)$$

For $\varepsilon > 0$ we call a solution y of (R^k) ε-optimal if $p^\top y \ge (1 - \varepsilon)q^*$, where q^* is the optimal value of (R^k). Convex problems of type (R^k) can be solved ε-optimally in polynomial time by interior points methods [19].

Algorithm 4: Randomized rounding with partial enumeration

---

1 $H_\delta \leftarrow \{i \in [n] : \exists k \in [r] \text{ with } w_{ii}^k > \delta c^k\}$;
2 $z_\delta \leftarrow$ optimal solution of (P^k) with $x_i = 0 \, \forall i \in [n] \setminus H_\delta$ and $|N_1(x)| \le \frac{r}{\delta}$ (via enumeration) ;
3 $y_\delta \leftarrow$ approximate solution of (P^k) with $x_i = 0 \, \forall i \in H_\delta$ computed by randomized rounding (Algorithm 3) with $\alpha = \alpha_\delta$;
4 **return** $\operatorname{argmax}_{x \in \{y_\delta, z_\delta\}} p^\top x$

We proceed to derive an approximation algorithm based on solving (R^k). For some fixed value $\delta \in (0,1)$, we call items i with $w_{ii}^k \le \delta c^k$ for all $k \in [r]$ δ-light. All other items are called δ-heavy. We first assume that all items are δ-light and devise a randomized constant-factor approximation algorithm for (P^k) based on randomized rounding; see Algorithm 3. To that end, for some vector $y \in [0,1]^n$, denote by $\mathrm{Ber}(y)$ the vector of stochastically independent binary random variables $X = (X_1, \ldots, X_n)^\top$ with the property $\mathbb{P}[X_i = 1] = y_i$ and $\mathbb{P}[X_i = 0] = 1 - y_i$, for $i \in [n]$.

Lemma 6. *Let $\delta \in (0,1)$ and assume that all items $i \in [n]$ are δ-light. Let $\varepsilon \in (0,1)$, p^* be the optimal value of (P^k), y be an ε-optimal solution of (R^k), $\alpha \in (0,1)$, and $X = \mathrm{Ber}(\alpha y)$. Then, $\mathbb{E}[p^\top X \mid X \text{ is feasible}] \ge f(\alpha, \delta)(1 - \varepsilon)p^*$, where $f(\alpha, \delta) = \alpha \big(1 - g(\alpha, \delta)\big)^r$ and $g(\alpha, \delta) = \alpha \big(1 + (1 + \delta^{\frac{1}{3}})^3\big) + (1 - \alpha)\delta$.*

Standard calculus shows that Lemma 6 provides the best approximation guarantee for $\alpha_\delta := (1-\delta)/(r+1)(1-\delta+(1+\sqrt[3]{\delta})^3)$. The approximation guarantee approaches $\frac{1}{2(r+1)}\left(\frac{r}{r+1}\right)^r \ge \frac{1}{2e(r+1)}$ as ε and δ go to zero. We proceed to show that for this α, the probability that the random vector $X = \mathrm{Ber}(\alpha y)$ produced by Algorithm 3 is infeasible for (P^k) can be bounded from above by $\frac{1}{2}$.

Lemma 7. *Let y be an optimal solution to (R^k), $\alpha \in (0,1)$, and $X = \mathrm{Ber}(\alpha y)$. Then $\mathbb{P}[X \text{ infeasible for } (P^k)] \le r(\alpha^2 + \alpha)$. In particular, if $\alpha = \alpha_\delta$, then $\mathbb{P}[X \text{ infeasible for } (P^k)] \le \frac{1}{2}$.*

To finish the proof, we show that for any constant $\delta \in (0,1)$, any optimal solution to (P^k) contains a constant number of δ-heavy items only.

Lemma 8. *Let x^* be an optimal solution to problem (P^k), let $\delta \in (0,1)$, and let $H^* := \{i \in [n] : x_i^* = 1 \text{ and } i \text{ is } \delta\text{-heavy}\}$. Then $|H^*| \leq \frac{r}{\delta}$.*

We are now in position to devise a randomized constant-factor approximation algorithm for Problem (P^k). The algorithm first enumerates all solutions using only heavy items, then computes a solution with randomized rounding involving only the light items, and returns the better of the two solutions; see Algorithm 4.

Theorem 4. *For every $\bar{\varepsilon} > 0$, there are $\varepsilon > 0$ and $\delta > 0$ such that Algorithm 4 yields an $(\alpha + \bar{\varepsilon})$-approximation for (P^k) where $\alpha = \frac{1}{1+2(r+1)(\frac{r+1}{r})^r} \geq \frac{1}{1+2e(r+1)}$.*

7   Approximation Hardness

We finally note that packing problems with convex quadratic constraints of type (P) are APX-hard.

Theorem 5. *It is NP-hard to approximate packing problems with convex quadratic constraints by a factor of $\frac{91}{92} + \varepsilon$, for any $\varepsilon > 0$.*

Appendix A – Computational results

We apply our algorithms to gas transportation as described in Example 1, using the GasLib-134 instance [25], see Fig. 2. Sources and sinks are denoted by S and T, resp. Every $t \in T$ has a demand of q_t units of gas. To ensure network robustness in the sense of [14], we assume that all sinks between s_1 and s_2 are supplied by s_1, all sinks between s_3 and t_{45} by s_3, and all other sinks by s_2. Let T_i be the sinks supplied by s_i. For simplicity, we assume that for every $t \in T$, the economic welfare p_t of transporting q_t units of gas to t equals θq_t for $\theta > 0$.

The goal is to choose a welfare-maximal feasible subset of transportations, while the pressures at the first sink s_1 and the last source t_{45} are within their feasible interval. Let $\bar{E}$ denote the path from s_1 to t_{45}, and for every $t \in T_i$ denote by E_t the set of edges on the unique path from s_i to t, $i \in [3]$. Let $p = (p_t)_{t \in T}$, $W = (w_{t,t'})_{t,t' \in T}$, with $w_{t,t'} = \sum_{e \in \bar{E} \cap E_t \cap E_{t'}} \beta_e \, q_t \, q_{t'}$, and let $c = \bar{\pi}_{s_1} - \underline{\pi}_{t_{45}}$, where $\bar{\pi}_v$ and $\underline{\pi}_v$ denote the upper and lower bound on the squared pressure at node v, respectively. Finally, let $x = (x_t)_{t \in T} \in \{0,1\}^T$, where $x_t = 1$ if and only if sink t is supplied. This results in a formulation as (P); see Example 1.

The GasLib-134 instance contains different scenarios, where each scenario provides demands $\hat{q}_t$ for every sink $t \in T$. In order to make the optimization problem non-trivial, we increase the node demands by setting $q_t = \gamma \, \hat{q}_t$, for $\gamma \in \Gamma := \{5, 10, 50, 100\}$. We apply the golden ratio, greedy, and randomized rounding algorithm to the first 100 scenarios, using each $\gamma \in \Gamma$. The algorithms are executed using k initial elements in partial enumeration for each $k \in \{0, 1, 3\}$.

Randomized rounding is run with α chosen uniformly at random from $[0,1]$. Instead of a single feasible realization, we generate 100 feasible realizations of $\text{Ber}(\phi y)$ and return the one with the highest profit. For the golden ratio algorithm, instead of scaling the optimal solution y of (R_1) by ϕ, we scale it by the

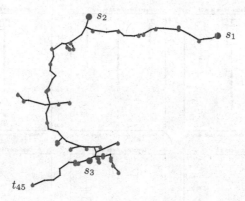

Fig. 2. The Gaslib-134 instance. Sources are shown in blue, sinks in red. (Color figure online)

Table 1. Mean and standard deviation (SD) of the approximation ratio of the greedy algorithm, the golden ratio algorithm, and randomized rounding. Each algorithm has been executed with partial enumeration of k elements.

	$k = 0$		$k = 1$		$k = 3$	
	Mean	SD	Mean	SD	Mean	SD
Greedy	0.927	0.0814	**0.985**	0.0247	**0.999**	0.0034
Golden ratio	0.870	0.1290	0.944	0.0769	0.976	0.0464
Rand. rounding	**0.950**	0.0437	0.984	0.0235	0.995	0.0124

largest number $\lambda \in [\phi, 1]$ such that λy is feasible for (R_2), using binary search. The result of each algorithm is compared to an optimal solution computed with a standard MIP solver. The computations were executed on a 4-core Intel Core i5-2520M processor with 2.5 GHz. The code is implemented in Python 3.6 and we use the SLSQP algorithm of the SciPy `optimize` package to solve (R_1). The results are shown in Table 1 and as box plots in Fig. 3.

The greedy algorithm achieves the best approximation ratios on average when combined with partial enumeration and is at least 20 times faster than the other algorithms, because the latter rely on solving (R_1) first. The approximation ratio of all three algorithms is on average much higher than their proven worst case lower bounds. However, the quality of the solutions produced by the golden ratio algorithm is subject to strong fluctuations. By running the algorithm with partial enumeration with $k = 3$ initial items, the ratio is at least ϕ for every instance, as guaranteed by Theorem 1.

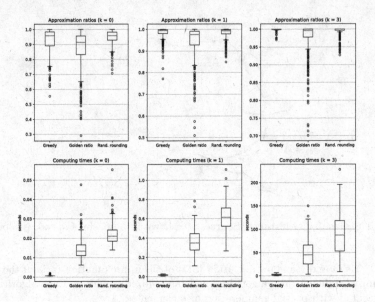

Fig. 3. Approximation ratios (top row) and computation times (bottom row) of the three algorithms when executed with partial enumeration of $k = 0, 1, 3$ elements. The red line indicates the median. (Color figure online)

References

1. Aggarwal, G., Hartline, J.D.: Knapsack auctions. In: Proceedings of 17th Annual ACM-SIAM Symposium Discrete Algorithms (SODA), pp. 1083–1092 (2006)
2. Bansal, N., Kimbrel, T., Pruhs, K.: Dynamic speed scaling to manage energy and temperature. In: Proceedings of 45th Annual IEEE Symposium Foundations Computer Science (FOCS), pp. 520–529 (2004)
3. Berman, A., Shaked-Monderer, N.: Completely Positive Matrices. World Scientific Publishing, Singapore (2003)
4. Briest, P., Krysta, P., Vöcking, B.: Approximation techniques for utilitarian mechanism design. SIAM J. Comput. **40**, 1587–1622 (2011)
5. Chau, C.-K., Elbassioni, K.M., Khonji, M.: Truthful mechanisms for combinatorial allocation of electric power in alternating current electric systems for smart grid. ACM Trans. Econ. Comput. **5** (2016). Art. nr. 7
6. Elbassioni, K.M., Nguyen, T.T.: Approximation algorithms for binary packing problems with quadratic constraints of low cp-rank decompositions. Discrete Appl. Math. **230**, 56–70 (2017)
7. Gallo, G., Hammer, P.L., Simeone, B.: Quadratic knapsack problems. Math. Program. Study **12**, 132–149 (1980)
8. Gibbard, A.: Manipulation of voting schemes: a general result. Econometrica **41**, 587–601 (1973)
9. Håstad, J.: Clique is hard to approximate within $n^{1-\epsilon}$. Acta Mathematica **182**(1), 105–142 (1999)
10. Ibarra, O.H., Kim, C.E.: Fast approximation algorithms for the knapsack and sum of subsets problems. J. ACM **22**, 463–468 (1975)

11. Irani, S., Pruhs, K.R.: Algorithmic problems in power management. SIGACT News **36**(2), 63–76 (2005)
12. Karp, R.M.: Reducibility among combinatorial problems. In: Miller, R.E., Thatcher, J.W., Bohlinger, J.D. (eds.) Complexity of Computer Computations. The IBM Research Symposia Series. Springer, Boston (1972). https://doi.org/10.1007/978-1-4684-2001-2_9
13. Klimm, M., Pfetsch, M.E., Raber, R., Skutella, M.: Packing under convex quadratic constraints. Preprint (2019). arXiv:1912.00468 [math.OC]
14. Klimm, M., Pfetsch, M.E., Raber, R., Skutella, M.: On the robustness of potential-based flow networks. Preprint (2020). https://opus4.kobv.de/opus4-trr154/frontdoor/index/index/docId/309
15. Kozlov, M.K., Tarasov, S.P., Khachiyan, L.G.: The polynomial solvability of convex quadratic programming. USSR Comput. Math. Math. Phys. **20**(5), 223–228 (1980)
16. McCabe, K.A., Rassenti, S.J., Smith, V.L.: Designing 'smart' computer-assisted markets: an experimental auction for gas networks. Eur. J. Polit. Econ. **5**, 259–283 (1989)
17. Mu'alem, A., Nisan, N.: Truthful approximation mechanisms for restricted combinatorial auctions. Games Econ. Behav. **64**, 612–631 (2008)
18. Myerson, R.B.: Optimal auction design. Math. Oper. Res. **6**, 58–73 (1981)
19. Nesterov, Y., Nemirovskii, A.: Interior-Point Polynomial Algorithms in Convex Programming, vol. 13. SIAM (1994)
20. Newbery, D.M.: Network capacity auctions: promise and problems. Utilities Policy **11**, 27–32 (2002)
21. Pferschy, U., Schauer, J.: Approximation of the quadratic knapsack problem. INFORMS J. Comput. **28**, 308–318 (2016)
22. Rader Jr., D.J., Woeginger, G.J.: The quadratic 0–1 knapsack problem with series-parallel support. Oper. Res. Lett. **30**, 159–166 (2002)
23. Rassenti, S.J., Reynolds, S.S., Smit, V.L.: Cotenancy and competition in an experimental auction market for natural gas pipeline networks. Econ. Theory **4**, 41–65 (1994)
24. Sahni, S.: Approximate algorithms for the 0/1 knapsack problem. J. ACM **22**(1), 115–124 (1975)
25. Schmidt, M., et al.: GasLib - a library of gas network instances. Data **2**(4) (2017). Article 40
26. Sviridenko, M.: A note on maximizing a submodular set function subject to a knapsack constraint. Oper. Res. Lett. **32**, 41–43 (2004)
27. Weymouth, T.R.: Problems in natural gas engineering. Trans. Am. Soc. Mech. Eng. **34**, 185–231 (1912)
28. Wierman, A., Andrew, L.L.H., Tang, A.: Power-aware speed scaling in processor sharing systems: optimality and robustness. Perform. Eval. **69**, 601–622 (2012)
29. Woeginger, G.J.: When does a dynamic programming formulation guarantee the existence of a fully polynomial time approximation scheme (FPTAS)? INFORMS J. Comput. **12**, 57–74 (2000)

Weighted Triangle-Free 2-Matching Problem with Edge-Disjoint Forbidden Triangles

Yusuke Kobayashi[✉][iD]

Research Institute for Mathematical Sciences, Kyoto University,
Kyoto 606-8502, Japan
yusuke@kurims.kyoto-u.ac.jp

Abstract. The weighted $\mathcal{T}$-free 2-matching problem is the following
problem: given an undirected graph G, a weight function on its edge set,
and a set $\mathcal{T}$ of triangles in G, find a maximum weight 2-matching containing no triangle in $\mathcal{T}$. When $\mathcal{T}$ is the set of all triangles in G, this problem
is known as the weighted triangle-free 2-matching problem, which is a
long-standing open problem. A main contribution of this paper is to give
a first polynomial-time algorithm for the weighted $\mathcal{T}$-free 2-matching
problem under the assumption that $\mathcal{T}$ is a set of edge-disjoint triangles.
In our algorithm, a key ingredient is to give an extended formulation
representing the solution set, that is, we introduce new variables and represent the convex hull of the feasible solutions as a projection of another
polytope in a higher dimensional space. Although our extended formulation has exponentially many inequalities, we show that the separation
problem can be solved in polynomial time, which leads to a polynomial-
time algorithm for the weighted $\mathcal{T}$-free 2-matching problem.

Keywords: Triangle-free 2-matchings · b-factors · Extended
formulation · Polynomial-time algorithm

1 Introduction

1.1 2-Matchings Without Short Cycles

In an undirected graph, an edge set M is said to be a 2-*matching*[1] if each vertex
is incident to at most two edges in M. Finding a 2-matching of maximum size is
a classical combinatorial optimization problem, which can be solved efficiently

---

[1] Although such an edge set is often called a *simple 2-matching* in the literature, we
call it a 2-*matching* to simplify the description.

Supported by JSPS KAKENHI Grant Numbers JP16K16010, 16H03118, JP18H05291,
and JP19H05485, Japan.

D. Bienstock and G. Zambelli (Eds.): IPCO 2020, LNCS 12125, pp. 280–293, 2020.
https://doi.org/10.1007/978-3-030-45771-6_22

by using a matching algorithm. By imposing restrictions on 2-matchings, various extensions have been introduced and studied in the literature. Among them, the problem of finding a maximum 2-matching without short cycles has attracted attentions, because it has applications to approximation algorithms for TSP and its variants. We say that a 2-matching M is $C_{\leq k}$-*free* if M contains no cycle of length k or less, and the $C_{\leq k}$-*free* 2-*matching problem* is to find a $C_{\leq k}$-free 2-matching of maximum size in a given graph. When $k \leq 2$, every 2-matching without self-loops and parallel edges is $C_{\leq k}$-free, and hence the $C_{\leq k}$-free 2-matching problem can be solved in polynomial time. On the other hand, when $n/2 \leq k \leq n - 1$, where n is the number of vertices in the input graph, the $C_{\leq k}$-free 2-matching problem is NP-hard, because it decides the existence of a Hamiltonian cycle. These facts motivate us to investigate the borderline between polynomially solvable cases and NP-hard cases of the problem. Hartvigsen [12] gave a polynomial-time algorithm for the $C_{\leq 3}$-free 2-matching problem, and Papadimitriou showed that the problem is NP-hard when $k \geq 5$ (see [6]). The polynomial solvability of the $C_{\leq 4}$-free 2-matching problem is still open, whereas some positive results are known for special cases. For the case when the input graph is restricted to be bipartite, Hartvigsen [13], Király [18], and Frank [10] gave min-max theorems, Hartvigsen [14] and Pap [26] designed polynomial-time algorithms, Babenko [1] improved the running time, and Takazawa [28] showed decomposition theorems. Recently, Takazawa [29,30] extended these results to a generalized problem. When the input graph is restricted to be subcubic, i.e., the maximum degree is at most three, Bérczi and Végh [4] gave a polynomial-time algorithm for the $C_{\leq 4}$-free 2-matching problem. Relationship between $C_{\leq k}$-free 2-matchings and jump systems is studied in [3,8,22].

There are a lot of studies also on the weighted version of the $C_{\leq k}$-free 2-matching problem. In the weighted problem, an input consists of a graph and a weight function on the edge set, and the objective is to find a $C_{\leq k}$-free 2-matching of maximum total weight. Király proved that the weighted $C_{\leq 4}$-free 2-matching problem is NP-hard even if the input graph is restricted to be bipartite (see [10]), and a stronger NP-hardness result was shown in [3]. Under the assumption that the weight function satisfies a certain property called *vertex-induced on every square*, Makai [24] gave a polyhedral description and Takazawa [27] designed a combinatorial polynomial-time algorithm for the weighted $C_{\leq 4}$-free 2-matching problem in bipartite graphs. The case of $k = 3$, which we call the *weighted triangle-free* 2-*matching problem*, is a long-standing open problem. For the weighted triangle-free 2-matching problem in subcubic graphs, Hartvigsen and Li [15] gave a polyhedral description and a polynomial-time algorithm, followed by a slight generalized polyhedral description by Bérczi [2] and another polynomial-time algorithm by Kobayashi [19]. Relationship between $C_{\leq k}$-free 2-matchings and discrete convexity is studied in [19,20,22].

1.2   Our Results

The previous papers on the weighted triangle-free 2-matching problem [2,15,19] deal with a generalized problem in which we are given a set T of forbidden triangles as an input in addition to a graph and a weight function. The objective is to find a maximum weight 2-matching that contains no triangle in $\mathcal{T}$, which we call the *weighted $\mathcal{T}$-free 2-matching problem*. In this paper, we focus on the case when $\mathcal{T}$ is a set of edge-disjoint triangles, i.e., no pair of triangles in $\mathcal{T}$ shares an edge in common. A main contribution of this paper is to give a first polynomial-time algorithm for the weighted $\mathcal{T}$-free 2-matching problem under the assumption that $\mathcal{T}$ is a set of edge-disjoint triangles. Note that we impose an assumption only on $\mathcal{T}$, and no restriction is required for the input graph. We now describe the formal statement of our result.

Let $G = (V, E)$ be an undirected graph with vertex set V and edge set E, which might have self-loops and parallel edges. For a vertex set $X \subseteq V$, let $\delta_G(X)$ denote the set of edges between X and $V \setminus X$. For $v \in V$, $\delta_G(\{v\})$ is simply denoted by $\delta_G(v)$. For $v \in V$, let $\dot{\delta}_G(v)$ denote the multiset of edges incident to $v \in V$, that is, a self-loop incident to v is counted twice. We omit the subscript G if no confusion may arise. For $b \in \mathbf{Z}_{\geq 0}^V$, an edge set $M \subseteq E$ is said to be a *b-matching* (resp. *b-factor*) if $\left| M \cap \dot{\delta}(v) \right| \leq b(v)$ (resp. $\left| M \cap \dot{\delta}(v) \right| = b(v)$) for every $v \in V$. If $b(v) = 2$ for every $v \in V$, a b-matching and a b-factor are called a *2-matching* and a *2-factor*, respectively. Let $\mathcal{T}$ be a set of triangles in G, where a *triangle* is a cycle of length three. For a triangle T, let $V(T)$ and $E(T)$ denote the vertex set and the edge set of T, respectively. An edge set $M \subseteq E$ is said to be *$\mathcal{T}$-free* if $E(T) \not\subseteq M$ for every $T \in \mathcal{T}$. For a vertex set $S \subseteq V$, let $E[S]$ denote the set of all edges with both endpoints in S. For an edge weight vector $w \in \mathbf{R}^E$, we consider the problem of finding a $\mathcal{T}$-free b-matching (resp. b-factor) maximizing $w(M)$, which we call the *weighted $\mathcal{T}$-free b-matching (resp. b-factor) problem*. Note that, for a set A and a vector $c \in \mathbf{R}^A$, we denote $c(A) = \sum_{a \in A} c(a)$.

Our main result is formally stated as follows.

Theorem 1. *There exists a polynomial-time algorithm for the following problem: given a graph $G = (V, E)$, $b(v) \in \mathbf{Z}_{\geq 0}$ for each $v \in V$, a set $\mathcal{T}$ of edge-disjoint triangles, and a weight $w(e) \in \mathbf{R}$ for each $e \in E$, find a $\mathcal{T}$-free b-factor $M \subseteq E$ that maximizes the total weight $w(M)$.*

A proof of this theorem is given in Sect. 4. Since finding a maximum weight $\mathcal{T}$-free b-matching can be reduced to finding a maximum weight $\mathcal{T}$-free b-factor by adding dummy vertices and zero-weight edges, Theorem 1 implies the following corollary.

Corollary 1. *There exists a polynomial-time algorithm for the following problem: given a graph $G = (V, E)$, $b(v) \in \mathbf{Z}_{\geq 0}$ for each $v \in V$, a set $\mathcal{T}$ of edge-disjoint triangles, and a weight $w(e) \in \mathbf{R}$ for each $e \in E$, find a $\mathcal{T}$-free b-matching $M \subseteq E$ that maximizes the total weight $w(M)$.*

In particular, we can find a $\mathcal{T}$-free 2-matching (or 2-factor) $M \subseteq E$ that maximizes the total weight $w(M)$ in polynomial time if $\mathcal{T}$ is a set of edge-disjoint triangles.

1.3   Key Ingredient: Extended Formulation

A natural strategy to solve the maximum weight $\mathcal{T}$-free b-factor problem is to give a polyhedral description of the $\mathcal{T}$-free b-factor polytope as Hartvigsen and Li [15] did for the subcubic case. However, as we will see in Example 1, giving a system of inequalities that represents the $\mathcal{T}$-free b-factor polytope seems to be quite difficult even when $\mathcal{T}$ is a set of edge-disjoint triangles. A key idea of this paper is to give an extended formulation of the $\mathcal{T}$-free b-factor polytope, that is, we introduce new variables and represent the $\mathcal{T}$-free b-factor polytope as a projection of another polytope in a higher dimensional space.

Extended formulations of polytopes arising from various combinatorial optimization problems have been intensively studied in the literature, and the main focus in this area is on the number of inequalities that are required to represent the polytope. If a polytope has an extended formulation with polynomially many inequalities, then we can optimize a linear function in the original polytope by the ellipsoid method (see e.g. [11]). On the other hand, even if a linear function on a polytope can be optimized in polynomial time, the polytope does not necessarily have an extended formulation of polynomial size. In this context, the existence of a polynomial size extended formulation has attracted attentions. See survey papers [5,17] for previous work on extended formulations.

In this paper, under the assumption that $\mathcal{T}$ is a set of edge-disjoint triangles, we give an extended formulation of the $\mathcal{T}$-free b-factor polytope that has exponentially many inequalities (Theorem 2). In addition, we show that the separation problem for the extended formulation is solvable in polynomial time, and hence we can optimize a linear function on the $\mathcal{T}$-free b-factor polytope by the ellipsoid method in polynomial time. This yields a first polynomial-time algorithm for the weighted $\mathcal{T}$-free b-factor (or b-matching) problem. Note that it is rare that the first polynomial-time algorithm was designed with the aid of an extended formulation. To the best of our knowledge, the *weighted linear matroid parity problem* was the only such problem before this paper (see [16]).

1.4   Organization of the Paper

The rest of this paper is organized as follows. In Sect. 2, we introduce an extended formulation of the $\mathcal{T}$-free b-factor polytope. The outline of the correctness proof is given in Sect. 3. In Sect. 4, we give a polynomial-time algorithm for the weighted $\mathcal{T}$-factor problem and prove Theorem 1. Finally, we conclude this paper with remarks in Sect. 5. Some of the proofs are omitted due to the space constraint and given in the full version [21].

2   Extended Formulation of the $\mathcal{T}$-free b-factor Polytope

Let $G = (V, E)$ be a graph, $b \in \mathbf{Z}_{\geq 0}^{V}$ be a vector, and $\mathcal{T}$ be a set of forbidden triangles. Throughout this paper, we only consider the case when triangles in $\mathcal{T}$ are mutually edge-disjoint.

284   Y. Kobayashi

For an edge set $M \subseteq E$, define its characteristic vector $x_M \in \mathbf{R}^E$ by

$$x_M(e) = \begin{cases} 1 & \text{if } e \in M, \\ 0 & \text{otherwise.} \end{cases} \tag{1}$$

The $\mathcal{T}$-*free* b-*factor polytope* is defined as $\text{conv}\{x_M \mid M$ is a $\mathcal{T}$-free b-factor in $G\}$, where conv denotes the convex hull of vectors, and the b-*factor polytope* is defined similarly. Edmonds [9] shows that the b-factor polytope is determined by the following inequalities.

$$x(\dot{\delta}(v)) = b(v) \qquad\qquad (v \in V) \tag{2}$$
$$0 \leq x(e) \leq 1 \qquad\qquad (e \in E) \tag{3}$$
$$\sum_{e \in F_0} x(e) + \sum_{e \in F_1} (1 - x(e)) \geq 1 \qquad\qquad ((S, F_0, F_1) \in \mathcal{F}) \tag{4}$$

Here, $\mathcal{F}$ is the set of all triples (S, F_0, F_1) such that $S \subseteq V$, (F_0, F_1) is a partition of $\delta(S)$, and $b(S) + |F_1|$ is odd. Note that $x(\dot{\delta}(v)) = \sum_{e \in \dot{\delta}(v)} x(e)$ and $x(e)$ is added twice if e is a self-loop incident to v.

In order to deal with $\mathcal{T}$-free b-factors, we consider the following constraint in addition to (2)–(4).

$$x(E(T)) \leq 2 \qquad\qquad (T \in \mathcal{T}) \tag{5}$$

However, as we will see in Example 1, the system of inequalities (2)–(5) does not represent the $\mathcal{T}$-free b-factor polytope. Note that when we consider uncapacitated 2-factors, i.e., we are allowed to use two copies of the same edge, it is shown by Cornuejols and Pulleyblank [7] that the $\mathcal{T}$-free uncapacitated 2-factor polytope is represented by $x(e) \geq 0$ for $e \in E$, $x(\dot{\delta}(v)) = 2$ for $v \in V$, and (5).

Example 1. Consider the graph $G = (V, E)$ in Fig. 1. Let $b(v) = 2$ for every $v \in V$ and $\mathcal{T}$ be the set of all triangles in G. Then, G has no $\mathcal{T}$-free b-factor, i.e., the $\mathcal{T}$-free b-factor polytope is empty. For $e \in E$, let $x(e) = 1$ if e is drawn as a blue line in Fig. 1 and let $x(e) = \frac{1}{2}$ otherwise. Then, we can easily check that x satisfies (2), (3), and (5). Furthermore, since x is represented as a linear combination of two b-factors M_1 and M_2 shown in Figs. 2 and 3, x satisfies (4).

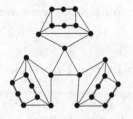
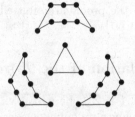
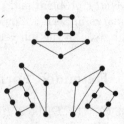

Fig. 1. Graph $G = (V, E)$ (Color figure online)   **Fig. 2.** b-factor M_1 **Fig. 3.** b-factor M_2

In what follows in this section, we introduce new variables and give an extended formulation of the $\mathcal{T}$-free b-factor polytope. For $T \in \mathcal{T}$, we denote $\mathcal{E}_T = \{J \subseteq E(T) | J \neq E(T)\}$. For $T \in \mathcal{T}$ and $J \in \mathcal{E}_T$, we introduce a new variable $y(T, J)$. Roughly, $y(T, J)$ denotes the (fractional) amount of b-factors M satisfying $M \cap E(T) = J$. In particular, when x and y are integral, $y(T, J) = 1$ if and only if the b-factor M corresponding to (x, y) satisfies $M \cap E(T) = J$. We consider the following inequalities.

$$\sum_{J \in \mathcal{E}_T} y(T, J) = 1 \qquad\qquad (T \in \mathcal{T}) \qquad\qquad (6)$$

$$\sum_{e \in J \in \mathcal{E}_T} y(T, J) = x(e) \qquad\qquad (T \in \mathcal{T},\ e \in E(T)) \qquad\qquad (7)$$

$$y(T, J) \geq 0 \qquad\qquad (T \in \mathcal{T},\ J \in \mathcal{E}_T) \qquad\qquad (8)$$

If T is clear from the context, $y(T, J)$ is simply denoted by $y(J)$. Since triangles in $\mathcal{T}$ are edge-disjoint, this causes no ambiguity unless $J = \emptyset$. In addition, for $\alpha, \beta \in E(T)$, $y(\{\alpha\})$, $y(\{\alpha, \beta\})$, and $y(\emptyset)$ are simply denoted by y_α, $y_{\alpha\beta}$, and $y_\emptyset$, respectively.

We now strengthen (4) by using y. For $(S, F_0, F_1) \in \mathcal{F}$, let $\mathcal{T}_S = \{T \in \mathcal{T} | E(T) \cap \delta(S) \neq \emptyset\}$. For $T \in \mathcal{T}_S$ with $E(T) = \{\alpha, \beta, \gamma\}$ and $E(T) \cap \delta(S) = \{\alpha, \beta\}$, we define

$$q^*(T) = \begin{cases} y_\alpha + y_{\alpha\gamma} & \text{if } \alpha \in F_0 \text{ and } \beta \in F_1, \\ y_\beta + y_{\beta\gamma} & \text{if } \beta \in F_0 \text{ and } \alpha \in F_1, \\ y_\emptyset + y_\gamma & \text{if } \alpha, \beta \in F_1, \\ y_{\alpha\beta} & \text{if } \alpha, \beta \in F_0. \end{cases}$$

Note that this value depends on $(S, F_0, F_1) \in \mathcal{F}$ and y, but it is simply denoted by $q^*(T)$ for a notational convenience. We consider the following inequality.

$$\sum_{e \in F_0} x(e) + \sum_{e \in F_1} (1 - x(e)) - \sum_{T \in \mathcal{T}_S} 2q^*(T) \geq 1 \qquad ((S, F_0, F_1) \in \mathcal{F}) \qquad (9)$$

For $T \in \mathcal{T}_S$ with $E(T) = \{\alpha, \beta, \gamma\}$ and $E(T) \cap \delta(S) = \{\alpha, \beta\}$, the contribution of α, β, and T to the left-hand side of (9) is equal to the amount of b-factors M such that $|M \cap \{\alpha, \beta\}| \not\equiv |F_1 \cap \{\alpha, \beta\}| \pmod 2$ by the following observations.

- If $\alpha \in F_0$ and $\beta \in F_1$, then (6) and (7) show that $x(\alpha) = y_\alpha + y_{\alpha\beta} + y_{\alpha\gamma}$ and $1 - x(\beta) = 1 - (y_\beta + y_{\alpha\beta} + y_{\beta\gamma}) = y_\emptyset + y_\alpha + y_\gamma + y_{\alpha\gamma}$. Therefore, $x(\alpha) + (1 - x(\beta)) - 2q^*(T) = y_\emptyset + y_\gamma + y_{\alpha\beta}$, which denotes the amount of b-factors M such that $|M \cap \{\alpha, \beta\}|$ is even.
- If $\beta \in F_0$ and $\alpha \in F_1$, then (6) and (7) show that $(1 - x(\alpha)) + x(\beta) - 2q^*(T) = y_\emptyset + y_\gamma + y_{\alpha\beta}$, which denotes the amount of b-factors M such that $|M \cap \{\alpha, \beta\}|$ is even.
- If $\alpha, \beta \in F_1$, then (6) and (7) show that $(1 - x(\alpha)) + (1 - x(\beta)) - 2q^*(T) = y_\alpha + y_\beta + y_{\alpha\gamma} + y_{\beta\gamma}$, which denotes the amount of b-factors M such that $|M \cap \{\alpha, \beta\}|$ is odd.

– If $\alpha, \beta \in F_0$, then (6) and (7) show that $x(\alpha) + x(\beta) - 2q^*(T) = y_\alpha + y_\beta + y_{\alpha\gamma} + y_{\beta\gamma}$, which denotes the amount of b-factors M such that $|M \cap \{\alpha, \beta\}|$ is odd.

Let P be the polytope defined by

$$P = \{(x, y) \in \mathbf{R}^E \times \mathbf{R}^Y | x \text{ and } y \text{ satisfy } (2), (3), \text{and } (5) - (9)\},$$

where $Y = \{(T, F)|T \in \mathcal{T}, \ F \in \mathcal{E}_T\}$. Note that we do not need (4), because it is implied by (9). Define the projection of P onto E as

$$\text{proj}_E(P) = \{x \in \mathbf{R}^E | \text{There exists } y \in \mathbf{R}^Y \text{ such that } (x, y) \in P\}.$$

Our aim is to show that $\text{proj}_E(P)$ is equal to the $\mathcal{T}$-free b-factor polytope. It is not difficult to see that the $\mathcal{T}$-free b-factor polytope is contained in $\text{proj}_E(P)$.

Lemma 1. *The $\mathcal{T}$-free b-factor polytope is contained in* $\text{proj}_E(P)$.

Proof. Suppose that $M \subseteq E$ is a $\mathcal{T}$-free b-factor in G and define $x_M \in \mathbf{R}^E$ by (1). For $T \in \mathcal{T}$ and $J \in \mathcal{E}_T$, define

$$y_M(T, J) = \begin{cases} 1 & \text{if } M \cap E(T) = J, \\ 0 & \text{otherwise.} \end{cases}$$

We can easily see that (x_M, y_M) satisfies (2), (3), and (5)–(8). Thus, it suffices to show that (x_M, y_M) satisfies (9). Assume to the contrary that (9) does not hold for $(S, F_0, F_1) \in \mathcal{F}$. Then, $x_M(e) = 0$ for every $e \in F_0 \backslash \bigcup_{T \in \mathcal{T}_S} E(T)$ and $x_M(e) = 1$ for every $e \in F_1 \backslash \bigcup_{T \in \mathcal{T}_S} E(T)$. Furthermore, since the contribution of $E(T) \cap \delta(S)$ and T to the left-hand side of (9) is equal to 1 if and only if $|M \cap E(T) \cap \delta(S)| \not\equiv |F_1 \cap E(T)| \pmod 2$, we obtain $|M \cap E(T) \cap \delta(S)| \equiv |F_1 \cap E(T)| \pmod 2$ for every $T \in \mathcal{T}_S$. Then,

$$|M \cap \delta(S)| = |(M \cap \delta(S)) \backslash \bigcup_{T \in \mathcal{T}_S} E(T)| + \sum_{T \in \mathcal{T}_S} |M \cap E(T) \cap \delta(S)|$$

$$\equiv |F_1 \backslash \bigcup_{T \in \mathcal{T}_S} E(T)| + \sum_{T \in \mathcal{T}_S} |F_1 \cap E(T)| = |F_1|.$$

Since M is a b-factor, it holds that $|M \cap \delta(S)| \equiv b(S) \pmod 2$, which contradicts that $b(S) + |F_1|$ is odd. $\square$

To prove the opposite inclusion (i.e., $\text{proj}_E(P)$ is contained in the $\mathcal{T}$-free b-factor polytope), we consider a relaxation of (9). For $T \in \mathcal{T}_S$ with $E(T) = \{\alpha, \beta, \gamma\}$ and $E(T) \cap \delta(S) = \{\alpha, \beta\}$, we define

$$q(T) = \begin{cases} y_\alpha + y_{\alpha\gamma} & \text{if } \alpha \in F_0 \text{ and } \beta \in F_1, \\ y_\beta + y_{\beta\gamma} & \text{if } \beta \in F_0 \text{ and } \alpha \in F_1, \\ y_\gamma & \text{if } \alpha, \beta \in F_1, \\ 0 & \text{if } \alpha, \beta \in F_0. \end{cases}$$

Since $q(T) \leq q^*(T)$ for every $T \in \mathcal{T}_S$, the following inequality is a relaxation of (9).

$$\sum_{e \in F_0} x(e) + \sum_{e \in F_1} (1 - x(e)) - \sum_{T \in \mathcal{T}_S} 2q(T) \geq 1 \qquad ((S, F_0, F_1) \in \mathcal{F}) \qquad (10)$$

Define a polytope Q and its projection onto E as

$$Q = \{(x, y) \in \mathbf{R}^E \times \mathbf{R}^Y \mid x \text{ and } y \text{ satisfy } (2), (3), (5) - (8), \text{ and } (10)\},$$

$$\text{proj}_E(Q) = \{x \in \mathbf{R}^E \mid \text{There exists } y \in \mathbf{R}^Y \text{ such that } (x, y) \in Q\}.$$

Since (10) is implied by (9), we have that $P \subseteq Q$ and $\text{proj}_E(P) \subseteq \text{proj}_E(Q)$. In addition, we show the following proposition whose proof outline is given in Sect. 3.

Proposition 1. $\text{proj}_E(Q)$ *is contained in the $\mathcal{T}$-free b-factor polytope.*

By Lemma 1, Proposition 1, and $\text{proj}_E(P) \subseteq \text{proj}_E(Q)$, we obtain the following theorem.

Theorem 2. *Let $G = (V, E)$ be a graph, $b(v) \in \mathbf{Z}_{\geq 0}$ for each $v \in V$, and let $\mathcal{T}$ be a set of edge-disjoint triangles. Then, both $\text{proj}_E(P)$ and $\text{proj}_E(Q)$ are equal to the $\mathcal{T}$-free b-factor polytope.*

We remark here that we do not know how to prove directly that $\text{proj}_E(P)$ is contained in the $\mathcal{T}$-free b-factor polytope. Introducing $\text{proj}_E(Q)$ and considering Proposition 1, which is a stronger statement, is a key idea in our proof. We also note that our algorithm in Sect. 4 is based on the fact that the $\mathcal{T}$-free b-factor polytope is equal to $\text{proj}_E(P)$. In this sense, both $\text{proj}_E(P)$ and $\text{proj}_E(Q)$ play important roles in this paper.

Example 2. Suppose that $G = (V, E)$, $b \in \mathbf{Z}_{\geq 0}^V$, and $x \in \mathbf{R}^E$ are as in Example 1. Let T be the central triangle in G and let $E(T) = \{\alpha, \beta, \gamma\}$. If $y \in \mathbf{R}^Y$ satisfies (6) and (8), then $y_{\alpha\beta} + y_{\beta\gamma} + y_{\alpha\gamma} \leq 1$. Thus, without loss of generality, we may assume that $y_{\alpha\beta} \leq \frac{1}{3}$ by symmetry. Let S be a vertex set with $\delta(S) = \{\alpha, \beta\}$. Then, (10) does not hold for $(S, \{\alpha\}, \{\beta\}) \in \mathcal{F}$, because $x(\alpha) + (1 - x(\beta)) - 2q(T) = 1 - x(\alpha) - x(\beta) + 2y_{\alpha\beta} \leq \frac{2}{3} < 1$. Therefore, x is not in $\text{proj}_E(Q)$.

3   Outline of the Proof of Proposition 1

In this section, we describe the outline of the proof of Proposition 1. In our proof, we use the following lemma whose proof is given in Appendix A.

Lemma 2. *Let x be an extreme point of $\text{proj}_E(Q)$. Then, one of the following holds.*

(i) $x = x_M$ for some $\mathcal{T}$-free b-factor $M \subseteq E$.
(ii) (5) is tight for some $T \in \mathcal{T}$.

(iii) There exists a vector $y \in \mathbf{R}^Y$ with $(x, y) \in Q$ such that (10) is tight for some $(S, F_0, F_1) \in \mathcal{F}$ with $\mathcal{T}_S^+ \neq \emptyset$, where we define $\mathcal{T}_S^+ = \{T \in \mathcal{T} | E(T) \cap \delta(S) \cap F_1 \neq \emptyset\}$.

We prove Proposition 1 by induction on $|\mathcal{T}|$. If $|\mathcal{T}| = 0$, then y does not exist and (10) is equivalent to (4). Thus, $\mathrm{proj}_E(Q)$ is the b-factor polytope, which shows the base case of the induction.

Fix an instance $(G, b, \mathcal{T})$ with $|\mathcal{T}| \geq 1$ and assume that Proposition 1 holds for instances with smaller $|\mathcal{T}|$. Suppose that $Q \neq \emptyset$, which implies that $b(V)$ is even as $(V, \emptyset, \emptyset) \notin \mathcal{F}$ by (10). Pick up an extreme point x of $\mathrm{proj}_E(Q)$ and let $y \in \mathbf{R}^Y$ be a vector with $(x, y) \in Q$. Our aim is to show that x is contained in the $\mathcal{T}$-free b-factor polytope.

We apply Lemma 2 to obtain one of (i), (ii), and (iii). If (i) holds, that is, $x = x_M$ for some $\mathcal{T}$-free b-factor $M \subseteq E$, then x is obviously in the $\mathcal{T}$-free b-factor polytope. If (ii) holds, that is, (5) is tight for some $T \in \mathcal{T}$, then we replace T with a certain graph and apply the induction. If (iii) holds, that is, (10) is tight for some $(S, F_0, F_1) \in \mathcal{F}$ with $\mathcal{T}_S^+ \neq \emptyset$, then we divide G into two graphs, where one corresponds to S and the other corresponds to $V \backslash S$, apply the induction for each graph, and merge them. See the full version [21] for a complete proof.

4   Algorithm

In this section, we give a polynomial-time algorithm for the weighted $\mathcal{T}$-free b-factor problem and prove Theorem 1. Our algorithm is based on the ellipsoid method using the fact that the $\mathcal{T}$-free b-factor polytope is equal to $\mathrm{proj}_E(P)$ (Theorem 2). In order to apply the ellipsoid method, we need a polynomial-time algorithm for the separation problem. That is, for $(x, y) \in \mathbf{R}^E \times \mathbf{R}^Y$, we need a polynomial-time algorithm that concludes $(x, y) \in P$ or returns a violated inequality.

Let $(x, y) \in \mathbf{R}^E \times \mathbf{R}^Y$. We can easily check whether (x, y) satisfies (2), (3), and (5)–(8) or not in polynomial time. In order to solve the separation problem for (9), we use the following theorem, which implies that the separation problem for (4) can be solved in polynomial time.

Theorem 3 (Padberg-Rao [25] (see also [23])). *Suppose we are given a graph $G' = (V', E')$, $b' \in \mathbf{Z}_{\geq 0}^{V'}$, and $x' \in [0, 1]^{E'}$. Then, in polynomial time, we can compute $S' \subseteq V'$ and a partition (F_0', F_1') of $\delta_{G'}(S')$ that minimize $\sum_{e \in F_0'} x'(e) + \sum_{e \in F_1'} (1 - x'(e))$ subject to $b'(S') + |F_1'|$ is odd.*

In what follows, we reduce the separation problem for (9) to that for (4) and utilize Theorem 3. Suppose that $(x, y) \in \mathbf{R}^E \times \mathbf{R}^Y$ satisfies (2), (3), and (5)–(8). For each triangle $T \in \mathcal{T}$, we remove $E(T)$ and add a vertex r_T together with three new edges $e_1 = r_T v_1$, $e_2 = r_T v_2$, and $e_3 = r_T v_3$ (Fig. 4). Let $E_T' = \{e_1, e_2, e_3\}$ and define $x'(e_1) = x(\alpha) + x(\gamma) - 2y_{\alpha\gamma}$, $x'(e_2) = x(\alpha) + x(\beta) - 2y_{\alpha\beta}$, and $x'(e_3) = x(\beta) + x(\gamma) - 2y_{\beta\gamma}$. Let $G' = (V', E')$ be the graph obtained from G

by applying this procedure for every $T \in \mathcal{T}$. Define $b' \in \mathbf{Z}_{\geq 0}^{V'}$ as $b'(v) = b(v)$ for $v \in V$ and $b'(v) = 0$ for $v \in V' \backslash V$. By setting $x'(e) = x(e)$ for $e \in E' \cap E$ and by defining $x'(e)$ as above for $e \in E' \backslash E$, we obtain $x' \in [0, 1]^{E'}$. Then, we can show the following lemma whose proof is given in Appendix B.

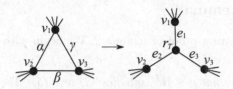

Fig. 4. Replacement of a triangle $T \in \mathcal{T}$

Lemma 3. *Suppose that* $(x, y) \in \mathbf{R}^E \times \mathbf{R}^Y$ *satisfies (2), (3), and (5)–(8). Define* $G' = (V', E')$, b', *and* x' *as above. Then,* (x, y) *violates (9) for some* $(S, F_0, F_1) \in \mathcal{F}$ *if and only if there exist* $S' \subseteq V'$ *and a partition* (F_0', F_1') *of* $\delta_{G'}(S')$ *such that* $b'(S') + |F_1'|$ *is odd and* $\sum_{e \in F_0'} x'(e) + \sum_{e \in F_1'} (1 - x'(e)) < 1$.

Since our proof of Lemma 3 is constructive, given $S' \subseteq V'$ and $F_0', F_1' \subseteq E'$ such that (F_0', F_1') is a partition of $\delta_{G'}(S')$, $b'(S') + |F_1'|$ is odd, and $\sum_{e \in F_0'} x'(e) + \sum_{e \in F_1'} (1 - x'(e)) < 1$, we can construct $(S, F_0, F_1) \in \mathcal{F}$ for which (x, y) violates (9) in polynomial time. By combining this with Theorem 3, it holds that the separation problem for P can be solved in polynomial time. Therefore, the ellipsoid method can maximize a linear function on P in polynomial time (see e.g. [11]), and hence we can maximize $\sum_{e \in E} w(e)x(e)$ subject to $x \in \text{proj}_E(P)$. By perturbing the objective function if necessary, we can obtain a maximizer x^* that is an extreme point of $\text{proj}_E(P)$. Since each extreme point of $\text{proj}_E(P)$ corresponds to a $\mathcal{T}$-free b-factor by Theorem 2, x^* is a characteristic vector of a maximum weight $\mathcal{T}$-free b-factor. This completes the proof of Theorem 1.

5   Concluding Remarks

This paper gives a first polynomial-time algorithm for the weighted $\mathcal{T}$-free b-matching problem where $\mathcal{T}$ is a set of edge-disjoint triangles. A key ingredient is an extended formulation of the $\mathcal{T}$-free b-factor polytope with exponentially many inequalities. As we mentioned in Sect. 1.3, it is rare that the first polynomial-time algorithm was designed with the aid of an extended formulation. This approach has a potential to be used for other combinatorial optimization problems for which no polynomial-time algorithm is known.

Some interesting problems remain open. Since the algorithm proposed in this paper relies on the ellipsoid method, it is natural to ask whether we can design a combinatorial polynomial-time algorithm. It is also open whether our approach can be applied to the weighted $C_{\leq 4}$-free b-matching problem in general graphs under the assumption that the forbidden cycles are edge-disjoint and the weight is vertex-induced on every square. In addition, the weighted $C_{\leq 3}$-free 2-matching problem and the $C_{\leq 4}$-free 2-matching problem are big open problems in this area.

A   Proof of Lemma 2

In this section, we give a proof of Lemma 2. We begin with the following easy lemma.

Lemma 4. *Suppose that $x \in \mathbf{R}^E$ satisfies (3) and (5). Then, there exists $y \in \mathbf{R}^Y$ that satisfies (6)–(8).*

Proof. Let $T \in \mathcal{T}$ be a triangle with $E(T) = \{\alpha, \beta, \gamma\}$ and $x(\alpha) \geq x(\beta) \geq x(\gamma)$. For $J \in \mathcal{E}_T$, we define $y(T, J)$ as follows.

- If $x(\alpha) \geq x(\beta) + x(\gamma)$, then $y_{\alpha\beta} = x(\beta)$, $y_{\alpha\gamma} = x(\gamma)$, $y_\emptyset = 1 - x(\alpha)$, $y_\alpha = x(\alpha) - x(\beta) - x(\gamma)$, and $y_\beta = y_\gamma = y_{\beta\gamma} = 0$.
- If $x(\alpha) < x(\beta) + x(\gamma)$, then $y_{\alpha\beta} = \frac{1}{2}(x(\alpha) + x(\beta) - x(\gamma))$, $y_{\alpha\gamma} = \frac{1}{2}(x(\alpha) + x(\gamma) - x(\beta))$, $y_{\beta\gamma} = \frac{1}{2}(x(\beta) + x(\gamma) - x(\alpha))$, $y_\emptyset = 1 - \frac{1}{2}(x(\alpha) + x(\beta) + x(\gamma))$, and $y_\alpha = y_\beta = y_\gamma = 0$.

Then, y satisfies (6)–(8). $\qquad\square$

By using this lemma, we can prove Lemma 2.

Proof (Proof of Lemma 2). We prove (i) by assuming that (ii) and (iii) do not hold. Since (10) is not tight for any $(S, F_0, F_1) \in \mathcal{F}$ with $\mathcal{T}_S^+ \neq \emptyset$, x is an extreme point of

$$\{x \in \mathbf{R}^E | \text{There exists } y \in \mathbf{R}^Y \text{ such that } (x, y) \text{ satisfies } (2) - (8)\},$$

because (4) is a special case of (10) in which $\mathcal{T}_S^+ = \emptyset$. By Lemma 4, this polytope is equal to $\{x \in \mathbf{R}^E | x \text{ satisfies } (2) - (5)\}$. Since (5) is not tight for any $T \in \mathcal{T}$, x is an extreme point of $\{x \in \mathbf{R}^E | x \text{ satisfies } (2) - (4)\}$, which is the b-factor polytope. Thus, x is a characteristic vector of a b-factor. Since x satisfies (5), it holds that $x = x_M$ for some $\mathcal{T}$-free b-factor $M \subseteq E$. $\qquad\square$

B   Proof of Lemma 3

In this section, we give a proof of Lemma 3.

First, to show the "only if" part, assume that (x, y) violates (9) for some $(S, F_0, F_1) \in \mathcal{F}$. Recall that $\mathcal{T}_S = \{T \in \mathcal{T} | E(T) \cap \delta_G(S) \neq \emptyset\}$. Define $S' \subseteq V'$ by $S' = S \cup \{r_T | T \in \mathcal{T}, |V(T) \cap S| \geq 2\}$. Then, for each $T \in \mathcal{T}_S$, $E'_T \cap \delta_{G'}(S')$ consists of a single edge, which we denote e_T. Define F'_0 and F'_1 as follows:

$$F'_0 = (F_0 \cap E') \cup \{e_T | T \in \mathcal{T}_S, |E(T) \cap F_1| = 0 \text{ or } 2\},$$
$$F'_1 = (F_1 \cap E') \cup \{e_T | T \in \mathcal{T}_S, |E(T) \cap F_1| = 1\}.$$

It is obvious that (F_0', F_1') is a partition of $\delta_{G'}(S')$ and $b'(S') + |F_1'| \equiv b(S) + |F_1| \equiv 1 \pmod{2}$.

To show that $\sum_{e \in F_0'} x'(e) + \sum_{e \in F_1'} (1 - x'(e)) < 1$, we evaluate $x'(e_T)$ or $1 - x'(e_T)$ for each $T \in \mathcal{T}_S$. Let $T \in \mathcal{T}_S$ be a triangle such that $E(T) = \{\alpha, \beta, \gamma\}$ and $E(T) \cap \delta_G(S) = \{\alpha, \beta\}$. Then, we obtain the following by the definition of $q^*(T)$.

- If $T \in \mathcal{T}_S$ and $\alpha, \beta \in F_0$, then $x(\alpha) + x(\beta) - 2q^*(T) = x'(e_T)$.
- If $T \in \mathcal{T}_S$ and $\alpha, \beta \in F_1$, then $(1 - x(\alpha)) + (1 - x(\beta)) - 2q^*(T) = x'(e_T)$.
- If $T \in \mathcal{T}_S$, $\alpha \in F_0$, and $\beta \in F_1$, then $x(\alpha) + (1 - x(\beta)) - 2q^*(T) = y_\emptyset + y_\gamma + y_{\alpha\beta} = 1 - x'(e_T)$.
- If $T \in \mathcal{T}_S$, $\beta \in F_0$, and $\alpha \in F_1$, then $(1 - x(\alpha)) + x(\beta) - 2q^*(T) = y_\emptyset + y_\gamma + y_{\alpha\beta} = 1 - x'(e_T)$.

With these observations, we obtain

$$\sum_{e \in F_0'} x'(e) + \sum_{e \in F_1'} (1 - x'(e)) = \sum_{e \in F_0} x(e) + \sum_{e \in F_1} (1 - x(e)) - \sum_{T \in \mathcal{T}_S} 2q^*(T) < 1,$$

which shows the "only if" part.

We next show the "if" part. For edge sets $F_0', F_1' \subseteq E'$, we denote $g(F_0', F_1') = \sum_{e \in F_0'} x'(e) + \sum_{e \in F_1'} (1 - x'(e))$ to simplify the notation. Let (S', F_0', F_1') be a minimizer of $g(F_0', F_1')$ subject to (F_0', F_1') is a partition of $\delta_{G'}(S')$ and $b'(S') + |F_1'|$ is odd. Among minimizers, we choose (S', F_0', F_1') so that $F_0' \cup F_1'$ is inclusion-wise minimal. To derive a contradiction, assume that $g(F_0', F_1') < 1$. We can show the following claim by a case analysis (see [21] for a proof).

Claim. Let $T \in \mathcal{T}$ be a triangle as shown in Fig. 4 and denote $\hat{F}_0 = F_0' \cap E_T'$ and $\hat{F}_1 = F_1' \cap E_T'$. Then, we obtain the following.

(i) If $v_1, v_2, v_3 \notin S'$, then $r_T \notin S'$.
(ii) If $v_1, v_2, v_3 \in S'$, then $r_T \in S'$.
(iii) If $v_1 \in S'$, $v_2, v_3 \notin S'$, and $\left| \hat{F}_1 \right|$ is even, then $g(\hat{F}_0, \hat{F}_1) = x'(e_1) = x(\alpha) + x(\gamma) - 2y_{\alpha\gamma}$.
(iv) If $v_1 \in S'$, $v_2, v_3 \notin S'$, and $\left| \hat{F}_1 \right|$ is odd, then $g(\hat{F}_0, \hat{F}_1) = 1 - x'(e_1) = y_\emptyset + y_\beta + y_{\alpha\gamma}$.

Note that each $T \in \mathcal{T}$ satisfies exactly one of (i)–(iv) by changing the labels of v_1, v_2, and v_3 if necessary. In what follows, we construct $(S, F_0, F_1) \in \mathcal{F}$ for which (x, y) violates (9). We initialize (S, F_0, F_1) as

$$S = S' \cap V, \qquad F_0 = F_0' \cap E, \qquad F_1 = F_1' \cap E,$$

and apply the following procedures for each triangle $T \in \mathcal{T}$.

- If T satisfies the condition (i) or (ii), then we do nothing.
- If T satisfies the condition (iii), then we add α and γ to F_0.

– If T satisfies the condition (iv), then we add α to F_0 and add γ to F_1.

Then, we obtain that (F_0, F_1) is a partition of $\delta_G(S)$, $b(S) + |F_1| \equiv b'(S') + |F_1'| \equiv 1 \pmod 2$, and

$$\sum_{e \in F_0} x(e) + \sum_{e \in F_1} (1 - x(e)) - \sum_{T \in \mathcal{T}_S} 2q^*(T) = \sum_{e \in F_0'} x'(e) + \sum_{e \in F_1'} (1 - x'(e)) < 1$$

by the above claim. This shows that (x, y) violates (9) for $(S, F_0, F_1) \in \mathcal{F}$, which completes the proof of the "if" part.   □

References

1. Babenko, M.A.: Improved algorithms for even factors and square-free simple b-matchings. Algorithmica **64**(3), 362–383 (2012). https://doi.org/10.1007/s00453-012-9642-6
2. Bérczi, K.: The triangle-free 2-matching polytope of subcubic graphs. Technical report TR-2012-2, Egerváry Research Group (2012)
3. Bérczi, K., Kobayashi, Y.: An algorithm for $(n - 3)$-connectivity augmentation problem: jump system approach. J. Comb. Theory Ser. B **102**(3), 565–587 (2012). https://doi.org/10.1016/j.jctb.2011.08.007
4. Bérczi, K., Végh, L.A.: Restricted b-matchings in degree-bounded graphs. In: Eisenbrand, F., Shepherd, F.B. (eds.) IPCO 2010. LNCS, vol. 6080, pp. 43–56. Springer, Heidelberg (2010). https://doi.org/10.1007/978-3-642-13036-6_4
5. Conforti, M., Cornuéjols, G., Zambelli, G.: Extended formulations in combinatorial optimization. 4OR **8**(1), 1–48 (2010). https://doi.org/10.1007/s10288-010-0122-z
6. Cornuéjols, G., Pulleyblank, W.: A matching problem with side conditions. Discrete Math. **29**(2), 135–159 (1980). https://doi.org/10.1016/0012-365x(80)90002-3
7. Cornuejols, G., Pulleyblank, W.R.: Perfect triangle-free 2-matchings. In: Rayward-Smith, V.J. (ed.) Mathematical Programming Studies, vol. 13, pp. 1–7. Springer, Heidelberg (1980). https://doi.org/10.1007/bfb0120901
8. Cunningham, W.H.: Matching, matroids, and extensions. Math. Program. **91**(3), 515–542 (2001). https://doi.org/10.1007/s101070100256
9. Edmonds, J.: Maximum matching and a polyhedron with 0, 1-vertices. J. Res. Nat. Bur. Stand. B **69**, 125–130 (1965)
10. Frank, A.: Restricted t-matchings in bipartite graphs. Discrete Appl. Math. **131**(2), 337–346 (2003). https://doi.org/10.1016/s0166-218x(02)00461-4
11. Grötschel, M., Lovász, L., Schrijver, A.: Geometric Algorithms and Combinatorial Optimization. Algorithms and Combinatorics, vol. 2. Springer, Heidelberg (1988). https://doi.org/10.1007/978-3-642-78240-4
12. Hartvigsen, D.: Extensions of Matching Theory. Ph.D. thesis, Carnegie Mellon University (1984)
13. Hartvigsen, D.: The square-free 2-factor problem in bipartite graphs. In: Cornuéjols, G., Burkard, R.E., Woeginger, G.J. (eds.) IPCO 1999. LNCS, vol. 1610, pp. 234–241. Springer, Heidelberg (1999). https://doi.org/10.1007/3-540-48777-8_18
14. Hartvigsen, D.: Finding maximum square-free 2-matchings in bipartite graphs. J. Comb. Theor. Ser. B **96**(5), 693–705 (2006). https://doi.org/10.1016/j.jctb.2006.01.004

15. Hartvigsen, D., Li, Y.: Polyhedron of triangle-free simple 2-matchings in subcubic graphs. Math. Program. **138**(1–2), 43–82 (2012). https://doi.org/10.1007/s10107-012-0516-0

16. Iwata, S., Kobayashi, Y.: A weighted linear matroid parity algorithm. In: Proceedings of the 49th Annual ACM SIGACT Symposium on Theory of Computing - STOC 2017, pp. 264–276. ACM Press (2017). https://doi.org/10.1145/3055399.3055436

17. Kaibel, V.: Extended formulations in combinatorial optimization. Technical report (2011). arXiv:1104.1023

18. Király, Z.: C_4-free 2-factors in bipartite graphs. Technical report TR-2012-2, Egerváry Research Group (1999)

19. Kobayashi, Y.: A simple algorithm for finding a maximum triangle-free 2-matching in subcubic graphs. Discrete Optim. **7**(4), 197–202 (2010). https://doi.org/10.1016/j.disopt.2010.04.001

20. Kobayashi, Y.: Triangle-free 2-matchings and M-concave functions on jump systems. Discrete Appl. Math. **175**, 35–42 (2014). https://doi.org/10.1016/j.dam.2014.05.016

21. Kobayashi, Y.: Weighted triangle-free 2-matching problem with edge-disjoint forbidden triangles (2019). arXiv:1911.06436

22. Kobayashi, Y., Szabó, J., Takazawa, K.: A proof of Cunningham's conjecture on restricted subgraphs and jump systems. J. Comb. Theor. Ser. B **102**(4), 948–966 (2012). https://doi.org/10.1016/j.jctb.2012.03.003

23. Letchford, A.N., Reinelt, G., Theis, D.O.: Odd minimum cut sets and b-matchings revisited. SIAM J. Discrete Math. **22**(4), 1480–1487 (2008). https://doi.org/10.1137/060664793

24. Makai, M.: On maximum cost $K_{t,t}$-free t-matchings of bipartite graphs. SIAM J. Discrete Math. **21**(2), 349–360 (2007). https://doi.org/10.1137/060652282

25. Padberg, M.W., Rao, M.R.: Odd minimum cut-sets and b-matchings. Math. Oper. Res. **7**(1), 67–80 (1982)

26. Pap, G.: Combinatorial algorithms for matchings, even factors and square-free 2-factors. Math. Program. **110**(1), 57–69 (2007). https://doi.org/10.1007/s10107-006-0053-9

27. Takazawa, K.: A weighted $K_{t,t}$-free t-factor algorithm for bipartite graphs. Math. Oper. Res. **34**(2), 351–362 (2009). https://doi.org/10.1287/moor.1080.0365

28. Takazawa, K.: Decomposition theorems for square-free 2-matchings in bipartite graphs. Discrete Appl. Math. **233**, 215–223 (2017). https://doi.org/10.1016/j.dam.2017.07.035

29. Takazawa, K.: Excluded t-factors in bipartite graphs: a unified framework for non-bipartite matchings and restricted 2-matchings. In: Eisenbrand, F., Koenemann, J. (eds.) IPCO 2017. LNCS, vol. 10328, pp. 430–441. Springer, Cham (2017). https://doi.org/10.1007/978-3-319-59250-3_35

30. Takazawa, K.: Finding a maximum 2-matching excluding prescribed cycles in bipartite graphs. Discrete Optim. **26**, 26–40 (2017). https://doi.org/10.1016/j.disopt.2017.05.003

Single Source Unsplittable Flows
with Arc-Wise Lower and Upper Bounds

Sarah Morell and Martin Skutella[✉]

Combinatorial Optimization & Graph Algorithms Group, Institut für Mathematik,
Technische Universität Berlin, Berlin, Germany
{morell,skutella}@math.tu-berlin.de

Abstract. In a digraph with a source and several destination nodes
with associated demands, an unsplittable flow routes each demand along
a single path from the common source to its destination. Given some
flow x that is not necessarily unsplittable but satisfies all demands, it is
a natural question to ask for an unsplittable flow y that does not deviate
from x by too much, i.e., $y_a \approx x_a$ for all arcs a. Twenty years ago, in a
landmark paper, Dinitz, Garg, and Goemans [3] proved that there is an
unsplittable flow y such that $y_a \leq x_a + d_{\max}$ for all arcs a, where $d_{\max}$
denotes the maximum demand value.

Our first contribution is a considerably simpler one-page proof for this
classical result, based upon an entirely new approach. Secondly, using a
subtle variant of this approach, we obtain a new result: There is an
unsplittable flow y such that $y_a \geq x_a - d_{\max}$ for all arcs a. Finally,
building upon an iterative rounding technique previously introduced by
Kolliopoulos and Stein [10] and Skutella [15], we prove existence of
an unsplittable flow that simultaneously satisfies the upper and lower
bounds for the special case when demands are integer multiples of each
other. For arbitrary demand values, we prove the slightly weaker simultaneous bounds $x_a/2 - d_{\max} \leq y_a \leq 2x_a + d_{\max}$ for all arcs a.

1 Introduction

Ever since the seminal work of Ford and Fulkerson [5] network flows belong to the
most important and fundamental class of problems in combinatorial optimization
and mathematical programming. We refer to the classical textbook [1] by Ahuja,
Magnanti, and Orlin as well as the very recent new textbook [16] by Williamson
on the topic.

Problem Setting and Notation. Let $D = (V, A)$ be a directed acyclic graph
with source node $s \in V$ and k commodities with destination nodes $t_1, \ldots, t_k \in V$
and associated demands $d_1, \ldots, d_k \in \mathbb{R}_{>0}$. A flow $x \in \mathbb{R}_{\geq 0}^A$ *satisfies the given*

Partially supported by DFG Priority Programme 1736 (grant SK 58/10-2).

D. Bienstock and G. Zambelli (Eds.): IPCO 2020, LNCS 12125, pp. 294–306, 2020.
https://doi.org/10.1007/978-3-030-45771-6_23

demands if it simultaneously sends d_i units of flow from s to t_i, for all $i = 1, \ldots, k$. That is, x must satisfy the following flow conservation constraints:

$$x(\delta^{\text{in}}(v)) - x(\delta^{\text{out}}(v)) = \begin{cases} d_i & \text{if } v = t_i \text{ for some } i \in \{1, \ldots, k\}, \\ -\sum_{i=1}^{k} d_i & \text{if } v = s, \\ 0 & \text{otherwise.} \end{cases} \quad (1)$$

Here, $\delta^{\text{in}}(v)$ and $\delta^{\text{out}}(v)$ denote the set of incoming and outgoing arcs of node v, respectively; for $B \subseteq A$, let $x(B) := \sum_{a \in B} x_a$. In the following, whenever we refer to a *flow*, we mean a flow satisfying the given demands, i.e., Constraints (1), unless stated otherwise.

The following classical integrality property of network flows follows, for example, from the fact that the node-arc incidence matrix, which implicitly occurs on the left-hand side of (1), is totally unimodular.

Theorem 1. *If the demands $d_1, \ldots, d_k$ are all integral, then any flow x can be written as a convex combination of integral flows such that each such integral flow $y \in \mathbb{Z}_{\geq 0}^A$ satisfies*

$$\lfloor x_a \rfloor \leq y_a \leq \lceil x_a \rceil \qquad \text{for all } a \in A.$$

In particular, there exists an integral flow obeying these upper and lower bounds.

Single Source Unsplittable Flows. In 1996, Kleinberg [6] introduced *single source unsplittable flows*. A flow is called *unsplittable* if the entire demand of each commodity is routed along one path from the source to its destination node. That is, an unsplittable flow y can be specified as follows: for all $i = 1, \ldots, k$, there is one s-t_i-path P_i^y in D such that

$$y_a = \sum_{i : a \in P_i^y} d_i \qquad \text{for all } a \in A.$$

In order to emphasize the fact that a particular flow x is not necessarily unsplittable, we sometimes refer to x as a *fractional* flow in this case. Notice that for the special case of unit demands, a flow is unsplittable if and only if it is integral.

Related Literature. Single source unsplittable flows constitute a special case of more general unsplittable flows where each commodity has its own source and destination node. General unsplittable flows have been well studied in the literature as an interesting extension of disjoint paths. For instance, if we are given arc capacities and demands for each commodity and look for an unsplittable flow of minimum congestion, i.e., of minimum overload of arc capacities, Raghavan and Thompson [12,13] present an approximation algorithm based on their randomized rounding technique. We refer to the survey [9] by Kolliopoulos for further results on general unsplittable flows.

The problem of finding a single source unsplittable flow in a directed graph with capacities on the arcs contains several well-known NP-complete problems as special cases, e.g., Partition, Bin Packing, or even scheduling parallel machines with makespan objective; we refer to [6] for more details and other special cases.

Kleinberg [6], Dinitz, Garg, and Goemans [3], Kolliopoulos and Stein [10], and Skutella [15] present approximation algorithms for various optimization versions of the single source unsplittable flow problem. Du and Kolliopoulos [4] have implemented and empirically tested several of those approximation algorithms.

Baier et al. [2] introduce the following relaxation of unsplittable flows. For a given $k \geq 1$, a *k-splittable flow* must route each commodity along at most k paths. In particular, 1-splittable flows are unsplittable flows. It follows from the classical flow decomposition theorem that k-splittability is not a meaningful restriction for $k \geq |A|$. Single source k-splittable flows are studied, e.g., by Kolliopoulos [8], Koch, Skutella, and Spenke [7], and Salazar and Skutella [14].

Let $d_{\max} := \max\{d_1, \ldots, d_k\}$ denote the maximum demand value. The central result in the seminal paper of Dinitz, Garg, and Goemans [3] is the following theorem on single source unsplittable flows:

Theorem 2. *For a given flow x, there is an unsplittable flow y such that*

$$y_a \leq x_a + d_{\max} \qquad for \ all \ a \in A. \tag{2}$$

The proof of this theorem given in [3] is in fact algorithmic, that is, the authors provide an efficient algorithm that turns the given flow x into an unsplittable flow y satisfying the arc-wise upper bounds (2). Starting from the fractional flow x, the algorithm repeatedly augments flow along cycles featuring a special property until all commodities have been routed unsplittably.

Contribution and Outline. In Sect. 2 we present a considerably simpler one-page proof of Theorem 2, based upon an entirely new approach. Dinitz, Garg, and Goemans start with the fractional flow $y := x$ and then iteratively modify y, always maintaining Property (2), until y is unsplittable. In contrast, we start with an arbitrary unsplittable flow y violating Property (2) and then iteratively modify y, always maintaining an unsplittable flow, until y meets Property (2).

In Sect. 3, using a similar approach, we derive the following new covering analogue of the packing result in Theorem 2:

Theorem 3. *For a given flow x, there is an unsplittable flow y such that*

$$y_a \geq x_a - d_{\max} \qquad for \ all \ a \in A. \tag{3}$$

To prove this result, we again iteratively turn an arbitrary unsplittable flow into one that meets Property (3). In contrast to the packing result, however, the proof of the covering bounds in (3) turns out to be somewhat more intricate, requiring several additional insights and arguments.

Section 4 considers upper bounds (2) and lower bounds (3) simultaneously. Using techniques introduced by Kolliopoulos and Stein [10], Skutella [15], and Martens et al. [11], one can obtain the following generalization of Theorem 1:

Theorem 4. *If the demands are multiples of each other, i.e., $d_1 \mid d_2 \mid \ldots \mid d_k$, then any flow x can be written as a convex combination of unsplittable flows such that each such unsplittable flow y satisfies*

$$x_a - d_{\max} \leq y_a \leq x_a + d_{\max} \quad \text{for all } a \in A. \tag{4}$$

In particular, there does exist an unsplittable flow y obeying (4).

Finally, for arbitrary demand values, we obtain the following slightly weaker bounds:

Theorem 5. *For a given flow x, there is an unsplittable flow y such that*

$$\frac{x_a}{2} - d_{\max} \leq y_a \leq 2\,x_a + d_{\max} \quad \text{for all } a \in A. \tag{5}$$

We conclude in Sect. 5 by pointing out several interesting open problems and stating a stronger version of Goemans' unsplittable flow conjecture.

Preliminaries. We assume throughout this paper that, without loss of generality, each node $v \in V$ lies on an s-t_i-path for some $i \in \{1, \ldots, k\}$. Paths are considered to be subsets of the given arc set A. For a path Q and two nodes $v, w \in V$ lying on path Q (with v being visited first), the v-w-subpath of Q is denoted by $Q|_{[v,w]}$. Finally, for a subset of nodes $X \subseteq V \setminus \{s\}$, let $d(X) := \sum_{i:t_i \in X} d_i$.

2   A Short Proof of the Dinitz-Garg-Goemans Theorem

Our novel proof of Theorem 2 relies on a simple augmentation step, called *upper bound preserving (UBP)* augmentation step. For a given flow x and an arbitrary unsplittable flow y, we say that a node v is *UBP-reachable w.r.t. y* if there is an s-v-path Q in D such that

$$y_a \leq x_a \quad \text{for all } a \in Q.$$

A UBP augmentation step for an unsplittable flow y is defined as follows: Given a node v that is UBP-reachable w.r.t. y along path Q and a commodity i such that node v lies on path P_i^y, reroute commodity i from s to v along path Q. This results in a new unsplittable flow y' using a new s-t_i-path $P_i^{y'}$ with $P_i^{y'}|_{[s,v]} = Q$; see Fig. 2 in Appendix A. Notice that

$$y_a' \leq x_a + d_i \quad \text{for all } a \in P_i^{y'}|_{[s,v]},$$

which explains why the augmentation step is called *upper bound preserving*.

If an unsplittable flow y' results from y by a finite sequence of UBP augmentation steps, we write $y \overset{\text{UBP}}{\leadsto} y'$. We say that a node v is *eventually UBP-reachable w.r.t. y* if there is an unsplittable flow y' with $y \overset{\text{UBP}}{\leadsto} y'$ such that v is UBP-reachable w.r.t. y'.

Lemma 1. *For any unsplittable flow y, all nodes in V are eventually UBP-reachable w.r.t. y.*

Proof. Assume that there is an unsplittable flow y such that the set X_y of eventually UBP-reachable nodes w.r.t. y is a proper subset of V, i.e., $X_y \subsetneq V$. Applying UBP augmentation steps to y cannot enlarge the set of eventually UBP-reachable nodes. Hence, if we choose y such that X_y is inclusion-wise minimal, then $X_y = X_{y'} =: X$ for any unsplittable flow y' with $y \overset{\text{UBP}}{\rightsquigarrow} y'$. We prove three important properties of y and X:

(P1) $y_a > x_a$ for all $a \in \delta^{\text{out}}(X)$.

Indeed, if there is an arc $a = (v, w) \in \delta^{\text{out}}(X)$ with $y_a \leq x_a$, let y' be an unsplittable flow resulting from y by a shortest possible sequence of UBP augmentation steps such that v is UBP-reachable w.r.t. y'. As long as node v is not UBP-reachable, flow on arc a cannot increase during a UBP augmentation step. Hence, $y'_a \leq y_a \leq x_a$ which implies that not only v but also w is UBP-reachable w.r.t. y'. Hence, $w \in X$, a contradiction.

(P2) $y(\delta^{\text{in}}(X)) = \sum_{a \in \delta^{\text{in}}(X)} y_a > 0$.

Indeed, $\delta^{\text{out}}(X) \neq \emptyset$, as otherwise there is no path from source $s \in X$ to the nodes in $V \setminus X$. Both flows x and y satisfy the same set of demands, i.e., $y(\delta^{\text{out}}(X)) - y(\delta^{\text{in}}(X)) = d(V \setminus X) = x(\delta^{\text{out}}(X)) - x(\delta^{\text{in}}(X))$. But, by Property (P1), $y(\delta^{\text{out}}(X)) > x(\delta^{\text{out}}(X))$. Hence, $y(\delta^{\text{in}}(X)) > 0$.

(P3) $y'(\delta^{\text{in}}(X)) \leq y(\delta^{\text{in}}(X))$ for each unsplittable flow y' with $y \overset{\text{UBP}}{\rightsquigarrow} y'$.

Let a UBP augmentation step reroute commodity i from source s to some node $v \in X$ along an s-v-path Q. By definition of X, path Q must remain in X. Moreover, $P_i^y|_{[s,v]}$ may either remain in X, implying that the total income flow of y does not change; or it may exit and re-enter X, implying that $y(\delta^{\text{in}}(X))$ decreases by (a multiple of) demand d_i.

With a view to Property (P3), we choose y with $y(\delta^{\text{in}}(X))$ minimal. By Property (P2), there is an arc $a = (w, v) \in \delta^{\text{in}}(X)$ such that $y_a > 0$. Since $v \in X$, there is an unsplittable flow y' with $y \overset{\text{UBP}}{\rightsquigarrow} y'$ such that v is UBP-reachable w.r.t. y' via an s-v-path Q. Flow on arc a remains unchanged, i.e., $y'_a = y_a$, and there is a commodity i with $a \in P_i$. Rerouting commodity i from s to v along Q decreases $y'(\delta^{\text{in}}(X)) = y(\delta^{\text{in}}(X))$, contradicting its minimality. $\square$

In order to prove Theorem 2, we start with an arbitrary unsplittable flow y_0. Apply Lemma 1 successively to each of the destination nodes t_i with $i = 1, \ldots, k$: Once t_i is UBP-reachable w.r.t. some unsplittable flow via an s-t_i-path Q, we apply a UBP augmentation step, rerouting demand d_i along path Q. The resulting flow y_i satisfies the desired arc-wise upper bound on all arcs lying on the new s-t_i-path $P_i^{y_i} = Q$. Notice that whenever a commodity i is routed properly, it remains so, independently of which UBP augmentation steps may follow. This implies for y_i that all commodities $1, \ldots, i$ are now properly routed. Hence, the final flow y_k satisfies the arc-wise upper bound on all arcs. This concludes the proof of Theorem 2.

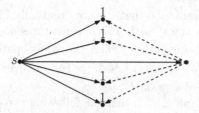

Fig. 1. An instance with k demands of value 1. Flow x is given as follows: solid arcs have flow value $\frac{k}{k+1}$ and dashed arcs carry flow value $\frac{1}{k+1}$. Notice that any unsplittable flow sends zero flow on some solid arc.

3 Unsplittable Flows with Arc-Wise Lower Bounds

Notice that the lower bound (3) in Theorem 3 is tight in the following sense: For each $\varepsilon > 0$, there is a digraph D together with a fractional flow x such that no unsplittable flow y satisfies $y_a \geq x_a - d_{\max} + \varepsilon$, for all $a \in A$; see the instance depicted in Fig. 1 with $k = \lceil 1/\varepsilon \rceil$.

We point out that the techniques provided in [3] are not adapted for handling arc-wise lower bounds, see Appendix B for further details. The proof of Theorem 3 is based on a similar approach as our proof of the Dinitz-Garg-Goemans Theorem in Sect. 2, yet turns out to be somewhat more intricate. Since we are no longer interested in upper bounds but in lower bounds, we now use *lower bound preserving (LBP)* augmentation steps: for an unsplittable flow y, we say that node v is *LBP-reachable w.r.t. y* if there is a commodity i whose s-t_i-path P_i^y passes through node v and the s-v-subpath $P_i^y|_{[s,v]}$ satisfies

$$y_a \geq x_a \qquad \text{for all } a \in P_i^y|_{[s,v]}.$$

To emphasize the role of commodity i, we also say that node v is *LBP-reachable for commodity i w.r.t. y* in this case.

An LBP augmentation step for an unsplittable flow y is defined as follows: Given a node v that is LBP-reachable for commodity i w.r.t. y, reroute commodity i from source s to node v along an arbitrary s-v-path Q. This results in a new unsplittable flow y' using a new s-t_i-path $P_i^{y'}$ with $P_i^{y'}|_{[s,v]} = Q$; see Fig. 3 in Appendix A. Notice that for each $a \in P_i^y|_{[s,v]}$ we have $y'_a \geq x_a - d_i$, which explains why the augmentation step is called *lower bound preserving*.

If an unsplittable flow y' results from y by a finite sequence of LBP augmentation steps, we write $y \overset{\text{LBP}}{\leadsto} y'$. We say that a node v is *eventually LBP-reachable (for commodity i) w.r.t. y* if there is an unsplittable flow y' with $y \overset{\text{LBP}}{\leadsto} y'$ such that v is LBP-reachable (for commodity i) w.r.t. y'.

Proposition 1. *Let y be an unsplittable flow and v a node on path P_i^y for some commodity i such that v is eventually LBP-reachable w.r.t. y' for all y' with $y \overset{\text{LBP}}{\leadsto} y'$. Then v is eventually LBP-reachable for commodity i w.r.t. y.*

Proof. Assume by contradiction that v is not eventually LBP-reachable for commodity i w.r.t. y. Then any y' with $y \overset{\text{LBP}}{\rightsquigarrow} y'$ constitutes another counterexample. In particular, node v is on path $P_i^{y'}$ and the s-v-subpath $P_i^{y'}|_{[s,v]}$ contains at least one arc a such that $y'_a < x_a$. Let $a^{y'} = (u^{y'}, w^{y'})$ be such arc closest to v. Let $d^{y'}$ be the number of arcs on $P_i^{y'}|_{[w^{y'},v]}$.

Choose counterexample y such that the following two criteria are met in the given order: (i) d^y is maximal, (ii) y_{a^y} is maximal.

Consider a sequence of unsplittable flows $y = y_0, y_1, \ldots, y_q$, where each y_j results from y_{j-1} via an LBP augmentation step, such that there is a node w on $P_i^y|_{[w^y,v]}$ that is LBP-reachable w.r.t. y_q but no node on $P_i^y|_{[w^y,v]}$ is LBP-reachable w.r.t. y_j for any $j < q$. Such a sequence exists since node v is eventually LBP-reachable w.r.t. y. As no node on $P_i^y|_{[w^y,v]}$ is LBP-reachable w.r.t. y_j for any $j < q$, flow on the arcs of $P_i^y|_{[u^y,v]}$ is never decreased during these LBP augmentation steps and $P_i^y|_{[u^y,v]} = P_i^{y_j}|_{[u^y,v]}$ for all $j \leq q$. In particular, due to (i) and (ii), $d^{y_j} = d^y$, $a^{y_j} = a^y$, and the flow on a^y remains unchanged for all $j \leq q$. Node w is LBP-reachable for some commodity w.r.t y_q. Rerouting that commodity along $P_i^y|_{[s,w]}$ strictly increases flow on a^y, contradicting the choice of y in terms of (i) and (ii). □

Lemma 2. *For any unsplittable flow y, all nodes in V are eventually LBP-reachable for some commodity w.r.t. y.*

Proof. Assume that there is an unsplittable flow y such that the set X_y of eventually LBP-reachable nodes w.r.t. y is a proper subset of V, i.e., $X_y \subsetneq V$. Applying LBP augmentation steps to y cannot enlarge the set of eventually LBP-reachable nodes. Hence, if we choose y such that X_y is inclusion-wise minimal, then $X_y = X_{y'} =: X$ for any unsplittable flow y' with $y \overset{\text{LBP}}{\rightsquigarrow} y'$. Notice that an LBP augmentation step cannot decrease the outgoing flow $y(\delta^{\text{out}}(X))$. Indeed, if a node $v \in X$ is LBP-reachable for a commodity i w.r.t. y, then all nodes on the s-v-subpath $P_i^y|_{[s,v]}$ are LBP-reachable for commodity i w.r.t y. Hence, $P_i^y|_{[s,v]}$ remains inside of the set X_y and rerouting commodity i cannot decrease $y(\delta^{\text{out}}(X))$. Therefore, $y'(\delta^{\text{out}}(X)) \geq y(\delta^{\text{out}}(X))$ for each unsplittable flow y' with $y \overset{\text{LBP}}{\rightsquigarrow} y'$. With a view to this property, choose y with $y(\delta^{\text{out}}(X))$ maximal. We prove two important properties of y and X:

(P1') $\delta^{\text{in}}(X) = \emptyset$.

If there is an arc $(w,v) \in \delta^{\text{in}}(X)$, let y' with $y \overset{\text{LBP}}{\rightsquigarrow} y'$ be such that node $v \in X$ is LBP-reachable for some commodity i w.r.t. y'. Rerouting commodity i along an arbitrary s-v-path Q with $(w,v) \in Q$ increases flow on (w,v) as well as on some arc in $\delta^{\text{out}}(X)$, contradicting the maximality of $y(\delta^{\text{out}}(X))$.

(P2') There is an arc $a \in \delta^{\text{out}}(X)$ with $y_a > 0$ and $y_a \geq x_a$.

Consider a node $v \in V \setminus X$. By assumption, there is a v-t_i-path in D for some commodity i. Since $\delta^{\text{in}}(X) = \emptyset$ by Property (P1'), node t_i must also lie in $V \setminus X$. Therefore, again by Property (P1'), we have

$$y(\delta^{\text{out}}(X)) = x(\delta^{\text{out}}(X)) = d(V \setminus X) \geq d_i > 0.$$

This implies the existence of arc a with the properties stated above.

Consider an arc $a = (v, w) \in \delta^{\text{out}}(X)$ as in (P2') and a commodity i with $a \in P_i^y$. Since $w \in V \setminus X$ and $\delta^{\text{in}}(X) = \emptyset$ by (P1'), it is impossible for any sequence of LBP augmentations to add or delete flow on arc a. In particular, $y_a' = y_a \geq x_a$ and $a \in P_i^{y'}$ for each unsplittable flow y' with $y \overset{\text{LBP}}{\rightsquigarrow} y'$. By Proposition 1, node v is eventually LBP-reachable for commodity i w.r.t. y. Hence, node w is eventually LBP-reachable w.r.t. y, a contradiction to $w \notin X$. $\qquad\square$

Lemma 3. *For any unsplittable flow y and any arc a, there is an unsplittable flow y' with $y \overset{\text{LBP}}{\rightsquigarrow} y'$ such that $y_a' \geq x_a$.*

Proof. Assume that there is an unsplittable flow y and an arc $a = (v, w) \in A$ such that $y_a' < x_a$ for any y' with $y \overset{\text{LBP}}{\rightsquigarrow} y'$. Notice that flow on arc a can never be decreased by an LBP augmentation step. We may choose y in such a way that y_a is maximal. By Lemma 2, node w is LBP-reachable for some commodity i w.r.t. an unsplittable flow y' with $y \overset{\text{LBP}}{\rightsquigarrow} y'$. Rerouting i along any s-w-path Q with $a \in Q$ increases the flow on arc a, hence contradicting its maximality. $\qquad\square$

We can now prove Theorem 3: Let y be an arbitrary unsplittable flow. By Lemma 3, for each arc $a \in A$ there is an unsplittable flow y' with $y \overset{\text{LBP}}{\rightsquigarrow} y'$ such that $y_a' \geq x_a$. Notice that any flow resulting from y' by a sequence of LBP augmentation has a flow value on a of at least $x_a - d_{\max}$. Going through all arcs successively leads to a final unsplittable flow with the desired properties.

Unsplittable Flows Satisfying Lower Capacities on Arcs

In the context of unsplittable flows respecting arc-wise upper bounds, several interesting problem variants have been considered in the literature; see, e.g., [6]. Theorem 2 immediately implies approximation results for the *minimum congestion problem* whose objective is to bound the violation of given upper bounds (arc capacities). Another prominent problem, the *minimum number of rounds problem*, asks for a partition of the set of commodities into a minimum number of subsets (rounds) such that each subset can be routed unsplittably without violating given arc capacities. Finally, the *maximum routable demand problem* asks for a feasible (w.r.t. arc capacities) unsplittable routing of a subset of commodities of maximum total demand. We refer to [3] for further details.

With a view to these optimization problems, the (fractional) flow x in Theorem 2 plays the role of a solution to a fractional relaxation obeying given arc capacities. Similarly, our new result in Theorem 3 is relevant for unsplittable flow problems with lower capacities on the arcs. If we assume that a given (fractional) flow x obeys lower arc capacities, a meaningful question in this context is, how many copies of the commodities are needed such that one can find an unsplittable routing y of *all* copies such that $y_a \geq x_a$ for all arcs $a \in A$.

Corollary 1. *Given a fractional flow x, let $\alpha \geq 1$ be such that $\alpha \cdot x_a \geq \max_i d_i$ for all $a \in A$, where the maximum is taken over all i with arc a lying on an s-t_i-path. Then, at most $\lceil 1 + \alpha \rceil$ copies of commodities are necessary in order to find an unsplittable routing y of all those copies such that $y_a \geq x_a$ for all $a \in A$.*

Proof. Let $\tilde{x}$ be given by $\tilde{x}_a = \lceil 1+\alpha \rceil \cdot x_a$ for each $a \in A$, hence satisfying $\lceil 1+\alpha \rceil$ copies of each demand d_i for $i = 1, \ldots, k$. As a consequence of Theorem 3, there is an unsplittable flow y satisfying $\lceil 1+\alpha \rceil$ copies of each demand d_i, $i = 1, \ldots, k$, such that $y_a \geq \tilde{x}_a - \max_i d_i$ for all $a \in A$, where the maximum is taken over all commodities i with arc a lying on an s-t_i-path. Since $\alpha \cdot x_a \geq \max_i d_i$ for all $a \in A$, and by definition of $\tilde{x}$, we get $y_a \geq \lceil 1 + \alpha \rceil \cdot x_a - \alpha \cdot x_a \geq x_a$ for all $a \in A$. $\square$

4   Combining Lower and Upper Bounds

In this section we obtain results on unsplittable flows that simultaneously satisfy arc-wise upper and lower bounds with respect to the given fractional flow x. We first consider the special case where demands are multiples of each other, i.e., $d_1 \mid d_2 \mid \ldots \mid d_k$, and give a rough sketch how Theorem 4 can be obtained via methods introduced by Kolliopoulos and Stein [10], Skutella [15], and Martens et al. [11]. To make the terminology precise, for $a, b \in \mathbb{R}_{>0}$ we write $a \mid b$ if there is an integer $c \in \mathbb{Z}$ such that $a \cdot c = b$. In this case we say that b is a-*integral*.

Proof Sketch of Theorem 4. In the considered case, the demands are all $d_{\min}$-integral where $d_{\min}$ is the minimum demand value. By scaling the demand values and the flow x accordingly, we may assume that $d_{\min} = 1$ such that all demands are integral. Therefore, by Theorem 1, the given fractional flow x is a convex combination of $d_{\min}$-integral flows whose flow values on the arcs differ from x by at most $d_{\min}$. Thus, it suffices to show that any $d_{\min}$-integral flow can be written as a convex combination of unsplittable flows whose flow values differ by at most $d_{\max} - d_{\min}$.

Notice that a $d_{\min}$-integral flow can be interpreted to route all commodities i of value $d_i = d_{\min}$ unsplittably by iteratively choosing a flow-carrying s-t_i-path P_i for each such commodity. We can thus decrease the flow along P_i by $d_i = d_{\min}$ and delete commodity i from the instance. This leaves us with a $d_{\min}$-integral flow satisfying all demands strictly larger than $d_{\min} =: \tilde{d}_1$. We are thus left to deal with $\tilde{d}_2$-integral demands where $\tilde{d}_2$ denotes the smallest remaining demand value. Again applying Theorem 1, our $\tilde{d}_1$-integral flow satisfying $\tilde{d}_2$-integral demands can be written as a convex combination of $\tilde{d}_2$-integral flows whose flow values differ by at most $\tilde{d}_2 - \tilde{d}_1$. Thus, overall, the flow values of the $\tilde{d}_2$-integral flows differ from x by at most $\tilde{d}_1 + (\tilde{d}_2 - \tilde{d}_1) = \tilde{d}_2$. Applying this chain of arguments iteratively finally yields Theorem 4 as well as an efficient algorithm to compute the desired convex combination of unsplittable flows. We refer to [11, 15] for further details.

Proof Sketch of Theorem 5. The basic idea of the proof is to round down the demand values such that the rounded demands satisfy the conditions of Theorem 4. The flow x has to be modified accordingly in a careful way in order to finally satisfy the desired lower and upper bounds (5). Then, we apply Theorem 4 to the modified flow x which yields an unsplittable flow satisfying the rounded demands. Finally, we increase flow on the paths of this unsplittable flow to meet the original demand values; see Algorithm 1.

Algorithm 1:

Input : A flow x^0 on D satisfying demands $d_1, \ldots, d_k$;
Output: An unsplittable flow y given by an s-t_i-path for $i = 1, \ldots, k$;
For $i = 1, \ldots, k$, set $\bar{d}_i := d_{\min} \cdot 2^{\lfloor \log(d_i/d_{\min}) \rfloor}$;
Compute a flow $\bar{x}$ satisfying demands $\bar{d}_1, \ldots, \bar{d}_k$ with $\bar{x}_a \geq \frac{x_a}{2}$ for all $a \in A$;
Apply Theorem 4 to $\bar{x}$, yielding an unsplittable flow $\bar{y}$ for demands $\bar{d}_1, \ldots, \bar{d}_k$;
Return unsplittable flow y for original demands with $P_i^y = P_i^{\bar{y}}$, for $i = 1, \ldots, k$;

Lemma 4. *Algorithm 1 computes an unsplittable flow y satisfying* (5).

We refer to Appendix C for the proof of Lemma 4.

5 Conclusion

We conclude by pointing out interesting open problems and conjectures related to the results presented in this paper.

While the original proof of Theorem 2 in [3] comes with an *efficient* algorithm for computing an unsplittable flow y satisfying (2), our proof seems to only give rise to an exponential time algorithm. We conjecture, however, that there always exists a sequence of UBP augmentation steps leading to an unsplittable flow y satisfying (2) whose length is polynomially bounded. We also conjecture that a polynomially bounded sequence of LBP augmentation steps exists in the context of Theorem 3.

With respect to the combination of upper and lower bounds discussed in Sect. 4, we conjecture that for a given flow x, there always exists an unsplittable flow y such that both upper bounds (2) and lower bounds (3) hold. Even more, we conjecture that any flow x can be written as a convex combination of unsplittable flows y satisfying (2) and (3). This can be equivalently stated as follows (cf., e.g., [11]):

Conjecture 1. Given arbitrary cost $c = (c_a)_{a \in A}$ on the arcs and a (fractional) flow x, there is an unsplittable flow y such that $c(y) \leq c(x)$ and

$$x_a - d_{\max} \leq y_a \leq x_a + d_{\max} \quad \text{for all } a \in A.$$

This conjecture is a strengthening of a famous, yet still unresolved conjecture of Goemans (see [15]) which does not take the lower bounds on y into account. Theorem 4 implies that Conjecture 1 is true for the special case of demands that are multiples of each other (also see Theorem 1 for the case of unit demands).

We hope that the new techniques and methods presented in Sects. 2 and 3 will turn out to be of further use and stimulate progress towards these open problems. They might also be useful in the context of k-splittable flows considered, e.g., in [8,14].

Acknowledgements. The authors would like to thank Rico Zenklusen as well as Mohammed Majthoub Almoghrabi and Philipp Warode for interesting discussions on the topic of this paper.

A   Illustration of UBP and LBP Augmentation Steps

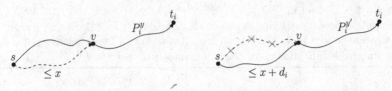

Fig. 2. For a given unsplittable flow y, let v be UBP-reachable w.r.t. y along a path Q (illustrated dashed on the left) and let i be a commodity such that v lies on path P_i^y. A UBP augmentation step reroutes commodity i from s to v along path Q. The resulting unsplittable flow y' is illustrated on the right.

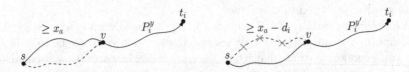

Fig. 3. For a given unsplittable flow y, let v be LBP-reachable for some commodity i w.r.t. y and Q be an arbitrary s-v-path (illustrated dashed on the left). An LBP augmentation step reroutes commodity i from s to v along path Q. The resulting unsplittable flow y' is illustrated on the right.

B   Counterexample

In Fig. 4 we give an example showing that the algorithm of Dinitz, Garg, and Goemans [3] (that was designed for the problem with arc-wise upper bounds) is not adapted for handling arc-wise lower bounds. Their violation can be arbitrarily large, as we can see by adding commodities and expanding the graph in Fig. 4. Consequently, even though the two problems regarding arc-wise upper and lower bounds seem to be similar in spirit, we need completely new tools in order to solve the above mentioned problem.

C   Proof of Lemma 4

In order to prove the correctness of Algorithm 1 and the lower (resp. upper) bound, we use the following well-known results on splittable flows, also known as the *cut condition*; see, e.g., [1]: Given an arc-capacitated directed graph $D = (V, A)$ with a source node s and k commodities with corresponding destination nodes $t_1, \ldots, t_k$ and demands $d_1, \ldots, d_k \in \mathbb{R}_{>0}$, there is a feasible flow satisfying all demands if and only if, for any subset $T \subset V \setminus \{s\}$, the sum of capacities of arcs in the directed cut $(V \setminus T, T)$ is at least $d(T)$.

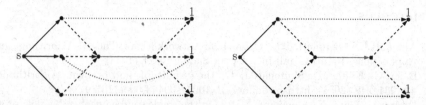

Fig. 4. Let the digraph have three commodities with demand 1, as illustrated on the left, and let the fractional flow x be given as follows: $x_{\text{solid}} = 1$, $x_{\text{dashed}} = 1 - \epsilon$, $x_{\text{dotted}} = \epsilon$, and $x_{\text{red}} = 3 - 3\epsilon$ for some $\epsilon > 0$. The algorithm in [3] first moves one of the demands to the source node along its corresponding dotted arc, hence decreasing flow on the red arc by $1 - \epsilon$. The remaining instance with two commodities and $x'_{\text{red}} = 2 - 2\epsilon$ is illustrated on the right. Any further augmentation step decreases flow on the red arc by $1 - \epsilon$, hence violating the desired lower bound on the red arc. (Color figure online)

By definition, $\bar{d}_i \leq d_i < 2\bar{d}_i$ for all $i = 1, \ldots, k$. We thus get $d_i - \bar{d}_i < \frac{d_i}{2}$ for all $i = 1, \ldots, k$. We first show that there is a flow $\bar{x}$ satisfying demands $\bar{d}_1, \ldots, \bar{d}_k$ such that $\bar{x}_a \geq \frac{x_a}{2}$ for all $a \in A$. By the cut condition for flow x, it holds that

$$\sum_{a \in \delta^{\text{out}}(T)} x_a \geq d(T) \qquad \text{for each } T \subseteq V \setminus \{s\}.$$

Therefore,

$$\sum_{a \in \delta^{\text{out}}(T)} \frac{x_a}{2} \geq \frac{d(T)}{2} > d(T) - \bar{d}(T).$$

By the cut condition, there is a flow satisfying demands $d_i - \bar{d}_i$ for $i = 1, \ldots, k$ and obeying arc capacities $\frac{x_a}{2}$ for all $a \in A$. Subtract this flow from the original flow x to get a flow $\bar{x}$ satisfying demands $\bar{d}_1, \ldots, \bar{d}_k$. The flow $\bar{x}$ actually satisfies

$$\bar{x}_a \geq x_a - \frac{x_a}{2} = \frac{x_a}{2} \quad \text{for all } a \in A.$$

Applying Theorem 4 to $\bar{x}$ leads to an unsplittable flow $\bar{y}$ such that

$$\bar{x}_a - \bar{d}_{\max} \leq \bar{y}_a \leq \bar{x}_a + \bar{d}_{\max} \quad \text{for all } a \in A.$$

By construction of flow y, we obtain the lower bound

$$y_a \geq \bar{y}_a \geq \bar{x}_a - \bar{d}_{\max} \geq \frac{x_a}{2} - d_{\max} \quad \text{for all } a \in A.$$

In order to prove the upper bound, let d_{i_a} be a commodity with maximal demand that is routed across an arc $a \in A$. Notice that $\bar{y}$ even satisfies a slightly stronger upper bound (also see [15]): we have $\bar{y}_a \leq \bar{x}_a + \bar{d}_{i_a}$ for each arc a. Therefore,

$$y_a = \sum_{i: a \in P_i} d_i \leq d_{i_a} + 2 \sum_{\substack{i: a \in P_i, \\ i \neq i_a}} \bar{d}_i \leq d_{i_a} + 2\bar{x}_a \leq 2x_a + d_{\max}.$$

This concludes the proof. □

References

1. Ahuja, R.K., Magnanti, T.L., Orlin, J.B.: Network Flows: Theory, Algorithms, and Applications. Prentice-Hall Inc., Upper Saddle River (1993)
2. Baier, G., Köhler, E., Skutella, M.: On the k-splittable flow problem. Algorithmica **42**, 231–248 (2005). https://doi.org/10.1007/s00453-005-1167-9
3. Dinitz, Y., Garg, N., Goemans, M.X.: On the single source unsplittable flow problem. Combinatorica **19**, 17–41 (1999). https://doi.org/10.1007/s004930050043
4. Du, J., Kolliopoulos, S.: Implementing approximation algorithms for the single-source unsplittable flow problem. J. Exp. Algorithmics **10**, 2–3 (2005)
5. Ford, L.R., Fulkerson, D.R.: Flows in Networks. Princeton University Press, Princeton (1962)
6. Kleinberg, J.M.: Approximation algorithms for disjoint paths problems. Ph.D. thesis, M.I.T. (1996)
7. Koch, R., Skutella, M., Spenke, I.: Maximum k-splittable s, t-flows. Theor. Comput. Syst. **43**, 56–66 (2008). https://doi.org/10.1007/s00224-007-9068-8
8. Kolliopoulos, S.G.: Minimum-cost single-source 2-splittable flow. Inf. Process. Lett. **94**, 15–18 (2005)
9. Kolliopoulos, S.G.: Edge-disjoint paths and unsplittable flow. In: Gonzalez, T.F. (ed.) Handbook of Approximation Algorithms and Metaheuristics. Chapman and Hall/CRC, Boca Raton (2007)
10. Kolliopoulos, S.G., Stein, C.: Approximation algorithms for single-source unsplittable flow. SIAM J. Comput. **31**, 919–946 (2002)
11. Martens, M., Salazar, F., Skutella, M.: Convex combinations of single source unsplittable flows. In: Arge, L., Hoffmann, M., Welzl, E. (eds.) ESA 2007. LNCS, vol. 4698, pp. 395–406. Springer, Heidelberg (2007). https://doi.org/10.1007/978-3-540-75520-3_36
12. Raghavan, P.: Probabilistic construction of deterministic algorithms: approximating packing integer programs. J. Comput. Syst. Sci. **37**, 130–143 (1988)
13. Raghavan, P., Thompson, C.D.: Randomized rounding: a technique for provably good algorithms and algorithmic proofs. Combinatorica **7**, 365–374 (1987). https://doi.org/10.1007/BF02579324
14. Salazar, F., Skutella, M.: Single-source k-splittable min-cost flows. Oper. Res. Lett. **37**, 71–74 (2009)
15. Skutella, M.: Approximating the single source unsplittable min-cost flow problem. Math. Program. **91**(3), 493–514 (2001). https://doi.org/10.1007/s101070100260
16. Williamson, D.P.: Network Flow Algorithms. Cambridge University Press, Cambridge (2019)

Maximal Quadratic-Free Sets

Gonzalo Muñoz[1]([✉])[iD] and Felipe Serrano[2][iD]

[1] Universidad de O'Higgins, Rancagua, Chile
gonzalo.munoz@uoh.cl
[2] Zuse Institute Berlin, Berlin, Germany
serrano@zib.de

Abstract. The intersection cut paradigm is a powerful framework that facilitates the generation of valid linear inequalities, or cutting planes, for a potentially complex set S. The key ingredients in this construction are a simplicial conic relaxation of S and an S-free set: a convex zone whose interior does not intersect S. Ideally, such S-free set would be maximal inclusion-wise, as it would generate a deeper cutting plane. However, maximality can be a challenging goal in general. In this work, we show how to construct maximal S-free sets when S is defined as a general quadratic inequality. Our maximal S-free sets are such that efficient separation of a vertex in LP-based approaches to quadratically constrained problems is guaranteed. To the best of our knowledge, this work is the first to provide maximal quadratic-free sets.

Keywords: Non-convex quadratic · Intersection cut · S-free sets

1 Introduction

Cutting planes have been at the core of the development of tractable computational techniques for integer-programming for decades. Their rich theory and remarkable empirical performance have constantly caught the attention of the optimization community, and has recently seen renewed efforts on their extensions to the nonlinear setting. Consider a generic optimization problem, which we assume to have linear objective without loss of generality:

$$\min \quad c^\mathsf{T} x \tag{1a}$$
$$\text{s.t.} \quad x \in S \subseteq \mathbb{R}^n. \tag{1b}$$

A common framework for finding strong approximations to this problem is to first find $\bar{x}$, an extreme point optimal solution of an LP relaxation of (1), and check if $\bar{x} \in S$. If so, then (1) is solved. Otherwise, we try to find an inequality separating $\bar{x}$ from S. Such inequality can be used to refine the LP relaxation of (1).

One way of finding such a cutting plane is through the *intersection cut* framework [4,23,39]. We refer the reader to [17] for background on this procedure.

© Springer Nature Switzerland AG 2020
D. Bienstock and G. Zambelli (Eds.): IPCO 2020, LNCS 12125, pp. 307–321, 2020.
https://doi.org/10.1007/978-3-030-45771-6_24

For the purposes of this article, it suffices to know that to compute an intersection cut we need $\bar{x} \notin S$ as above, a simplicial conic relaxation of S with apex $\bar{x}$, and an S-free set C—a convex set satisfying $\text{int}(C) \cap S = \emptyset$—such that $\bar{x} \in \text{int}(C)$. In this work, we assume that $\bar{x}$ and the simplicial cone are given, and thus only focus on the construction of the S-free sets. A particularly important case is obtained when (1) is a quadratic problem, i.e.,

$$S = \{x \in \mathbb{R}^n : x^\mathsf{T} Q_i x + b_i^\mathsf{T} x + c_i \leq 0, \, i = 1, \ldots, m\}$$

for certain $n \times n$ matrices Q_i, not necessarily positive semi-definite. Note that if $\bar{x} \notin S$, then there exists $i \in \{1, \ldots, m\}$ such that

$$\bar{x} \notin S_i := \{x \in \mathbb{R}^n : x^\mathsf{T} Q_i x + b_i^\mathsf{T} x + c_i \leq 0\},$$

and constructing an S_i-free set containing $\bar{x}$ would suffice to ensure separation. Thus, slightly abusing notation, given $\bar{x}$ we focus on a systematic way of constructing S-free sets containing $\bar{x}$, where S is defined using a single quadratic inequality:

$$S = \{x \in \mathbb{R}^n : x^\mathsf{T} Q x + b^\mathsf{T} x + c \leq 0\}.$$

As a final note, if we consider the simplest form of intersection cuts, where the cuts are computed using the intersection points of the S-free set and the extreme rays of the simplicial conic relaxation of S (i.e., using the gauge function), then the larger the S-free set the better the cut. In other words, if two S-free sets C_1, C_2 are such that $C_1 \subsetneq C_2$, the intersection cut derived from C_2 is stronger than the one derived from C_1 [16]. Therefore, we aim at computing *maximal* S-free sets.

Contribution. The main contribution of this paper is an explicit construction of maximal S-free sets, when S is defined using a non-convex quadratic inequality (Theorems 3 and 4). While these maximal S-free sets are constructed using semi-infinite representations, we show equivalent and simple closed-form representations of them. In order to construct these sets, we also derive maximal S-free sets for sets S defined as the intersection of a homogeneous quadratic inequality intersected with a linear *homogeneous inequality*. These are an important intermediate step in our construction, but they are of independent interest as well.

In order to show our results, we state and prove a criterion for maximality of S-free sets which generalizes a criterion proven by Dey and Wolsey (the 'only if' of [20, Proposition A.4]) in the case of maximal *lattice-free* sets (Definition 2 and Theorem 1). We also develop a new criterion that can handle a special phenomenon that arises in our setting and also in nonlinear integer programming: the boundary of a maximal S-free set may not even intersect S. Instead, the intersection might be at "infinity". We formalize this in Definition 3 and show the criterion in Theorem 1.

1.1 Literature Review

The history of intersection cuts and S-free sets dates back to the 60's. They were originally introduced in the nonlinear setting by Tuy [39] for the problem of minimizing a concave function over a polytope. Later on, they were introduced in integer programming by Balas [4]. The more modern form of intersection cuts deduced from an arbitrary convex S-free set is due to Glover [23], although the term S-free was coined by Dey and Wolsey [20]. While the origin of intersection cuts was in nonlinear optimization, most developments have been in the mixed integer linear programming literature. See e.g. [6,16,18] for in-depth analyses of the relation of intersection cuts using maximal $\mathbb{Z}^n$-free sets and the generation of facets of $\mathrm{conv}(S)$. We also refer the reader to [2,3,5,14,19,25] and references therein. For extensions of this approach to the mixed-integer conic case see [1, 8,9,26,27,32,33,37,40].

Lately, there has been several developments of intersection cuts in a nonlinear setting. Fischetti et al. [21] applied intersection cuts to bilevel optimization. Bienstock et al. [10,11] studied outer-product-free sets; these can be used for generating intersection cuts for polynomial optimization when using an extended formulation. Serrano [36] showed how to construct a concave underestimator of any factorable function and from them one can build intersection cuts for factorable mixed integer nonlinear programs. Fischetti and Monaci [22] constructed bilinear-free sets through a bound disjunction and underestimating a bilinear term with McCormick inequalities [30].

Of all these approaches for constructing intersection cuts in a nonlinear setting, the only one that ensures maximality of the corresponding S-free sets is the work of Bienstock et al. [10,11]. While their approach can also be used to generate cuts in our setting, the space where they are computed and definition of S differ significantly: in an n-dimensional quadratic setting, the approach of Bienstock et al. would use a matrix-based extended formulation of dimension proportional to n^2 [28,29,38], define S as the set of positive semi-definite and rank 1 matrices and aim for maximality using this notion. Our approach can construct a maximal S-free set in the original space. These differences make the approaches incomparable at this point: the quadratic dimension increase can be a drawback, however stronger cuts can be derived from extended formulations in some cases [12]. A comparison is subject of future work.

If one were to rely on a second-order cone instead of a simplicial cone to compute a cut, one can use the results of [15] to obtain valid inequalities. The relation between these approaches is subject of future work. We refer to the survey [13] for other efforts of extending cutting planes to the nonlinear setting.

1.2   Notation

We mostly follow standard notation. $\|\cdot\|$ is the euclidean norm in $\mathbb{R}^n$. $D_r(x)$ is the boundary of the euclidean ball centered at x of radius r, i.e., $D_r(x) = \{y \in \mathbb{R}^n : \|y-x\| = r\}$. Given a vector v and a set C, we denote the distance between v and C as $\mathrm{dist}(v,C) = \inf_{x \in C} \|v - x\|$. We denote the set $\{v + x : x \in C\}$ as $v + C$. We denote the transpose operator as $(\cdot)^\mathsf{T}$. For a set of vectors $\{v^1, \ldots, v^k\} \subseteq \mathbb{R}^n$, we denote $\langle v^1, \ldots, v^k \rangle$ the subspace generated by them. Given a set $C \subseteq \mathbb{R}^n$, $\mathrm{conv}(C)$ and $\mathrm{int}(C)$ denotes the convex hull and interior of C, respectively. We denote an inequality $\alpha^\mathsf{T} x \leq \beta$ by (α, β). If $\beta = 0$ we denote it as well as α.

2   Preliminaries

As we mentioned above, our main object of study is the set $S = \{x \in \mathbb{R}^p : q(x) \leq 0\} \subseteq \mathbb{R}^p$, where q is a quadratic function. To make the analysis easier, we can work on $\mathbb{R}^{p+1}$ and consider the cone generated by $S \times \{1\}$, namely, $\{(x, z) \in \mathbb{R}^{p+1} : z^2 q(\frac{x}{z}) \leq 0, \ z \geq 0\}$. To recover the original S, however, we must intersect the cone with $z = 1$. Since we are interested in maximal S-free sets, this motivates the following definition, see also [5].

Definition 1. *Given $S, C, H \subseteq \mathbb{R}^n$ where S is closed, C is closed and convex and H is an affine hyperplane, we say that C is S-free with respect to H if $C \cap H$ is $S \cap H$-free w.r.t the induced topology in H. We say C is* maximal *S-free with respect to H, if for any $C' \supseteq C$ that is S-free with respect to H it holds that $C' \cap H \subseteq C \cap H$.*

The next definition is motivated by the sufficient (and necessary) condition for a full dimensional $\mathbb{Z}^n$-free set to be maximal: it must be a polyhedron containing a point of $\mathbb{Z}^n$ in the relative interior of each facet [31, Theorem 6.18].

Definition 2. *Given a convex set $C \subseteq \mathbb{R}^n$ and a valid inequality $\alpha^\mathsf{T} x \leq \beta$, we say that a point $x_0 \in \mathbb{R}^n$* exposes *(α, β) with respect to C or that (α, β) is* exposed *by x_0 if: $\alpha^\mathsf{T} x_0 = \beta$ and if $\gamma^\mathsf{T} x \leq \delta$ is any other non-trivial valid inequality for C such that $\gamma^\mathsf{T} x_0 = \delta$, then there exists a $\mu > 0$ such that $\gamma = \mu\alpha$ and $\beta = \mu\delta$. We omit saying "with respect to C" if it is clear from context.*

Points that expose inequalities are also called *smooth* points [24]. A related concept is that of *blocking points* [7].[1]

The reader might already anticipate our criteria: if all valid inequalities of an S-free set are exposed by elements of S, then the set should be maximal S-free. A result like this can be shown (see [35]), but there are cases where a maximal S-free set has inequalities that are not exposed by any point in S. This phenomenon occurs in nonlinear integer programming as well [34], but does not occur in the integer *linear* setting [6]. This leads to the following definition.

Definition 3. *Given a convex set $C \subseteq \mathbb{R}^n$ with non-empty recession cone and a valid inequality $\alpha^\mathsf{T} x \leq \beta$, we say that a sequence $(x_n)_n \subseteq \mathbb{R}^n$* exposes *$(\alpha, \beta)$ at infinity with respect to C if (1) $\|x_n\| \to \infty$, (2) $\frac{x_n}{\|x_n\|} \to d \in rec(C)$, (3) d exposes $\alpha^\mathsf{T} x \leq 0$ with respect to $rec(C)$, and (4) there exists y such that $\alpha^\mathsf{T} y = \beta$ such that $dist(x_n, y + \langle d \rangle) \to 0$. As before, we omit saying "with respect to C" if it is clear from context.*

The next theorem summarizes how we use exposing points and exposing sequences to determine maximality of a convex S-free set.

---

[1] For the reader familiar with the notion of *exposed point* from convex analysis, we would like to point that if C is convex with 0 in its interior and the valid inequality $\alpha^\mathsf{T} x \leq 1$ has an *exposing point*, then α is an *exposed point* of the polar of C.

Theorem 1. *Let $S \subseteq \mathbb{R}^n$ be a closed set, H an affine hyperplane and $C \subseteq \mathbb{R}^n$ a convex S-free set. Assume that $C = \{x \in \mathbb{R}^n : \alpha^T x \leq \beta, \forall (\alpha, \beta) \in \Gamma\}$.*

If for every inequality (α, β) there is, either an $x \in S \cap C$ that exposes (α, β), or a sequence $(x_n)_n \subseteq S$ that exposes (α, β) at infinity, then C is maximal S-free.

Similarly, if for every inequality (α, β) there is, either an $x \in S \cap C \cap H$ that exposes (α, β), or a sequence $(x_n)_n \subseteq S \cap H$ that exposes (α, β) at infinity, then C is maximal S-free with respect to H.

Finally, to easily find exposing points we rely on the following result. We recall that a function is sublinear if and only if it is convex and positive homogeneous.

Lemma 1. *Let $\phi : \mathbb{R}^n \to \mathbb{R}$ be a sublinear function, $\lambda \in D_1(0)$, and let*

$$C = \{(x, y) : \phi(y) \leq \lambda^T x\}.$$

Let $(\bar{x}, \bar{y}) \in C$ be such that ϕ is differentiable at $\bar{y}$ and $\phi(\bar{y}) = \lambda^T \bar{x}$. Then $(\bar{x}, \bar{y})$ exposes the valid inequality $-\lambda^T x + \nabla \phi(\bar{y})^T y \leq 0$.

These definitions and results summarize our main tools. For the sake of brevity we omit some technical details and intermediate steps which the interested reader can find in the full-length version of this paper [35].

3   Homogeneous Quadratics

In this section we construct maximal S^h-free sets that contain a vector $\bar{x} \notin S^h$ for $S^h = \{x \in \mathbb{R}^p : x^T Q x \leq 0\}$. This is our building block towards maximality in the general case. After a change of variable, we can assume that

$$S^h = \{(x, y, z) \in \mathbb{R}^{n+m+l} : \sum_{i=1}^{n} x_i^2 - \sum_{i=1}^{m} y_i^2 \leq 0\} \tag{2a}$$

$$= \{(x, y) \in \mathbb{R}^{n+m} : \|x\| - \|y\| \leq 0\} \times \mathbb{R}^l. \tag{2b}$$

Thus, we focus on $S^h = \{(x, y) \in \mathbb{R}^{n+m} : \|x\| - \|y\| \leq 0\}$ and assume we are given $(\bar{x}, \bar{y})$ such that $\|\bar{x}\| > \|\bar{y}\|$.

Remark 1. Using (2b) instead of (2a) seems irrelevant, but it plays a subtle role in our pursuit of maximality. We elaborate this in [35]. Additionally, the transformation used to bring S^h to the "diagonal" form (2a) is, in general, not unique. Nonetheless, maximality of the S^h-free sets is preserved, as there always exist such transformation which is linear and one-to-one. In [35] we also discuss this further.

The construction of maximal S^h-free sets is simple, and follows from the approach of Serrano [36]. The function $f(x, y) = \|x\| - \|y\|$ has the following *concave* underestimator at $\bar{x} \neq 0$: $\frac{\bar{x}^T x}{\|\bar{x}\|} - \|y\|$. Letting $\lambda = \frac{\bar{x}}{\|\bar{x}\|}$, we get the following S^h-free set containing $(\bar{x}, \bar{y})$ in its interior

$$C_\lambda = \{(x, y) \in \mathbb{R}^{n+m} : \lambda^T x \geq \|y\|\} \tag{3}$$

$$= \{(x, y) \in \mathbb{R}^{n+m} : -\lambda^T x + \beta^T y \leq 0, \; \forall \beta \in D_1(0)\}.$$

Theorem 2. *Let $S^h = \{(x,y) \in \mathbb{R}^{n+m} : \|x\| \le \|y\|\}$ and $C_\lambda = \{(x,y) \in \mathbb{R}^{n+m} : \lambda^\mathsf{T}x \ge \|y\|\}$ for $\lambda = \frac{\bar{x}}{\|\bar{x}\|}$. Then, C_λ is a maximal S^h-free set and contains $(\bar{x}, \bar{y})$ in its interior.*

Proof (sketch). To show that C_λ is maximal, we use the norm function $\|\cdot\|$ as ϕ in Lemma 1. We can show that for each β with $\|\beta\| = 1$ the point $(\lambda, \beta) \in S^h \cap C_\lambda$ exposes the valid inequality $-\lambda^\mathsf{T}x + \beta^\mathsf{T}y \le 0$. With Theorem 1 we conclude.

4   Including a Single Homogeneous Linear Constraint

In general, any S defined using a quadratic function can be described as

$$\{(x,y,z) \in \mathbb{R}^{n+m+l} : \|x\| \le \|y\|, a^\mathsf{T}x + d^\mathsf{T}y + h^\mathsf{T}z = -1\}. \tag{4}$$

via a linear one-to-one transformation. This preserves maximality of S-free sets. Note that the case $h \ne 0$ can be tackled directly using Sect. 3: in this case $C \times \mathbb{R}^l$ is maximal S-free, where C is any maximal S^h-free. Thus, we consider

$$S = \{(x,y) \in \mathbb{R}^{n+m} : \|x\| \le \|y\|, a^\mathsf{T}x + d^\mathsf{T}y = -1\}. \tag{5}$$

The set S above is our goal. However, at this point, a simpler set to study is

$$S_{\le 0} = \{(x,y) \in \mathbb{R}^{n+m} : \|x\| \le \|y\|, a^\mathsf{T}x + d^\mathsf{T}y \le 0\}.$$

In this section we construct maximal $S_{\le 0}$-free sets that contain $(\bar{x}, \bar{y})$ satisfying $\|\bar{x}\| > \|\bar{y}\|$, $a^\mathsf{T}\bar{x} + d^\mathsf{T}\bar{y} \le 0$. We distinguish the following cases.

$$\|a\| \le \|d\| \wedge m > 1 \;(\textbf{Case 1}) \quad \vee \quad \|a\| \ge \|d\| \;(\textbf{Case 2}).$$

We omit $m = 1 \wedge \|a\| < \|d\|$, as it can be shown that $S_{\le 0}$ is convex in such case.

4.1   Case 1: $\|a\| \le \|d\| \wedge M > 1$

The set C_λ is clearly $S_{\le 0}$-free. The strategy for proving maximality of C_λ with respect to S^h was to show that $(\lambda, \beta) \in S^h \cap C_\lambda$ exposes each valid inequality $-\lambda^\mathsf{T}x + \beta^\mathsf{T}y \le 0$. However, these inequalities may not all have exposing points in $S_{\le 0} \cap C_\lambda$, as we do not necessarily have $a^\mathsf{T}\lambda + d^\mathsf{T}\beta \le 0$. To handle this issue, let

$$G(\lambda) = \{\beta : \|\beta\| = 1, a^\mathsf{T}\lambda + d^\mathsf{T}\beta \le 0\}.$$

and define

$$C_{G(\lambda)} = \{(x,y) \in \mathbb{R}^{n+m} : -\lambda^\mathsf{T}x + \beta^\mathsf{T}y \le 0, \; \forall \beta \in G(\lambda)\}.$$

Intuitively, $C_{G(\lambda)}$ is obtained from enlarging C_λ by removing all inequalities that do not have an exposing point in $S_{\le 0}$. It is reasonable to expect maximality, but we need to be careful with losing $S_{\le 0}$-freeness. Indeed,

Proposition 1. *Let $(\bar{x}, \bar{y}) \notin S_{\le 0}$ such that $a^\mathsf{T}\bar{x} + d^\mathsf{T}\bar{y} \le 0$ and $\lambda = \frac{\bar{x}}{\|\bar{x}\|}$. If $\|d\| \ge \|a\|$ and $m > 1$, then $C_{G(\lambda)}$ is maximal $S_{\le 0}$-free and contains $(\bar{x}, \bar{y})$.*

We would like to emphasize that the assumptions $\|d\| \ge \|a\| \wedge m > 1$ play a key role, as otherwise the set $C_{G(\lambda)}$ is not $S_{\le 0}$-free in general. Figure 1 shows an example of this (for explicit details see Example 1 in the appendix).

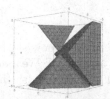 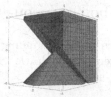

(a) $S_{\leq 0}$ in Example 1 (orange) and the corresponding C_λ set (green). The latter is $S_{\leq 0}$-free but not maximal.

(b) $S_{\leq 0}$ in Example 1 (orange) and the corresponding $C_{G(\lambda)}$ set (green). The latter is not $S_{\leq 0}$-free.

Fig. 1. Sets C_λ and $C_{G(\lambda)}$ in Example 1 for the case $\|a\| > \|d\|$.(Color figure online)

4.2 Case 2: $\|a\| \geq \|d\|$

As $C_{G(\lambda)}$ is not necessarily $S_{\leq 0}$-free in this case, we devise another strategy from an equivalent description of C_λ. Note that $\|y\| = \max\{\lambda^\mathsf{T}\beta : \|\beta\| \leq \|y\|\}$ for any $\lambda \in \mathcal{D}_1(0)$. Thus,

$$C_\lambda = \{(x,y) \in \mathbb{R}^{n+m} : \max_\beta\{\lambda^\mathsf{T}\beta : (\beta,y) \in S^h\} \leq \lambda^\mathsf{T}x\}.$$

This motivates the following definitions

$$\phi_\lambda(y) = \max_x\{\lambda^\mathsf{T}x : (x,y) \in S_{\leq 0}\} \tag{6}$$

$$C_{\phi_\lambda} = \{(x,y) : \phi_\lambda(y) \leq \lambda^\mathsf{T}x\}. \tag{7}$$

We now proceed to describe ϕ_λ and its properties. Note that in this case $\|a\| = 0$ implies $S_{\leq 0} = S^h$. Thus, we assume $\|a\| = 1$.

Proposition 2. *Let $\lambda, a \in \mathcal{D}_1(0) \subseteq \mathbb{R}^n$ and $d \in \mathbb{R}^m$ such that $\|d\| \leq 1$. Then,*

$$\phi_\lambda(y) = \begin{cases} \|y\|, & \text{if } \lambda^\mathsf{T}a\|y\| + d^\mathsf{T}y \leq 0 \\ \sqrt{(\|y\|^2 - (d^\mathsf{T}y)^2)(1 - (\lambda^\mathsf{T}a)^2)} - d^\mathsf{T}y\lambda^\mathsf{T}a, & \text{otherwise.} \end{cases}$$
$$\tag{8}$$

Furthermore, ϕ_λ is sublinear and: if $\|d\| = 1 \wedge m > 1$, then ϕ_λ is differentiable in $\mathbb{R}^m \setminus d\mathbb{R}_+$; otherwise ϕ_λ is differentiable in $\mathbb{R}^m \setminus \{0\}$.

In this case, the fact that $\|a\| \geq \|d\|$ plays a key role in ϕ_λ being convex. From here we can show the maximality of C_{ϕ_λ}.

Proposition 3. *Consider $(\bar{x}, \bar{y}) \notin S_{\leq 0}$ such that $a^\mathsf{T}\bar{x} + d^\mathsf{T}\bar{y} \leq 0$ and let $\lambda = \frac{\bar{x}}{\|\bar{x}\|}$. Let $\phi_\lambda(y)$ and C_{ϕ_λ} defined as in (6) and (7), respectively. If $\|d\| \leq \|a\| = 1$, then $C_{\phi_\lambda} = \{(x,y) : \phi_\lambda(y) \leq \lambda^\mathsf{T}x\}$ is maximal $S_{\leq 0}$-free and contains $(\bar{x}, \bar{y})$ in its interior.*

Proof (sketch). As ϕ_λ is sublinear and has good differentiability properties, we can use Lemma 1. We show that $(x_\beta, \beta) \in S_{\leq 0} \cap C_{\phi_\lambda}$, $\beta \in \mathcal{D}_1(0)$, exposes $-\lambda^\mathsf{T}x + \nabla\phi_\lambda(\beta)^\mathsf{T}y \leq 0$, where x_β is the optimal solution of $\phi_\lambda(\beta)$ in (6).

5   Non-homogeneous Quadratics

As discussed at the beginning of the previous section, we now study a general non-homogeneous quadratic which can be written as S in (5). We assume we are given $(\bar{x}, \bar{y})$ such that $\|\bar{x}\| > \|\bar{y}\|$, $a^{\mathsf{T}}\bar{x} + d^{\mathsf{T}}\bar{y} = -1$ and distinguish the following cases:

$$\|a\| \leq \|d\| \wedge m > 1 \text{ (Case 1)} \quad \vee \quad \|a\| > \|d\| \text{ (Case 2)}.$$

Much like in Sect. 4, we are omitting a simple case. Indeed, it can be proven that when $\|a\| \leq \|d\| \wedge m = 1$, the set S is convex. Since $S \subsetneq S_{\leq 0}$, then $C_{G(\lambda)}$ (C_{ϕ_λ}) is S-free in Case 1 (Case 2) as per Sect. 4. It is natural to wonder whether these sets are maximal already.

5.1   Case 1: $\|a\| \leq \|d\| \wedge M > 1$

To prove maximality of $C_{G(\lambda)}$ with respect to $S_{\leq 0}$ we exploit that $C_{G(\lambda)}$ is directly defined as the inequalities of C_λ exposed by elements in $S_{\leq 0}$. Consider an inequality of $C_{G(\lambda)}$ with coefficients $(-\lambda, \beta)$ such that $a^{\mathsf{T}}\lambda + d^{\mathsf{T}}\beta < 0$. Its exposing point $(\lambda, \beta) \in S_{\leq 0}$ can be scaled by $\mu = \frac{-1}{a^{\mathsf{T}}\lambda + d^{\mathsf{T}}\beta}$ to the exposing point $\mu(\lambda, \beta) \in S$. Thus, almost every inequality of $C_{G(\lambda)}$ is exposed by points of S. Thanks to $m > 1$, we can remove inequalities that correspond to β such that $a^{\mathsf{T}}\lambda + d^{\mathsf{T}}\beta = 0$ without changing the set $C_{G(\lambda)}$. This strategy shows the following.

Theorem 3. Let $\lambda = \frac{\bar{x}}{\|\bar{x}\|}$, $H = \{(x, y) \in \mathbb{R}^{n+m} : a^{\mathsf{T}}x + d^{\mathsf{T}}y = -1\}$, and

$$S_{\leq 0} = \{(x, y)\mathbb{R}^{n+m} : \|x\| \leq \|y\|, \ a^{\mathsf{T}}x + d^{\mathsf{T}}y \leq 0\},$$

where $\|a\| \leq \|d\| \wedge m > 1$. Then, $C_{G(\lambda)}$ is maximal $S_{\leq 0}$-free with respect to H and contains $(\bar{x}, \bar{y})$ in its interior.

This theorem states that obtaining a maximal S-free set in this case amounts to simply use the maximal $S_{\leq 0}$-free set $C_{G(\lambda)}$, and then intersect it with H.

5.2   Case 2: $\|a\| > \|d\|$

Since $\|a\| > 0$, we can assume that $\|a\| = 1$ after rescaling the variables. Unfortunately, in this case the maximality of C_{ϕ_λ} with respect to $S_{\leq 0}$ does not carry over to S. In Fig. 2 we show this issue (for details see Example 2 in the appendix). This figure displays an interesting feature though: the inequalities defining C_{ϕ_λ} seem to have the correct "slope" and just need to be translated. We conjecture, then, that in order to find a maximal S-free set, we only need to adequately relax the inequalities of C_{ϕ_λ}. Recall that

$$C_{\phi_\lambda} = \{(x, y) : \phi_\lambda(y) \leq \lambda^{\mathsf{T}}x\} = \{(x, y) : -\lambda^{\mathsf{T}}x + \nabla\phi_\lambda(\beta)^{\mathsf{T}}y \leq 0, \forall \beta \in D_1(0)\}.$$

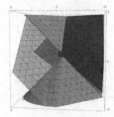 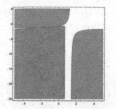

(a) $S_{\leq 0}$ (orange), H (green) and C_{ϕ_λ} (blue).

(b) Projection onto (x_1, x_2) of $S_{\leq 0} \cap H$ (orange) and $C_{\phi_\lambda} \cap H$ (blue).

Fig. 2. Plots of $S_{\leq 0}$, H and C_{ϕ_λ} as defined in Example 2 showing that C_{ϕ_λ} is not necessarily maximal $S_{\leq 0}$-free with respect to H in the case $\|a\| > \|d\|$. (Color figure online)

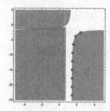

Fig. 3. Projection onto (x_1, x_2) of $S_{\leq 0} \cap H$ (orange), $C_{\phi_\lambda} \cap H$ (blue), and the sequence $(z_n)_{n \in \mathbb{N}}$ in Example 2 for several values of n (red). The sequence diverges "downwards". (Color figure online)

We thus search for $r(\beta)$ such that

$$C = \{(x,y) \, : \, -\lambda^\mathsf{T} x + \nabla \phi_\lambda(\beta)^\mathsf{T} y \leq r(\beta), \forall \beta \in D_1(0)\}, \tag{9}$$

is S-free. When β satisfies $\lambda^\mathsf{T} a + d^\mathsf{T} \beta < 0$, it should be that $r(\beta) = 0$, as in this case the inequality is the same in C_{ϕ_λ} and $C_{G(\lambda)}$ (since $\nabla \phi_\lambda(\beta) = \beta$), and has an exposing point in S. Thus, in the following we find $r(\beta)$ when $\lambda^\mathsf{T} a + d^\mathsf{T} \beta \geq 0$.

Construction of $r(\beta)$: Let $\beta \in D_1(0)$ be such that $\lambda^\mathsf{T} a + d^\mathsf{T} \beta \geq 0$. The valid inequality $-\lambda^\mathsf{T} x + \nabla \phi_\lambda(\beta)^\mathsf{T} y \leq 0$ is not exposed by any point in $S \cap C_{\phi_\lambda}$ (see [35] for a proof). However, it is exposed by $(x_\beta, \beta) \in S_{\leq 0}$, where x_β is the optimal solution of (6) (see proof sketch of Proposition 3). Note that $(x_\beta, \beta) \in H_0 = \{(x,y) \, : \, a^\mathsf{T} x + d^\mathsf{T} y = 0\}$, as otherwise we can scale it so that it belongs to S.

To find $r(\beta)$ we first find a sequence of points, $(x_n, y_n)_{n \in \mathbb{N}}$, in $S_{\leq 0}$ that converge to (x_β, β). We take $y_n = \beta$ and $x_n \in \langle \lambda, a \rangle$ such that $\|x_n\| = 1$, $a^\mathsf{T} x_n + d^\mathsf{T} \beta < 0$, and $x_n \to x_\beta$. The requirement $x_n \in \langle \lambda, a \rangle$ is fairly technical and involves also assuming $\lambda \neq \pm a$. However, we can show that such requirement can always be assumed without loss of generality. We can also show this sequence always exists when $\|d\| < \|a\| = 1$. Then, we scale the sequence: $z_n = -\frac{(x_n, \beta)}{a^\mathsf{T} x_n + d^\mathsf{T} \beta} \in S$. This last scaled sequence diverges, as the denominator goes to 0. The idea is that the violation $(-\lambda, \nabla \phi_\lambda(\beta))^\mathsf{T} z_n$ given by this sequence will give us, in the limit, the maximum relaxation that will ensure S-freeness. In Fig. 3 we illustrate what such a sequence looks like in the case of Fig. 2b. Thus, we define

$$r(\beta) = - \lim_{n \to \infty} \frac{-\lambda^\mathsf{T} x_n + \nabla \phi_\lambda(\beta)^\mathsf{T} \beta}{a^\mathsf{T} x_n + d^\mathsf{T} \beta}. \tag{10}$$

With these considerations, we can prove the following.

Lemma 2. *Let $\beta \in D_1(0)$ such that $\lambda^\mathsf{T} a + d^\mathsf{T} \beta \geq 0$. For any sequence $(x_n)_{n \in \mathbb{N}}$ as described above and $r(\beta)$ defined as (10), it holds that*

$$r(\beta) = \frac{d^\mathsf{T} \beta + \lambda^\mathsf{T} a \phi_\lambda(\beta)}{\phi_\lambda(\beta) + d^\mathsf{T} \beta \lambda^\mathsf{T} a}.$$

Finally, we show that our construction yields the desired result.

Theorem 4. *Let $S_{\leq 0}$, H and λ be defined as in Theorem 3, but with $\|a\| > \|d\|$ instead. Define $r(\beta)$ as in Lemma 2 for $\lambda^\mathsf{T} a + d^\mathsf{T} \beta \geq 0$ and 0 otherwise. Then, C defined in (9) is maximal $S_{\leq 0}$-free with respect to H and contains $(\bar{x}, \bar{y})$. Furthermore,*

$$C = \left\{ (x,y) : \begin{array}{ll} \phi_\lambda(y) \leq \lambda^\mathsf{T} x & \text{if } \lambda^\mathsf{T} a \|y\| + d^\mathsf{T} y \leq 0 \\ \phi_\lambda\left(y - \dfrac{d}{1 - \|d\|^2} \right) \leq \lambda^\mathsf{T} \left(x + \dfrac{a}{1 - \|d\|^2} \right) & \text{otherwise} \end{array} \right\}. \tag{11}$$

Proof (sketch). The proof relies on both expressions for $r(\beta)$ given in (10) and in Lemma 2. The latter is used for proving S-freeness of C, and the former is used for proving maximality. The sequence $(z_n)_{n \in \mathbb{N}}$ we used in defining $r(\beta)$ in (10) exposes the corresponding inequality *at infinity*, as per Definition 3. With Theorem 1 we conclude. To obtain a closed-form expression of C, we conjecture that, since C is obtained from C_{ϕ_λ} by "translating" its inequalities, a translation of (7) might yield C. After some algebraic manipulation we obtain (11).

The last theorem shows an explicit formula for obtaining maximal S-free sets (recall that $S = S_{\leq 0} \cap H$). In Fig. 5 (appendix) we show similar plots to Fig. 2 with C instead of C_{ϕ_λ}.

6   Conclusions

In this work we have shown how to construct *maximal* quadratic-free sets. Using the *intersection cut* framework, these sets can be used for generating deep cutting planes for quadratically constrained problems. The maximal quadratic-free sets we construct in this work allow for an *efficient* computation of the corresponding intersection cuts, as all of them have a closed-form expressions: see (3), (7), (11) for the closed-form expressions of C_λ, C_{ϕ_λ} and C, respectively, and (8) for the explicit description of the ϕ_λ function. The case of $C_{G(\lambda)}$ can be found in [35].

We strongly believe that, by carefully laying a theoretical framework for quadratic-free sets, this work provides an important contribution to the understanding and future computational development of non-convex quadratically constrained optimization problems. The empirical performance of these intersection cuts remains to be seen, and it is ongoing work. Future work includes a careful comparison of the resulting intersection cuts and the related approaches mentioned in the literature review, in particular the work of [10] and [15].

Acknowledgements. We are indebted to Franziska Schlösser for several inspiring conversations. We would like to thank Stefan Vigerske, Antonia Chmiela, Ksenia Bestuzheva and Nils-Christian Kempke for helpful discussions. We would also like to thank the three anonymous reviewers for their valuable feedback. Lastly, we would like to acknowledge the support of the IVADO Institute for Data Valorization for their support through the IVADO Post-Doctoral Fellowship program and to the IVADO-ZIB academic partnership. The described research activities are funded by the German Federal Ministry for Economic Affairs and Energy within the project EnBA-M (ID: 03ET1549D). The work for this article has been (partly) conducted within the Research Campus MODAL funded by the German Federal Ministry of Education and Research (BMBF grant number 05M14ZAM).

Appendix

Example 1 (Homogeneous case). Consider the set $S_{\leq 0}$, defined as

$$S_{\leq 0} = \{(x_1, x_2, y) \in \mathbb{R}^3 \ : \ \|x\| \leq |y|, \quad a^\mathsf{T} x + dy \leq 0\}$$

with $a = (-1/\sqrt{2}, 1/\sqrt{2})^\mathsf{T}$ and $d = 1/\sqrt{2}$, and $(\bar{x}, \bar{y}) = (-1, -1, 0)^\mathsf{T}$. This point satisfies the linear inequality in $S_{\leq 0}$, but it is not in $S_{\leq 0}$. It is not too hard to check that $G(\lambda) = \{-1\}$. In Fig. 1 we show $S_{\leq 0}$, the $S_{\leq 0}$-free set given by C_λ and the set $C_{G(\lambda)}$. In this case $\|a\| = 1 > 1/\sqrt{2} = |d|$, so we have no guarantee on the $S_{\leq 0}$-freeness of $C_{G(\lambda)}$. Even more, it is not $S_{\leq 0}$-free.

Moving forward, since $\lambda^\mathsf{T} a = 0$ we have

$$\lambda^\mathsf{T} a \|y\| + d^\mathsf{T} y \leq 0 \Longleftrightarrow y \leq 0.$$

A simple calculation using (8) yields

$$\phi_\lambda(y) = \begin{cases} -y, & \text{if } y \leq 0 \\ \frac{y}{\sqrt{2}} & \text{if } y > 0 \end{cases}$$

In Fig. 4 we show the set C_{ϕ_λ}, which is maximal $S_{\leq 0}$-free.

Example 2 (Non-homogeneous case). We continue with $S_{\leq 0}$ defined in Example 1, but we now look for maximality with respect to

$$H = \{(x, y) \in \mathbb{R}^{n+m} \ : \ a^\mathsf{T} x + d^\mathsf{T} y = -1\}.$$

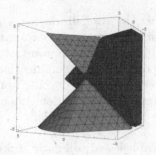

Fig. 4. $S_{\leq 0}$ in Example 1 (orange) and C_{ϕ_λ} set (blue). The latter is maximal $S_{\leq 0}$-free. (Color figure online)

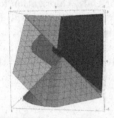 

(a) $S_{\leq 0}$ (orange), H (green) and C_1 (blue). In this case C_1 is no longer $S_{\leq 0}$-free.

(b) Projection onto (x_1, x_2) of $S_{\leq 0} \cap H$ (orange) and $C_1 \cap H$ (blue).

Fig. 5. Plots of $S_{\leq 0}$, H and C_1 as defined in Example 2 showing that C_1 is maximal $S_{\leq 0}$-free with respect to H.

In this case, $C_{\phi_\lambda} \cap H$ is not maximal $S_{\leq 0} \cap H$-free, as shown in Fig. 2. Since $\lambda = \frac{1}{\sqrt{2}}(-1, -1)^\mathsf{T}$, we see that

$$C_{\phi_\lambda} = \{(x, y) : \frac{1}{\sqrt{2}}(x_1 + x_2) - y \leq 0, \tag{12a}$$

$$\frac{1}{\sqrt{2}}(x_1 + x_2) + \frac{1}{\sqrt{2}}y \leq 0\}. \tag{12b}$$

It is not hard to check that $-(\frac{1}{\sqrt{2}}, \frac{1}{\sqrt{2}}, \sqrt{2}) \in S_{\leq 0} \cap H \cap C_{\phi_\lambda}$ exposes inequality (12a). This is the tangent point in Fig. 2b. On the other hand, (12b), which is obtained from $\beta = 1$, does not have an exposing point in $S_{\leq 0} \cap H \cap C_{\phi_\lambda}$, and corresponds to an inequality we should relax as per our discussion. This inequality, however, is exposed by $(x_\beta, \beta) = (0, -1, 1) \in S_{\leq 0} \cap C_{\phi_\lambda}$. Consider now the sequence defined as

$$(x_n, \beta) = \left(\frac{1}{\sqrt{n^2 + 1}}, -\frac{n}{\sqrt{n^2 + 1}}, 1\right) \in S_{\leq 0}.$$

Clearly the limit of this sequence is $(0, -1, 1)$ and

$$a^\mathsf{T} x_n + d^\mathsf{T} \beta = \frac{1}{\sqrt{2}} \left(-\frac{1}{\sqrt{n^2 + 1}} - \frac{n}{\sqrt{n^2 + 1}} + 1 \right) < 0.$$

Now we let

$$z_n = -\frac{(x_n, \beta)}{a^\mathsf{T} x_n + d^\mathsf{T} \beta} \in S_{\leq 0} \cap H.$$

which diverges. In Fig. 3, we plot the first two components of the sequence $(z_n)_{n \in \mathbb{N}}$ along with $S_{\leq 0} \cap H$ and $C_{\phi_\lambda} \cap H$. The sequence $(z_n)_{n \in \mathbb{N}}$ moves along the boundary of $S_{\leq 0} \cap H$ towards an "asymptote" from where we deduce $r(\beta)$.

In this case $r(-1) = 0$, and it can be checked that $r(1) = 1$ using its formula. Now, let

$$C_1 = \{(x, y) : -\lambda^\mathsf{T} x + \nabla \phi_\lambda(\beta)^\mathsf{T} y \leq r(\beta), \text{ for all } \beta \in D_1(0)\}$$

$$= \{(x, y) : \frac{1}{\sqrt{2}}(x_1 + x_2) - y \leq 0, \frac{1}{\sqrt{2}}(x_1 + x_2) + \frac{1}{\sqrt{2}} y \leq 1\}.$$

Figure 5 shows the same plots as Fig. 2 with C_1 instead of C_{ϕ_λ}.

References

1. Andersen, K., Jensen, A.N.: Intersection cuts for mixed integer conic quadratic sets. In: Goemans, M., Correa, J. (eds.) IPCO 2013. LNCS, vol. 7801, pp. 37–48. Springer, Heidelberg (2013). https://doi.org/10.1007/978-3-642-36694-9_4
2. Andersen, K., Louveaux, Q., Weismantel, R.: An analysis of mixed integer linear sets based on lattice point free convex sets. Math. Oper. Res. **35**(1), 233–256 (2010)
3. Andersen, K., Louveaux, Q., Weismantel, R., Wolsey, L.A.: Inequalities from two rows of a simplex tableau. In: Fischetti, M., Williamson, D.P. (eds.) IPCO 2007. LNCS, vol. 4513, pp. 1–15. Springer, Heidelberg (2007). https://doi.org/10.1007/978-3-540-72792-7_1
4. Balas, E.: Intersection cuts–a new type of cutting planes for integer programming. Oper. Res. **19**(1), 19–39 (1971). https://doi.org/10.1287/opre.19.1.19
5. Basu, A., Conforti, M., Cornuéjols, G., Zambelli, G.: Maximal lattice-free convex sets in linear subspaces. Math. Oper. Res. **35**(3), 704–720 (2010). https://doi.org/10.1287/moor.1100.0461
6. Basu, A., Conforti, M., Cornuéjols, G., Zambelli, G.: Minimal inequalities for an infinite relaxation of integer programs. SIAM J. Discrete Math. **24**(1), 158–168 (2010). https://doi.org/10.1137/090756375
7. Basu, A., Dey, S.S., Paat, J.: Nonunique lifting of integer variables in minimal inequalities. SIAM J. Discrete Math. **33**(2), 755–783 (2019). https://doi.org/10.1137/17m1117070
8. Belotti, P., Góez, J.C., Pólik, I., Ralphs, T.K., Terlaky, T.: On families of quadratic surfaces having fixed intersections with two hyperplanes. Discrete Appl. Math. **161**(16–17), 2778–2793 (2013)
9. Belotti, P., Góez, J.C., Pólik, I., Ralphs, T.K., Terlaky, T.: A conic representation of the convex hull of disjunctive sets and conic cuts for integer second order cone optimization. In: Al-Baali, M., Grandinetti, L., Purnama, A. (eds.) Numerical Analysis and Optimization. SPMS, vol. 134, pp. 1–35. Springer, Cham (2015). https://doi.org/10.1007/978-3-319-17689-5_1

10. Bienstock, D., Chen, C., Munoz, G.: Outer-product-free sets for polynomial optimization and oracle-based cuts. Math. Program. 1–44 (2020)
11. Bienstock, D., Chen, C., Muñoz, G.: Intersection cuts for polynomial optimization. In: Lodi, A., Nagarajan, V. (eds.) IPCO 2019. LNCS, vol. 11480, pp. 72–87. Springer, Cham (2019). https://doi.org/10.1007/978-3-030-17953-3_6
12. Bodur, M., Dash, S., Günlük, O.: Cutting planes from extended LP formulations. Math. Program. **161**(1–2), 159–192 (2017)
13. Bonami, P., Linderoth, J., Lodi, A.: Disjunctive cuts for mixed integer nonlinear programming problems. In: Progress in Combinatorial Optimization, pp. 521–541 (2011). (chapter 18)
14. Borozan, V., Cornuéjols, G.: Minimal valid inequalities for integer constraints. Math. Oper. Res. **34**(3), 538–546 (2009). https://doi.org/10.1287/moor.1080.0370
15. Burer, S., Kılınç-Karzan, F.: How to convexify the intersection of a second order cone and a nonconvex quadratic. Math. Program. **162**(1–2), 393–429 (2017)
16. Conforti, M., Cornuéjols, G., Daniilidis, A., Lemaréchal, C., Malick, J.: Cut-generating functions and S-free sets. Math. Oper. Res. **40**(2), 276–391 (2015). https://doi.org/10.1287/moor.2014.0670
17. Conforti, M., Cornuéjols, G., Zambelli, G.: Corner polyhedron and intersection cuts. Surv. Oper. Res. Manage. Sci. **16**(2), 105–120 (2011). https://doi.org/10.1016/j.sorms.2011.03.001
18. Cornuéjols, G., Wolsey, L., Yıldız, S.: Sufficiency of cut-generating functions. Math. Program. **152**(1–2), 643–651 (2015)
19. Dey, S.S., Wolsey, L.A.: Lifting integer variables in minimal inequalities corresponding to lattice-free triangles. In: Lodi, A., Panconesi, A., Rinaldi, G. (eds.) IPCO 2008. LNCS, vol. 5035, pp. 463–475. Springer, Heidelberg (2008). https://doi.org/10.1007/978-3-540-68891-4_32
20. Dey, S.S., Wolsey, L.A.: Constrained infinite group relaxations of MIPs. SIAM J. Optim. **20**(6), 2890–2912 (2010). https://doi.org/10.1137/090754388
21. Fischetti, M., Ljubić, I., Monaci, M., Sinnl, M.: Intersection cuts for bilevel optimization. In: Louveaux, Q., Skutella, M. (eds.) IPCO 2016. LNCS, vol. 9682, pp. 77–88. Springer, Cham (2016). https://doi.org/10.1007/978-3-319-33461-5_7
22. Fischetti, M., Monaci, M.: A branch-and-cut algorithm for mixed-integer bilinear programming. Eur. J. Oper. Res. (2019). https://doi.org/10.1016/j.ejor.2019.09.043
23. Glover, F.: Convexity cuts and cut search. Oper. Res. **21**(1), 123–134 (1973). https://doi.org/10.1287/opre.21.1.123
24. Goberna, M., González, E., Martínez-Legaz, J., Todorov, M.: Motzkin decomposition of closed convex sets. J. Math. Anal. Appl. **364**(1), 209–221 (2010). https://doi.org/10.1016/j.jmaa.2009.10.015
25. Gomory, R.E., Johnson, E.L.: Some continuous functions related to corner polyhedra. Math. Program. **3**–3(1), 23–85 (1972). https://doi.org/10.1007/bf01584976
26. Kılınç-Karzan, F.: On minimal valid inequalities for mixed integer conic programs. Math. Oper. Res. **41**(2), 477–510 (2015)
27. Kılınç-Karzan, F., Yıldız, S.: Two-term disjunctions on the second-order cone. Math. Program. **154**(1–2), 463–491 (2015)
28. Lasserre, J.B.: Global optimization with polynomials and the problem of moments. SIAM J. Optim. **11**(3), 796–817 (2001)
29. Laurent, M.: Sums of squares, moment matrices and optimization over polynomials. In: Putinar, M., Sullivant, S. (eds.) Emerging Applications of Algebraic Geometry. The IMA Volumes in Mathematics and its Applications, vol. 149, pp. 157–270. Springer, New York (2009). https://doi.org/10.1007/978-0-387-09686-5_7

30. McCormick, G.P.: Computability of global solutions to factorable nonconvex programs: part I—convex underestimating problems. Math. Program. **10**(1), 147–175 (1976). https://doi.org/10.1007/bf01580665
31. Conforti, M., Cornuejols, G., Zambelli, G.: Integer Programming. Springer, Heidelberg (2014). https://doi.org/10.1007/978-3-319-11008-0
32. Modaresi, S., Kılınç, M.R., Vielma, J.P.: Split cuts and extended formulations for mixed integer conic quadratic programming. Oper. Res. Lett. **43**(1), 10–15 (2015)
33. Modaresi, S., Kılınç, M.R., Vielma, J.P.: Intersection cuts for nonlinear integer programming: convexification techniques for structured sets. Math. Program. **155**(1–2), 575–611 (2016)
34. Morán, D., Dey, S.S.: On maximal S-free convex sets. SIAM J. Discrete Math. **25**(1), 379–393 (2011). https://doi.org/10.1137/100796947
35. Muñoz, G., Serrano, F.: Maximal quadratic-free sets. arXiv preprint arXiv:1911.12341 (2019)
36. Serrano, F.: Intersection cuts for factorable MINLP. In: Lodi, A., Nagarajan, V. (eds.) IPCO 2019. LNCS, vol. 11480, pp. 385–398. Springer, Cham (2019). https://doi.org/10.1007/978-3-030-17953-3_29
37. Shahabsafa, M., Góez, J.C., Terlaky, T.: On pathological disjunctions and redundant disjunctive conic cuts. Oper. Res. Lett. **46**(5), 500–504 (2018)
38. Shor, N.Z.: Quadratic optimization problems. Sov. J. Comput. Syst. Sci. **25**, 1–11 (1987)
39. Tuy, H.: Concave programming with linear constraints. In: Doklady Akademii Nauk, vol. 159, pp. 32–35. Russian Academy of Sciences (1964)
40. Yıldız, S., Kılınç-Karzan, F.: Low-complexity relaxations and convex hulls of disjunctions on the positive semidefinite cone and general regular cones. Optim. Online (2016)

On Generalized Surrogate Duality
in Mixed-Integer Nonlinear Programming

Benjamin Müller[1]([✉])[ID], Gonzalo Muñoz[2][ID], Maxime Gasse[3][ID],
Ambros Gleixner[1][ID], Andrea Lodi[3][ID], and Felipe Serrano[1][ID]

[1] Zuse Institute Berlin, Berlin, Germany
{benjamin.mueller,gleixner,serrano}@zib.de
[2] Universidad de O'Higgins, Rancagua, Chile
gonzalo.munoz@uoh.cl
[3] Polytechnique Montréal, Montréal, Canada
{maxime.gasse,andrea.lodi}@polymtl.ca

Abstract. Due to both theoretical and practical considerations, relaxations of MINLPs are usually required to be convex. Nonetheless, current optimization solvers can often successfully handle a moderate presence of nonconvexities, which opens the door for the use of potentially tighter nonconvex relaxations. In this work, we exploit this fact and make use of a nonconvex relaxation obtained via aggregation of constraints: a *surrogate* relaxation. These relaxations were actively studied for linear integer programs in the 70s and 80s, but they have been scarcely considered since. We revisit these relaxations in an MINLP setting and show the computational benefits and challenges they can have. Additionally, we study a generalization of such relaxation that allows for multiple aggregations simultaneously and present the first algorithm that is capable of computing the best set of aggregations. We propose a multitude of computational enhancements for improving its practical performance and evaluate the algorithm's ability to generate strong dual bounds through extensive computational experiments.

Keywords: Surrogate relaxation · MINLP · Nonconvex optimization

1 Introduction

We consider a mixed-integer nonlinear program (MINLP) of the form

$$\min_{x \in X} \left\{ c^\mathsf{T} x : g_i(x) \leq 0 \text{ for all } i \in \mathcal{M} \right\}, \tag{1}$$

where $X := \{ x \in \mathbb{R}^{n-p} \times \mathbb{Z}^p : Ax \leq b \}$ is a compact mixed-integer linear set, each $g_i : \mathbb{R}^n \to \mathbb{R}$ is a factorable continuous function [36], and $\mathcal{M} := \{1, \ldots, m\}$. Many real-world applications are inherently nonlinear, and can be formulated as a MINLP. See, e.g., [26] for an overview.

The state-of-the-art algorithm for solving nonconvex MINLPs to global ϵ-optimality is spatial branch and bound, see, e.g., [28, 42, 43], whose performance

© Springer Nature Switzerland AG 2020
D. Bienstock and G. Zambelli (Eds.): IPCO 2020, LNCS 12125, pp. 322–337, 2020.
https://doi.org/10.1007/978-3-030-45771-6_25

highly depends on the tightness of the relaxations used. Those relaxations are typically convex. As a result of the rapid progress during the last decades, current solvers can often handle a moderate presence of nonconvex constraints efficiently. This progress opens the door for practical use of potentially tighter *nonconvex* relaxations in MINLP solvers. In this paper, we explore a nonconvex relaxation referred to as *surrogate relaxation* [19].

Definition 1 (Surrogate Relaxation). *For a given $\lambda \in \mathbb{R}_+^m$, we call the following optimization problem a* surrogate relaxation *of* (1):

$$S(\lambda) := \min_{x \in X} \left\{ c^\mathsf{T} x : \sum_{i \in M} \lambda_i g_i(x) \leq 0 \right\}. \tag{2}$$

Denote by $\mathcal{F} \subseteq \mathbb{R}^n$ the feasible region of (1), and $S_\lambda \subseteq \mathbb{R}^n$ that of (2). Clearly, $\mathcal{F} \subseteq S_\lambda$ holds for every $\lambda \in \mathbb{R}_+^m$, and as such (2) provides a valid lower bound of (1). Moreover, solving (2) might be computationally more convenient than solving (1), since there is only one nonconvex nonlinear constraint in $S(\lambda)$.

The quality of the bound provided by $S(\lambda)$ may be highly dependent on λ, and therefore it is natural to consider the *surrogate dual* problem.

Definition 2 (Surrogate Dual). *We call the following optimization problem the* surrogate dual *of* (1):

$$\sup_{\lambda \in \mathbb{R}_+^m} S(\lambda). \tag{3}$$

The function S is only lower semi-continuous [25], thus, there might be no λ such that $S(\lambda)$ is equal to (3). The surrogate dual is closely related to the *Lagrangian dual*, but provides a bound that is at least as good [25,33]. In contrast to the Lagrangian, which is always concave, S is only quasi-concave [25].

Contribution. In this paper, we revisit surrogate duality in the context of mixed-integer nonlinear programming. To the best of our knowledge, surrogate relaxations have never been considered for solving general MINLPs. The first contribution of the paper is an algorithm capable of solving a *generalized* surrogate dual that allows for multiple aggregations of the nonlinear constraints simultaneously. We also prove its convergence guarantees. Secondly, we present computational enhancements to make the algorithm practical. Our developed algorithm allows us to compare the performance of the classic surrogate relaxation with the generalized one. Finally, we provide a broad computational analysis using publicly available benchmark instances, and we show the practical interest in our proposed approach.

2   Background

Surrogate constraints were first introduced by Glover [19] in the context of zero-one linear integer programming problems. Extensions were provided by Balas [7] and Geoffrion [18], which were then analyzed from a unified perspective by

Glover [20]. A theoretical analysis of surrogate duality in a nonlinear setting was first presented by Greenberg and Pierskalla [25]. They also proposed a generalization using multiple disjoint aggregation constraints. A similar generalization allowing multiple aggregations was later proposed by Glover [21]. These were proposed without computational evaluation.

The first algorithmic method for solving (3) is attributed to Banerjee [9]. In the context of integer linear programming, he proposed a Benders' approach similar to the one considered by us. Karwan [32] expanded on this approach, including a refinement of that of Banerjee and subgradient-based methods. Independently, Dyer [15] proposed similar methods to those of Karwan. Karwan and Rardin [33,34] further developed these approaches. Other search procedures involve consecutive Lagrangian dual searches [35,44] and heuristics [17].

Surrogate constraints have been used in various applications: in primal heuristics for IPs [22], knapsack problems with a quadratic objective function [14], the job shop problem [16], generalized assignment problems [40], among others. We refer the reader to [5,23] for reviews on surrogate duality methods.

To the best of our knowledge, the efforts for practical implementations of multiplier search methods have mainly focused on *linear* integer programs. This focus can be explained by the maturity of computational optimization tools available during the time where most of these implementations have been developed. We are only aware of two exceptions: the entropy approach to nonlinear programming (see [48,53]) which uses an entropy-based surrogate reformulation instead of a weighted sum of the constraints; and the work by Nakagawa [39], who considered *separable* nonlinear integer programming and presented a novel, albeit expensive, algorithm for solving the surrogate dual. Regarding the generalization of the surrogate dual that considers multiple aggregated constraints, we are not aware of any work considering a multiplier search method with provable guarantees or a computational implementation of a heuristic approach.

3 Generalized Surrogate Duality

We consider a generalization of surrogate relaxations that was introduced by Glover [21]. Instead of a single aggregation, as in (2), the generalization allows for $K \in \mathbb{N}$ aggregations. These are encoded by the nonnegative vector $\lambda = (\lambda^1, \lambda^2, \ldots, \lambda^K) \in \mathbb{R}_+^{Km}$

$$\sum_{i \in \mathcal{M}} \lambda_i^k g_i(x) \leq 0, \quad k \in \{1, \ldots, K\}. \tag{4}$$

Similar to S_λ, for a vector $\lambda \in \mathbb{R}_+^{Km}$ the feasible region of the K *-surrogate relaxation* is given by the intersection $S_\lambda^K := \bigcap_{k=1}^{K} S_{\lambda^k}$, where S_{λ^k} is the feasible region of the surrogate relaxation $S(\lambda^k)$ for $\lambda^k \in \mathbb{R}_+^m$. It clearly follows that S_λ^K is a relaxation for (1). The best dual bound for (1) generated by a K-surrogate relaxation is given by

$$\sup_{\lambda \in \mathbb{R}_+^{Km}} S^K(\lambda), \tag{5}$$

which we call the K-*surrogate dual*. Note that scaling each $\lambda^k \in \mathbb{R}_+^m$ individually by a positive scalar does not affect the value of $S^K(\lambda)$, therefore, it is possible to impose additional normalization constraints $\left\|\lambda^k\right\|_1 \leq 1$ for each k. Using the proof idea of the case $K = 1$ by Glover [21], we can prove the following.

Proposition 1. *If g_i is continuous for every $i \in \mathcal{M}$ and X is compact, then $S^K : \mathbb{R}_+^{Km} \rightarrow \mathbb{R}$ is lower semi-continuous for any choice of K.*

One important difference to the classic surrogate dual is that $S^K(\lambda)$ is no longer quasi-concave. Due to this, subgradient-based methods, as in [32], do not solve (5) to global optimality. Even though (5) is substantially more difficult to solve than (3), it may also provide considerably stronger bounds for (1):

Proposition 2. *The following inequality holds for any $K \in \mathbb{N}$:*

$$\sup_{\lambda \in \mathbb{R}_+^{Km}} S^K(\lambda) \leq \sup_{\lambda \in \mathbb{R}_+^{(K+1)m}} S^{K+1}(\lambda). \tag{6}$$

In Sect. 6, we analyze computationally how big the gap between different values of K can be.

4 An Algorithm for the K-surrogate Dual

We now show how to solve the surrogate dual with a Benders' approach. This type of algorithm was presented independently for the $K = 1$ case by Banerjee [9], Karwan [32], and Dyer [15]. Here, we generalize it for arbitrary K and prove its convergence guarantees. To the best of our knowledge, our generalization is the first algorithm in the literature that solves (5).

4.1 A Benders' Approach

We use an iterative approach that alternates between a master- and sub-problem. Assume that we have a solution $\bar{x}$ of $S^K(\lambda)$. The master problem searches for the next multipliers ensuring that $\bar{x}$ is not considered in later iterations, i.e., it computes a vector $\tilde{\lambda}$ that satisfies $\sum_{i \in \mathcal{M}} \tilde{\lambda}_i^k g_i(\bar{x}) > 0$ for some $k \in \{1, \ldots, K\}$. One way of achieving this is through the following disjunctive program:

$$\max \quad \Psi$$
$$\text{s.t.} \quad \bigvee_{k=1}^{K} \left(\sum_{i \in \mathcal{M}} \lambda_i^k g_i(\bar{x}) \geq \Psi \right) \quad \text{for all } \bar{x} \in \mathcal{X}, \tag{7}$$
$$\left\|\lambda^k\right\|_1 \leq 1, \lambda^k \geq 0 \quad \text{for all } k \in \{1, \ldots, K\},$$

where $\mathcal{X} \subseteq X$ is the set of generated points of the sub-problems. The sub-problem solves $S^K(\tilde{\lambda})$ and thus provides a valid dual bound. The resulting scheme is formalized in Algorithm 1.

Algorithm 1: Algorithm for the K-surrogate dual.

Input: MINLP of the form (1), $K \in \mathbb{N}$, threshold $\epsilon > 0$
Output: dual bound $D \in \mathbb{R}$ for (1)

1 initialize $\lambda := 0 \in \mathbb{R}_+^{Km}$, $\Psi := \infty$, $\mathcal{X} := \emptyset$, $D := -\infty$
2 **while** $\Psi \geq \epsilon$ **do**
3     $\bar{x} := \mathrm{argmin}_x \{ c^\mathsf{T} x : x \in S_\lambda^K \}$
4     $D := \max\{D, c^\mathsf{T}\bar{x}\}$
5     $\mathcal{X} := \mathcal{X} \cup \{\bar{x}\}$
6     $(\lambda, \Psi) :=$ optimal solution of (7) for $\mathcal{X}$

7 **return** D

In order to solve (7) we use the following big-M formulation:

$$\max \ \Psi$$

$$\text{s.t.} \sum_{i \in \mathcal{M}} \lambda_i^k g_i(\bar{x}) \geq \Psi - M(1 - z_k^{\bar{x}}) \quad \text{for all } k \in \{1, \ldots, K\}, \bar{x} \in \mathcal{X},$$

$$\sum_{k=1}^K z_k^{\bar{x}} = 1 \qquad\qquad\qquad \text{for all } \bar{x} \in \mathcal{X}, \tag{8}$$

$$z_k^{\bar{x}} \in \{0, 1\} \qquad\qquad\qquad \text{for all } k \in \{1, \ldots, K\}, \bar{x} \in \mathcal{X},$$

$$\left\| \lambda^k \right\|_1 \leq 1, \ \lambda^k \geq 0 \qquad\quad \text{for all } k \in \{1, \ldots, K\},$$

where M is a large constant. A binary variable $z_k^{\bar{x}}$ indicates if the k-th disjunction of (7) is used to cut off the point $\bar{x} \in \mathcal{X}$. It is well known that big-M formulations are not considered strong in MILPs, given their usual weak LP relaxations. Other formulations in extended spaces can yield better theoretical guarantees (e.g., [8,10,49]). However, these formulations typically require to add copies of variables depending on the number of disjunctions, which in our setting is rapidly increasing. Furthermore, as we discuss below, we do not require a tight LP relaxation of (7), and thus, we opted to use (8).

4.2 Convergence

The fact that the dual bounds obtained by Algorithm 1 converge to the optimal value of the K-surrogate dual is summarized in the following theorem.

Theorem 1. *Denote by $(\lambda^t, \Psi^t)_{t \in \mathbb{N}}$ the sequence of values obtained after solving (7) in Algorithm 1 for $\epsilon = 0$. The algorithm either **(a)** terminates in T steps, i.e., $\Psi^T = 0$, in which case $\max_{1 \leq t \leq T} S^K(\lambda^t)$ is equal to (5), or **(b)** $\sup_{t \geq 1} S^K(\lambda^t)$ is equal to (5).*

The result for $K = 1$ was originally proven in [34], which we generalize to $K > 1$. The key in the proof is to show that $(\Psi^t)_{t \in \mathbb{N}}$ converges to zero in Case (b). For this, we use the compactness of the feasible regions of both the master and sub-problems and the continuity of the constraint functions g_i. In addition, this shows that Algorithm 1 terminates after finitely many steps for any $\epsilon > 0$.

5 Computational Enhancements

In what follows, we present computational enhancements that speed up Algorithm 1 and improve the quality of the dual bound that can be achieved in practice.

Refined MILP Relaxation. Instead of only using $Ax \leq b$ of (1), we exploit an LP relaxation of (1) that is available in LP-based spatial branch and bound. This LP contains constraints that have been derived from, e.g., integrality restrictions of variables (e.g., MIR cuts [41] and Gomory cuts [24]), gradient cuts [30], RLT cuts [46], SDP cuts [47], or underestimators for each g_i. We also make use of objective cutoff information. If there is a feasible solution x^* to (1), then the relaxation can be strengthened by the inequality $c^\mathsf{T} x \leq c^\mathsf{T} x^*$. This inequality preserves all optimal solutions of (1) and can improve the optimal value of (3).

Early Stopping in the Sub-problem. One crucial ingredient to speed up Algorithm 1, proposed in [32] and [15] independently for $K = 1$, is an early stopping criterion for $S^K(\lambda)$. If Algorithm 1 has proven a dual bound D in some previous iteration, we can stop the solving process of $S^K(\lambda)$ if some $\bar{x} \in S_\lambda^K$ with $c^\mathsf{T}\bar{x} \leq D$ has been found. The point $\bar{x}$ provides a new inequality for (8) and shows that λ will not lead to a better dual bound. All convergence/correctness statements remain valid, and a good quality $\bar{x}$ can often be found fast by heuristics for MINLPs. Furthermore, we can apply the same idea with other choices of D, in which case D would act as a *target* dual bound that we want to prove.

Early Stopping of the Master Problem. While the optimal value of Ψ is needed to decide whether the algorithm terminated, to ensure progress of Algorithm 1 it is enough to compute a feasible point $(\Psi, \lambda^1, \ldots, \lambda^K)$ of (8) with $\Psi > 0$. We balance these two opposing forces with the following early stopping method. While solving (8), we have access to valid dual and primal bounds. Let Ψ_p^t and Ψ_d^t be the primal and dual bounds obtained from the master problem in iteration t of Algorithm 1. We stop the master problem in iteration $t+1$ as soon as $\Psi_p^{t+1} \geq \alpha \Psi_d^t$ holds for a fixed $\alpha \in (0,1]$. The parameter α controls the trade-off between proving a good dual bound Ψ_d^{t+1} and saving time for solving the master problem. In our experiments, we observed that setting α to 0.2 performs well.

Support Stabilization. Much like in column generation approaches, Algorithm 1 can suffer from convergence issues. Deriving "stabilization" techniques that can avoid oscillations of the λ variables and tailing-off effects, among others, are common goals for improving performance, see, e.g., [4,6,37]. Our following *support stabilization* technique was developed to handle some of these issues. Specifically, once Algorithm 1 finds a multiplier vector that improves the overall dual bound, we restrict the support to that of the improving dual multiplier. This step restricts the search space and improves solution times of (8) considerably. Once stalling is detected, we remove the support restriction until another multiplier vector that improves the dual bound is found.

Trust-Region Stabilization. Even using the previous stabilization technique, the non-zero entries of the λ vectors can (and do, in practice) vary significantly from iteration to iteration. To remedy this, we incorporated a classic stabilization technique: a *box trust-region* stabilization [13]. Given a reference solution $(\hat{\lambda}^1, \ldots, \hat{\lambda}^k)$, we impose $\|(\lambda^1, \ldots, \lambda^k) - (\hat{\lambda}^1, \ldots, \hat{\lambda}^k)\|_\infty \leq \delta$ in (8) for some parameter δ. This prevents the λ variables from oscillating excessively, and carefully updating $(\hat{\lambda}^1, \ldots, \hat{\lambda}^k)$ and δ can maintain the convergence guarantees of the algorithm. In our implementation, we maintain a fixed $(\hat{\lambda}^1, \ldots, \hat{\lambda}^k)$ until we obtain an improvement or the algorithm stalls. When any of this happens, we remove the box and compute a new $(\hat{\lambda}^1, \ldots, \hat{\lambda}^k)$ with (8) without any stabilization added.

Other Enhancements. For the sake of completeness, we briefly discuss some natural ideas that can be considered but that we identified to not work in our setting. These are not included in our final implementation.

- *Dual objective cutoff in the sub-problem:* The sequence of dual bounds D provided by $c^\mathsf{T} x^*$ in Step 3 might not be monotone. Thus, one could think of adding a *dual objective cutoff* $c^\mathsf{T} x \geq D$ to the sub-problem $S(\lambda)$. However, this increases degeneracy and results in an overall negative effect. We confirmed this with extensive computational experiments.
- *Multiplier symmetry breaking:* Permuting the continuous λ variables in (8) yields equivalent solutions. With this in mind, we experimented with different symmetry-breaking constraints to speed up Algorithm 1. However, all inequalities that we used had insignificant impact.
- *Constraint filtering:* We tested different filtering heuristics to preselect nonlinear constraints to use in the aggregations and alleviate the computational burden of solving (8). We used the violation of the constraints for an optimal solution of the LP, MILP, and convex NLP relaxation of the MINLP, as measures of "importance" of nonlinear constraints, among others. After extensive testing, unfortunately, we could not identify a good filtering rule that selects few nonlinear constraints and results in strong bounds for (5).

6   Computational Experiments

In this section, we present a computational study on 1671 publicly available instances of the MINLPLib [38]. We conduct three experiments in order to answer different questions:

1. ROOTGAP: How much of the root gap with respect to the MILP relaxation can be closed by using the K-surrogate dual?
2. BENDERS: How much do the ideas of Sect. 5 improve the performance of Algorithm 1?
3. DUALBOUND: Can Algorithm 1 improve on the dual bounds obtained by the MINLP solver SCIP?

Our methods are implemented in the MINLP solver SCIP [45]. We refer to [2] for an overview of the general solving algorithm and to [50,51] for the particular MINLP features of SCIP.

6.1   Experimental Setup

For the ROOTGAP experiment, we run Algorithm 1 for one hour for each choice of $K \in \{1, 2, 3\}$ after the root node has been completely processed by SCIP. To measure how much more root gap can be closed by using $K + 1$ instead of K, we use the best found aggregation vector of K as an initial point for $K + 1$.

The goal of the BENDERS experiment is to analyze the impact of the presented computational enhancements. We compare the following settings, each of them using a time limit of one hour and $K = 3$:

- DEFAULT: Algorithm 1 with all techniques described in Sect. 5.
- PLAIN: A plain version of Algorithm 1 without enhancements.
- NOSTAB: Same as DEFAULT but without trust-region and support stabilization.
- NOSUPP: Same as DEFAULT but without support stabilization.
- NOEARLY: Same as DEFAULT but without early termination for the master problem.

In the DUALBOUND experiment we proceed as follows. First, we collect the dual bounds for all instances that could not be solved by SCIP within three hours. Afterwards, we apply Algorithm 1 for $K = 3$ with a time limit of three hours, and set a target dual bound (see Sect. 6) of $D + (P - D) \cdot 0.2$, where D is the dual bound obtained by default SCIP and P the best known primal bound reported. This means we aim for a gap closed reduction of at least 20%.

Test Set. We use the publicly available instances of the MINLPLib [38]. We selected all instances that were available in OSiL format and consisted of nonlinear expressions that could be handled by SCIP. In total these are 1671 instances.

Performance Evaluation. We use the *gap closed* measure to compare dual bounds relative to a given primal bound. If $d_1, d_2 \in \mathbb{R}$ are two dual bounds for (1) with $d_1 \le d_2$ and $p \in \mathbb{R}$ a reference primal bound, then the function $GC(p, d_1, d_2) = \frac{d_2 - d_1}{p - d_1}$ measures the *gap closed* improvement of d_2 with respect to d_1.

To evaluate algorithmic performance over a large test set of benchmark instances, we use *shifted geometric means*. This avoids results being dominated by outliers with large absolute values (as is the case for the arithmetic mean) and avoids an over-representation of differences among very small values. See also the discussion in [2,3,27]. As shift values we use 10 s for averaging over running time and 5% for averaging over gap closed values.

Hardware and Software. The computing environment is a cluster of 64bit Intel Xeon X5672 CPUs at 3.2 GHz with 12 MB cache and 48 GB main memory. To safeguard against a mutual slowdown of parallel processes, we run only one job per node at a time. We use a development version of SCIP with CPLEX 12.8.0.0 as LP solver [29], the algorithmic differentiation code CppAD 20180000.0 [11], the graph automorphism package bliss 0.73 [31] for detecting MILP symmetry, and Ipopt 3.12.11 with Mumps 4.10.0 [1] as NLP solver [12,52].

Table 1. Results for the ROOTGAP experiment. A row $m \geq x$ considers all instances that have at least x nonlinear constraints. The second part of the table considers instances for which at least one setting closes at least 1% of the root gap.

Group	# instances	$K = 1$	$K = 2$	$K = 3$
ALL	633	18.4%	21.4%	23.4%
$m \geq 10$	528	14.6%	16.9%	18.4%
$m \geq 50$	229	7.1%	7.9%	8.5%
AFFECTED	469	35.0%	42.2%	46.9%
$m \geq 10$	370	30.1%	36.0%	40.1%
$m \geq 50$	115	23.9%	28.0%	30.8%

6.2 Computational Results

ROOTGAP Experiment. From all instances of MINLPLib, we filter those for which SCIP's MILP relaxation proves optimality in the root node, no primal solution is known, or SCIP aborted due to numerical issues. This leaves 633 instances.

Aggregated results are reported in Table 1. First, we observe an average gap reduction of 18.4% for $K = 1$, 21.4% for $K = 2$, and 23.4% for $K = 3$, respectively. The same tendency is true when considering instances grouped by their number of nonlinear constraints, and the results are even more dramatic when filtering-out unaffected instances. From these results, we see that using surrogate relaxations has a tremendous impact on reducing the root gap, even more so when using the generalized surrogate dual for $K = 2$ and $K = 3$.

BENDERS Experiment. Table 2 reports aggregated results for the BENDERS experiment. First, we observe that the DEFAULT performs significantly better than PLAIN. Only on 36 instances PLAIN closes more gap, but over all instances it closes on average 13.1% less gap than DEFAULT. On instances with a larger number of nonlinear constraints, DEFAULT performs even better: on the 107 instances with at least 50 nonlinear constraints, PLAIN closes 25.8% less gap than DEFAULT.

Table 2 also shows that DEFAULT dominates NOSTAB, NOSUPP, and NOEARLY. The most relevant computational enhancement is the early termination of the master problem. Even though Table 2 suggests that the trust-region and support stabilization are not crucial for closing a significant portion of the root gap, both techniques are important to exploit the λ space in a more structured way. Due to space limitations, we relegate this discussion to the Appendix.

DUALBOUND Experiment. For this experiment, we include all instances which could not be solved by SCIP with default settings within three hours, have a final gap of at least 10%, terminate without an error, and contain at least four nonlinear constraints. This leaves 209 instances. To compute gaps, we use the best known primal bounds from the MINLPLib as reference values. Table 4 in the Appendix reports detailed results on the subset of instances for which Algorithm 1 was able to improve on the bound obtained by SCIP, which was the case for 53 of the 209 instances. On these instances, the average gap of 284.3%

Table 2. Results for the BENDERS experiment. The column "M"/"L" reports the number of instances for which DEFAULT could close at least 1% more/less root gap than the settings of the corresponding column. Column "rgc" reports the average root gap closed relative to our default settings (in %). Instances for which no setting could close at least 1% of the root gap are filtered out.

Group	# instances	PLAIN			NOSTAB			NOSUPP			NOEARLY		
		M	L	rgc	M	L	rgc	M	L	rgc	M	L	rgc
ALL	457	100	36	86.9	40	31	98.3	41	27	98.4	94	40	88.2
$m \geq 10$	346	90	29	84.1	34	27	98.7	38	24	98.5	85	33	85.4
$m \geq 50$	107	35	1	74.2	13	7	98.9	14	12	98.4	32	2	76.3

for SCIP could be reduced to an average gap of 142.8%. Generalized surrogate duality works particularly well on difficult nonconvex MINLPs: for example, for all polygon* instances and four facloc* instances we find better bounds than the *best known dual bounds* from the MINLPLib, as shown in Table 3 of the Appendix.

7   Conclusion

In this article, we studied theoretical and computational aspects of surrogate relaxations for MINLPs. We developed the first algorithm able to solve exactly a generalization of the surrogate dual problem and proved its convergence guarantees. Our extensive computational study on the heterogeneous set of publicly available instances of the MINLPLib showed that exploiting surrogate duality can lead to significantly better dual bounds than standard LP-based spatial branch-and-bound. Additionally, our experiments showed that the presented computational enhancements are key to obtaining strong dual bounds for problems with a larger number of nonlinear constraints. Finally, we showed that our Algorithm can yield an average gap reduction from 284.3% to 142.8% on challenging instances. This includes instances of MINLPLib where we were able to significantly improve the best known dual bounds.

We hope our results not only help to revitalize surrogate duality and make it a viable alternative for challenging MINLPs, but also open the door for new developments using these techniques. Two important open questions are the following. First, in particular in the case of QCQPs, how much can be gained from requiring the surrogate relaxations to be convex? This could yield much more tractable sub-problems but might compromise the quality of the relaxation. Second, what is the best way of developing a surrogate-based spatial branch-and-bound approach? Solving a surrogate dual in each node of the branching tree most certainly will be impractical, but a coordinated scheme for sharing information between different surrogate duals across the tree could reduce relaxation time to a level at which search trees can be traversed to near-optimality.

Acknowledgements. We gratefully acknowledge support from the Research Campus MODAL (BMBF Grant 05M14ZAM) and the Institute for Data Valorization (IVADO) through an IVADO Postdoctoral Fellowship.

A   Appendix

A.1   The Effect of Stabilization

To visualize the importance of stabilization techniques, we use the instance `genpooling_lee1`. Figure 1 shows the achieved dual bounds after each iteration of Algorithm 1 for `DEFAULT` and `NOSTAB`. The achieved dual bound of -4775.26 with `DEFAULT` is significantly better than the dual bound of -5006.95 when using `NOSTAB`. More importantly, stabilization helps to reach the final dual bound much earlier.

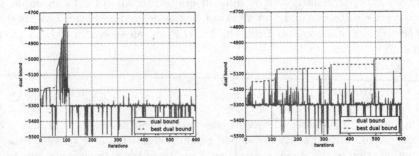

Fig. 1. Dual bounds for `DEFAULT` (left) and `NOSTAB` (right) on `genpooling_lee1` using $K = 3$. The red dashed curve shows the best found dual bound so far, whereas the blue curve shows the computed dual bound at each iteration. (Color figure online)

A.2   Detailed Results for the DUALBOUND Experiment

Table 3. Best known primal vs. the dual bounds computed by SCIP, Algorithm 1, and reported in MINLPLib for all `polygon*` and four `facloc*` instances.

Instance	Best primal	Dual bounds		
		MINLPLib	SCIP	DUALBOUND
`polygon25`	-0.78	-5.80	-4.24	-3.94
`polygon50`	-0.78	-15.27	-10.78	-8.72
`polygon75`	-0.78	-24.87	-16.82	-13.55
`polygon100`	-0.78	-34.00	-24.37	-19.03
`facloc1_3_95`	12.30	4.46	5.50	5.70
`facloc1_4_80`	7.88	0.16	0.09	0.41
`facloc1_4_90`	10.46	0.48	0.49	1.18
`facloc1_4_95`	11.18	0.79	1.40	2.40

Table 4. Detailed results for the DUALBOUND experiment on 53 instances for which Algorithm 1 could find a better dual bound in the root node.

Instance	m	n	p	Obj	Best primal	gap$_{SCIP}$	gap$_{Benders}$
case_1scv2	696	3377	25	Max	7888.57	2e+10	5e+09
crudeoil_pooling_ct1	37	362	71	Min	210538	27.1	21.6
crudeoil_pooling_ct3	182	966	108	Min	287000	27.3	21.6
eg_int_s	27	33968	3	Min	6.4531	3097.2	106.4
genpool10i	300	1294	187	Min	1.19809e+06	49.7	43.4
genpool10paper	33	727	187	Min	1.16851e+06	47.3	41.1
genpool15i	675	2759	352	Min	992088	34.5	25.8
heatexch_gen3	241	1627	60	Min	122519	125.4	118.8
multiplants_mtg1c	28	297	104	Max	683.971	498.9	362.2
multiplants_mtg6	65	410	140	Max	5314.43	31.0	18.3
ngone	4951	5344	0	Min	−0.0683939	1223.9	872.8
orth_d3m6_pl	66	303	0	Min	0.707107	88.3	58.2
orth_d4m6_pl	41	223	0	Min	0.649519	138.1	65.1
polygon100	4951	19899	0	Min	−0.785056	3004.5	2324.6
polygon25	301	1224	0	Min	−0.779741	444.3	405.5
polygon50	1226	4949	0	Min	−0.783875	1275.8	1012.9
polygon75	2776	11174	0	Min	−0.784464	2044.2	1626.8
radar-3000-10-a-8_lat_7	3000	18000	3000	Min	1039	6590.4	5672.2
rsyn0810m03h	18	679	198	Max	2722.45	74.4	8.3
rsyn0815m03h	33	842	217	Max	2827.93	78.4	28.6
rsyn0820m03h	42	972	242	Max	2028.81	139.1	83.3
rsyn0820m04h	56	1311	339	Max	2450.77	147.6	91.7
rsyn0830m02h	40	801	173	Max	730.507	177.8	94.4
rsyn0830m03h	60	1229	289	Max	1543.06	140.0	98.6
rsyn0840m02h	56	991	199	Max	734.984	222.7	102.8
rsyn0840m03h	84	1507	333	Max	2742.65	72.2	38.4
rsyn0840m04h	112	2044	467	Max	2564.5	148.7	88.4
sfacloc1_3_80	15	338	62	Min	8.52307	524.1	499.9
sfacloc1_3_95	15	278	9	Min	12.3025	123.7	115.8
sfacloc1_4_80	15	415	62	Min	7.8791	8198.1	1814.0
sfacloc1_4_90	15	383	30	Min	10.4575	2018.7	785.3
sfacloc1_4_95	15	355	9	Min	11.1841	699.6	365.4
sonet23v4	23	297	273	Min	-22747.5	34.3	27.1
sonet24v5	24	299	297	Min	-34704	78.9	73.0
sssd15-06	18	168	108	Min	539635	88.8	79.4
sssd18-06	18	186	126	Min	397992	73.4	64.2
sssd20-04persp	12	156	92	Min	347691	16.3	9.1
sssd22-08	24	280	200	Min	508714	75.6	69.6

(continued)

Table 4. (*continued*)

Instance	m	n	p	Obj	Best primal	gap$_{SCIP}$	gap$_{Benders}$
sssd25-04persp	12	176	112	Min	300177	22.0	12.8
syn20m03h	42	518	82	Max	2646.95	21.1	10.7
syn20m04h	56	695	115	Max	3532.74	50.7	15.1
syn30m02h	40	501	77	Max	399.684	219.0	14.6
syn30m03h	60	763	129	Max	654.155	234.3	54.4
syn30m04h	80	1022	181	Max	865.723	274.3	109.1
syn40m02h	56	691	103	Max	388.773	362.0	60.1
syn40m03h	84	1053	173	Max	395.149	492.8	239.7
tls12	12	575	419	Min	108.8	6769.5	3415.9
tls6	6	201	159	Min	15.3	87.2	48.5
tls7	7	306	250	Min	15	829.4	203.8
tspn10	11	212	45	Min	225.126	80.0	22.0
tspn12	13	302	66	Min	262.647	123.4	39.8
waterful2	57	798	56	Min	1012.61	162.4	50.2
watersym1	29	406	28	Min	913.776	114.4	32.1

$m/n/p$ — total number of nonlinear onstraints/variables/integer variables
obj — objective sense
best primal — reference primal bound
gap$_{SCIP}$ — remaining gap for SCIP with default settings (in %)
gap$_{Benders}$ — remaining gap after applying Algorithm 1 in the root (in %)

References

1. MUMPS: Multifrontal massively parallel sparse direct solver. http://mumps.enseeiht.fr
2. Achterberg, T.: Constraint integer programming. Ph.D. thesis, Technische Universität Berlin (2007). https://doi.org/10.14279/depositonce-1634. URN:nbn:de:kobv:83-opus-16117
3. Achterberg, T., Wunderling, R.: Mixed integer programming: analyzing 12 years of progress. In: Jünger, M., Reinelt, G. (eds.) Facets of Combinatorial Optimization, pp. 449–481. Springer, Heidelberg (2013). https://doi.org/10.1007/978-3-642-38189-8_18
4. van Ackooij, W., Frangioni, A., de Oliveira, W.: Inexact stabilized benders' decomposition approaches with application to chance-constrained problems with finite support. Comput. Optim. Appl. **65**(3), 637–669 (2016). https://doi.org/10.1007/s10589-016-9851-z
5. Alidaee, B.: Zero duality gap in surrogate constraint optimization: a concise review of models. Eur. J. Oper. Res. **232**(2), 241–248 (2014). https://doi.org/10.1016/j.ejor.2013.04.023
6. Amor, H.M.B., Desrosiers, J., Frangioni, A.: On the choice of explicit stabilizing terms in column generation. Discrete Appl. Math. **157**(6), 1167–1184 (2009). https://doi.org/10.1016/j.dam.2008.06.021
7. Balas, E.: Discrete programming by the filter method. Oper. Res. **15**(5), 915–957 (1967). https://doi.org/10.1287/opre.15.5.915

8. Balas, E.: Disjunctive programming: properties of the convex hull of feasible points. Discrete Appl. Math. **89**(1–3), 3–44 (1998). https://doi.org/10.1016/s0166-218x(98)00136-x
9. Banerjee, K.: Generalized Lagrange multipliers in dynamic programming. Ph.D. thesis, University of California, Berkeley (1971)
10. Bonami, P., Lodi, A., Tramontani, A., Wiese, S.: On mathematical programming with indicator constraints. Math. Program. **151**(1), 191–223 (2015). https://doi.org/10.1007/s10107-015-0891-4
11. COIN-OR: CppAD, a package for differentiation of C++ algorithms. http://www.coin-or.org/CppAD
12. COIN-OR: Ipopt, Interior point optimizer. http://www.coin-or.org/Ipopt
13. Conn, A.R., Gould, N.I.M., Toint, P.L.: Trust Region Methods. Society for Industrial and Applied Mathematics, Philadelphia (2000). https://doi.org/10.1137/1.9780898719857
14. Djerdjour, M., Mathur, K., Salkin, H.M.: A surrogate relaxation based algorithm for a general quadratic multi-dimensional knapsack problem. Oper. Res. Lett. **7**(5), 253–258 (1988). https://doi.org/10.1016/0167-6377(88)90041-7
15. Dyer, M.E.: Calculating surrogate constraints. Math. Program. **19**(1), 255–278 (1980). https://doi.org/10.1007/bf01581647
16. Fisher, M., Lageweg, B., Lenstra, J., Kan, A.: Surrogate duality relaxation for job shop scheduling. Discrete Appl. Math. **5**(1), 65–75 (1983). https://doi.org/10.1016/0166-218x(83)90016-1
17. Gavish, B., Pirkul, H.: Efficient algorithms for solving multiconstraint zero-one knapsack problems to optimality. Math. Program. **31**(1), 78–105 (1985). https://doi.org/10.1007/bf02591863
18. Geoffrion, A.M.: Implicit enumeration using an imbedded linear program. Technical report, May 1967. https://doi.org/10.21236/ad0655444
19. Glover, F.: A multiphase-dual algorithm for the zero-one integer programming problem. Oper. Res. **13**(6), 879–919 (1965). https://doi.org/10.1287/opre.13.6.879
20. Glover, F.: Surrogate constraints. Oper. Res. **16**(4), 741–749 (1968). https://doi.org/10.1287/opre.16.4.741
21. Glover, F.: Surrogate constraint duality in mathematical programming. Oper. Res. **23**(3), 434–451 (1975). https://doi.org/10.1287/opre.23.3.434
22. Glover, F.: Heuristics for integer programming using surrogate constraints. Decis. Sci. **8**(1), 156–166 (1977). https://doi.org/10.1111/j.1540-5915.1977.tb01074.x
23. Glover, F.: Tutorial on surrogate constraint approaches for optimization in graphs. J. Heuristics **9**(3), 175–227 (2003). https://doi.org/10.1023/a:1023721723676
24. Gomory, R.E.: An algorithm for the mixed integer problem. Technical report. P-1885, The RAND Corporation, June 1960
25. Greenberg, H.J., Pierskalla, W.P.: Surrogate mathematical programming. Oper. Res. **18**(5), 924–939 (1970). https://doi.org/10.1287/opre.18.5.924
26. Grossmann, I.E., Sahinidis, N.V.: Special issue on mixed integer programmingand its application to engineering, part I. Optim. Eng. **3**(4), 52–76 (2002)
27. Hendel, G.: Empirical analysis of solving phases in mixed integer programming. Master's thesis, Technische Universität Berlin, August 2014. URN:nbn:de: http://nbn-resolving.de/urn:nbn:de:0297-zib-54270
28. Horst, R., Tuy, H.: Global Optimization. Springer, Berlin Heidelberg (1996). DOI: https://doi.org/10.1007/978-3-662-03199-5
29. ILOG, I.: ILOG CPLEX: High-performance software for mathematical programming and optimization. http://www.ilog.com/products/cplex/

30. Kelley Jr., J.E.: The cutting-plane method for solving convex programs. J. Soc. Ind. Appl. Math. **8**(4), 703–712 (1960). https://doi.org/10.1137/0108053

31. Junttila, T., Kaski, P.: Bliss: a tool for computing automorphism groups and canonical labelings of graphs. (2012). http://www.tcs.hut.fi/Software/bliss/

32. Karwan, M.H.: Surrogate constraint duality and extensions in integer programming. Ph.D. thesis, Georgia Institute of Technology, January 1976

33. Karwan, M.H., Rardin, R.L.: Some relationships between Lagrangian and surrogate duality in integer programming. Math. Program. **17**(1), 320–334 (1979). https://doi.org/10.1007/bf01588253

34. Karwan, M.H., Rardin, R.L.: Surrogate dual multiplier search procedures in integer programming. Oper. Res. **32**(1), 52–69 (1984). https://doi.org/10.1287/opre.32.1.52

35. Kim, S.L., Kim, S.: Exact algorithm for the surrogate dual of an integer programming problem: subgradient method approach. J. Optim. Theory Appl. **96**(2), 363–375 (1998). https://doi.org/10.1023/a:1022622231801

36. McCormick, G.P.: Computability of global solutions to factorable nonconvex programs: Part i – convex underestimating problems. Math. Program. **10**(1), 147–175 (1976). https://doi.org/10.1007/bf01580665

37. du Merle, O., Villeneuve, D., Desrosiers, J., Hansen, P.: Stabilized column generation. Discrete Math. **194**(1–3), 229–237 (1999). https://doi.org/10.1016/s0012-365x(98)00213-1

38. MINLP library. http://www.minlplib.org/

39. Nakagawa, Y.: An improved surrogate constraints method for separable nonlinear integer programming. J. Oper. Res. Soc. Jpn **46**(2), 145–163 (2003). https://doi.org/10.15807/jorsj.46.145

40. Narciso, M.G., Lorena, L.A.N.: Lagrangean/surrogate relaxation for generalized assignment problems. Eur. J. Oper. Res. **114**(1), 165–177 (1999). https://doi.org/10.1016/s0377-2217(98)00038-1

41. Nemhauser, G.L., Wolsey, L.A.: A recursive procedure to generate all cuts for 0–1 mixed integer programs. Math. Program. **46**(1–3), 379–390 (1990). https://doi.org/10.1007/bf01585752

42. Quesada, I., Grossmann, I.E.: A global optimization algorithm for linear fractional and bilinear programs. J. Glob. Optim. **6**(1), 39–76 (1995). https://doi.org/10.1007/bf01106605

43. Ryoo, H., Sahinidis, N.: Global optimization of nonconvex NLPs and MINLPs with applications in process design. Comput. Chem. Eng. **19**(5), 551–566 (1995). https://doi.org/10.1016/0098-1354(94)00097-2

44. Sarin, S., Karwan, M.H., Rardin, R.L.: A new surrogate dual multiplier search procedure. Naval Res. Logistics **34**(3), 431–450 (1987). https://doi.org/10.1002/1520-6750(198706)34:3⟨431::aid-nav3220340309⟩3.0.co;2-p

45. SCIP - Solving Constraint Integer Programs. http://scip.zib.de

46. Sherali, H.D., Adams, W.P.: A Reformulation-Linearization Technique for Solving Discrete and Continuous Nonconvex Problems. Springer, New York (1999). https://doi.org/10.1007/978-1-4757-4388-3

47. Sherali, H.D., Fraticelli, B.M.P.: Enhancing RLT relaxations via a new class of semidefinite cuts. J. Glob. Optim. **22**(1/4), 233–261 (2002). https://doi.org/10.1023/a:1013819515732

48. Templeman, A.B., Xingsi, L.: A maximum entropy approach to constrained nonlinear programming. Eng. Optim. **12**(3), 191–205 (1987). https://doi.org/10.1080/03052158708941094

49. Vielma, J.P.: Small and strong formulations for unions of convex sets from the Cayley embedding. Math. Program. **177**(1–2), 21–53 (2018). https://doi.org/10.1007/s10107-018-1258-4

50. Vigerske, S.: Decomposition in multistage stochastic programming and a constraint integer programming approach to mixed-integer nonlinear programming. Ph.D. thesis, Humboldt-Universität zu Berlin, Mathematisch-Naturwissenschaftliche Fakultät II (2013). URN:nbn:de:kobv:11-100208240

51. Vigerske, S., Gleixner, A.: SCIP: global optimization of mixed-integer nonlinear programs in a branch-and-cut framework. Optim. Methods Softw. **33**(3), 563–593 (2017). https://doi.org/10.1080/10556788.2017.1335312

52. Wächter, A., Biegler, L.T.: On the implementation of an interior-point filter line-search algorithm for large-scale nonlinear programming. Math. Program. **106**(1), 25–57 (2005). https://doi.org/10.1007/s10107-004-0559-y

53. Xingsi, L.: An aggregate constraint method for non-linear programming. J. Oper. Res. Soc. **42**(11), 1003–1010 (1991). https://doi.org/10.1057/jors.1991.190

The Integrality Number of an Integer Program

Joseph Paat, Miriam Schlöter[(✉)], and Robert Weismantel

Department of Mathematics, ETH Zürich, Zürich, Switzerland
{joseph.paat,miriam.schloeter,robert.weismantel}@ifor.math.ethz.ch

Abstract. We introduce the *integrality number* of an integer program (IP) in inequality form. Roughly speaking, the integrality number is the smallest number of integer constraints needed to solve an IP via a mixed integer (MIP) relaxation. One notable property of this number is its invariance under unimodular transformations of the constraint matrix. Considering the largest minor Δ of the constraint matrix, our analysis allows us to make statements of the following form: there exist numbers $\tau(\Delta)$ and $\kappa(\Delta)$ such that an IP with $n \geq \tau(\Delta)$ many variables and $n + \kappa(\Delta) \cdot \sqrt{n}$ many inequality constraints can be solved via a MIP relaxation with fewer than n integer constraints. A special instance of our results shows that IPs defined by only n constraints can be solved via a MIP relaxation with $O(\sqrt{\Delta})$ many integer constraints.

1 Introduction

Let $A \in \mathbb{Z}^{m \times n}$ satisfy $\mathrm{rank}(A) = n$ and $c \in \mathbb{Z}^n$. We denote the integer linear program parameterized by right hand side $b \in \mathbb{Z}^m$ by

$$\mathrm{IP}_{A,c}(b) := \max\{c^\mathsf{T} x : Ax \leq b \text{ and } x \in \mathbb{Z}^n\}.$$

See [8] for more on parametric integer programs.

We are interested in solving $\mathrm{IP}_{A,c}(b)$ by relaxing it to have fewer integer constraints. In special cases we can solve $\mathrm{IP}_{A,c}(b)$ with zero integer constraints by only solving its linear relaxation

$$\mathrm{LP}_{A,c}(b) := \max\{c^\mathsf{T} x : Ax \leq b\}.$$

However, these special cases require the underlying polyhedron to have optimal integral vertices, which occurs for instance when A is totally unimodular. Our target is to consider general matrices A and relaxations in the form of a mixed integer program:

$$W\text{-MIP}_{A,c}(b) := \max\{c^\mathsf{T} x : Ax \leq b, \ Wx \in \mathbb{Z}^k, \text{ and } x \in \mathbb{R}^n\},$$

where $k \in \mathbb{Z}_{\geq 0}$ and $W \in \mathbb{Z}^{k \times n}$ satisfies $\mathrm{rank}(W) = k$.

© Springer Nature Switzerland AG 2020
D. Bienstock and G. Zambelli (Eds.): IPCO 2020, LNCS 12125, pp. 338–350, 2020.
https://doi.org/10.1007/978-3-030-45771-6_26

One sufficient condition for solving $\text{IP}_{A,c}(b)$ using a mixed integer relaxation is that the vertices of $W\text{-MIP}_{A,c}(b)$ are integral. The *vertices* of $W\text{-MIP}_{A,c}(b)$ are the vertices of the polyhedron

$$\text{conv}\left(\{x \in \mathbb{R}^n : Ax \le b, \ Wx \in \mathbb{Z}^k\}\right).$$

If W can be chosen such that every vertex of $W\text{-MIP}_{A,c}(b)$ is integral, then an optimal valued vertex of $W\text{-MIP}_{A,c}(b)$ also solves $\text{IP}_{A,c}(b)$. Moreover, if W has this property, then $\text{IP}_{A,c}(b)$ has a feasible solution if and only if $W\text{-MIP}_{A,c}(b)$ does too. Lenstra's algorithm combined with the ellipsoid method can find an optimal valued vertex of $W\text{-MIP}_{A,c}(b)$ in polynomial time when k is fixed (see [15], also [7,14]). This leads us to consider the smallest k for which such a W exists:

$$i_A(b) := \min\left\{k \in \mathbb{Z}_{\ge 0} : \begin{array}{l} \exists \, W \in \mathbb{Z}^{k \times n} \text{ such that all vertices} \\ \text{of } W\text{-MIP}_{A,c}(b) \text{ are integral} \end{array}\right\}$$

We refer to $i_A(b)$ as the *integrality number* of $\text{IP}_{A,c}(b)$. The integrality number can be interpreted as the number of integer constraints needed to solve $\text{IP}_{A,c}(b)$ with a relaxation of the form $W\text{-MIP}_{A,c}(b)$. Observe that $i_A(b)$ is independent of c because we look for W that describe all vertices of $W\text{-MIP}_{A,c}(b)$. Moreover, $i_A(b)$ is always finite. If $\text{IP}_{A,c}(b)$ is infeasible, then $i_A(b) = 0$ holds vacuously because there are no vertices of $\text{IP}_{A,c}(b)$. On the other hand, if $\text{IP}_{A,c}(b)$ is feasible, then $i_A(b) \le n$ because we can always choose $W = \mathbb{I}^n$. Our goal is to find situations (besides when $\text{IP}_{A,c}(b)$ is infeasible) in which $i_A(b) < n$.

Our first main result bounds $i_A(b)$ using two well-studied data parameters. The first of these parameters is the largest full rank minor of A. The largest full rank minor of a matrix $C \in \mathbb{R}^{d \times \ell}$ is denoted by

$$\Delta = \Delta(C) := \max\{|\det(B)| : B \text{ is a } (\text{rank}(C) \times \text{rank}(C)) \text{ submatrix of } C\}.$$

We use Δ to denote the value $\Delta(A)$ unless explicitly stated otherwise. Solving $\text{IP}_{A,c}(b)$ for bounded values of $\Delta(A)$ has been studied extensively over the years with recent renewed interest [1,2,18]. The second parameter is the cardinality of the column set of an integer-valued matrix. For $r, \Delta \in \mathbb{Z}_{\ge 1}$ define

$$c(r, \Delta) := \max\left\{d : \begin{array}{l} \exists \, B \in \mathbb{Z}^{r \times d} \text{ with } d \text{ distinct columns,} \\ \text{rank}(B) = r, \text{ and } \Delta(B) \le \Delta \end{array}\right\}.$$

Heller [12] and Glanzer et al. [10] showed that

$$c(r, \Delta) \le \begin{cases} r^2 + r + 1 & \text{if } \Delta = 1 \\ \Delta^{2 + \log_2 \log_2(\Delta)} \cdot r^2 + 1 & \text{if } \Delta \ge 2. \end{cases} \tag{1}$$

Theorem 1. *Let* $A \in \mathbb{Z}^{m \times n}$, $c \in \mathbb{Z}^n$, *and* $b \in \mathbb{Z}^m$. *Suppose*

$$A = \begin{pmatrix} A^1 \\ A^2 \end{pmatrix},$$

where $r := \text{rank}(A^2)$ *and* $A^1 \in \mathbb{Z}^{n \times n}$ *is a full rank matrix with* $\delta := |\det(A^1)|$.

(a) If $r = 0$, then $i_A(b) \leq 6\delta^{1/2} + \log_2(\delta)$.

(b) If $r \geq 1$, then $i_A(b) \leq [6\delta^{1/2} + \log_2(\delta)] \cdot \min\{c(r, \Delta(A^2)), c(r, \Delta)\}$.

Combining Theorems 1 and (1) yields the following corollary.

Corollary 1. *There exist constants* $\tau(\Delta), \kappa(\Delta) > 0$ *that satisfy the following: if* $IP_{A,c}(b)$ *has* $n \geq \tau(\Delta)$ *many variables and at most* $n + \kappa(\Delta) \cdot \sqrt{n}$ *many constraints, then it can be reformulated using fewer than* n *integrality constraints.*

Our proof of Theorem 1 is inspired by the notion of affine TU decompositions introduced by Bader et al. [3]. An affine TU decomposition of A is an equation $A = A_0 + UW$ such that $[A_0^\mathsf{T} \ W^\mathsf{T}]$ is totally unimodular and U is an integral matrix. It can be shown that such a decomposition implies that

$$\operatorname{conv}(\{x \in \mathbb{R}^n : Ax \leq b, \ Wx \in \mathbb{Z}^k\})$$

is an integral polyhedron. Integral vertices are preserved under unimodular maps, while the property of total unimodularity is not. This implies that affine TU decompositions are not robust under unimodular transformations of A. In contrast, the integrality number is preserved under unimodular maps (see Lemma 2). We use this fact to construct a new homogeneous matrix decomposition that produces integral vertices and is invariant under unimodular maps. Furthermore, our construction is a method for creating relaxations of $IP_{A,c}(b)$ for general A while Bader et al. only describe relaxations for certain examples and prove hardness results [3, Section 3]. It is worth mentioning that the examples in [3] can be modified to show that Theorem 1 (a) is tight. See also Hupp [13] who investigated computational benefits of affine TU decompositions.

The proof of Theorem 1 is given in Sect. 2. Given any basis matrix A^1 (one of which can be found efficiently), the matrix W underlying the result can be constructed in polynomial time even if we do not know Δ a priori or Δ is large. The number of distinct columns $c(r, \cdot)$ plays an important role in bounding $i_A(b)$ because we aggregate equal columns and represent them with a single integer constraint. This is analogous to common aggregation techniques in one-row knapsack problems. In Theorem 1 (b) the value $c(r, \cdot)$ can be replaced by the number of distinct columns of A^2, but we present the results in terms of $c(r, \cdot)$ because (1) then allows us to bound $i_A(b)$ in terms of r and $\Delta(A^2)$. The next lemma shows that the number of distinct columns of A^2 is also bounded by $c(r, \Delta)$ even if $\Delta(A^2)$ is significantly larger than Δ; hence, we can also bound $i_A(b)$ in terms of r and Δ only. The matrix A^2 is one choice of YA^1 in the next lemma, but we state the result in generality as it may be of independent interest in future research regarding $c(r, \cdot)$.

Lemma 1. *Let* Δ, δ, *and* A^1 *as in Theorem 1. If* $Y \in \mathbb{R}^{r \times n}$ *satisfies* $\operatorname{rank}(Y) = r$, $\Delta(Y) \leq \Delta/\delta$, *and* $YA^1 \in \mathbb{Z}^{r \times n}$, *then* Y *has at most* $c(r, \Delta)$ *many distinct columns.*

The bounds in Theorem 1 grow larger than n when r is larger than $\sqrt{n}$. However, it turns out that *most* problems of the form $IP_{A,c}(b)$ have redundant constraints making it possible to bound $i_A(b)$ by a function of only Δ. To formalize this, we define the *density of a set* $\mathcal{A} \subseteq \mathbb{Z}^m$ to be

$$\Pr(\mathcal{A}) := \liminf_{t \to \infty} \frac{|\{-t, \ldots, t\}^m \cap \mathcal{A}|}{|\{-t, \ldots, t\}^m|}.$$

The value $\Pr(\mathcal{A})$ can be interpreted as the likelihood that the family $\{\mathrm{IP}_{A,c}(b) : b \in \mathcal{A}\}$ occurs within $\{\mathrm{IP}_{A,c}(b) : b \in \mathbb{Z}^m\}$. However, the functional is not formally a probability measure but rather a lower density function. The functional $\Pr(\cdot)$ has been used before to study sparse solutions of $\mathrm{IP}_{A,c}(b)$ [5,16]. Other significant asymptotic results were given by Gomory [11] and Wolsey [19], who showed that the value function of $\mathrm{IP}_{A,c}(b)$ has periodic asymptotic behavior.

Theorem 2. *It holds that* $\Pr(\mathcal{G}^I \cup \mathcal{G}^L) = 1$, *where*

$$\begin{aligned}
\mathcal{G}^I &:= \{b \in \mathbb{Z}^m : i_A(b) \in O(\Delta^{1/2}) \text{ and } \mathrm{IP}_{A,c}(b) \text{ is feasible}\}, \text{ and} \\
\mathcal{G}^L &:= \{b \in \mathbb{Z}^m : \mathrm{LP}_{A,c}(b) \text{ is infeasible}\}.
\end{aligned} \tag{2}$$

The proof of Theorem 2 is given in Sect. 3, and it relies on the fact that most feasible regions $\mathrm{IP}_{A,c}(b)$ are either infeasible or simplicial polytopes. In the proof we find a set $\mathcal{G} \subseteq \mathbb{Z}^m$ such that $\Pr(\mathcal{G}) = 1$ and for every $b \in \mathcal{G}$ we provide an efficient construction of a matrix $W \in \mathbb{Z}^{k \times n}$ such that $k \in O(\Delta^{1/2})$ and $W\text{-MIP}_{A,c}(b)$ has integer vertices. This allows us to conclude that almost all integer programs in n variables can be solved as a mixed integer problem with only $O(\Delta^{1/2})$ many integer constraints.

Corollary 2. *Let* $A \in \mathbb{Z}^{m \times n}$ *and* $c \in \mathbb{Z}^n$. *There exists* $\mathcal{G} \subseteq \mathbb{Z}^m$ *such that* $\Pr(\mathcal{G}) = 1$, *and for every* $b \in \mathcal{G}$, $\mathrm{IP}_{A,c}(b)$ *can be solved as a mixed integer program with only* $O(\Delta^{1/2})$ *many integer constraints.*

A consequence of Corollary 2 is that almost all problems can be solved in polynomial time provided Δ is constant. This consequence can also be derived from a classic dynamic programming result by Gomory involving the so-called group relaxation [11] as well as from the dynamic programs presented in [1] or [9]. The running time of the latter has lesser dependence on Δ than our approach in Corollary 2. However, these proof ideas use dynamic programming rather than mixed integer relaxations. To the best of our knowledge, Corollary 2 cannot be derived from these works.

Notation. Denote the largest minor of $C \in \mathbb{R}^{m \times n}$ by

$$\Delta^{\max}(C) := \max\{|\det(B)| : B \text{ a submatrix of } A\}.$$

Denote the i-th row of C by C_i. For $I \subseteq \{1, \ldots, m\}$ let C_I denote the $|I| \times n$ matrix consisting of the rows $\{C_i : i \in I\}$. A matrix $U \in \mathbb{Z}^{n \times n}$ is *unimodular* if $|\det(U)| = 1$. A matrix $W \in \mathbb{Z}^{k \times n}$ is *totally unimodular* if $\Delta^{\max}(W) \leq 1$. For $K^1, K^2 \subseteq \mathbb{R}^n$ define $K^1 + K^2 := \{x + y : x \in K^1, y \in K^2\}$. Denote the $d \times d$ identity matrix by $\mathbb{I}^d$ and the $d \times k$ all zero matrix by $\mathbb{0}^{d \times k}$.

2   The Proof of Theorem 1

Throughout this section, we assume that A^1 is a given $n \times n$ invertible submatrix of A. The first step in our proof of Theorem 1 is to perform a suitable unimodular transformation to A. The next result, which is proven in the appendix, states that the integrality number is preserved under unimodular transformations.

Lemma 2. Let $c \in \mathbb{Z}^n$, $b \in \mathbb{Z}^m$, $A \in \mathbb{Z}^{m \times n}$, and $U \in \mathbb{Z}^{n \times n}$ be unimodular. Then $i_A(b) = i_{AU}(b)$.

The particular unimodular transformation that we want to apply is the one that transforms A into *Hermite Normal Form*. Notice that if we permute the rows of $(A\ b)$, then the optimization problem $\text{IP}_{A,c}(b)$ remains the same. After reordering the constraints of $\text{IP}_{A,c}(b)$, there exists a unimodular matrix U such that AU is in Hermite Normal Form (see, e.g., [17]):

$$A = \begin{pmatrix} A^1 \\ A^2 \end{pmatrix}, \text{ where } A^1 = \begin{pmatrix} \mathbb{I}^{n-\ell} & \mathbb{0}^{(n-\ell)\times\ell} \\ & A_I^1 \end{pmatrix} \text{ and } A_I^1 = \begin{pmatrix} * & \cdots & * & \alpha_1 & & \\ \vdots & & \vdots & \vdots & \ddots & \\ * & \cdots & * & * & * & \alpha_\ell \end{pmatrix}, \quad (3)$$

$\alpha_1, \ldots, \alpha_\ell \in \mathbb{Z}_{\geq 2}$, $A_{i,j}^1 \leq \alpha_i - 1$ if $j < i$ and $A_{i,j}^1 = 0$ if $j > i$, and $\delta = |\det(A^1)|$. In light of Lemma 2, we assume that A is in Hermite Normal Form for the rest of the section.

We proceed by solving instances of the following problem:

Given $C \in \mathbb{R}^{p \times n}$, find a totally unimodular matrix $W \in \mathbb{Z}^{k \times n}$
and $V \in \mathbb{R}^{p \times k}$ such that $C = VW$. $\qquad(4)$

Our use of (4) for bounding $i_A(b)$ is justified by the following lemma.

Lemma 3. Let $W \in \mathbb{Z}^{k \times n}$ be totally unimodular satisfying $\text{rank}(W) = k$.

$$\text{If } W \text{ satisfies (4) for } C = \begin{pmatrix} A_I^1 \\ A^2 \end{pmatrix}, \text{ then } i_A(b) \leq k. \quad (5)$$

Proof. If W-$\text{MIP}_{A,c}(b)$ is infeasible, then the result is vacuously true because there are no vertices. A vertex z^* of W-$\text{MIP}_{A,c}(b)$ has the form

$$z^* = \begin{pmatrix} A_J \\ W \end{pmatrix}^{-1} \begin{pmatrix} b_J \\ y \end{pmatrix}, \quad (6)$$

where $y \in \mathbb{Z}^k$, $J \subseteq \{1, \ldots, m\}$, and $|J| = n - k$. The rows of A_J are linearly independent of the rows of W. By (4) and (5) we know there is a matrix V satisfying $C = VW$, so the only rows of A that are linearly independent of W are those not belonging to C. Hence, the rows of A_J are a subset of the first $n - \ell$ rows defining A^1 in (3) and they form a partial-identity matrix. Thus, the matrix in (6) is unimodular because W is totally unimodular. By Cramer's Rule we have $z^* \in \mathbb{Z}^n$. Every vertex of W-$\text{MIP}_{A,c}(b)$ is integral, so $i_A(b) \leq k$.   □

It remains to discuss how to find a totally unimodular matrix W satisfying (4) for $C \in \mathbb{R}^{p \times n}$. We could choose $W = \mathbb{I}^n$, but we want k to be as small as possible. In order to find W with fewer rows, we write C differently. Consider a finite nonempty set $B \subseteq \mathbb{R}^p$ and a finite (possibly empty) set $T \subseteq \mathbb{R}^p \setminus \{0\}$ satisfying

$$\text{the columns of } C \text{ are contained in } B + (T \cup \{0\}). \tag{7}$$

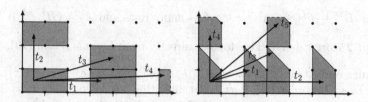

Fig. 1. Let $C \in \mathbb{Z}^{20 \times 2}$ have columns $(\{0, \ldots, 5\} \times \{0, \ldots, 2\}) \cup \{[6,0]^\mathsf{T}, [0,3]^\mathsf{T}\}$. Two choices of B and T satisfying (7) are shown. Each finite set B and its translations are shaded. The vectors in T are denoted by t_i.

Fig. 1 gives examples of B and T.

Every column $u \in C$ can be written as $u = v + t$ for some $v \in B$ and $t \in T \cup \{0\}$. If many representations exist, then choose one. We can write C as

$$C = (B \ T)W, \quad \text{where} \quad W := \begin{pmatrix} W^B \\ W^T \end{pmatrix}, \tag{8}$$

$W^B \in \{0,1\}^{|B| \times n}$, and $W^T \in \{0,1\}^{|T| \times n}$. Note W^T has $|T|$ rows rather than $|T|+1$ because any column of C that is in B can be represented without T. The benefit of creating W using (8) is that it only has $|B| + |T|$ rows. Thus, if B and T have only a few elements, then W has few rows. We refer to the construction of B and T using an oracle called $\text{COVER}(C)$. In what follows, any sets B and T constructed using $\text{COVER}(C)$ will always be finite. We ensure W has full row rank by removing linearly dependent rows. Lemma 4 shows that W constructed in this way is totally unimodular.

Lemma 4. Let $C \in \mathbb{R}^{p \times n}$. If $W \in \mathbb{Z}^{k \times n}$ is constructed as in (8) using $(B, T) = \text{COVER}(C)$, then W is totally unimodular.

In order to prove Theorem 1 we use two specific constructions of B and T. The first construction solves (4) for $C = A_I^1$.

Construction 1. For $i \in \{1, \ldots, \ell\}$ let $k_i \in \{0, \ldots, \alpha_i - 1\}$. Let z^i denote the column of A_I^1 whose i-th component is α_i. Define

$$B = (\{0, \ldots, k_1\} \times \ldots \times \{0, \ldots, k_\ell\}) \cup \{z^1, \ldots, z^\ell\}$$
$$\text{and} \quad T = \{z \in \mathbb{Z}^\ell : z_i \in \{0, k_i + 1, \ldots, \beta_i \cdot (k_i + 1)\} \ \forall \ i \in \{1, \ldots, \ell\}\} \setminus \{0\}, \tag{9}$$

where

$$\beta_i := \left\lfloor \frac{\alpha_i - 1}{k_i + 1} \right\rfloor \quad \text{for all } i \in \{1, \ldots, \ell\}.$$

Lemma 5. *The sets B and T defined in (9) satisfy (7) for $C = A_I^1$. Also, $k_1, \ldots, k_\ell$ can be chosen such that $|B| + |T| \leq 6\delta^{1/2} + \log_2(\delta)$, where $\delta = \prod_{i=1}^\ell \alpha_i$.*

The proofs of Lemmas 4 and 5 appear in the appendix. For the rest of this section we use W^I to denote the totally unimodular matrix derived from Construction 1 and Lemma 5. We also use B^I and T^I to denote the corresponding finite set and translation set. By Lemma 5

$$W^I \text{ has } |B^I| + |T^I| \leq 6\Delta^{1/2} + \log_2(\Delta) \text{ many rows and } A_I^1 = (B^I \ T^I)W^I. \quad (10)$$

Proof (of Theorem 1(a)). This follows directly from Lemma 3 and (10). $\qquad \square$

We turn our attention to proving $i_A(b) \leq [6\delta^{1/2} + \log_2(\delta)] \cdot c(r, \Delta(A^2))$, which we refer to as Theorem 1(b) Part 1. To show this, we solve (4) for $C = A^2$. Our construction is a simple enumeration of the distinct columns of A^2.

Construction 2. *Let $C \in \mathbb{R}^{p \times n}$. Choose B to be the set of columns of C and $T = \emptyset$. By design, $|B| + |T| = |B|$ equals the number of distinct columns in C.*

It is not difficult to combine Constructions 1 and 2 in order to solve (4) for C in (5). We state this as a lemma without proof.

Lemma 6. *For $i \in \{1, 2\}$ let $C^i \in \mathbb{R}^{p_i \times n}$ and construct $W^i \in \mathbb{Z}^{k_i \times n}$ with $(B^i, T^i) = \text{Cover}^i(C^i)$ (the oracle depends on i). Set*

$$B := B^1 \times B^2, \quad T := (T^1 \cup \{0\}) \times (T^2 \cup \{0\}) \setminus \{0\}, \text{ and } C := \begin{pmatrix} C^1 \\ C^2 \end{pmatrix}.$$

If $W \in \mathbb{Z}^{k \times n}$ is constructed using (8) and (B, T), then W solves (4) for C. Note $k \leq |B^1| \cdot |B^2| + (|T^1| + 1) \cdot (|T^2| + 1) - 1$.

We are now prepared to prove Theorem 1(b) Part 1.

Proof (of Theorem 1(b) Part 1). Let B^I, T^I, and W^I be as in (10). Let $\overline{A}^2 \in \mathbb{Z}^{r \times n}$ be any submatrix of A^2 with $\text{rank}(\overline{A}^2) = r$. Thus, there exists $V^2 \in \mathbb{R}^{(m-n) \times r}$ such that $A^2 = V^2 \overline{A}^2$. The number of distinct columns of $\overline{A}^2$ is bounded by $c(r, \Delta(A^2))$. Construction 2 yields a nonempty set B^2 of cardinality at most $c(r, \Delta(A^2))$, an empty translation set T^2, and a totally unimodular $W^2 \in \mathbb{Z}^{k \times n}$ with $k \leq |B^2| + |T^2| \leq c(r, \Delta(A^2))$ such that $\overline{A}^2 = (B^2 \ T^2)W^2$. Thus, $A^2 = V^2(B^2 \ T^2)W^2$. By Lemma 6, we can combine W^I and W^2 to create a totally unimodular $W \in \mathbb{Z}^{k \times n}$ satisfying Lemma 3 and

$$i_A(b) \leq k \leq |B^I| \cdot |B^2| + (|T^I| + 1)(|T^2| + 1) - 1 = |B^I| \cdot |B^2| + |T^I|$$
$$\leq (|B^I| + |T^I|) \cdot (|B^2| + |T^2|) \leq [6\delta^{1/2} + \log_2(\delta)] \cdot c(r, \Delta(A^2)). \quad \square$$

If $r = \text{rank}(A^2)$ is at least n, then $\Delta(A^2)$ is the maximum over all $n \times n$ determinants of A^2 and thus $\Delta(A^2) \leq \Delta(A) = \Delta$. However, if $r < n$, then $\Delta(A^2)$ may be significantly larger than Δ. This follows from the fact that there exists a unique matrix Y satisfying $A^2 = YA^1$ (because A^1 is invertible) and an

$r \times r$ determinant of A^2 is a linear combination of $r \times r$ determinants of A from the Cauchy-Binet formula on the system $A^2 = YA^1$. Nevertheless, the number of distinct columns of A^2 can still be bounded by $c(r, \Delta)$ rather than $c(r, \Delta(A^2))$. To motivate why this is true, we note that the parallelepiped generated by the rows of A^1 induces a group on $\mathbb{Z}^n$ of size δ, and Y contains the coordinates mapping these group elements to the rows of A^2. This mapping allows us to view distinct columns of A^2 as distinct columns of Y. We can bound distinct columns of Y using the group structure induced by A^1. This is Lemma 1.

Proof (of Lemma 1). Define

$$\Pi := \{g \in [0,1)^n : g^\mathsf{T} A^1 \in \mathbb{Z}^n\}.$$

The set $\{g^\mathsf{T} A^1 : g \in \Pi\}$ is the additive quotient group of $\mathbb{Z}^n$ factored by the rows of A^1, and Π is isomorphic to this group. The identity element of Π is 0. The group operation of Π is addition modulo 1. It is known that $|\Pi| = \Delta$ and for all $z \in \mathbb{Z}^n$ there exists a unique $g \in \Pi$ and $v \in \mathbb{Z}^n$ such that $z^\mathsf{T} = (g+v)^\mathsf{T} A^1$ (see, e.g., [4, §VII]).

Recall $YA^1 \in \mathbb{Z}^{r \times n}$. Thus, there exist matrices $G \in \mathbb{R}^{n \times r}$ and $V \in \mathbb{Z}^{n \times r}$ such that the columns of G are in Π and $YA^1 = (G+V)^\mathsf{T} A^1$. Because A^1 is invertible we have $Y = (G+V)^\mathsf{T}$. The columns $G_1, \ldots, G_r$ of G form a sequence of nested subgroups

$$\{0\} \subseteq \langle\{G_1\}\rangle \subseteq \ldots \subseteq \langle\{G_1, \ldots, G_r\}\rangle,$$

where $\langle\Omega\rangle := \{\sum_{h \in \Omega} \lambda_h h \bmod 1 : \lambda \in \mathbb{Z}^\Omega\}$ for $\Omega \subseteq \Pi$. For each $i \in \{1, \ldots, r\}$ the smallest positive integer α_i such that $\alpha_i G_i \bmod 1 \in \langle\{G_1, \ldots, G_{i-1}\}\rangle$ is the so-called index of $\langle\{G_1, \ldots, G_{i-1}\}\rangle$ in $\langle\{G_1, \ldots, G_i\}\rangle$, i.e.,

$$\alpha_1 = |\langle\{G_1\}\rangle| \quad \text{and} \quad \alpha_i = \frac{|\langle\{G_1, \ldots, G_i\}\rangle|}{|\langle\{G_1, \ldots, G_{i-1}\}\rangle|} \quad \forall i \in \{2, \ldots, r\}.$$

The definition of α_i implies that there are integers $\beta_i^1, \ldots, \beta_i^{i-1}$ such that

$$\alpha_i G_i - \sum_{j=1}^{i-1} \beta_i^j G_j \in \mathbb{Z}^n.$$

We create a lower-triangular matrix $E \in \mathbb{Z}^{r \times r}$ from these linear forms as follows: the i-th row of E is $[-\beta_i^1, \ldots, -\beta_i^{i-1}, \alpha_i, 0, \ldots, 0]$. By design $EG^\mathsf{T} \in \mathbb{Z}^{r \times n}$, so $E(G+V)^\mathsf{T} = EY \in \mathbb{Z}^{r \times n}$. Also,

$$\det(E) = \prod_{i=1}^r \alpha_i = |\langle\{G_1\}\rangle| \cdot \prod_{i=2}^r \frac{|\langle\{G_1, \ldots, G_i\}\rangle|}{|\langle\{G_1, \ldots, G_{i-1}\}\rangle|} = |\langle\{G_1, \ldots, G_r\}\rangle| \leq \delta,$$

where the last inequality follows from Lagrange's Theorem and the fact that $\langle\{G_1, \ldots, G_r\}\rangle$ is a subgroup of Π whose order is δ. Furthermore, $\mathrm{rank}(EY) = \mathrm{rank}(Y) = r$ and an $r \times r$ submatrix of EY is of the form EF for an $r \times r$ submatrix F of Y. The assumption $\Delta(Y) \leq \Delta/\delta$ implies $|\det(F)| \leq \Delta/\delta$. Hence, $|\det(EF)| = |\det(E)| \cdot |\det(F)| \leq \Delta$ and so $\Delta(EY) \leq \Delta$.

Columns of Y are distinct if and only if the corresponding columns of integer-valued EY are distinct because E is invertible. The function $c(r, \cdot)$ is nondecreasing. Hence, EY has at most $c(r, \Delta(EY)) \leq c(r, \Delta)$ many distinct columns. □

We now show Theorem 1(b) Part 2, i.e., $i_A(b) \leq [6\delta^{1/2} + \log_2(\delta)] \cdot c(r, \Delta)$.

Proof (of Theorem 1(b) Part 2). Recall that we assume that A is in Hermite Normal Form (3). We construct W satisfying Lemma 3. If we remove linearly dependent rows of A^2, then W will still satisfy the conditions of the lemma. Thus, we assume $A^2 \in \mathbb{Z}^{r \times n}$ and $\mathrm{rank}(A^2) = r$. The matrix A^1 is invertible, so there exist $R \in \mathbb{R}^{r \times (n-\ell)}$ and $Q \in \mathbb{R}^{r \times \ell}$ such that $A^2 = [R\ Q]A^1 = [R\ \mathbb{0}^{r \times \ell}] + QA_I^1$. Using R and Q we can also rewrite A as

$$A = \begin{pmatrix} A^1 \\ A^2 \end{pmatrix} = \begin{pmatrix} \mathbb{I}^n \\ R\ Q \end{pmatrix} A^1.$$

Note that

$$\Delta^{\max}\left(\begin{pmatrix} \mathbb{I}^n \\ R\ Q \end{pmatrix}\right) \leq \frac{\Delta}{\delta}. \tag{11}$$

If (11) is false, then there exists a submatrix D of the matrix in (11) with $|\det(D)| > \Delta/\delta$. The matrix in (11) contains $\mathbb{I}^n$, so we can extend D and assume $D \in \mathbb{R}^{n \times n}$. Notice DA^1 is an $n \times n$ submatrix of A with $|\det(DA^1)| > (\Delta/\delta) \cdot |\det(A^1)| = \Delta$. However, this contradicts the definition of Δ and proves (11).

By Lemma 1 $[R\ Q]$ has at most $c(r, \Delta)$ many distinct columns. The matrix $[R\ Q\ \mathbb{0}^{r \times \ell}]$ also has at most $c(r, \Delta)$ many distinct columns because $\Delta([R\ Q]) = \Delta([R\ Q\ \mathbb{0}^{r \times \ell}])$. We can apply Construction 2 with $C = [R\ \mathbb{0}^{r \times \ell}]$ to obtain a nonempty finite set B^R, an empty translation set T^R, and a totally unimodular matrix $W^R \in \mathbb{Z}^{\tau \times n}$ such that $|B^R| = \tau \leq c(r, \Delta)$ and $[R\ \mathbb{0}^{r \times \ell}] = (B^R\ T^R)W^R$. Applying Lemma 6 to W^I and W^R yields a nonempty finite set B, a translation set T, and a totally unimodular matrix $W \in \mathbb{Z}^{k \times n}$ such that

$$\begin{pmatrix} A_I^1 \\ R\ \mathbb{0}^{r \times \ell} \end{pmatrix} = (B\ T)W.$$

The latter condition implies there is a submatrix $V \in \mathbb{R}^{\ell \times k}$ of $(B\ T)$ such that $A_I^1 = VW$. Using this and the displayed identity, we see that

$$\begin{pmatrix} A_I^1 \\ A^2 \end{pmatrix} = \begin{pmatrix} A_I^1 \\ [R\ \mathbb{0}^{r \times \ell}] + QA_I^1 \end{pmatrix} = \left[(B\ T) + \begin{pmatrix} \mathbb{0}^{\ell \times k} \\ QV \end{pmatrix} \right] W.$$

Hence, W satisfies the assumptions of Lemma 3 and

$$i_A(b) \leq k \leq |B^I| \cdot |B^R| + (|T^I| + 1) \cdot (|T^R| + 1) - 1 = |B^I| \cdot |B^R| + |T^I|$$
$$\leq (|B^I| + |T^I|) \cdot (|B^R| + |T^R|) \leq [6\delta^{1/2} + \log_2(\delta)] \cdot c(r, \Delta). \qquad \square$$

3   The Proof of Theorem 2

Let $A \in \mathbb{Z}^{m \times n}$ have $\mathrm{rank}(A) = n$. Our approach to bound $i_A(b)$ starts by finding an $\mathrm{LP}_{A,c}(b)$ basis matrix A_I, if one exists. The matrix A_I is square, so we can apply Theorem 1 (a) to find a suitable W for which $W\text{-MIP}_{A_I,c}(b_I)$ has integer vertices. These vertices may violate constraints $A_j x \leq b_j$ for $j \in \{1,\ldots,m\} \setminus I$. Intuitively, this means the vertices of $W\text{-MIP}_{A_I,c}(b_I)$ are close to other facets of $\text{LP}_{A,c}(b)$. The next lemma gives a bound on the coefficients b_j that ensures the vertices are valid for $\text{LP}_{A,c}(b)$ and $\text{IP}_{A,c}(b)$. The proof is in the appendix. A set $I \subseteq \{1,\ldots,m\}$ is a *basis* if $|I| = n$ and $\text{rank}(A_I) = n$, and I is feasible if $(A_I)^{-1} b_I$ is a feasible solution for $\text{LP}_{A,c}(b)$. Set $\Delta^{\max} := \Delta^{\max}(A)$.

Lemma 7. *Let $b \in \mathbb{Z}^m$ be such that $\text{LP}_{A,c}(b)$ is feasible and $I \subseteq \{1,\ldots,m\}$ a feasible $\text{LP}_{A,c}(b)$ basis. Let $W \in \mathbb{Z}^{k\times n}$ and assume that z^* is a feasible vertex of $W\text{-MIP}_{A_I,c}(b_I)$. If $A_j A_I^{-1} b_I + (n\Delta^{\max})^2 < b_j$ for all $j \in \{1,\ldots,m\} \setminus I$, then z^* is also feasible for $W\text{-MIP}_{A,c}(b)$.*

Lemma 7 inspires the following definition of $\mathcal{G}$ for Theorem 2:

$$\mathcal{G} := \left\{ b \in \mathbb{Z}^m : \begin{array}{c} A_j A_I^{-1} b_I + (n\Delta^{\max})^2 < b_j \text{ for all feasible bases } I \\ \text{of } \text{LP}_{A,c}(b) \text{ and } j \in \{1,\ldots,m\} \setminus I \end{array} \right\}. \quad (12)$$

Recall $\mathcal{G}^I$ and $\mathcal{G}^L$ from (2). Lemma 7 can be combined with Theorem 1 (a) to argue $\mathcal{G} \subseteq \mathcal{G}^I \cup \mathcal{G}^L$. We now show $\Pr(\mathbb{Z}^m \setminus \mathcal{G}) = 0$ by showing that $\mathbb{Z}^m \setminus \mathcal{G}$ is contained in a finite union of hyperplanes in $\mathbb{Z}^m$. The proof of the following lemma is in the appendix.

Lemma 8. *For each basis $I \subseteq \{1,\ldots,m\}$, set $\Delta_I := |\det(A_I)|$. It follows that*

$$\mathbb{Z}^m \setminus \mathcal{G} \subseteq \bigcup_{\substack{I \subseteq \{1,\ldots,m\} \\ I \text{ basis}}} \bigcup_{j \notin I} \bigcup_{r=0}^{\Delta_I (n\Delta^{\max})^2} \{b \in \mathbb{Z}^m : \Delta_I b_j = \Delta_I A_j A_I^{-1} b_I + r\}.$$

The value $\Pr(\mathbb{Z}^m \setminus \mathcal{G})$ is zero because $\mathbb{Z}^m \setminus \mathcal{G}$ is contained in a finite union of hyperplanes, each of which has measure zero in $\mathbb{R}^m$.

Proof (of Theorem 2). If we prove

$$\lim_{t \to \infty} \frac{|\{-t,\ldots,t\}^m \cap (\mathbb{Z}^m \setminus \mathcal{G})|}{|\{-t,\ldots,t\}^m|} = 0, \quad (13)$$

then we will have proven $\Pr(\mathcal{G})$ is defined by a true limit and $\Pr(\mathcal{G}) = 1$. The denominator of (13) is $(2t + 1)^m$; we show the numerator is in $O((2t + 1)^{m-1})$.

Lemma 8 implies that $|\{-t,\ldots,t\}^m \cap (\mathbb{Z}^m \setminus \mathcal{G})|$ is at most

$$\sum_{\substack{I \subseteq \{1,\ldots,m\} \\ I \text{ basis}}} \sum_{j \notin I} \sum_{r=0}^{\Delta_I (n\Delta^{\max})^2} |\{b \in \{-t,\ldots,t\}^m : \Delta_I b_j = \Delta_I A_j A_I^{-1} b_I + r\}|.$$

Consider a basis I, an index $j \notin I$, and a value $r \in \{0,\ldots,\Delta_I(n\Delta^{\max})^2\}$. If $b \in \{-t,\ldots,t\}^m$ and $\Delta_I b_j = \Delta_I A_j A_I^{-1} b_I + r$, then b_j is fixed and

$$|\{b \in \{-t,\ldots,t\}^m : \Delta_I b_j = \Delta_I A_j A_I^{-1} b_I + r\}| \leq \prod_{i \neq j}^m |\{-t,\ldots,t\}| = (2t + 1)^{m-1}.$$

The previous two inequalities imply that $|\{-t,\ldots,t\}^m \cap (\mathbb{Z}^m \setminus \mathcal{G})|$ is at most

$$\binom{m}{n}(m-n)(n^2(\Delta^{\max})^3 + 1)(2t+1)^{m-1} \in O((2t+1)^{m-1}). \qquad \square$$

Acknowledgements. The authors wish to thank Helene Weiß and Stefan Weltge for their help that led to major improvements of the manuscript. We are also grateful to the anonymous referees for their comments that improved the presentation of the material. The third author acknowledges the support from the Einstein Foundation Berlin.

Appendix

Proof (of Lemma 2). Let $W \in \mathbb{Z}^{k \times n}$ be a matrix with minimal k such that all vertices of W-$\mathrm{MIP}_{A,c}(b)$ are integral. Set $\overline{A} := AU$, $\overline{c} := c^\mathsf{T}U$, and $\overline{W} := WU$. The matrix U^{-1} maps the vertices of W-$\mathrm{MIP}_{A,c}(b)$ to those of $\overline{W}$-$\mathrm{MIP}_{\overline{A},\overline{c}}(b)$, and U^{-1} maps $\mathbb{Z}^n$ to $\mathbb{Z}^n$. Thus, $\overline{W} \in \mathbb{Z}^{k \times n}$ and the vertices of $\overline{W}$-$\mathrm{MIP}_{\overline{A},\overline{c}}(b)$ are integral. Hence, $i_A(b) \geq i_{\overline{A}}(b)$. To see why the reverse inequality holds, it is enough to notice that U^{-1} is also unimodular. $\qquad \square$

Proof (of Lemma 4). By Ghouila-Houri (see, e.g.,[17, §19]) it is enough to show

$$y := \sum_{w \in \widehat{W} \cap W^B} -w + \sum_{w \in \widehat{W} \cap W^T} w \in \{-1,0,1\}^n$$

for a subset $\widehat{W}$ of the rows of W. Recall that every column u of $C = (B \ \ T)W$ can be written as $u = v + t$ for some $v \in B$ and $t \in T \cup \{0\}$. Hence, a column of W has at most two non-zero entries, where a non-zero entry equals 1. One of these entries is in the rows of W^B while the other is in W^T. This shows $y \in \{-1,0,1\}^n$. $\qquad \square$

Proof (of Lemma 5). Let z be a column of A_I^1. If $z \in \{z^1,\ldots,z^\ell\}$, then $z = z + 0 \in B + (T \cup \{0\})$. Else, $z \in \{0,\ldots,\alpha_1 - 1\} \times \ldots \times \{0,\ldots,\alpha_\ell - 1\}$. Define

$$t_i = \left\lfloor \frac{z_i}{k_i+1} \right\rfloor \cdot (k_i+1) \text{ and } v_i = z_i - t_i \text{ for all } i \in \{0,\ldots,\ell\}.$$

We have $v_i \in \{0,\ldots,k_i\}$ for each $i \in \{0,\ldots,\ell\}$, so $v := (v_0,\ldots,v_\ell)^\mathsf{T} \in B$. To see that $(t_1,\ldots,t_\ell)^\mathsf{T} \in T$ note that $t_i \leq \beta_i \cdot (k_i+1)$ for each $i \in \{1,\ldots,\ell\}$.

It is left to choose $k_1,\ldots,k_\ell$ such that $|B| + |T| \leq 6\delta^{1/2} + \log_2(\delta)$. Note $\ell \leq \log_2(\delta)$ as $\alpha_1,\ldots,\alpha_\ell \geq 2$. By permuting rows and columns we assume $\alpha_1 \leq \alpha_2 \leq \ldots \leq \alpha_\ell$. We consider two cases.

Case 1. Assume $\alpha_\ell = \delta^\tau$ for $\tau \geq 1/2$. This implies $\prod_{i=1}^{\ell-1} \alpha_i = \delta^{1-\tau} \leq \delta^{1/2}$. Let $\sigma \geq 0$ such that $1 - \tau + \sigma = 1/2$. For each $i \in \{1,\ldots,\ell-1\}$ define $k_i := \alpha_i - 1$ and set $k_\ell := \lceil \delta^\sigma \rceil$. The value β_ℓ in (9) satisfies

$$\beta_\ell = \left\lfloor \frac{\alpha_\ell - 1}{\lceil \delta^\sigma \rceil + 1} \right\rfloor = \left\lfloor \frac{\delta^\tau - 1}{\lceil \delta^\sigma \rceil + 1} \right\rfloor \leq \lceil \delta^{1/2} \rceil \leq \delta^{1/2} + 1.$$

Define $B = \overline{B} \cup \{z^1, \ldots, z^\ell\}$, where $\overline{B} := \{0, \ldots, k_1\} \times \ldots \times \{0, \ldots, k_\ell\}$, and the set T via (9). A direct computation reveals that $|B| + |T| = |\overline{B}| + |T| + \ell$ is upper bounded by

$$\delta^{1-\tau}(\lceil \delta^\sigma \rceil + 1) + (\beta_\ell + 1) + \log_2(\delta) \le 6\delta^{1/2} + \log_2(\delta).$$

Case 2. Assume $\alpha_\ell < \delta^{1/2}$, which implies $\delta^{1/2} < \prod_{i=1}^{\ell-1} \alpha_i$. Let $j \in \{1, \ldots, \ell-2\}$ be the largest index with $\gamma := \prod_{i=1}^{j} \alpha_i \le \delta^{1/2}$. Let $\sigma \ge 0$ be such that $\gamma \cdot \delta^\sigma = \delta^{1/2}$ and $\tau < 1/2$ be such that $\alpha_{j+1} = \delta^\tau$. Note that $0 \le \sigma < \tau$ and $\delta^{\tau-\sigma} \cdot \prod_{i=j+2}^{\ell} \alpha_i = \delta^{1/2}$. For each $i \in \{1, \ldots, j\}$ define $k_i := 0$, for each $i \in \{j+2, \ldots, \ell\}$ define $k_i := \alpha_i - 1$, and set $k_{j+1} := \lceil \delta^{\tau-\sigma} \rceil$. The value β_{j+1} in (9) satisfies

$$\beta_{j+1} = \left\lfloor \frac{\alpha_{j+1} - 1}{\lceil \delta^{\tau-\sigma} \rceil + 1} \right\rfloor = \left\lfloor \frac{\delta^\tau - 1}{\lceil \delta^{\tau-\sigma} \rceil + 1} \right\rfloor \le \lceil \delta^\sigma \rceil.$$

Define $B = \overline{B} \cup \{z^1, \ldots, z^\ell\}$, where $\overline{B} := \{0, \ldots, k_1\} \times \ldots \times \{0, \ldots, k_\ell\}$, and the set T via (9). In this case we have $|B| + |T| = |\overline{B}| + |T| + l$ is upper bounded by

$$\left(\prod_{i=j+2}^{\ell} \alpha_i\right)(\delta^{\tau-\sigma} + 2) + \gamma(\delta^\sigma + 2) + \log_2(\delta) \le 6\delta^{1/2} + \log_2(\delta). \qquad \square$$

Proof (of Lemma 7). Let $x^* := A_I^{-1} b_I$ be the feasible vertex solution to $\mathrm{LP}_{A,c}(b)$ with respect to the basis I. Applying Theorem 1 in [6] to the simplicial problems $\mathrm{LP}_{A_I,c}(b)$ and $\mathrm{IP}_{A_I,c}(b)$ shows that z^* satisfies $\|z^* - x^*\|_\infty \le n\Delta^{\max}$. Thus, for every $j \in \{1, \ldots, m\} \setminus I$ we have

$$|A_j z^* - A_j x^*| \le \|A_j\|_1 \cdot \|z^* - x^*\|_\infty \le n^2 \|A_j\|_\infty \cdot \Delta^{\max} \le (n\Delta^{\max})^2.$$

The assumption $A_j A_I^{-1} b_I + (n\Delta^{\max})^2 < b_j$ implies

$$A_j z^* \le A_j x^* + |A_j z^* - A_j x^*| \le A_j x^* + (n\Delta^{\max})^2 = A_j A_I^{-1} b_I + (n\Delta^{\max})^2 < b_j.$$

Thus, z^* is feasible for $\mathrm{W\text{-}MIP}_{A,c}(b)$. $\qquad \square$

Proof (of Lemma 8). Let $b \in \mathbb{Z}^m \setminus \mathcal{G}$. Therefore, there exists a feasible $\mathrm{LP}_{A,c}(b)$ basis $I \subseteq \{1, \ldots, m\}$ and $j \in \{1, \ldots, m\} \setminus I$ such that $b_j \le A_j A_I^{-1} b_I + (n\Delta^{\max})^2$. Recall $A_j A_I^{-1} b_I \le b_j$ because $A_I^{-1} b_I$ is feasible for $\mathrm{LP}_{A,c}(b)$. Thus, $\mathbb{Z}^m \setminus \mathcal{G}$ is in

$$\{b \in \mathbb{Z}^m : \exists \text{ a basis } I \text{ and } j \in \{1, \ldots, m\} \setminus I \text{ with } b_j \le A_j A_I^{-1} b_I + (n\Delta^{\max})^2\}.$$

Cramer's Rule implies that $\Delta_I \cdot A_j A_I^{-1} b_I \in \mathbb{Z}$ for all $j \in \{1, \ldots, m\} \setminus I$. Thus,

$$\{b \in \mathbb{Z}^m : \exists \text{ a basis } I \text{ and } j \notin I \text{ with } \Delta_I b_j \le \Delta_I A_j A_I^{-1} b_I + \Delta_I(n\Delta^{\max})^2\}$$

$$\subseteq \bigcup_{\substack{I \subseteq \{1, \ldots, m\} \\ I \text{ basis}}} \bigcup_{j \notin I} \bigcup_{r=0}^{\Delta_I(n\Delta^{\max})^2} \{b \in \mathbb{Z}^m : \Delta_I b_j = \Delta_I A_j A_I^{-1} b_I + r\}. \qquad \square$$

References

1. Artmann, S., Eisenbrand, F., Glanzer, C., Oertel, T., Vempala, S., Weismantel, R.: A note on non-degenerate integer programs with small sub-determinants. Oper. Res. Lett. **44**(5), 635–639 (2016)
2. Artmann, S., Weismantel, R., Zenklusen, R.: A strongly polynomial algorithm for bimodular integer linear programming. In: Proceedings of the 49th Annual ACM SIGACT Symposium on Theory of Computing, pp. 1206–1219 (2017)
3. Bader, J., Hildebrand, R., Weismantel, R., Zenklusen, R.: Mixed integer reformulations of integer programs and the affine TU-dimension of a matrix. Math. Programm. **169**(2), 565–584 (2017). https://doi.org/10.1007/s10107-017-1147-2
4. Barvinok, A.: A Course in Convexity. Graduate Studies in Mathematics, vol. 54. American Mathematical Society, Providence (2002)
5. Bruns, W., Gubeladze, J.: Normality and covering properties of affine semigroups. J. für die reine und angewandte Mathematik **510**, 151–178 (2004)
6. Cook, W., Gerards, A., Schrijver, A., Tardos, E.: Sensitivity theorems in integer linear programming. Math. Programm. **34**, 251–264 (1986). https://doi.org/10.1007/BF01582230
7. Dadush, D., Peikert, C., Vempala, S.: Enumerative lattice algorithms in any norm via M-ellipsoid coverings. In: 2011 IEEE 52nd Annual Symposium on Foundations of Computer Science, pp. 580–589 (2011)
8. Eisenbrand, F., Shmonin, G.: Parametric integer programming in fixed dimension. Math. Oper. Res. **33**(4), 839–850 (2008). https://doi.org/10.1287/moor.1080.0320
9. Eisenbrand, F., Weismantel, R.: Proximity results and faster algorithms for integer programming using the Steinitz lemma. In: Proceedings of the Twenty-Ninth Annual ACM-SIAM Symposium on Discrete Algorithms, pp. 808–816 (2018)
10. Glanzer, C., Weismantel, R., Zenklusen, R.: On the number of distinct rows of a matrix with bounded subdeterminants. SIAM J. Discrete Math. **32**, 1706–1720 (2018)
11. Gomory, R.E.: On the relation between integer and noninteger solutions to linear programs. Proc. Nat. Acad. Sci. **53**(2), 260–265 (1965). https://doi.org/10.1073/pnas.53.2.260
12. Heller, I.: On linear systems with integral valued solutions. Pac. J. Math. **7**, 1351–1364 (1957)
13. Hupp, L.M.: Integer and Mixed-Integer Reformulations of Stochastic Resource-Constrained, and Quadratic Matching Problems. Ph.D. thesis. Friedrich-Alexander-Universitat Erlangen-Nurnberg (2017)
14. Kannan, R.: Minkowski's convex body theorem and integer programming. Math. Oper. Res. **12**(3), 415–440 (1987)
15. Lenstra, H.: Integer programming with a fixed number of variables. Math. Oper. Res. **8**, 538–548 (1983)
16. Oertel, T., Paat, J., Weismantel, R.: Sparsity of integer solutions in the average case. In: Lodi, A., Nagarajan, V. (eds.) IPCO 2019. LNCS, vol. 11480, pp. 341–353. Springer, Cham (2019). https://doi.org/10.1007/978-3-030-17953-3_26
17. Schrijver, A.: Theory of Linear and Integer Programming. Wiley, New York (1986)
18. Veselov, S., Chirkov, A.: Integer programming with bimodular matrix. Discrete Optim. **6**, 220–222 (2009)
19. Wolsey, L.: The b-hull of an integer program. Discrete Appl. Math. **3**(3), 193–201 (1981). https://doi.org/10.1016/0166-218X(81)90016-0. http://www.sciencedirect.com/science/article/pii/0166218X81900160

Persistency of Linear Programming Relaxations for the Stable Set Problem

Elisabeth Rodríguez-Heck[1], Karl Stickler[1], Matthias Walter[2(✉)],
and Stefan Weltge[3]

[1] Lehrstuhl für Operations Research, RWTH Aachen University, Aachen, Germany
`rodriguez-heck@or.rwth-aachen.de`, `karl.stickler@rwth-aachen.de`
[2] Department of Applied Mathematics, University of Twente,
Enschede, The Netherlands
`m.walter@utwente.nl`
[3] Department of Mathematics, Technical University of Munich, Munich, Germany
`weltge@tum.de`

Abstract. The Nemhauser-Trotter theorem states that the standard linear programming (LP) formulation for the stable set problem has a remarkable property, also known as (weak) *persistency*: for every optimal LP solution that assigns integer values to some variables, there exists an optimal integer solution in which these variables retain the same values. While the standard LP is defined by only non-negativity and edge constraints, a variety of stronger LP formulations have been studied and one may wonder whether any of them has the this property as well. We show that any stronger LP formulation that satisfies mild conditions cannot have the persistency property on all graphs, unless it is always equal to the stable-set polytope.

Keywords: Persistency · Integer linear programming · Stable set

1 Introduction

Given an undirected graph G with node set $V(G)$ and edge set $E(G)$, and node weights $w \in \mathbb{R}^{V(G)}$, the (weighted) stable-set problem asks for finding a stable set S in G that maximizes $\sum_{v \in S} w_v$, where a set S is called stable if G contains no edge with both endpoints in S. While the stable-set problem is NP-hard, it is a common approach to maximize $w^\mathsf{T} x$ over the *edge relaxation*

$$R_{\text{stab}}^{\text{edge}}(G) := \left\{ x \in [0,1]^{V(G)} \mid x_v + x_w \leq 1 \text{ for each edge } \{v,w\} \in E(G) \right\}$$

and use optimal (fractional) solutions to gain insights about optimal 0/1-solutions. Note that the 0/1-points in the edge relaxation are precisely the characteristic vectors of stable sets in G, and that maximizing a linear objective over the edge relaxation is a linear program that can be solved efficiently. Given an optimal solution of this linear program, its objective value is clearly an upper

© Springer Nature Switzerland AG 2020
D. Bienstock and G. Zambelli (Eds.): IPCO 2020, LNCS 12125, pp. 351–363, 2020.
https://doi.org/10.1007/978-3-030-45771-6_27

bound on the value of any 0/1-solution and its entries may guide initial decisions in a branch-and-bound algorithm. While this is also the case for general polyhedral relaxations, it turns out that optimal solutions of the edge relaxation have a remarkable property that allows to reduce the size of the problem by fixing some variables to provable optimal integer values.

Definition 1 (Persistency). *We say that a polytope $P \subseteq [0,1]^n$ has the persistency property if for every objective vector $c \in \mathbb{R}^n$ and every c-maximal point $x \in P$, there exists a c-maximal integer point $y \in P \cap \{0,1\}^n$ such that $x_i = y_i$ for each $i \in \{1,2,\ldots,n\}$ with $x_i \in \{0,1\}$.*

Proposition 1 (Nemhauser and Trotter [8]). *The edge relaxation $R_{stab}^{edge}(G)$ has the persistency property for every graph G.*

In other words, the result of Nemhauser and Trotter [8] states that if $x^\star$ is an optimal solution for the edge relaxation, then there exists an optimal stable set $S^\star$ satisfying $V_1 \subseteq S^\star \subseteq V(G) \setminus V_0$, where $V_i := \{v \in V(G) \mid x_v^\star = i\}$ for $i = 0, 1$. In this case, the nodes in $V_0 \cup V_1$ can be deleted and the search only has to be performed on the remaining graph. Clearly, this reduction is significant if $x^\star$ assigns integer values to many nodes.

For the maximum cardinality stable set problem it has been shown that the probability of obtaining a single integer component when solving the LP relaxation is very low for large random graphs [11]. However, persistencies have been proved to be very useful in a different context, when dealing with highly structured instances arising in the field of computer vision. More precisely, Hammer, Hansen and Simeone [4] provided a reduction of (Unconstrained) Quadratic Binary Programming (QBP) to the stable set problem and showed that weak persistency holds for (QBP) as well. Boros et al. [1] provided an algorithm to compute the largest possible set of variables to fix via persistencies in a quadratic binary program in polynomial time, which has been successfully used in practice to solve very large image restoration problems [3,5,7].

In general, dual bounds obtained from the edge relaxation are quite weak, and several families of additional inequalities have been studied in order to strengthen this formulation. Examples are the clique inequalities [10], (lifted) odd-cycle inequalities [10,15] and clique-family inequalities [9]. Most of these families were discovered by systematically studying the facets of the *stable-set polytope* $P_{stab}(G)$, which is the convex hull of the characteristic vectors of stable sets in G. The stable-set polytope itself is known to be a complicated polytope. In particular, one cannot expect to be able to completely characterize its facial structure [6]. Thus, the following question is natural.

Do there exist stronger linear programming formulations for the stable set problem that also have the persistency property for every graph G?

In this paper, we answer the question negatively. More precisely, we show that an LP formulation (satisfying mild conditions) that is stronger than the edge formulation cannot have the persistency property on all graphs, unless it always yields the stable-set polytope.

Outline. The paper is structured as follows. We start by introducing the conditions we impose on the LP formulation in Sect. 2. Our main result and its consequences are presented in Sect. 3. Section 4 is dedicated to the proof of the main result. Our preprint [12] provides running examples that illustrate the steps of the proof.

2   LP Formulations for Stable Set

It is clear that, for a *single* non-bipartite graph G, one can artificially construct polytopes strictly between $R_{\text{stab}}^{\text{edge}}(G)$ and $P_{\text{stab}}(G)$ that have the persistency property. For instance, if $x \in R_{\text{stab}}^{\text{edge}}(G) \setminus P_{\text{stab}}(G)$ is any point that has only fractional coordinates, then the polytope $\text{conv}(P_{\text{stab}}(G) \cup x)$ has the persistency property for trivial reasons. In this work, however, we consider relaxations defined for *every* graph that arise in a more structured way.

To this end, let $\mathcal{G}$ denote the set of finite undirected simple graphs. We regard an LP *formulation* for the stable set problem as a map that assigns to every graph $G \in \mathcal{G}$ a polytope $R_{\text{stab}}(G) \supseteq P_{\text{stab}}(G)$. As an example, the edge formulation assigns $R_{\text{stab}}^{\text{edge}}(G)$ to every graph G. Next, let us specify some natural conditions that are satisfied by many prominent formulations and under which our main result holds. Each of these conditions is defined for a formulation R_{stab}.

First, we require that the formulation R_{stab} is at least as strong as the edge formulation. Formally,

$$\text{for each } G \in \mathcal{G}, \text{ we have } P_{\text{stab}}(G) \subseteq R_{\text{stab}}(G) \subseteq R_{\text{stab}}^{\text{edge}}(G). \tag{A}$$

Second, the inequalities defining R_{stab} must be derived from facets of P_{stab}:

for each $G \in \mathcal{G}$, each inequality with support $U \subseteq V(G)$ that is facet-   (B)
defining for $R_{\text{stab}}(G)$ is also facet-defining for $P_{\text{stab}}(G[U])$,

where $G[U]$ denotes the subgraph induced by U. Note that inequalities need to define facets only on their support graph, hence also the generally not facet-defining odd-cycle inequalities (see [10]) satisfy (B). However, a formulation consisting of only rank inequalities (see [2]) does not satisfy (B).

Third, for every graph $G \in \mathcal{G}$, validity of facet-defining inequalities of $R_{\text{stab}}(G)$ shall be inherited by induced subgraphs. Formally,

for each $G \in \mathcal{G}$, each inequality with support $U \subseteq V(G)$ that is facet-   (C)
defining for $R_{\text{stab}}(G)$ is valid (although not necessarily facet-defining)
for $R_{\text{stab}}(G[U])$.

This requirement ensures that if an (irredundant) inequality arises for some graph then it must (at least implicitly) occur for all induced subgraphs for which it is defined. The reverse implication is imposed by the fourth condition, although in a more structured way. For this, we need the following definitions.

Let $G_1, G_2 \in \mathcal{G}$ and let $v_1 \in V(G_1)$, $v_2 \in V(G_2)$. Then the 1-*sum of* G_1 *and* G_2 *at* v_1 *and* v_2, denoted by $G_1 \oplus_{v_2}^{v_1} G_2$ is the graph obtained from the disjoint union of G_1 and G_2 by identifying v_1 with v_2. Moreover, let $P \subseteq \mathbb{R}^m$ and $Q \subseteq \mathbb{R}^n$ be polytopes and let $i \in \{1, 2, \ldots, m\}$ and $j \in \{1, 2, \ldots, n\}$. The 1-*sum of* P *and* Q *at coordinates* i *and* j, denoted by $P \oplus_j^i Q$, is defined as the projection of $\mathrm{conv}(\{(x, y) \in P \times Q \mid x_i = y_j\})$ onto all variables except for y_j. Notice that this projection is an isomorphism from the convex hull to its image since the variables x_i and y_j are equal.

Our fourth condition requires that for every pair of graphs $G_1, G_2 \in \mathcal{G}$, validity of inequalities is acquired by their 1-sum. Formally,

$$R_{\mathrm{stab}}(G_1 \oplus_{v_2}^{v_1} G_2) = R_{\mathrm{stab}}(G_1) \oplus_{v_2}^{v_1} R_{\mathrm{stab}}(G_2) \text{ holds for all } G_1, G_2 \in \mathcal{G} \qquad (D)$$
$$\text{and all nodes } v_1 \in V(G_1) \text{ and } v_2 \in V(G_2).$$

Also this condition is very natural since every inequality that is valid for $R_{\mathrm{stab}}(G_1)$ is also valid for $P_{\mathrm{stab}}(G_1 \oplus_{v_2}^{v_1} G_2)$, and hence its participation in $R_{\mathrm{stab}}(G_1 \oplus_{v_2}^{v_1} G_2)$ is reasonable.

3   Results

We say that two formulations R_{stab}^1 and R_{stab}^2 are *equivalent* if $R_{\mathrm{stab}}^1(G) = R_{\mathrm{stab}}^2(G)$ holds for every $G \in \mathcal{G}$, in which case we write $R_{\mathrm{stab}}^1 \equiv R_{\mathrm{stab}}^2$. We can now state our main result.

Theorem 1. *Let R_{stab} be a formulation satisfying* (A)–(D). *Then $R_{stab}(G)$ has the persistency property for all graphs $G \in \mathcal{G}$ if and only if $R_{stab} \equiv R_{stab}^{edge}$ or $R_{stab} \equiv P_{stab}$.*

Sufficiency follows from Proposition 1 and from the fact that $P_{\mathrm{stab}}(G)$ is an integral polytope for every $G \in \mathcal{G}$. Before we prove necessity in Sect. 4, let us mention some direct implications of Theorem 1 for known relaxations.

Corollary 1. *The clique relaxation*

$$R_{stab}^{clq}(G) = \left\{ x \in \mathbb{R}^{V(G)} \mid x(V(C)) \leq 1 \text{ for each clique } C \text{ of } G \right\}$$

does not have the persistency property for all graphs $G \in \mathcal{G}$.

Proof. It is easy to see that $R_{\mathrm{stab}}^{\mathrm{clq}}$ satisfies Properties (A) and (D). For Properties (B) and (C), consider a clique C of some graph $G \in \mathcal{G}$. Clearly, C is also a clique of $G[V(C)]$ and the inequality is known to be facet-defining for $P_{\mathrm{stab}}(G[V(C)])$ (see Theorem 2.4 in [10]). □

Also the relaxation based on odd-cycle inequalities satisfies these properties, although the inequalities are generally not facet-defining.

Corollary 2. *The odd-cycle relaxation*

$$R_{stab}^{oc}(G) = \left\{ x \in R_{stab}^{edge}(G) \mid x(V(C)) \leq \frac{|V(C)|-1}{2} \text{ for each odd cycle } C \text{ of } G \right\}$$

does not have the persistency property for all graphs $G \in \mathcal{G}$.

Proof. It is easy to see that R_{stab}^{oc} satisfies Properties (A) and (D). For Properties (B) and (C), consider an odd cycle C of some graph $G \in \mathcal{G}$. To induce a facet, C must be chordless, and the odd-cycle inequality is facet-defining for $P_{stab}(G[V(C)])$ (see Theorem 3.3 in [10]). □

4 Proof of the Main Result

Let us fix any formulation R_{stab} over $\mathcal{G}$ satisfying Properties (A)–(D). To prove the "only if" implication of Theorem 1 we have to verify that if $R_{stab} \neq R_{stab}^{edge}$ and $R_{stab} \neq P_{stab}$, then $R_{stab}(G)$ does not have the persistency property for all graphs $G \in \mathcal{G}$. Equivalently, we have to prove the following:

> If there exist graphs $G_1, G_2 \in \mathcal{G}$ with $R_{stab}(G_1) \neq R_{stab}^{edge}(G_1)$ and
> $R_{stab}(G_2) \neq P_{stab}(G_2)$, then there exists a graph $G^\star$ for which the   (◇)
> polytope $R_{stab}(G^\star)$ does not have the persistency property.

Given G_1 and G_2, we will provide an explicit construction of $G^\star$ and show that $R_{stab}(G^\star)$ does not have the persistency property. To see the latter, we will give an objective vector $c^\star \in \mathbb{R}^{V(G^\star)}$ such that every $c^\star$-maximal solution over $R_{stab}(G^\star)$ has a certain coordinate equal to zero while every $c^\star$-maximal stable set in $G^\star$ contains the corresponding node.

The graph $G^\star$ will consist of an "inner" graph G^{in} with $R_{stab}(G^{in}) \neq R_{stab}^{edge}(G^{in})$ and $|V(G^{in})| - 1$ copies of an "outer" graph G^{out} with $R_{stab}(G^{out}) \neq P_{stab}(G^{out})$. Each copy of G^{out} is attached to a node of G^{in} via the 1-sum operation. The only node of G^{in} that does *not* have a copy of G^{out} attached corresponds precisely to the coordinate showing that $R_{stab}(G^\star)$ does not have the persistency property. Note that such graphs G^{in}, G^{out} exist due to the hypothesis of (◇). Among all such graphs, we will make particular choices satisfying some additional properties that we specify in the next sections.

4.1 The Graph G^{out}

In the definition of the auxiliary graph G^{out} we will make use of the following lemma. In what follows, for a polytope $P \subseteq \mathbb{R}^n$ and a vector $c \in \mathbb{R}^n$, let us denote the optimal face of P induced by c by $opt(P, c) := \arg\max\{c^\mathsf{T}x \mid x \in P\}$.

Lemma 1. *Let $P, Q \subseteq \mathbb{R}^n$ be polytopes. If there exists a vector $c \in \mathbb{R}^n$ such that $\dim(opt(Q, c)) < \dim(opt(P, c))$, then there exists a vector $c' \in \mathbb{R}^n$ such that $opt(Q, c')$ is a vertex of Q, while $opt(P, c')$ is not a vertex of P.*

The lemma is proved in Appendix A. The graph G^{out} is now defined through the following statement.

Claim 1. Assuming the hypothesis of $(\Diamond)$, there exists a graph $G^{\text{out}} \in \mathcal{G}$, a vector $c^{\text{out}} \in \mathbb{R}^{V(G^{\text{out}})}$ and a node $v^{\text{out}} \in V(G^{\text{out}})$ such that $\text{opt}(R_{\text{stab}}(G^{\text{out}}), c^{\text{out}}) = \{\hat{x}\}$ holds with $\hat{x}_{v^{\text{out}}} \geq \frac{1}{2}$ and such that $\text{opt}(P_{\text{stab}}(G), c^{\text{out}})$ contains a vertex $\bar{x} \in \{0,1\}^{V(G^{\text{out}})}$ with $\bar{x}_{v^{\text{out}}} = 0$.

Proof. Let $G \in \mathcal{G}$ be such that $R_{\text{stab}}(G) \neq P_{\text{stab}}(G)$. Such a graph exists by hypothesis of $(\Diamond)$. By Property (A), there exists an inequality $a^\intercal x \leq \delta$ that is facet-defining for $P_{\text{stab}}(G)$, but not valid for $R_{\text{stab}}(G)$.

We claim that the face $\text{opt}(R_{\text{stab}}(G), a)$ is not a facet of $R_{\text{stab}}(G)$. Assume for a contradiction that $\text{opt}(R_{\text{stab}}(G), a)$ is a facet of $R_{\text{stab}}(G)$ and define $\delta' := \max\{a^\intercal x \mid x \in R_{\text{stab}}(G)\}$. Since $a^\intercal x \leq \delta$ is not valid for $R_{\text{stab}}(G)$, we have $\delta' > \delta$. Property (B) implies that $a^\intercal x \leq \delta'$ is facet-defining for $P_{\text{stab}}(G[\text{supp}(a)])$, and in particular, equality holds for the characteristic vector of some stable set $S \subseteq V(G[\text{supp}(a)])$. Since S is also a stable set in G, this contradicts the assumption that $a^\intercal x \leq \delta$ is valid for $P_{\text{stab}}(G)$.

By Lemma 1, there exists a vector $c \in \mathbb{R}^n$ such that $\text{opt}(R_{\text{stab}}(G), c) = \{\hat{x}\}$ and $\text{opt}(P_{\text{stab}}(G), c)$ has (at least) two vertices $\bar{x}^1, \bar{x}^2 \in \{0,1\}^{V(G)}$. Since $\bar{x}^1 \neq \bar{x}^2$, there exists a coordinate $u \in V(G)$ at which they differ and we can assume $\bar{x}_u^1 = 0$ and $\bar{x}_u^2 = 1$ without loss of generality. If $\hat{x}_u \geq \frac{1}{2}$, we can choose $G^{\text{out}} := G$, $c^{\text{out}} := c$ and $v^{\text{out}} := u$. Together with $\hat{x}$ and $\bar{x}^1$, they satisfy the requirements of the lemma.

Otherwise, let G' be the graph G with an additional edge $\{u, u'\}$ attached at u. Formally, let G'' be the graph consisting of a single edge $\{u, u'\}$ and let $G' := G \oplus_u^u G''$. By Property (D), $R_{\text{stab}}(G') = R_{\text{stab}}(G) \oplus_u^u R_{\text{stab}}(G'')$ holds. Since G'' is a single edge, $R_{\text{stab}}^{\text{edge}}(G'') = P_{\text{stab}}(G'')$ holds. Thus, $R_{\text{stab}}(G')$ is described by all inequalities that are valid for $R_{\text{stab}}(G)$ together with $x_{u'} \geq 0$ and $x_u + x_{u'} \leq 1$. Hence, for a sufficiently small $\varepsilon > 0$ and the objective vector $c' \in \mathbb{R}^{V(G')}$ with $c'_{u'} = \varepsilon$, $c'_u = c_u + 2\varepsilon$ and $c'_v = c_v$ for all $v \in V(G) \setminus \{u\}$, the maximization of c' over $R_{\text{stab}}(G')$ yields a unique optimum $\hat{x}' \in \mathbb{R}^{V(G')}$ with $\hat{x}'_v = \hat{x}_v$ for all $v \in V(G)$ and $\hat{x}'_{u'} = 1 - \hat{x}'_u > \frac{1}{2}$, while the maximization of c' over $P_{\text{stab}}(G')$ admits an optimum $\bar{x}' \in \mathbb{R}^{V(G')}$ with $\bar{x}'_u = 1$ and $\bar{x}'_{u'} = 0$. Now, $G^{\text{out}} := G'$, $c^{\text{out}} := c'$ and $v^{\text{out}} := u'$ together with $\hat{x}'$ and $\bar{x}'$ satisfy the requirements of the lemma. $\qquad\square$

4.2   The Graph G^{in}

Among all graphs $G \in \mathcal{G}$ with $R_{\text{stab}}(G) \neq R_{\text{stab}}^{\text{edge}}(G)$ we choose G^{in} to have a minimum number of nodes. Note that G^{in} exists by hypothesis of $(\Diamond)$. We assume $V(G^{\text{in}}) = \{1, 2, \ldots, n\}$. Let $Ax \leq b$ (with $A \in \mathbb{Z}^{m \times n}$ and $b \in \mathbb{Z}^m$) be the system containing inequalities for all facets of $R_{\text{stab}}(G^{\text{in}})$ that are not valid for $R_{\text{stab}}^{\text{edge}}(G^{\text{in}})$. Note that $m \geq 1$ and $n \geq 3$ hold by assumption on G^{in}.

Claim 2. $A_{i,j} \geq 1$ holds for every $i \in \{1, 2, \ldots, m\}$ and every $j \in \{1, 2, \ldots, n\}$.

Proof. It is a basic fact that every facet-defining inequality of a stable-set polytope that is not a nonnegativity constraint is of the form $a^{\mathsf{T}}x \leq \beta$ for some nonnegative vector $a \in \mathbb{R}^n$ (see Section 9.3 in [13]). Assume, $A_{i,j} = 0$ holds for some i, j. By Property (C), $A_{i,\star}x \leq b_i$ is valid for $R_{\text{stab}}(G[\text{supp}(A_{i,\star})])$, while it is not valid for $R_{\text{stab}}^{\text{edge}}(G^{\text{in}}[\text{supp}(A_{i,\star})])$, contradicting minimality of G^{in}. $\square$

4.3 The Graph $G^\star$

For each $j \in \{2, 3, \ldots, n\}$ let G^j be an isomorphic copy of G^{out} such that $V(G^j) \cap V(G^k) = \varnothing$ whenever $j \neq k$. Let $c^j \in \mathbb{R}^{V(G^j)}$ and $v^j \in V(G^j)$ be the vector and node corresponding to c^{out} and v^{out} in Claim 1, respectively. Now $G^\star$ is defined as the 1-sum of G^{in} with all G^j at the respective nodes $j \in V(G^{\text{in}})$ and $v^j \in V(G^j)$, i.e., $G^\star := G^{\text{in}} \oplus_{v^2}^2 G^2 \oplus_{v^3}^3 \cdots \oplus_{v^n}^n G^n$, where the $\oplus$-operator has to be applied from left to right. By Property (D) we have

$$R_{\text{stab}}(G^\star) = R_{\text{stab}}(G^{\text{in}}) \oplus_{v^2}^2 R_{\text{stab}}(G^2) \oplus_{v^3}^3 \cdots \oplus_{v^n}^n R_{\text{stab}}(G^n).$$

4.4 The Objective Vector

It remains to construct an objective vector $c^\star \in \mathbb{R}^{V(G^\star)}$ that shows that $R_{\text{stab}}(G^\star)$ does not have the persistency property. Let A, b be as in the previous section, and denote by $a := A_{1,\star}$ the first row of A. We will define $c^\star$ via

$$c_1^\star := \varepsilon \quad \text{and} \quad c_v^\star := a_j \cdot c_v^j \text{ for all } v \in V(G^j), \ j \in \{2, 3, \ldots, n\},$$

where $\varepsilon > 0$ is a positive constant that we will define later. Our first claim is independent of the specific choice of ε.

Claim 3. Every $c^\star$-maximal stable set in $G^\star$ contains node $1 \in V(G^{\text{in}})$.

Proof. By Claim 1 there exists, for each $j \in \{2, 3, \ldots, n\}$, a c^j-maximal stable set $S^j \subseteq V(G^j)$ that does not use v^j. Thus, the maximum objective value obtained on $V(G^\star \setminus \{1\})$ is $\sum_{j=2}^n a_j c^j(S^j)$, which is equal to the maximum objective value for all stable sets that do not contain node 1. Since $v^j \notin S^j$ for each j, the set $S^\star := \bigcup_{j=2}^n S^j \cup \{1\}$ is a stable set in $G^\star$ with objective value $\varepsilon + \sum_{j=2}^n a_j c^j(S^j) > \sum_{j=2}^n a_j c^j(S^j)$, which proves the claim. $\square$

To see that $R_{\text{stab}}(G^\star)$ does not have the persistency property, it suffices to establish the following claim, which then yields Theorem 1.

Claim 4. For $\varepsilon > 0$ small enough, each $c^\star$-optimal $x^\star \in R_{\text{stab}}(G^\star)$ satisfies $x_1^\star = 0$.

Let $x^\star$ be any $c^\star$-optimal point in $R_{\text{stab}}(G^\star)$. In order to understand the contributions of the variables corresponding to nodes $v \in V(G^j)$ to the total optimal value in terms of $x_{v^j}^\star$, let us introduce the function $f : [0, 1] \to \mathbb{R}$ defined via

$$f(y) := \max \left\{ c^{j\mathsf{T}}x \mid x \in R_{\text{stab}}(G^j) \text{ and } x_{v^j} = y \right\}.$$

Note that the definition is independent of j since all (G^j, c^j, v^j) are identical up to indexing. We observe that the restriction of $x^\star$ onto the coordinates corresponding to $V(G^{\mathrm{in}})$ is an optimal solution for

$$\max\left\{c'(x) \mid x \in R_{\mathrm{stab}}(G^{\mathrm{in}})\right\} = \max\left\{c'(x) \mid x \in R_{\mathrm{stab}}^{\mathrm{edge}}(G^{\mathrm{in}}),\, Ax \leq b\right\}, \quad (1)$$

where $c'(x) := \varepsilon x_1 + \sum_{j=2}^{n} a_j f(x_j)$. Thus, we see that Claim 4 immediately follows from the following result.

Claim 5. For $\varepsilon > 0$ small enough, each c'-optimal $x \in R_{\mathrm{stab}}(G^{\mathrm{in}})$ satisfies $x_1 = 0$.

We will consider the function $g : [0, \infty] \to \mathbb{R}$ defined via

$$g(z) := \max\left\{\sum_{j=2}^{n} a_j f(x_j) \mid a^\mathsf{T} x \leq z,\, x \in R_{\mathrm{stab}}^{\mathrm{edge}}(G^{\mathrm{in}})\right\}.$$

The intuition behind the proof of Claim 5 is the following: First, note that $c'(x)$ is the sum of εx_1 and the objective function defining g. Function $g(z)$ represents the contribution to the objective value of G^j for $j = 2, \ldots, n$ as a function of the right-hand side of the inequality $a^\mathsf{T} x \leq z$. We will soon prove that $g(z)$ is strictly increasing on the interval $z \in [0, b_1]$. Since $a_1 > 0$ and x_1 does not contribute to the maximum in the definition of g, the latter is attained only by solutions x with $x_1 = 0$. If we ignore, for a moment, the inequalities $Ax \leq b$, this shows that for sufficiently small ε, also every c'-maximal solution x satisfies $x_1 = 0$. The formal steps are as follows.

Claim 6. The functions f and g are concave. Moreover, g is strictly monotonically increasing on $[0, b_1]$.

Proof of Claim 5. Letting

$$\gamma := \min\left\{x_1 \mid x \text{ vertex of } R_{\mathrm{stab}}(G^\star) \text{ with } x_1 > 0\right\} \in (0, 1], \text{ and}$$
$$\lambda := \min\left\{\gamma/(A_{i,1} + \cdots + A_{i,n}) \mid i \in \{1, 2, \ldots, m\}\right\} \in (0, 1),$$

we claim that every choice of ε with

$$0 < \varepsilon < \lambda(g(b_1) - g(b_1 - a_1\gamma))$$

satisfies the assertion. First, we need to verify that the right-hand side of the inequality above is positive. To this end, note that $a_1 \leq b_1$ and hence $0 \leq b_1 - a_1\gamma < b_1$. By Claim 6 we have

$$g(b_1 - a_1\gamma) < g(b_1), \quad (2)$$

which yields positivity of the right-hand side.

Next, let ε be as above. For the sake of contradiction, assume that there exists a c'-optimal solution $x^\star \in R_{\mathrm{stab}}(G^{\mathrm{in}})$ with $x_1^\star > 0$. Note that $x^\star$ can be extended to a c^*-optimal solution over $R_{\text{stab}}(G^*)$, which we may assume to be a vertex of $R_{\text{stab}}(G^*)$, and hence $x_1^* \geq \gamma$. Let $\hat{x}^0 \in R_{\text{stab}}(G^{\text{in}})$ be equal to x^*, except for $\hat{x}_1^0 := 0$. Moreover, let $\hat{x}^1 \in \mathbb{R}^{V(G^{\text{in}})}$ be a maximizer of $g(b_1)$, which may not be contained in $R_{\text{stab}}(G^{\text{in}})$. Now consider the vector $\hat{x}^\lambda := (1 - \lambda)\hat{x}^0 + \lambda\hat{x}^1$. To obtain the desired contradiction, we will show that $\hat{x}^\lambda$ is contained in $R_{\text{stab}}(G^{\text{in}})$ and that $c'(\hat{x}^\lambda) > c'(x^*)$.

Since $\hat{x}^0$ and $\hat{x}^1$ both lie in $R_{\text{stab}}^{\text{edge}}(G^{\text{in}})$, also x^λ lies in $R_{\text{stab}}^{\text{edge}}(G^{\text{in}})$. Let $i \in \{1, 2, \ldots, m\}$. By Claim 2, $A_{i,1} \geq 1$ holds, which implies $A_{i,*}\hat{x}^0 \leq A_{i,*}x^* - \gamma \leq b_i - \gamma$. We obtain

$$A_{i,*}\hat{x}^\lambda = A_{i,*}\hat{x}^0 + \lambda A_{i,*}(\hat{x}^1 - \hat{x}^0) \leq b_i - \gamma + \lambda(A_{i,1} + \cdots + A_{i,n}) \leq b_i,$$

where the second inequality follows from the fact that each coordinate of $\hat{x}^1 - \hat{x}^0$ is bounded by 1, and the last inequality holds by the definition of λ. This shows that $\hat{x}^\lambda$ is contained in $R_{\text{stab}}(G^{\text{in}})$.

For the objective value of $\hat{x}^1$ we clearly have $c'(\hat{x}^1) \geq g(b_1)$. Moreover, since $\hat{x}_1^0 = 0$ we have

$$c'(\hat{x}^0) \leq g(a^\mathsf{T}\hat{x}^0) \leq g(b_1 - a_1\gamma) < g(b_1),$$

where the latter two inequalities again follow from Claim 6 and (2). Observe that concavity of f and nonnegativity of a imply concavity of $c'(x)$, which yields $c'(\hat{x}^\lambda) \geq (1 - \lambda)c'(\hat{x}^0) + \lambda c'(\hat{x}^1)$. We obtain

$$c'(x^*) - c'(\hat{x}^\lambda) \leq \big(\varepsilon + c'(\hat{x}^0)\big) - \big(c'(\hat{x}^0) - \lambda(c'(\hat{x}^0) - c'(\hat{x}^1))\big)$$
$$= \varepsilon + \lambda(c'(\hat{x}^0) - c'(\hat{x}^1)) \leq \varepsilon + \lambda(g(b_1 - a_1\gamma) - g(b_1)) < 0,$$

where the last inequality holds by definition of ε and due to (2).

$\square$

To conclude the proof of Theorem 1, it remains to prove Claim 6. The fact that f and g are concave is a simple consequence of the next basic lemma.

Lemma 2. *Let $P \subseteq \mathbb{R}^n$ be a non-empty polytope, let $c, a \in \mathbb{R}^n$ and let $\ell := \min\{a^\mathsf{T}x \mid x \in P\}$. The functions $h^=, h^\leq : [\ell, \infty) \to \mathbb{R}$ defined via $h^=(\beta) = \max\{c^\mathsf{T}x \mid x \in P, a^\mathsf{T}x = \beta\}$ and $h^\leq(\beta) = \max\{c^\mathsf{T}x \mid x \in P, a^\mathsf{T}x \leq \beta\}$ are concave. Moreover, there exists a number $\beta^* \in [\ell, \infty)$ such that $h^=$ and $h^\leq$ are identical and strictly monotonically increasing on the interval $[\ell, \beta^*]$, and $h^\leq$ is constant on the interval $[\beta^*, \infty)$.*

The lemma is proved in Appendix A. The proof of Claim 6 relies on the following result of Sewell.

Proposition 2 (Corollary 3.4.3 in [14]). *Let $\sum_{j=1}^n a_jx_j \leq b_1$ be a facet-defining inequality for the stable-set polytope of a graph on n nodes that is neither a bound nor an edge inequality. Then we have $a_1 \leq \sum_{j=1}^n a_j - 2b_1$.*

Proof of Claim 6. From Lemma 2 it is clear that f is concave. By rewriting

$$g(z) = \max\Big\{ \sum_{j=2}^{n} a_j \cdot \sum_{v \in V(G^j)} c_v^j x_v \mid \sum_{j=1}^{n} a_j x_j \le z,$$

$$x \in R_{\text{stab}}^{\text{edge}}(G^{\text{in}}) \oplus_{v^2}^2 R_{\text{stab}}(G^2) \oplus_{v^3}^3 \cdots \oplus_{v^n}^n R_{\text{stab}}(G^n) \Big\},$$

we also see that g is concave. Moreover, again by Lemma 2, there exists some $\beta^\star \ge 0$ such that g is strictly monotonically increasing on the interval $[0, \beta^\star]$, and constant on $[\beta^\star, \infty)$. It suffices to show that $\beta^\star \ge b_1$. To this end, let us get back to our initial definition of g, and let $\hat{x} \in R_{\text{stab}}^{\text{edge}}(G^{\text{in}})$ be a maximizer for $g(\infty)$. Note that $\beta^\star \ge a^\mathsf{T}\hat{x}$ by definition of $\beta^\star$, and hence we have to show that $\hat{x}$ satisfies $a^\mathsf{T}\hat{x} \ge b_1$.

Since the objective value of $\hat{x}$ does not depend on $\hat{x}_1$, we may assume that $\hat{x}_1 = 0$. By the construction of G^j and c^j, we know that f attains its unique maximum at $y^\star \ge \frac{1}{2}$. This implies $0 \le \hat{x}_j \le y^\star$ for $j = 2, 3, \ldots, n$. Moreover, we claim that also $\hat{x}_j \ge 1 - y^\star$ holds. Suppose not, then none of the edge inequalities involving x_j is tight. Then $\hat{x}_j < 1 - y^\star \le y^\star$ shows that increasing $\hat{x}_j$ would improve the objective value, which in turn contradicts optimality of $\hat{x}$. Consequently, even $1 - y^\star \le \hat{x}_j \le y^\star$ holds for $j = 2, 3, \ldots, n$.

Let $J(\alpha) := \{2 \le j \le n \mid \hat{x}_j = \alpha\}$ for $\alpha \in [1 - y^\star, y^\star]$. We will show that $a(J(\alpha)) \ge a(J(1 - \alpha))$ holds for all $\alpha \in (1/2, y^\star]$, where $a(J(\alpha))$ shall denote $\sum_{j \in J(\alpha)} a_j$. Note that this implies the claim since for each $\alpha \in (1/2, y^\star]$ we have

$$\sum_{j \in J(\alpha)} a_j \hat{x}_j + \sum_{j \in J(1-\alpha)} a_j \hat{x}_j = \sum_{j \in J(\alpha)} a_j \alpha + \sum_{j \in J(1-\alpha)} a_j (1 - \alpha)$$

$$= \alpha \cdot \underbrace{[a(J(\alpha)) - a(J(1 - \alpha))]}_{\ge 0} + a(J(1 - \alpha))$$

$$\ge \tfrac{1}{2} \cdot [a(J(\alpha)) - a(J(1 - \alpha))] + a(J(1 - \alpha))$$

$$= \sum_{j \in J(\alpha)} a_j \tfrac{1}{2} + \sum_{j \in J(1-\alpha)} a_j \tfrac{1}{2}$$

and hence

$$a^\mathsf{T}\hat{x} = \sum_{j=2}^{n} a_j \hat{x}_j = \sum_{j \in J(1/2)} a_j \hat{x}_j + \sum_{\alpha \in (1/2, y^\star]} \Big(\sum_{j \in J(\alpha)} a_j \hat{x}_j + \sum_{j \in J(1-\alpha)} a_j \hat{x}_j \Big)$$

$$\ge \sum_{j \in J(1/2)} a_j \tfrac{1}{2} + \sum_{\alpha \in (1/2, y^\star]} \Big(\sum_{j \in J(\alpha)} a_j \tfrac{1}{2} + \sum_{j \in J(1-\alpha)} a_j \tfrac{1}{2} \Big) \ge b_1,$$

where the last inequality follows from Proposition 2.

For the sake of contradiction, assume that $a(J(\alpha)) < a(J(1 - \alpha))$ holds for some $\alpha \in (1/2, y^\star]$. For a sufficiently small $\varepsilon' > 0$, the solution $\hat{x}' \in \mathbb{R}^{V(G^{\text{in}})}$ defined via

$$\hat{x}_j' := \begin{cases} \hat{x}_j + \varepsilon' & \text{if } j \in J(1 - \alpha) \\ \hat{x}_j - \varepsilon' & \text{if } j \in J(\alpha) \qquad \text{for } j = 1, 2, \ldots, n \\ \hat{x}_j & \text{otherwise} \end{cases}$$

is still contained in $R_{\mathrm{stab}}^{\mathrm{edge}}(G^{\mathrm{in}})$. To see this, observe that $\hat{x}_j' \geq 0$ holds for all $j \in V(G^{\mathrm{in}})$ since we only decrease entries that are at least $1/2$. Moreover, edge inequalities that are tight for $\hat{x}$ remain tight for $\hat{x}'$, since either none or both of its two node values are modified, where in the latter case, the value is increased by ε' for one node and decreased by ε' for the other. Finally, edge inequalities that are not tight for $\hat{x}$ will not be violated if we choose ε' sufficiently small. For the objective values we obtain

$$\sum_{j=2}^{n} a_j(f(\hat{x}_j') - f(\hat{x}_j)) = \sum_{j \in J(1-\alpha)} a_j(f(\hat{x}_j') - f(\hat{x}_j)) + \sum_{j \in J(\alpha)} a_j(f(\hat{x}_j') - f(\hat{x}_j))$$

$$= a(J(1-\alpha)) \cdot \big(f(1-\alpha+\varepsilon') - f(1-\alpha)\big) + a(J(\alpha)) \cdot \big(f(\alpha-\varepsilon') - f(\alpha)\big).$$

We also assume that ε' is small enough to guarantee $1-\alpha+\varepsilon' < \alpha-\varepsilon'$. Since f is concave and monotonically increasing in $[0, y^\star]$, we obtain $f(1-\alpha+\varepsilon')-f(1-\alpha) \geq f(\alpha)-f(\alpha-\varepsilon')$. Together with the assumption $a(J(1-\alpha)) > a(J(\alpha))$, this shows that the objective value of $\hat{x}'$ is strictly larger than that of $\hat{x}$, a contradiction to the optimality of $\hat{x}$. $\qquad\square$

Acknowledgements. We are grateful to four anonymous reviewers whose comments led to improvements of this manuscript.

A   Deferred proofs

Lemma 1. *Let $P, Q \subseteq \mathbb{R}^n$ be polytopes. If there exists a vector $c \in \mathbb{R}^n$ such that $\dim(\mathrm{opt}(Q, c)) < \dim(\mathrm{opt}(P, c))$, then there exists a vector $c' \in \mathbb{R}^n$ such that $\mathrm{opt}(Q, c')$ is a vertex of Q, while $\mathrm{opt}(P, c')$ is not a vertex of P.*

Proof. Let $c' \in \mathbb{R}^n$ be such that $\dim(\mathrm{opt}(Q, c')) < \dim(\mathrm{opt}(P, c'))$ holds, and among those, such that $\dim(\mathrm{opt}(Q, c'))$ is minimum. Clearly, c' is well-defined since $c' := c$ satisfies the conditions.

Assume, for the sake of contradiction, that $\dim(\mathrm{opt}(Q, c')) > 0$. Let $F := \mathrm{opt}(P, c')$ and $G := \mathrm{opt}(Q, c')$. Let $F_1, F_2, \ldots, F_k$ be the facets of F. By $n(F, F_i)$ we denote the set of vectors $w \in \mathbb{R}^n$ such that $\mathrm{opt}(F, w) \supseteq F_i$. Since F is a polytope, $\bigcup_{i \in \{1,2,\ldots,k\}} n(F, F_i)$ contains a basis U of $\mathbb{R}^n$. Moreover, not all vectors $u \in U$ can lie in $\mathrm{aff}(G)^\perp$, the orthogonal complement of $\mathrm{aff}(G)$, since then $\mathrm{aff}(G)^\perp = \mathbb{R}^n$ would hold, contradicting $\dim(G) > 0$. Let $u \in U \setminus \mathrm{aff}(G)^\perp$.

Now, for a sufficiently small $\varepsilon > 0$, $\mathrm{opt}(P, c' + \varepsilon u) \supseteq F_i$ for some $i \in \{1, 2, \ldots, k\}$, and $\mathrm{opt}(Q, c' + \varepsilon u)$ is a proper face of G. Thus, $c' + \varepsilon u$ satisfies the requirements at the beginning of the proof. However, $\dim(\mathrm{opt}(Q, c' + \varepsilon u)) < \dim(G)$ contradicts the minimality assumption, which concludes the proof. $\qquad\square$

Lemma 2. *Let $P \subseteq \mathbb{R}^n$ be a non-empty polytope, let $c, a \in \mathbb{R}^n$ and let $\ell := \min\{a^\intercal x \mid x \in P\}$. The functions $h^=, h^\leq : [\ell, \infty) \to \mathbb{R}$ defined via $h^=(\beta) = \max\{c^\intercal x \mid x \in P, a^\intercal x = \beta\}$ and $h^\leq(\beta) = \max\{c^\intercal x \mid x \in P, a^\intercal x \leq \beta\}$ are concave. Moreover, there exists a number $\beta^\star \in [\ell, \infty)$ such that $h^=$ and $h^\leq$ are identical and strictly monotonically increasing on the interval $[\ell, \beta^\star]$, and $h^\leq$ is constant on the interval $[\beta^\star, \infty)$.*

Proof. Let $Q := \{\left(\begin{smallmatrix} y_1 \\ y_2 \end{smallmatrix}\right) \mid \exists x \in P : a^\mathsf{T}x = y_1, \ c^\mathsf{T}x = y_2\} \subseteq \mathbb{R}^2$ be the projection of P along a and c. By construction, $h^{\leq}(\beta) = \max\{y_2 \mid y \in Q, \ y_1 \leq \beta\}$ holds. Considering that Q is a polytope of dimension at most 2, the claimed properties of $h^{\leq}$ and $h^{=}$ are obvious (see Fig. 1).   □

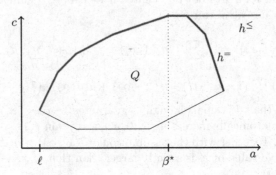

Fig. 1. Illustration of Lemma 2. The graph of $h^{\leq}$ is highlighted in red, while that of $h^{=}$ is highlighted in blue. (Color figure online)

References

1. Boros, E., Hammer, P.L., Sun, R., Tavares, G.: A max-flow approach to improved lower bounds for quadratic unconstrained binary optimization (QUBO). Disc. Optim. **5**(2), 501–529 (2008). In Memory of George B. Dantzig
2. Chvátal, V.: On certain polytopes associated with graphs. J. Combin. Theory, Ser. B **18**(2), 138–154 (1975)
3. Fix, A., Gruber, A., Boros, E., Zabih, R.: A hypergraph-based reduction for higher-order binary Markov random fields. IEEE Trans. Pattern Anal. Mach. Intell. **37**(7), 1387–1395 (2015)
4. Hammer, P.L., Hansen, P., Simeone, B.: Roof duality, complementation and persistency in quadratic 0–1 optimization. Math. Program. **28**(2), 121–155 (1984)
5. Ishikawa, H.: Transformation of general binary MRF minimization to the first-order case. IEEE Trans. Pattern Anal. Mach. Intell. **33**(6), 1234–1249 (2011)
6. Karp, R.M., Papadimitriou, C.H.: On linear characterizations of combinatorial optimization problems. SIAM J. Comput. **11**(4), 620–632 (1982)
7. Kolmogorov, V., Rother, C.: Minimizing nonsubmodular functions with graph cuts - a review. IEEE Trans. Pattern Anal. Mach. Intell. **29**(7), 1274–1279 (2007)
8. Nemhauser, G.L., Trotter, L.E.: Vertex packings: structural properties and algorithms. Math. Programm. **8**(1), 232–248 (1975)
9. Oriolo, G.: Clique family inequalities for the stable set polytope of quasi-line graphs. Disc. Appl. Math. **132**(1), 185–201 (2003). Stability in Graphs and Related Topics
10. Padberg, M.W.: On the facial structure of set packing polyhedra. Math. Program. **5**(1), 199–215 (1973)
11. Pulleyblank, W.R.: Minimum node covers and 2-bicritical graphs. Math. Program. **17**(1), 91–103 (1979)

12. Rodríguez-Heck, E., Stickler, K., Walter, M., Weltge, S.: Persistency of linear programming formulations for the stable set problem (2019). arXiv:1911.01478
13. Schrijver, A.: Theory of Linear and Integer Programming. John Wiley, New York (1986)
14. Sewell, E.C.: Stability critical graphs and the stable set polytope. Technical report, Cornell University Operations Research and Industrial Engineering (1990)
15. Trotter, L.E.: A class of facet producing graphs for vertex packing polyhedra. Disc. Math. **12**(4), 373–388 (1975)

Constructing Lattice-Free Gradient Polyhedra in Dimension Two

Joseph Paat[1], Miriam Schlöter[1(✉)], and Emily Speakman[2]

[1] Department of Mathematics, ETH Zürich, Zürich, Switzerland
miriam.schloeter@ifor.math.ethz.ch
[2] Department of Mathematical and Statistical Sciences,
University of Colorado Denver, Denver, USA

Abstract. Lattice-free gradient polyhedra are optimality certificates for mixed integer convex minimization models. We consider how to construct these polyhedra for unconstrained models with two integer variables. A classic result of Bell, Doignon, and Scarf states that a lattice-free gradient polyhedron exists with at most four facets. We show how to construct a sequence of (not necessarily lattice-free) gradient polyhedra, each of which has at most four facets, that finitely converges to a lattice-free gradient polyhedron. Each update requires constantly many gradient evaluations (By gradient evaluation, we refer to inner product evaluation using gradients. For our updates we require at most 18 gradient evaluations.). This update procedure imitates the gradient descent algorithm, and consequently, it yields a gradient descent type of algorithm for problems with two integer variables. An open question is to improve the convergence rates to obtain a minimizer or a lattice-free set.

1 Introduction

A polyhedron $P = \{(x, z) \in \mathbb{R}^d \times \mathbb{R}^n : A(x, z) \leq b\}$ is *lattice-free* if $\mathrm{intr}(P) \cap (\mathbb{R}^d \times \mathbb{Z}^n) = \emptyset$, where $\mathrm{intr}(P) := \{(x, z) \in \mathbb{R}^d \times \mathbb{R}^n : A(x, z) < b\}$.[1] Recent work on lattice-free sets has focused mainly on three topics. The first of these is the generation of valid inequalities for integer programs, see, e.g., [1,6,10,11,15]. The second is the classification of inclusion-wise maximal lattice-free sets [2,3,9, 16,22]. The third is the use of lattice-free sets as optimality certificates in mixed integer convex minimization. This paper focuses on the latter topic.

Let $f : \mathbb{R}^d \times \mathbb{R}^n \to \mathbb{R}$ be convex and differentiable with gradient ∇f. We assume oracle access to ∇f. Our results extend to non-differentiable functions via subgradients, but we assume differentiability for the sake of presentation. The unconstrained mixed integer convex minimization problem is

$$\min\{f(x, z) : (x, z) \in \mathbb{R}^d \times \mathbb{Z}^n\}. \tag{CM}$$

---

[1] If the inequality $\mathbf{0}^{\mathsf{T}}(x, z) \leq 0$ is in the system $A(x, z) \leq b$, then $\mathrm{intr}(P) = \emptyset$.

© Springer Nature Switzerland AG 2020
D. Bienstock and G. Zambelli (Eds.): IPCO 2020, LNCS 12125, pp. 364–377, 2020.
https://doi.org/10.1007/978-3-030-45771-6_28

Applications of (CM) include statistical regression and the closest vector problem. The *gradient polyhedron of a non-empty finite set* $\mathcal{U} \subseteq \mathbb{R}^d \times \mathbb{Z}^n$ is

$$\mathrm{GP}(\mathcal{U}) := \{(x, z) \in \mathbb{R}^d \times \mathbb{R}^n : \nabla f(\overline{x}, \overline{z})^\mathsf{T}(x - \overline{x}, z - \overline{z}) \leq 0 \;\forall (\overline{x}, \overline{z}) \in \mathcal{U}\}. \quad (1)$$

It follows from the definitions of $\mathrm{GP}(\mathcal{U})$ and ∇f that **if $\mathrm{GP}(\mathcal{U})$ is lattice-free, then there exists $(x^*, z^*) \in \mathcal{U}$ that is an optimal solution for** (CM). Consequently, if $\mathrm{GP}(\mathcal{U})$ is lattice-free, then it is an optimality certificate for (CM). The existence of a lattice-free gradient polyhedron $\mathrm{GP}(\mathcal{U})$ requires that (CM) has an optimal solution. We assume an optimal solution exists in the paper; our techniques do not immediately extend to detect if this assumption is violated.

Bell, Doignon, and Scarf [12,17,23] showed that if $n \geq 1$ and $d = 0$, then there exists $\mathcal{U} \subseteq \mathbb{Z}^n$ such that $|\mathcal{U}| \leq 2^n$ and $\mathrm{GP}(\mathcal{U})$ is lattice-free. Baes et al. [5] extended this to show that $|\mathcal{U}| \leq 2^n$ suffices for $d \geq 0$. Basu et al. [8] generalized this further to so-called S-free sets. Motivated by this existential result, we aim to algorithmically construct lattice-free gradient polyhedra with 2^n facets.

Constructing lattice-free gradient polyhedra is well studied when $n = 0$ and $d \geq 1$. Although rarely described as such, the gradient descent algorithm is a search algorithm for a lattice-free set when $n = 0$ (see [14, Chapter 9]). Gradient descent updates a point $x^i \in \mathbb{R}^d$ to x^{i+1} via $x^{i+1} = x^i - \alpha^i \nabla f(x^i)$, where $\alpha^i > 0$. The iterations $(x^i)_{i=1}^\infty$ converge to x^* with $\nabla f(x^*) = \mathbf{0}$. One can verify that

$$\nabla f(x^*) = \mathbf{0} \;\; \text{if and only if} \;\; \mathrm{intr}(\mathrm{GP}(\{x^*\})) \cap \mathbb{R}^d = \emptyset.$$

Thus, $(x^i)_{i=1}^\infty$ corresponds to a sequence of gradient polyhedra $(\mathrm{GP}(\{x^i\}))_{i=1}^\infty$ that 'converges' to a lattice-free gradient polyhedron $\mathrm{GP}(\{x^*\})$. One notable aspect of gradient descent is that it generates gradient polyhedra whose number of facets never exceeds the bound $2^n = 2^0 = 1$. Another notable aspect is that the initial x^1 can be chosen arbitrarily.

A lattice-free gradient polyhedron can also be constructed algorithmically if $n = 1$ and $d = 0$. In this case, $\mathcal{U} := \{\lfloor x^* \rfloor, \lceil x^* \rceil\}$ is the certifying set, where x^* is an optimal solution to the continuous relaxation of (CM). This $\mathcal{U}$ can be found by starting with an arbitrary $\mathcal{U}^i = \{z^i, z^i + 1\} \subseteq \mathbb{Z}$ and updating it as follows:

$$\mathcal{U}^{i+1} := \begin{cases} \{z^i - 1, z^i\} & \text{if } 0 < \nabla f(z^i) \\ \{z^i + 1, z^i + 2\} & \text{if } \nabla f(z^i + 1) < 0 \\ \mathcal{U}^i & \text{if } \nabla f(z^i) \leq 0 \leq \nabla f(z^i + 1). \end{cases}$$

The gradient comparisons ensure that $\mathcal{U}^i$ is updated by 'flipping' closer to x^*, and if $\mathcal{U}^{i+1} = \mathcal{U}^i$, then $\mathcal{U}^i = \mathcal{U}$ and $\mathrm{GP}(\mathcal{U}^i)$ is lattice-free. These updates are not the most efficient way of obtaining $\mathcal{U}$ (one can simply round x^*), but it does have the same properties as gradient descent: it generates a sequence $(\mathrm{GP}(\mathcal{U}^i))_{i=1}^\infty$ that yields a lattice-free set, the initial set $\mathcal{U}^1$ is arbitrary, and each update only requires a constant number of gradient evaluations. Also, the number of facets at each iterate $\mathrm{GP}(\mathcal{U}^i)$ does not exceed the bound $2^n = 2^1 = 2$. These updates and the rounding approach both generalize to the setting $n = 1$ and $d \geq 1$. However, if $n \geq 2$, then neither yields a lattice-free gradient polyhedron, in general.

Our first main result is an update procedure to create a lattice-free gradient polyhedron when $d = 0$ and $n = 2$. We say $\mathcal{U} \subseteq \mathbb{Z}^2$ is *unimodular* if

$$\mathcal{U} := \mathcal{U}(z, U) := \{z + Ue : e \in \{0,1\}^2\}, \tag{2}$$

for $z \in \mathbb{Z}^2$ and a matrix $U \in \mathbb{Z}^{2 \times 2}$ with $|\det(U)| = 1$. Our procedure updates any unimodular set and each update only needs constantly many gradient evaluations as opposed to solving 1- or 2-dimensional integer linear programs. We use two metrics of progress to ensure this procedure generates a sequence $(\mathcal{U}^i)_{i=1}^\infty = (\mathcal{U}(z^i, U^i))_{i=1}^\infty$ resulting in a lattice-free gradient polyhedron. The unimodularity of $\mathcal{U}$ ensures that during our update procedure for each iterate $\mathrm{GP}(\mathcal{U}^i)$ the bound of $2^2 = 4$ facets is not exceeded and that $\mathcal{U}$ always contains precisely 4 points.

The first progress measure is the minimum function value in $\mathcal{U}^i$, i.e.,

$$\min\{f(z) : z \in \mathcal{U}^i\}. \tag{3}$$

Our second measure is the distance from the optimal solution set of (CM) to $\mathcal{U}^i$ with respect to U^i:

$$\min\{\|(U^i)^{-1}(z^* - z)\|_1 : z \in \mathcal{U}^i \text{ and } z^* \text{ is optimal for CM}\}. \tag{4}$$

This measure is a notion of graphic distance of an optimal solution z^* to (CM). We discuss (4) more in Subsect. 2.1.

Our updates are such that (3) and (4) are both *non-increasing* in i. Furthermore, if neither measure strictly decreases from $\mathcal{U}^i$ to $\mathcal{U}^{i+1}$, then either $\mathcal{U}^i$ already contains a minimizer of (CM), or, one or both measures will strictly decrease when $\mathcal{U}^{i+1}$ is updated to $\mathcal{U}^{i+2}$.

Theorem 1. *Assume $d = 0$ and $n = 2$. Let $\mathcal{U}^i \subseteq \mathbb{Z}^2$ be a unimodular set. Any unimodular set $\mathcal{U}^i \subseteq \mathbb{Z}^2$ can be updated to a unimodular set $\mathcal{U}^{i+1} \subseteq \mathbb{Z}^2$ such that*

(a) the update only uses constantly many gradient evaluations,

(b) neither (3) nor (4) increases from $\mathcal{U}^i$ to $\mathcal{U}^{i+1}$, and

(c) if (3) and (4) do not decrease from $\mathcal{U}^i$ to $\mathcal{U}^{i+1}$ and $\mathcal{U}^{i+1}$ is updated to $\mathcal{U}^{i+2}$, then one of the following holds: (i) (3) or (4) decreases from $\mathcal{U}^{i+1}$ to $\mathcal{U}^{i+2}$ or (ii) $\mathcal{U}^i$ contains a minimizer of (CM).

Figure 1 illustrates how our update procedure iterates for a specific f and $\mathcal{U}^1$.

If $(\mathcal{U}^i)_{i=1}^\infty$ is constructed using updates from Theorem 1, then $\mathrm{GP}(\mathcal{U}^i)$ is guaranteed to eventually contain a minimizer of (CM). Our update procedure also guarantees that $\mathrm{GP}(\mathcal{U}^i)$ becomes lattice-free, and detecting this only needs constantly many gradient evaluations (see Lemma 3). It is worth reminding the reader that if we come across a unimodular set $\mathcal{U}^i$ that contains a point $z^* \in \mathbb{Z}^2$ satisfying $\nabla f(z^*) = \mathbf{0}$, then $\mathrm{GP}(\mathcal{U}^i)$ is lattice-free and z^* is a minimizer of (CM). We always assume this check is made and only update $\mathcal{U}^i$ if it fails.

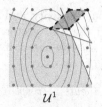
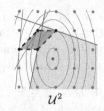
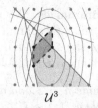
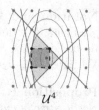

$$\mathcal{U}^1 \qquad\qquad \mathcal{U}^2 \qquad\qquad \mathcal{U}^3 \qquad\qquad \mathcal{U}^4$$

Fig. 1. A sequence $(\mathcal{U}^i)_{i=1}^4$ of unimodular sets generated by our updates for $f(x_1, x_2) :=$ $3x_1^2 + x_2^2 + x_1 + x_2$. The convex hull of each $\mathcal{U}^i$ is shaded in black, the hyperplanes defining $\mathrm{GP}(\mathcal{U}^i)$ are in red, and $\mathrm{GP}(\mathcal{U}^i)$ is shaded in red. Level curves of f are in gray. (Color figure online)

Theorem 2. *Let $(\mathcal{U}^i)_{i=1}^\infty$ be created using updates from Theorem 1. For some $T \in \mathbb{Z}$ the gradient polyhedron $\mathrm{GP}(\mathcal{U}^T)$ is lattice-free. Moreover, this can be checked using constantly many gradient evaluations.*

Theorem 2 implies that our update procedure provides a 'flipping' algorithm for solving (CM) when $n = 2$ and $d = 0$. We may start with any unimodular set $\mathcal{U}$, and then flip $\mathcal{U}$ (with a flip defined using only $\mathcal{U}$ and gradient information) until $\mathrm{GP}(\mathcal{U})$ is lattice-free. Other algorithms for (CM) use techniques such as branch and bound (see, e.g., [20,21]), outer approximations (see, e.g., [13,18]), convex separation [19, Theorem 6.7.10], or improvement oracles [4]. In contrast to the flipping algorithm, each of these algorithms use non-gradient information or create polyhedral relaxations with more than $2^2 = 4$ facets. Baes et al. [4] give a geometric algorithm for (CM) when $d = 0$ and $n = 2$ but explicitly use knowledge of a bounded set containing the minimum. We do not assume such knowledge. Our updates are conservative, but if f is L-Lipschitz continuous and c-strongly convex, then only $2(L/c+1)\|z^*\|_1$ many updates are needed to find an optimal solution z^*. An open question remains to find a 'best' update procedure.

Our updates can be extended to $d \geq 0$, provided we are able to exactly minimize $f_z(x) := f(x, z)$ over $x \in \mathbb{R}^d$ for each fixed $z \in \mathbb{Z}^2$. To see this, note that (CM) is the same as minimizing $f^{\min}(z) := \min\{f(x, z) : x \in \mathbb{R}^d\}$ over $\mathbb{Z}^2$.

Corollary 1. *The update procedure from Theorem 1 can be extended naturally to obtain an exact iterative algorithm to solve (CM) for functions $f : \mathbb{R}^d \times \mathbb{R}^2 \to \mathbb{R}$ provided we can exactly minimize $f_z(x)$ over $\mathbb{R}^d$.*

The update procedures from Theorem 1 and Corollary 1 require exact gradients to choose the correct updates. However, if f is strongly convex, then we can extend Theorem 1 to an update procedure that only requires *approximate* gradients. The major obstacle in extending Theorem 1 is that if one of the defining gradients is perturbed a lattice-free gradient polyhedra may no longer be lattice-free. We overcome this by identifying a subset of a gradient polyhedron that contains the minimizers of (CM) and is robust to gradient perturbations.

We omit some proofs in this abstract. See arXiv:2002.11076 for complete proofs and a discussion on extensions to the approximate gradient setting.

Notation and Preliminaries. We refer to [14] for more on convexity and gradients. We denote the i-th column of $U \in \mathbb{Z}^{2 \times 2}$ by u^i and the zero vector by $\mathbf{0}$. For $z, \overline{z} \in \mathbb{R}^2$ we say $\overline{z}$ *cuts* z if $\nabla f(\overline{z})^\intercal (z - \overline{z}) \geq 0$ and *strictly cuts* z if $\nabla f(\overline{z})^\intercal (z - \overline{z}) > 0$. The next result follows from the definition of convexity.

Proposition 1. *Assume $n = 2$ and $d = 0$. Let $z, \overline{z} \in \mathbb{R}^2$. If $\overline{z}$ cuts z (respectively, strictly cuts), then $f(z) \geq f(\overline{z})$ (respectively, $f(z) > f(\overline{z})$).*

The following properties of $\mathrm{GP}(\mathcal{U})$ can be shown using Proposition 1.

Lemma 1. *Assume $n = 2$ and $d = 0$. Let $\mathcal{U} \subseteq \mathbb{Z}^2$ be a non-empty finite set.*

(i) *If $z \notin \mathrm{intr}(\mathrm{GP}(\mathcal{U}))$, then $f(z) \geq \min\{f(\overline{z}) : \overline{z} \in \mathcal{U}\}$.*

(ii) *$\mathcal{U} \cap \mathrm{GP}(\mathcal{U}) \neq \emptyset$.*

(iii) *If $\mathcal{U}$ does not contain an optimal solution of (CM), then every optimal solution z^* of (CM) is in $\mathrm{intr}(\mathrm{GP}(\mathcal{U}))$.*

2   An Update Procedure for $n = 2$ and $d = 0$

Assume $n = 2$ and $d = 0$. Let $\mathcal{U} = \mathcal{U}(z, U)$ be unimodular as defined in (2). After multiplying u^1 and u^2 by ± 1 and relabeling the 'anchor' point $z \in \mathcal{U}$ to be another point in $\mathcal{U}$, we assume $\mathrm{GP}(\mathcal{U})$ fulfills *preprocessing* properties. The proof of Lemma 2 is in the appendix.

Lemma 2. *Let $\mathcal{U} = \mathcal{U}(z, U)$ be unimodular. We can preprocess $\mathrm{GP}(\mathcal{U})$ so that*

$$
\begin{aligned}
&(i) \ z \in \mathrm{GP}(\mathcal{U}), \\
&(ii) \ \textit{if } |\mathcal{U} \cap \mathrm{GP}(\mathcal{U})| = 2, \ \textit{then} \\
&\quad (a) \ \mathcal{U} \cap \mathrm{GP}(\mathcal{U}) = \{z, z + u^1\}, \ \textit{or} \\
&\quad (b) \ \mathcal{U} \cap \mathrm{GP}(\mathcal{U}) = \{z, z + u^1 + u^2\}, \ z \textit{ strictly cuts } z + u^1, \\
&\qquad \quad \textit{and } z + u^1 + u^2 \textit{ strictly cuts } z + u^2. \\
&(iii) \ \textit{if } |\mathcal{U} \cap \mathrm{GP}(\mathcal{U})| = 3, \ \textit{then } \mathcal{U} \cap \mathrm{GP}(\mathcal{U}) = \{z, z + u^1, z + u^2\}.
\end{aligned}
\tag{5}
$$

We update $\mathcal{U}$ by 'flipping' the columns of U to a new matrix $\overline{U}$ and preprocessing $(z, \overline{U})$ to satisfy (5). $\mathrm{FLIP}(U)$ denotes the matrix $\overline{U}$ obtained from this flipping, and $\mathrm{FLIP}(\mathcal{U})$ denotes the unimodular set obtained after preprocessing $(z, \overline{U})$. Table 1 defines $\mathrm{FLIP}(U)$. Certain flips rely on the following line segments:

$$
H^i := \{z + k \cdot u^1 + i \cdot u^2 \in \mathrm{intr}(\mathrm{GP}(\mathcal{U})) : k \in \mathbb{R}\} \quad \forall i \in \{-1, 1\}. \tag{6}
$$

We say that $\mathcal{U}$ is *connected* if $\mathcal{U} \cap \mathrm{GP}(\mathcal{U}) \supseteq \{z, z + u^1\}$ (**Cases 3–5**). Otherwise, we say $\mathcal{U}$ is *disconnected* (**Cases 1–2**). Note that $\mathcal{U}$ can be disconnected and not fit into **Cases 1** or **2**; this occurs when $\mathcal{U} = \{z\}$ but $\mathcal{U}$ does not meet the other conditions required to update in **Case 1**. We show in Theorem 3 that if $\mathcal{U}$ does not fall into one of these cases, then $\mathrm{GP}(\mathcal{U})$ is lattice-free.

Table 1. The different updates $\overline{U} = \text{FLIP}(U)$. Sample updates are drawn with $\mathcal{U}(z, U)$ and $\mathcal{U}(z, \overline{U})$ as dashed lines and $\text{GP}(\mathcal{U})$ in red.

Case 1: Assume $\mathcal{U} \cap \text{GP}(\mathcal{U}) = \{z\}$. For each $i \in \{1, 2\}$ define $\sigma^i := 1$ if $\nabla f(z)^\top(u^i) \leq 0$
and $\sigma^i := -1$ otherwise. **If** $(\sigma^1, \sigma^2) \neq (1, 1)$, **then** set $\text{FLIP}(U) := (\sigma^1 u^1, \sigma^2 u^2)$.

    Update to $\overline{U} := \text{FLIP}(U)$ $\longrightarrow$ 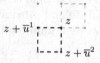

Case 2: Assume $\mathcal{U} \cap \text{GP}(\mathcal{U}) = \{z, z + u^1 + u^2\}$. Define $U' := (u^1, u^1 + u^2)$,
     $\mathcal{U}' := \mathcal{U}(z, U')$, $U'' := (-u^1, u^1 + u^2)$, and $\mathcal{U}'' := \mathcal{U}(z, U'')$.
     If $\mathcal{U}'$ is connected **or** $\text{GP}(\mathcal{U}') \cap \mathcal{U}' = \{z + 2u^1 + u^2\}$ **or**
       $z + u^1 + u^2$ strictly cuts $z - u^1$ and $z - u^1$ strictly cuts z,
       then set $\text{FLIP}(U) := U'$.
     Else if $\mathcal{U}''$ is connected **or** $\text{GP}(\mathcal{U}'') \cap \mathcal{U}'' = \{z - u^1\}$ **or**
       z strictly cuts $z + 2u^1 + u^2$ and $z + 2u^1 + u^2$ strictly cuts $z + u^1 + u^2$,
       then set $\text{FLIP}(U) := U''$.
     Else set $\text{FLIP}(U) := (-u^1, 2u^1 + u^2)$

    Update to $\overline{U} := \text{FLIP}(U)$ $\longrightarrow$ 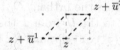

Case 3: Assume $\mathcal{U} \cap \text{GP}(\mathcal{U}) = \{z, z + u^1\}$ and $|H^i \cap \mathbb{Z}^2| = 1$ for some $i \in \{-1, 1\}$.
     Let $z + ku^1 + iu^2 \in H^i \cap \mathbb{Z}^2$. Set $\text{FLIP}(U) := (ku^1 + iu^2, -(k-1)u^1 - iu^2)$.

Case 4: Assume $\mathcal{U} \cap \text{GP}(\mathcal{U}) = \{z, z + u^1\}$ and $|H^i \cap \mathbb{Z}^2| \geq 2$ for some $i \in \{-1, 1\}$.
     Let $z + ku^1 + iu^2 \in H^i \cap \mathbb{Z}^2$ minimize $|k|$.
     If $k \geq 0$, **then** set $\text{FLIP}(U) := (u^1, ku^1 + iu^2)$.
     If $k \leq -1$, **then** set $\text{FLIP}(U) := (u^1, (k-1)u^1 + iu^2)$.

    Update to $\overline{U} := \text{FLIP}(U)$ $\longrightarrow$

Case 5: Assume $\mathcal{U} \cap \text{GP}(\mathcal{U}) = \{z, z + u^1, z + u^2\}$ and $|H^i \cap \mathbb{Z}^2| \geq 1$ for some $i \in \{-1, 1\}$.
     If $i = 1$, **then** set $\text{FLIP}(U) := (u^1, -u^1 + u^2)$. Here, $z - u^1 + u^2 \in H^1$.
     Else set $\text{FLIP}(U) := (u^1 - u^2, u^2)$. Here, $z + u^1 - u^2 \in H^{-1}$.

    Update to $\overline{U} := \text{FLIP}(U)$ $\longrightarrow$

2.1   Convergence Towards an Optimal Solution Of (CM): Theorem 1

Our first measure of progress for the update procedure is the smallest function value that we have seen thus far. For unimodular sets $\mathcal{U}, \mathcal{U}' \subseteq \mathbb{Z}^2$ we say

$$\mathcal{U} <_f \mathcal{U}' \quad \text{if } \min\{f(z) : z \in \mathcal{U}\} < \min\{f(z) : z \in \mathcal{U}'\}. \tag{7}$$

We define $\mathcal{U} \leq_f \mathcal{U}'$ and $\mathcal{U} =_f \mathcal{U}'$ similarly.

To motivate and provide intuition for our second measure of progress (4), let $\mathcal{U} = \mathcal{U}(z, U)$ be a unimodular set. The *orthants* corresponding to $\mathcal{U}$ are

$$\begin{aligned}
O_\mathcal{U}(z) &:= \{z + Ur : r \in \mathbb{Z}_{\leq 0} \times \mathbb{Z}_{\leq 0}\}, \\
O_\mathcal{U}(z + u^1) &:= \{(z + u^1) + Ur : r \in \mathbb{Z}_{\geq 0} \times \mathbb{Z}_{\leq 0}\}, \\
O_\mathcal{U}(z + u^2) &:= \{(z + u^2) + Ur : r \in \mathbb{Z}_{\leq 0} \times \mathbb{Z}_{\geq 0}\}, \text{ and} \\
O_\mathcal{U}(z + u^1 + u^2) &:= \{(z + u^1 + u^2) + Ur : r \in \mathbb{Z}_{\geq 0} \times \mathbb{Z}_{\geq 0}\}.
\end{aligned} \tag{8}$$

For every $x \in \mathbb{Z}^2$ and $w \in \mathcal{U}$, the difference vector $x - w$ is an integer combination of signed copies of u^1 and u^2 (recall U has $|\det(U)| = 1$ and see [7, §VII Corollary 2.2]). The number of signed copies is $\|U^{-1}(x - w)\|_1$. However, for every $x \in \mathbb{Z}^2$ there is a unique $w \in \mathcal{U}$ minimizing $\|U^{-1}(x - w)\|_1$. Denote this minimum value:

$$r_\mathcal{U}(x) := \min\{\|U^{-1}(x - w)\|_1 : w \in \mathcal{U}\}.$$

The function $r_\mathcal{U}$ is a distance measure from x to $\mathcal{U}$, and it can be visualized by considering the orthant $O_\mathcal{U}(\cdot)$ that contains x and reporting the distance with respect to the U from x the 'anchor' of that orthant. The value in (4) equals $\min\{r_\mathcal{U}(z^*) : z^* \text{ optimal for (CM)}\}$. Note $r_\mathcal{U}(z^*) = 0$ if and only if $z^* \in \mathcal{U}$.

Theorem 1 shows that (4) is non-increasing in i after updating by demonstrating that $r_{\mathcal{U}^i}(z^*)$ is non-increasing for every optimal z^* for (CM). Let $\mathcal{U}' \subseteq \mathbb{Z}^2$ be a unimodular set and $z^* \in \mathbb{Z}^2$ a fixed optimal solution to (CM). We write

$$\mathcal{U} <_r \mathcal{U}' \quad \text{if } r_\mathcal{U}(z^*) < r_{\mathcal{U}'}(z^*), \tag{9}$$

and define $\mathcal{U} \leq_r \mathcal{U}'$ and $\mathcal{U} =_r \mathcal{U}'$ similarly.

Theorem 1 *(a)* follows directly from Table 1. Theorem 1 *(b)* and *(c)* are implied by the next result, which is stated using our new notation. Theorem 3 also demonstrates that if no update occurs, then $\mathrm{GP}(\mathcal{U})$ is lattice-free. We emphasize that the $\leq_r$ used in Theorem 3 is satisfied for any choice of z^* used to define $\leq_r$. Hence, although (9) depends on z^*, the choice of z^* is arbitrary and $(\mathrm{FLIP}(\mathcal{U}^i))_{i=1}^\infty$ never diverges away from *any* optimal z^* with respect to $\leq_r$.

Theorem 3. *Assume $\mathcal{U}$ satisfies (5). If $\mathcal{U}$ does not satisfy a case in Table 1, then $\mathrm{GP}(\mathcal{U})$ is lattice-free and $\mathcal{U}$ contains an optimal solution of (CM). If $\mathcal{U}$ satisfies a case in Table 1 and does not contain an optimal solution of (CM), then at least one of the following holds:*

(a) $\mathrm{FLIP}(\mathcal{U}) <_f \mathcal{U}$ and $\mathrm{FLIP}(\mathcal{U}) \leq_r \mathcal{U}$

(b) $\mathrm{FLIP}(\mathcal{U}) \leq_f \mathcal{U}$ and $\mathrm{FLIP}(\mathcal{U}) <_r \mathcal{U}$

(c) $\mathrm{FLIP}(\mathcal{U}) \leq_f \mathcal{U}$, $\mathrm{FLIP}(\mathcal{U}) \leq_r \mathcal{U}$, and $\mathrm{FLIP}(\mathcal{U})$ is connected.

If $\mathcal{U}$ is also connected, then $\mathrm{FLIP}(\mathcal{U})$ satisfies (b).

It is important to note the advantage of the connected case in Theorem 3: when $\mathcal{U}$ is connected, we are able to quickly determine if $GP(\mathcal{U})$ is lattice-free.

Lemma 3. *Recall definition* (6). *If* $\mathcal{U} = \mathcal{U}(z, U)$ *is connected and satisfies* (5), *then* $GP(\mathcal{U})$ *is lattice-free if and only if* $H^{-1} \cap \mathbb{Z}^2 = \emptyset$ *and* $H^1 \cap \mathbb{Z}^2 = \emptyset$.

Table 1 does not consider $|\mathcal{U} \cap GP(\mathcal{U})| = 4$. The reason for this is that $GP(\mathcal{U})$ is lattice-free when $|\mathcal{U} \cap GP(\mathcal{U})| = 4$, a result that follows almost immediately from Lemma 3. The proofs of Lemma 3 and Corollary 2 are in the appendix.

Corollary 2. *If* $|\mathcal{U} \cap GP(\mathcal{U})| = 4$, *then* $GP(\mathcal{U})$ *is lattice-free.*

We first prove Theorem 3 when $\mathcal{U}$ does not fit in any case of Table 1.

Proof (of Theorem 3 when $\mathcal{U}$ does not fit into Table 1). Assume $\mathcal{U}$ does not fit in any case of Table 1. If $\mathcal{U}$ is connected, then Lemma 3 implies that $GP(\mathcal{U})$ is lattice-free. Assume that $\mathcal{U}$ is disconnected; we claim this yields a contradiction. If $\mathcal{U} \cap GP(\mathcal{U}) = \backslash \{z, z + u^1 + u^2\}$, then $\mathcal{U}$ fits into **Case 2**, which is a contradiction. Hence, we must have $\mathcal{U} \cap GP(\mathcal{U}) = \{z\}$, $\nabla f(z)^\intercal u^1 \leq 0$, and $\nabla f(z)^\intercal u^2 \leq 0$. This implies $\nabla f(z)^\intercal (u^1 + u^2) \leq 0$, so z does not strictly cut $z + u^1$, $z + u^2$, nor $z + u^1 + u^2$. For the identity $\mathcal{U} \cap GP(\mathcal{U}) = \{z\}$ to hold, it must then follow that $(\mathcal{U}\backslash\{z\}) \cap GP(\mathcal{U}\backslash\{z\}) = \emptyset$, which contradicts Lemma 1 *(ii)*. $\qquad\square$

We use the next observation and Lemma 4 to prove the remaining cases of Theorem 3. The lemma follows from the definitions in Table 1.

Observation 1. *Let* $x, v^1, v^2 \in \mathbb{Z}^2$ *and* $w = x + k_1 v^1 + k_2 v^2$ *for* $k_1, k_2 \geq 0$. *Then* $x \in \text{conv}\{w, x - v^1, x - v^2\}$.

Lemma 4. *If* $\mathcal{U} = \mathcal{U}(z, U)$ *satisfies* (5), *then* $\mathcal{U} \cap GP(\mathcal{U}) \subseteq \text{FLIP}(\mathcal{U})$. *Hence,* $\text{FLIP}(\mathcal{U}) \leq_f \mathcal{U}$ *by Lemma 1.*

In this abstract we only give a proof of Theorem 3 in **Case 2** as this is the only case that captures every possible outcome of Theorem 3.

*Proof (of Theorem 3 in **Case 2**).* The following sets closely related to the orthants in (8) will be helpful in the proof (see Fig. 2 *(i)*):

$$O_{\mathcal{U}}(z + u^1 + u^2)_B := \{(z + u^1 + u^2) + (u^1 + u^2)k_1 + u^2 k_2 : k_1, k_2 \in \mathbb{Z}_{\geq 0}\},$$
$$O_{\mathcal{U}}(z + u^1 + u^2)_A := O_{\mathcal{U}}(z + u^1 + u^2)\backslash O_{\mathcal{U}}(z + u^1 + u^2)_B,$$
$$O_{\mathcal{U}}(z)_B := \{z + k_1(u^1 + u^2) + k_2 u^2 : k_1, k_2 \in \mathbb{Z}_{\leq 0}\}, \text{ and}$$
$$O_{\mathcal{U}}(z)_A := O_{\mathcal{U}}(z)\backslash O_{\mathcal{U}}(z)_B.$$

Let z^* be optimal for (CM). We claim that $z^* \in O_{\mathcal{U}}(z)_A \cup O_{\mathcal{U}}(z + u^1 + u^2)_A$. We assume in Theorem 3 that $\mathcal{U}$ does not contain a minimizer of (CM), so $z^* \notin \mathcal{U}$. If $z^* \in O_{\mathcal{U}}(z + u^2)$, then $z + u^2 \in \text{conv}\{z^*, z + u^1 + u^2, z\}$ by Observation 1 with $x = z + u^2$, $v^1 = -u^1$, $v^2 = u^2$, and $w = z^*$. We have $\text{conv}\{z^*, z + u^1 + u^2, z\} \subseteq GP(\mathcal{U})$, so $z + u^2 \in GP(\mathcal{U})$. However, $z + u^2 \notin GP(\mathcal{U})$ by the definition of

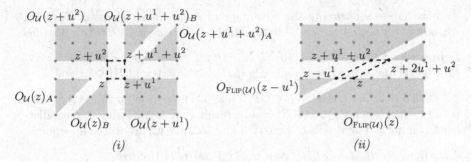

Fig. 2. *(i)* A depiction of the orthants and sets used to prove Theorem 3 in **Case 2**. *(ii)* A sample update for $\mathrm{FLIP}(U) = (-u^1, 2u^1 + u^2)$.

Case 2, which gives a contradiction. Thus, $z^* \notin O_{\mathcal{U}}(z + u^2)$. By symmetry, $z^* \notin O_{\mathcal{U}}(z + u^1)$. If $z^* \in O_{\mathcal{U}}(z + u^1 + u^2)_B \backslash \{z + u^1 + u^2\}$, then

$$z^* = (z + u^1 + u^2) + k_1(u^1 + u^2) + k_2 u^2, \text{ where } k_1, k_2 \in \mathbb{Z}_{\geq 0}.$$

We have $\nabla f(z + u^1 + u^2)^\mathsf{T}(-u^2) < 0$ by (5) and $\nabla f(z + u^1 + u^2)^\mathsf{T}(-u^1 - u^2) \leq 0$ because $z \in \mathrm{GP}(\mathcal{U})$. Thus, $z + u^1 + u^2$ cuts z^*, so $z^* \notin \mathrm{intr}(\mathrm{GP}(\mathcal{U}))$, which contradicts Lemma 1 *(iii)*. By symmetry, $z^* \notin O_{\mathcal{U}}(z)_B \backslash \{z\}$. We have argued that $z^* \in O_{\mathcal{U}}(z)_A \cup O_{\mathcal{U}}(z + u^1 + u^2)_A$.

Theorem 3 *(a)*, *(b)*, and *(c)* require that (7) and (9) do not increase after updating $\mathcal{U}$. We proved $\mathrm{FLIP}(\mathcal{U}) \leq_f \mathcal{U}$ in Lemma 4. It is left to show $\mathrm{FLIP}(\mathcal{U}) \leq_r \mathcal{U}$. We prove $\mathrm{FLIP}(\mathcal{U}) \leq_r \mathcal{U}$ in each of the three possible updates in **Case 2**.

Assume $\mathrm{FLIP}(U) = (-u^1, u^1 + u^2)$. If $z^* \in O_{\mathcal{U}}(z + u^1 + u^2)_A$, then

$$z^* = (z + u^1 + u^2) + k_1 u^1 + k_2 u^2 = (z + u^1 + u^2) + (k_1 - k_2)u^1 + k_2(u^1 + u^2),$$

where $k_1, k_2 \geq 0$ and $k_1 \geq k_2 + 1$. The latter equations imply $r_{\mathcal{U}}(x^*) = k_1 + k_2$ and $r_{\mathrm{FLIP}(\mathcal{U})}(z^*) = k_1 - k_2 + k_2$. Hence, $\mathrm{FLIP}(\mathcal{U}) \leq_r \mathcal{U}$. The case when $z^* \in O_{\mathcal{U}}(z)_A$ or $\mathrm{FLIP}(U) = (u^1, u^1 + u^2)$ can be proved by a similar argument.

Assume $\mathrm{FLIP}(U) = (-u^1, 2u^1 + u^2)$. If $z^* \in O_{\mathcal{U}}(z)_A$, then $z^* \in O_{\mathrm{FLIP}(\mathcal{U})}(z - u^1) \cup O_{\mathrm{FLIP}(\mathcal{U})}(z)$. This is indicated in Fig. 2 *(i)* and *(ii)*. The proof when $z^* \in O_{\mathrm{FLIP}(\mathcal{U})}(z)$ is similar to when $z^* \in O_{\mathrm{FLIP}(\mathcal{U})}(z - u^1)$, so we only consider $z^* \in O_{\mathrm{FLIP}(\mathcal{U})}(z)$. Here, there are $k_1, k_2 \in \mathbb{Z}_{\geq 0}$ such that

$$z^* = z + k_1(-u^1) + k_2(-u^2) = z + (-k_1 + 2k_2)u^1 + k_2(-2u^1 - u^2).$$

We have $k_1 > k_2$, or equivalently $k_2 > (-k_1 + 2k_2)$, because $z^* \in O_{\mathcal{U}}(z)_A$. Thus,

$$r_{\mathcal{U}}(z^*) = k_1 + k_2 > (-k_1 + 2k_2) + k_2 = r_{\mathrm{FLIP}(\mathcal{U})}(z^*).$$

Hence, $\mathrm{FLIP}(\mathcal{U}) <_r \mathcal{U}$. Notice that we have strict decrease of $<_r$ in this setting. The case $z^* \in O_{\mathcal{U}}(z + u^1 + u^2)_A$ can be handled by a symmetric argument. We have argued that $\mathrm{FLIP}(\mathcal{U}) \leq_r \mathcal{U}$ for each possible update of **Case 2**.

We have shown that $\text{FLIP}(\mathcal{U}) \leq_f \mathcal{U}$ and $\text{FLIP}(\mathcal{U}) \leq_r \mathcal{U}$. If either of the latter inequalities is strict, then *(a)* or *(b)* hold. It suffices to assume $\text{FLIP}(\mathcal{U}) =_f \mathcal{U}$ and $\text{FLIP}(\mathcal{U}) =_r \mathcal{U}$ and show $\text{FLIP}(\mathcal{U})$ is connected; this will prove *(c)*.

Assume neither *(a)* nor *(b)* hold. If we update U to $(-u^1, 2u^1 + u^2)$, then $\text{FLIP}(\mathcal{U}) <_r \mathcal{U}$ and *(b)* would hold. Hence, $\text{FLIP}(\mathcal{U}) = \mathcal{U}'$ or $\text{FLIP}(\mathcal{U}) = \mathcal{U}''$, where $\mathcal{U}'$, $\mathcal{U}''$, U', and U'' are defined in **Case 2** of Table 1. We show the proof when $\text{FLIP}(\mathcal{U}) = \mathcal{U}''$ as the other proof follows symmetrically. If $\mathcal{U}''$ is connected, then *(c)* holds. If $\mathcal{U}'' \cap \text{GP}(\mathcal{U}'') = \{z - u^1\}$, then the points in $\text{argmin}\{f(w) : w \in \mathcal{U}\}$ are all strictly cut by some point in $\mathcal{U}''$. This implies $\text{FLIP}(\mathcal{U}) <_f \mathcal{U}$, which is a contradiction. Thus, it is left to consider when z strictly cuts $z + 2u^1 + u^2$ and $z + 2u^1 + u^2$ strictly cuts $z + u^1 + u^2$. Here, $f(z) < f(z + u^1 + u^2)$ and $f(z) = \min\{f(w) : w \in \mathcal{U}\}$. The definition of **Case 2** and the preprocessing (5) on $\mathcal{U}$ imply that none of the points in $\mathcal{U}''\backslash\{z\} = \{z + u^2, z + u^1 + u^2, z - u^1\}$ are strictly cut by z. Hence, $|\,\text{GP}(\mathcal{U}'') \cap \mathcal{U}''| \neq 1$ as the points in $\mathcal{U}''\backslash\{z\}$ cannot strictly cut each other by Proposition 1. Thus, $|\,\text{GP}(\mathcal{U}'') \cap \mathcal{U}''| \geq 1$. One of these points must be z because $z = \min\{f(z) : z \in \mathcal{U}\}$. If $\mathcal{U}''$ was disconnected, then $\mathcal{U}'' \cap \text{GP}(\mathcal{U}'') = \{z, z + u^2\}$. However, $z + u^2$ is strictly cut by $z + u^1 + u^2$ by the preprocessing on $\mathcal{U}$. Hence, $\text{FLIP}(\mathcal{U})$ is connected. $\qquad\square$

Theorem 3 also implies the following theorem.

Theorem 4. *If $(\mathcal{U}^i)_{i=1}^{\infty}$ is constructed using the procedure in Theorem 1, then there exists $T_1 \in \mathbb{Z}$ such that $\mathcal{U}^{T_1}$ contains an optimal solution of (CM).*

2.2 Convergence Towards a Lattice-Free Set: Theorem 2

Let $\mathcal{U}^0$ be a unimodular set satisfying (5). For $i \in \mathbb{Z}_{\geq 0}$ set $\mathcal{U}^{i+1} = \text{FLIP}(\mathcal{U}^i)$ and preprocess $\mathcal{U}^{i+1}$ to satisfy (5). By Theorem 4 and Lemma 4, there exists $T_1 \in \mathbb{Z}_{\geq 1}$ such that $\text{GP}(\mathcal{U}^i) \cap \mathcal{U}^i$ contains an optimal solution of (CM) for all $i \geq T_1$. After relabeling indices, we assume $T_1 = 0$. It turns out that after at most one flip, each $\mathcal{U}^i$ is also connected. The proof appears in the appendix.

Lemma 5. *$\text{FLIP}(\mathcal{U}^i)$ is connected for all $i \geq 1$.*

To show that $(\mathcal{U}^i)_{i=1}^{\infty}$ finitely converges to a lattice-free gradient polyhedron, we now consider how the second minimum evolves as we update the unimodular sets. Set $f_2 := \infty$ and after $\mathcal{U}^{i+1} = \text{FLIP}(\mathcal{U}^i)$ is computed update f^2 as follows:

$$f_2 = \begin{cases} f_2 & \text{if } f_2 \geq \min\{f(w) : w \in \mathcal{U}' \cap \text{GP}(\mathcal{U}^{i+1})\} \\ \min\{f(w) : w \in \mathcal{U}' \cap \text{GP}(\mathcal{U}^{i+1})\} & \text{otherwise,} \end{cases}$$

where $\mathcal{U}' := \mathcal{U}^{i+1}\backslash\{z \in \mathcal{U}^{i+1} : f(z) = f(z^*)\}$. Note that $f_2 \geq f(z^*)$.

Lemma 6. *There exists $T_2 \in \mathbb{Z}$ such that f_2 is minimized at $\mathcal{U}^{T_2}$, or $\mathcal{U}^{T_2}$ does not fit into any case of Table 1 (and $\text{GP}(\mathcal{U}^{T_2})$ is lattice-free by Theorem 3).*

The proof of Lemma 6 is in the appendix. As was the case with z^*, it can be shown that any vector $z^2 \in \mathbb{Z}^2$ such that $f(z^2) = f_2$ is contained in $\mathcal{U}^i$ for all $i \geq T_2$.

Define f_3 analogously to f_2. By similar arguments, f_3 reaches its minimal value after finitely many additional flips. Once f_3 has reached its minimal possible value, the element $z^3 \in \mathcal{U}$ with $f(z^3) = f_3$ has to stay in $\mathrm{GP}(\mathcal{U})$ after flipping. This implies that after one additional flip we have reached a lattice-free gradient polyhedron by Corollary 2. This completes the proof of Theorem 2.

Acknowledgements. The authors would like to thank the anonymous reviewers for their thorough review and comments, which led to an improved presentation.

Appendix

Proof (of Lemma 2). Lemma 1 *(ii)* states that $\mathcal{U} \cap \mathrm{GP}(\mathcal{U}) \neq \emptyset$. If $|\mathcal{U} \cap \mathrm{GP}(\mathcal{U})| = 1$, then z can be relabeled so that Lemma 2 *(i)* holds. If $|\mathcal{U} \cap \mathrm{GP}(\mathcal{U})| = 3$, then z can be relabeled and the columns of U can be multiplied by ± 1 so that Lemma 2 *(iii)* holds. It remains to consider when $|\mathcal{U} \cap \mathrm{GP}(\mathcal{U})| = 2$. If the points in $\mathcal{U} \cap \mathrm{GP}(\mathcal{U})$ differ by $\pm u^1$ or $\pm u^2$, then z and U can be relabeled so that Lemma 2 *(ii -a)* holds.

We complete the proof by assuming the points in $\mathcal{U} \cap \mathrm{GP}(\mathcal{U})$ do not differ by $\pm u^1$ nor $\pm u^2$. Here, z and U can be relabeled so that $\mathcal{U} \cap \mathrm{GP}(\mathcal{U}) = \{z, z + u^1 + u^2\}$. Note that z cannot strictly cut both $z + u^1$ and $z + u^2$. Otherwise, $\nabla f(z)^\mathsf{T} u^1 > 0$ and $\nabla f(z)^\mathsf{T} u^2 > 0$ implying that z strictly cuts $z + u^1 + u^2$. However, $z + u^1 + u^2$ is assumed to be in $\mathrm{GP}(\mathcal{U})$ and cannot be strictly cut by z. Similarly, $z + u^1 + u^2$ cannot strictly cut both $z + u^1$ and $z + u^2$.

Also, $z + u^1$ cannot strictly cut $z + u^2$. Otherwise $\nabla f(z + u^1)^\mathsf{T}(u^2 - u^1) > 0$; this inequality, along with the fact that $z + u^1$ does not cut z because $z \in \mathcal{U} \cap \mathrm{GP}(\mathcal{U})$, implies $\nabla f(z + u^1)^\mathsf{T} u^2 > 0$. Consequently, $z + u^1$ strictly cuts $z + u^1 + u^2$, which is a contradiction. Similarly, $z + u^2$ cannot strictly cut $z + u^1$.

Collectively, the previous arguments imply that z and $z + u^1 + u^2$ must strictly cut $z + u^1$ and $z + u^2$. As neither z nor $z + u^1 + u^2$ can strictly cut both, we may assume that z strictly cuts $z + u^1$ and $z + u^1 + u^2$ strictly cuts $z + u^2$.   □

Proof (of Lemma 3). If $\mathrm{GP}(\mathcal{U})$ is lattice-free, then both $H^{-1} \cap \mathbb{Z}^2$ and $H^1 \cap \mathbb{Z}^2$ are empty. Assume to the contrary that $\mathrm{GP}(\mathcal{U})$ is not lattice-free but $H^{-1} \cap \mathbb{Z}^2 = \emptyset$ and $H^1 \cap \mathbb{Z}^2 = \emptyset$. Let $x \in \mathrm{intr}(\mathrm{GP}(\mathcal{U})) \cap \mathbb{Z}^2$ with $x = z + k_1 u^1 + k_2 u^2$ and $k_1, k_2 \in \mathbb{Z}$. If $|k_2| = 1$, then $x \in H^{-1}$ or $x \in H^1$, which is a contradiction. Thus, $|k_2| \geq 2$. The triangle $\mathrm{conv}\{z, z + u^1, x\}$ is contained in $\mathrm{GP}(\mathcal{U})$ as its three vertices are contained in $\mathrm{GP}(\mathcal{U})$ and $\mathrm{GP}(\mathcal{U})$ is convex. Notice that

$$|\det(z - x, (z + u^1) - x)| = |\det(-k_1 u^1 - k_2 u^2, -(k_1 - 1)u^1 - k_2 u^2)| = |k_2| \geq 2.$$

This implies $\mathrm{conv}\{z, z + u^1, x\} \backslash \{z, z + u^1, x\}$ contains an integer point $\overline{x} := z + k_1' u^1 + k_2' u^2$ with $k_1', k_2' \in \mathbb{Z}$ (see, e.g., [7, Page 291, Corollary (2.6)]). There are no such integer points in $\mathrm{conv}\{z, z + u^1\}$, so $0 < |k_2'| < |k_2|$. Hence, $\overline{x} \in \mathrm{intr}(\mathrm{GP}(\mathcal{U}))$. Repeating this with $\overline{x}$ eventually returns a point in $\mathrm{intr}(\mathrm{GP}(\mathcal{U})) \cap \mathbb{Z}^2$ that is also in H^{-1} or H^1, which is a contradiction.   □

Proof (of Corollary 2). If $|\mathcal{U} \cap \mathrm{GP}(\mathcal{U})| = 4$, then $\mathcal{U}$ is connected. Also, both $H^{-1} \cap \mathbb{Z}^2$ and $H^1 \cap \mathbb{Z}^2$ are empty otherwise one of the points in $\mathcal{U}$ is contained in $\mathrm{intr}(\mathrm{GP}(\mathcal{U}))$, which is a contradiction. Hence, $\mathrm{GP}(\mathcal{U})$ is lattice-free. $\qquad\square$

Proof (of Lemma 5). Let $i \in \mathbb{Z}_{\geq 0}$. We show that $\mathrm{FLIP}(\mathcal{U}^i)$ is connected. The possible cases are similar, and we only present the case when $\mathrm{GP}(\mathcal{U}^i) \cap \mathcal{U}^i = \{z, z + u^1, z + u^2\}$. We only consider the case $|H^{-1} \cap \mathbb{Z}^2| \geq 1$, i.e., $\mathrm{FLIP}(\mathcal{U}^i) = \{u^1 - u^2, u^1\}$. The argument for the case $|H^1 \cap \mathbb{Z}^2| \geq 1$ is symmetric. We have $z^* \in \mathrm{GP}(\mathcal{U}^i) \cap \mathcal{U}^i$ by Proposition 1, and $\mathrm{GP}(\mathcal{U}^i) \cap \mathcal{U}^i \subseteq \mathrm{GP}(\mathrm{FLIP}(\mathcal{U}^i)) \cap \mathrm{FLIP}(\mathcal{U}^i)$ by Lemma 4. The set $\mathrm{FLIP}(\mathcal{U}^i)$ is defined by **Case 5** in Table 1. Hence, $\mathrm{FLIP}(\mathcal{U}^i) = \{z, z + u^1, z + u^2, z + u^1 - u^2\}$ and $z + u^1 - u^2 \in \mathrm{intr}(\mathrm{GP}(\mathcal{U}^i))$. This implies that $z + u^1 - u^2$ is not cut by any of the points $z, z + u^2, z + u^1$. Thus, $\mathrm{GP}(\mathrm{FLIP}(\mathcal{U}^i)) \cap \mathrm{FLIP}(\mathcal{U}^i)$ contains at least two points: a minimizer z^* of (CM) contained in $\mathcal{U}^i$ and $z + u^1 - u^2$. Because of this, $\mathrm{FLIP}(\mathcal{U}^i)$ can only be disconnected if $z^* = z + u^2$. However, $z + u^2$ does not strictly cut z or $z + u^1$, therefore, this implies that $z + u^1 - u^2$ strictly cuts z and $z + u^1$, but not $z + u^2$. This is not possible. Thus, $\mathrm{FLIP}(\mathcal{U}^i)$ is connected. $\qquad\square$

Proof (of Lemma 6). By Lemma 5, $\mathcal{U}^i$ is connected for all $i \geq 1$. Hence, after at most one iteration f_2 becomes finite. Moreover, f_2 can only decrease a finite number of times because it is greater than or equal to the optimal value of (CM), which was assumed finite.

Let $i \in \mathbb{Z}_{\geq 1}$. We use the following implication:

If $f(w) = f_2$ for $w \in \mathcal{U}^i \cap \mathrm{GP}(\mathcal{U}^i)$, then either $w \in \mathrm{FLIP}(\mathcal{U}^i) \cap \mathrm{GP}(\mathcal{U}^i)$

or the value of f_2 strictly decreases after flipping $\mathcal{U}^i$ to $\mathcal{U}^{i+1} = \mathrm{FLIP}(\mathcal{U}^i)$.

$$(10)$$

To see this, assume $w \notin \mathrm{FLIP}(\mathcal{U}^i) \cap \mathrm{GP}(\mathcal{U}^i)$. Since $\mathcal{U}^i \cap \mathrm{GP}(\mathcal{U}^i) \subseteq \mathcal{U}^{i+1}$ by Lemma 4, this implies that w is strictly cut by one of the points in $\mathcal{U}^{i+1} \backslash \mathcal{U}^i$. Hence, f_2 strictly decreases by Proposition 1. This proves (10).

Let $i \in \mathbb{Z}_{\geq 1}$ and assume that f_2 is not minimized. Suppose $f(z_2) = f_2$ for $z_2 \in \mathcal{U}^i \cap \mathrm{GP}(\mathcal{U}^i)$. We show that f_2 decreases after a finite number of updates. This will prove the lemma because f_2 can only decrease a finite number of times. If f_2 decreases after updating $\mathcal{U}^i$ to $\mathcal{U}^{i+1}$, then we are done. Otherwise, $z_2 \in \mathcal{U}^{i+1} \cap \mathrm{GP}(\mathcal{U}^i)$ by (10), where $\mathcal{U}^{i+1} = \mathrm{FLIP}(\mathcal{U}^i)$. This implies that no points in $\mathcal{U}^{i+1} \backslash \mathcal{U}^i$ strictly cut z_2. The update definitions in Table 1 imply that

$$\text{if } \mathcal{U} \text{ is connected, then } (\mathrm{FLIP}(\mathcal{U}) \backslash \mathcal{U}) \cap \mathrm{intr}(\mathrm{GP}(\mathcal{U})) \neq \emptyset. \qquad (11)$$

It follows from (11) that there always exists a point in $\mathcal{U}^{i+1} \backslash \mathcal{U}^i$ that is not strictly cut by z^* or z_2. Also, $z_2 \in \mathcal{U}^{i+1} \cap \mathrm{GP}(\mathcal{U}^{i+1})$ and $S := (\mathcal{U}^{i+1} \cap \mathrm{GP}(\mathcal{U}^{i+1})) \backslash \{z^*, z_2\}$ is non-empty. Define $f_3 := \min\{f(w) : w \in S\}$ and let $z_3 \in S$ such that $f(z_3) = f_3$. In particular $f_3 < \infty$ and f_3 is lower bounded by f_2. We show that f_2 has to strictly decrease after at most $f_3 - f_2$ many additional updates.

Assume that after updating the set $\mathcal{U}^{i+1}$ to $\mathcal{U}^{i+2}$ the value of f_2 still has not decreased. Thus, $z_2 \in \mathcal{U}^{i+2}$. Also, if the value of f_3 has not decreased, then the point in $\mathcal{U}^{i+2} \backslash \mathcal{U}^{i+1} = \mathcal{U}^{i+2} \backslash \{z^*, z_2, z_3\}$ does not strictly cut (nor strictly cut by)

z^*, z_2 and z_3. Thus, $|\operatorname{GP}(\mathcal{U}^{i+2}) \cap \mathcal{U}^{i+2}| = 4$, $\operatorname{GP}(\mathcal{U}^{i+2})$ is already lattice-free by Corollary 2 and the update procedure terminates.

Thus, we can assume that the value of f_3 decreases after flipping from $\mathcal{U}^{i+1}$ to $\mathcal{U}^{i+2}$. Since f_3 can only decrease at most $f_3 - f_2$ times, this implies that after that many additional flips the value of f_2 has to decrease. Also note that during these intermediate flips the value f_2 does never increase. Iterating these arguments implies that f_2 has reached its minimal possible value after finitely many updates to $\mathcal{U}^1$. $\square$

References

1. Andersen, K., Louveaux, Q., Weismantel, R., Wolsey, L.A.: Inequalities from two rows of a simplex tableau. In: Fischetti, M., Williamson, D.P. (eds.) IPCO 2007. LNCS, vol. 4513, pp. 1–15. Springer, Heidelberg (2007). https://doi.org/10.1007/978-3-540-72792-7_1
2. Averkov, G., Krümpelmann, J., Weltge, S.: Notions of maximality for integral lattice-free polyhedra: the case of dimension three. Math. Oper. Res. **42**(4), 1035–1062 (2017)
3. Averkov, G., Wagner, C., Weismantel, R.: Maximal lattice-free polyhedra: finiteness and an explicit description in dimension three. Math. Oper. Res. **36**(4), 721–742 (2011)
4. Baes, M., Oertel, T., Wagner, C., Weismantel, R.: Mirror-descent methods in mixed-integer convex optimization. In: Jünger, M., Reinelt, G. (eds.) Facets of Combinatorial Optimization, pp. 101–131. Springer, Heidelberg (2013). https://doi.org/10.1007/978-3-642-38189-8_5
5. Baes, M., Oertel, T., Weismantel, R.: Duality for mixed-integer convex minimization. Math. Program. Ser. A **158**, 547–564 (2016)
6. Balas, E.: Intersection cuts - a new type of cutting planes for integer programming. Oper. Res. **19**(1), 19–39 (1971)
7. Barvinok, A.: A Course in Convexity, Graduate Studies in Mathematics, vol. 54. American Mathematical Society, Providence (2002)
8. Basu, A., Conforti, M., Cornuéjols, G., Weismantel, R., Weltge, S.: Optimality certificates for convex minimization and Helly numbers. Oper. Res. Lett. **45**, 671–674 (2017)
9. Basu, A., Conforti, M., Cornuéjols, G., Zambelli, G.: Maximal lattice-free convex sets in linear subspaces. Math. Oper. Res. **35**(3), 704–720 (2010)
10. Basu, A., Conforti, M., Di Summa, M.: A geometric approach to cut-generating functions. Math. Program. **151**(1), 153–189 (2015). https://doi.org/10.1007/s10107-015-0890-5
11. Basu, A., Cornuéjols, G., Köppe, M.: Unique minimal liftings for simplicial polytopes. Math. Oper. Res. **37**(2), 346–355 (2012)
12. Bell, D.: A theorem concerning the integer lattice. Stud. Appl. Math. **56**, 187–188 (1977)
13. Bonami, P., et al.: An algorithmic framework for convex mixed integer nonlinear programs. Discrete Optim. **5**(2), 186–204 (2008)
14. Boyd, S., Vandenberghe, L.: Convex Optimization. Cambridge University Press, Cambridge (2004)

15. Del Pia, A., Weismantel, R.: Relaxations of mixed integer sets from lattice-free polyhedra. Ann. Oper. Res. **240**(1), 95–117 (2015). https://doi.org/10.1007/s10479-015-2024-0

16. Dey, S.S., Wolsey, L.A.: Lifting integer variables in minimal inequalities corresponding to lattice-free triangles. In: Lodi, A., Panconesi, A., Rinaldi, G. (eds.) IPCO 2008. LNCS, vol. 5035, pp. 463–475. Springer, Heidelberg (2008). https://doi.org/10.1007/978-3-540-68891-4_32

17. Doignon, J.: Convexity in cristallographical lattices. J. Geom. **3**, 71–85 (1973)

18. Duran, M.A., Grossman, I.E.: An outer-approximation algorithm for a class of mixed-integer nonlinear programs. Math. Program. **36**(3), 307–339 (1986)

19. Grötschel, M., Lovász, L., Schrijver, A.: Geometric Algorithms and Combinatorial Optimization, Algorithms and Combinatorics, vol. 2, 1st edn. Springer, Heidelberg (1988). https://doi.org/10.1007/978-3-642-78240-4

20. Gupta, O., Ravindran, V.: Branch and bound experiments in convex nonlinear integer programming. Manag. Sci. **31**, 1533–1546 (1985)

21. Leyffer, S.: Integrating SQP and branch-and-bound for mixed integer nonlinear programming. Comput. Optim. Appl. **18**, 295–309 (2001)

22. Schnorr, C.P.: Geometry of numbers and integer programming. In: Cori, R., Wirsing, M. (eds.) STACS 1988. LNCS, vol. 294, pp. 1–7. Springer, Heidelberg (1988). https://doi.org/10.1007/BFb0035826

23. Scarf, H.: An observation on the structure of production sets with indivisibilities. Proc. Nat. Acad. Sci. U.S.A. **74**, 3637–3641 (1977)

Sequence Independent Lifting for the Set of Submodular Maximization Problem

Xueyu Shi, Oleg A. Prokopyev, and Bo Zeng[✉]

Department of Industrial Engineering, University of Pittsburgh, Pittsburgh, USA
{xus6,droleg,bzeng}@pitt.edu

Abstract. We study the polyhedral structure of a mixed 0-1 set arising in the submodular maximization problem, given by $P = \{(w, x) \in \mathbb{R} \times \{0,1\}^n : w \leq f(x), x \in \mathcal{X}\}$, where submodular function $f(x)$ is represented by a concave function composed with a linear function, and $\mathcal{X}$ is the feasible region of binary variables x. For $\mathcal{X} = \{0,1\}^n$, two families of facet-defining inequalities are proposed for the convex hull of P through restriction and lifting using submodular inequalities. When $\mathcal{X}$ is a partition matroid, we propose a new class of facet-defining inequalities for the convex hull of P through multidimensional sequence independent lifting. Our results enable us to unify and generalize the existing results on valid inequalities for the mixed 0-1 knapsack. Finally, we perform some preliminary computational experiments to illustrate the superiority of our facet-defining inequalities.

Keywords: Submodular function maximization · Sequence independent multidimensional lifting · Polyhedra

1 Introduction

For the ground set $N = \{1, \ldots, n\}$, consider a submodular maximization problem

$$\max_{S \subseteq N} \{f(S) : S \in \mathcal{I}\}, \tag{1}$$

where $\mathcal{I}$ is a collection of subsets of N, and $f : 2^N \to \mathbb{R}$ is a real-valued, submodular set function. Let $\rho_i(S) = f(S \cup \{i\}) - f(S)$ for $S \subseteq N$ and $i \in N \setminus S$. Function f is said to be *submodular* if $\rho_i(T) \geq \rho_i(S)$ for any $i \in N \setminus S$ and for all subsets $T \subseteq S \subseteq N$ [11].

In this paper, we are primarily interested in solving the submodular maximization problem exactly via mixed-integer programming. Specially, we focus on a class of submodular functions [1] that are represented by a concave function composed with a linear function, i.e.,

$$f(S) = g(a(S) + b), \tag{2}$$

where $g : \mathbb{R} \to \mathbb{R}$ is a concave function, $b \in \mathbb{R}$, vector $a \in \mathbb{R}^n$ whose components are either all nonnegative or all nonpositive, and $a(S) := \sum_{j \in S} a_j$. This class

© Springer Nature Switzerland AG 2020
D. Bienstock and G. Zambelli (Eds.): IPCO 2020, LNCS 12125, pp. 378–390, 2020.
https://doi.org/10.1007/978-3-030-45771-6_29

of functions is, in fact, very flexible and has been heavily used in number of domains, see, e.g., [4–6,8,9,12]. Without loss of generality, in the remainder of the paper we assume that $b = 0$ and $a \in \mathbb{R}^n_+$.

Let binary vectors x be the incidence vectors for subsets of N, then the submodular maximization problem (1) can be written as $\max\{w : w \leq f(x), x \in \mathcal{X} \subseteq \{0,1\}^n\}$, where $\mathcal{X}$ is the feasible region of x with respect to $\mathcal{I}$. Formally, we study the following mixed 0-1 submodular maximization set:

$$P = \{(w, x) \in \mathbb{R} \times \{0,1\}^n : w \leq g(a^T x), x \in \mathcal{X}\}. \tag{3}$$

In the remainder of the paper if $\mathcal{X} = \{0,1\}^n$, then we refer to P as P_0. Note that we can remove the assumption on the signs of a when $\mathcal{X} = \{0,1\}^n$ (as x_j can be replaced by $1 - x_j$ for all j such that $a_j < 0$).

The convex hull of P is a polyhedron since P is a union of a finite number of rays with same directions. Nevertheless, it is very challenging to have a complete characterization of this polyhedron; see, e.g., a discussion in [1]. Nemhauser and Wolsey [11] propose an approach with exponentially many inequalities to formulate P as a mixed-integer linear program. However, its linear program relaxation is quite weak and the traditional branch-and-bound methods are often ineffective [1]. Considering a situation where g is strictly increasing, concave and differentiable, Ahmed and Atamtürk [1] employ the lifting technique to derive strong lifted inequalities for P_0. Their study is mostly focused on continuous relaxation of the lifting function as it enables the use of the KKT conditions to derive its subadditive approximation. The numerical results in [1] demonstrate the effectiveness of this approximated lifting. In a subsequent paper, Yu and Ahmed [14] adopt this approximation idea to study set P, where $\mathcal{X}$ involves a single 0-1 knapsack constraint. Specifically, the authors consider its cardinality relaxation, and extend the subadditive approximate lifting developed in [1] to handle this somewhat more complex set.

Inspired by the fundamental results in [1,14], in this paper, we revisit the sequence independent (SI) lifting and its multidimensional extensions for conv(P). After developing a new class of subadditive functions, we recognize that the lifting functions of conv(P_0) given in [1] are naturally subadditive. We believe that this new finding is of a significant value for further systematic study of the submodular maximization problem and the associated set P.

Specifically, our technical results and contributions in this paper can be summarized as follows. First, instead of assuming that g is strictly concave, increasing and differentiable as in [1,14], we only assume that g is concave to ensure its submodularity. As mentioned earlier, in [1,14] the increasing and differentiable properties are exploited to apply the KKT conditions in order to derive the closed-form subadditive approximation of the lifting function. Our results do not rely on the KKT conditions. Clearly, with those key assumptions removed, a much broader class of concave functions can be considered.

Second, as an immediate application of our first point, the piecewise linear functions can be studied. In particular, we show in the following that the well-known mixed knapsack set is a special case of set P of the submodular maximization problem. Consider the mixed 0-1 knapsack set K, see [10], where

$$K = \{(\pi, x) \in \mathbb{R} \times \{0,1\}^n : a^T x \le b + \pi, \pi \ge 0, x \in \mathcal{X}\}.$$

Define $w = -\pi$, and piecewise concave function $g(z) = \min\{0, b - z\}$. Then we convert set K to the form of (3). We believe that these connections between the mixed 0-1 knapsack sets and submodular sets are novel and rather important, given that a number of computationally effective results have been derived in the literature for the mixed 0-1 knapsack sets. Furthermore, it is known [10] that the mixed 0-1 knapsack set K can be viewed as a relaxation of the popular single-node flow set, given by $F = \{(x, y) \in \{0,1\}^n \times \mathbb{R}_+^n : \sum_{j \in N^+} y_j - \sum_{j \in N^-} y_j \le b, y_j \le a_j x_j \ \forall j \in N, x \in \mathcal{X}\}$, where $N = N^+ \cup N^-$. This observation immediately implies that the valid inequalities for P are valid for the flow set F. Note that the connections between the submodularity and the flow models are also considered by Wolsey [13] and Atamtürk et al. [3].

Third, we develop several results to support lifting operations in the context of submodular optimization. In particular, new results on multidimensional SI lifting are derived, a new type of subadditive functions to ensure SI lifting is constructed. The latter result generalizes the existing ones in the literature.

Finally, in addition to strengthening the existing polyhedral results for P_0 we also derive new interesting results when P has a more complex structure. Specifically, for P_0, i.e., when $\mathcal{X} = \{0,1\}^n$, we strengthen the results in [1,14] and present two family of facets for $\text{conv}(P_0)$. Moreover, we consider set P when $\mathcal{X}$ is a partition matroid; such $\mathcal{X}$ is often encountered in the discrete optimization literature. For its convex hull, facet-defining inequalities are derived based on the multidimensional SI lifting.

Notation. Define set $[k] = \{1, 2, \ldots, k\}$ for any positive integer k. Let $\mathbf{e}_k$ be the unit vector whose entry value is 1 for kth component, and 0 for the others. Let $\bar{S} = N \setminus S$ for any $S \subseteq N$. Define $a(S) := \sum_{j \in S} a_j$ for $a \in \mathbb{R}^n$ and set $S \subseteq N$.

2   A New Class of Subadditive Function

In this section, we introduce a new class of subadditive functions, which is then exploited in Sect. 3.

Definition 1. *A function $\phi : \mathbb{R}^n \to \mathbb{R}$ is subadditive on $D \subseteq \mathbb{R}^n$ if $\phi(z_1) + \phi(z_2) \ge \phi(z_1 + z_2)$ whenever $z_1, z_2 \in D$ and $z_1 + z_2 \in D$. A function ϕ is called superadditive on D if $-\phi$ is subadditive on D.*

Theorem 1. *Given a sequence of values $a_1, a_2, \ldots$ such that $a_j \ge a_{j+1} \ge 0$ for all j, let $A_0 = 0$ and $A_k = \sum_{j=1}^k a_j$. Define a piecewise concave function $\phi : \mathbb{R}_+ \to \mathbb{R}$ as follows*

$$\phi(z) = \begin{cases} 0 & \text{if } z = 0, \\ g(z - A_k + v_k) + \phi(A_k) - g(v_k) & \text{if } A_k \le z \le A_{k+1}, k = 0, 1, \ldots, \end{cases} \tag{4}$$

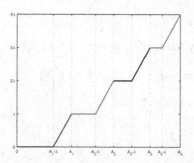

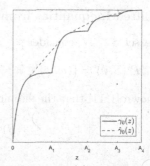

(a) The lifting function on the flow cover in-   (b) Lifting function $\gamma_0(z)$ and its approxima-
equality.   tion $\hat{\gamma}_0(z)$ in [1].

Fig. 1. Two subadditive functions in the form of ϕ, see Theorem 1.

*where g is a concave function and $\{v_k\}_{k=0}^{\infty}$ is a sequence of values such that
$v_k + a_{k+1} \leq v_{k+1} + a_{k+2}$. Then function ϕ is subadditive on $\mathbb{R}_+$.*

Proof. See Appendix A.   □

Remark. Note that subadditive function ϕ is of a general form as we do not
specify the representation of g. Hence, it provides an effective framework to
verify one function's subadditivity or to construct its subadditive approximation,
according to its particular structure. Moreover, if g is replaced by $-g$, which is
convex, then the resulting function ϕ becomes a superadditive function.

Example 1 (Adapted from [7]). Consider the single-node flow set F when $N^- = \emptyset$
and $\mathcal{X} = \{0, 1\}^n$. Given a flow cover $S = \{1, \ldots, s\}$ such that $\lambda = \sum_{j \in S} a_j - b > 0$
and $a_1 \geq \cdots \geq a_r > \lambda \geq a_{r+1} \geq \cdots \geq a_s$. Then the lifting function of the flow
cover inequality (depicted in Fig. 1(a)) is $\phi(0) = 0$ and

$$
\phi(z) = \begin{cases} g(z - A_k + a_1 - a_{k+1}) + \phi(A_k) & \text{if } A_k \leq z \leq A_{k+1}, k = 0, 1, \ldots, r - 2, \\ g(z - A_r + a_1) + \phi(A_{k-1}) & \text{if } z \geq A_{r-1}, \end{cases}
$$

where $\phi(A_k) = k\lambda$ and $g(z) = \max\{0, z - (a_1 - \lambda)\}$ is a convex function. Let
$v_k = a_1 - a_{k+1}$ for $k = 0, 1 \ldots, r-1$, then $g(v_k) = 0$ and the above lifting function
can be written in the form of (4). Thus, the lifting function is superadditive on
$\mathbb{R}_+$, which matches the corresponding result in [7].

3   Lifting for conv(P_0)

In this section, we study the lifting for the convex hull of the basic set P_0, i.e.,

$$
P_0 = \{(w, x) \in \mathbb{R} \times \{0, 1\}^n : w \leq g(a^T x)\}.
$$

We first derive a set of valid inequalities by exact lifting on variables fixed at
zeros. As we will see, the lifting is SI, and the resulting inequalities are facet-
defining for conv(P_0). Similarly, another family of facet-defining inequalities are
derived by SI lifting on variables fixed at ones.

3.1 Lifted Inequalities from Uplifting

Given a set $S \subseteq N$, consider set $P_0(\bar{S}, \emptyset)$ by fixing $x_j = 0$ for $j \in \bar{S}$ in P_0, i.e.,

$$P_0(\bar{S}, \emptyset) = \{(w, x) \in \mathbb{R} \times \{0,1\}^n : w \leq g(a^T x), x_j = 0 \ \forall j \in \bar{S}\}.$$

It is known [1,11] that the submodular inequality

$$w \leq f(S) - \sum_{j \in S} \rho_j(S \setminus j)(1 - x_j) \tag{5}$$

defines a facet of the convex hull of $P_0(\bar{S}, \emptyset)$. To perform lifting on inequality (5) for $\text{conv}(P_0)$, its lifting function is

$$\gamma_0(z) = \max w + \sum_{j \in S} \rho_j(S \setminus j)(1 - x_j) - f(S)$$

$$s.t. \ w \leq g(\sum_{j \in S} a_j x_j + z), \ x_j \in \{0,1\} \ \forall j \in S,$$

where $z \in \mathbb{R}_+$. With a similar strategy as in Proposition 5 in [1], we can derive the exactly same formula for $\gamma_0(z)$ and any general concave function g.

Proposition 1. *Suppose $S = \{1, \ldots, s\}$ and $a_1 \geq \cdots \geq a_s$, let $A_0 = 0$, $A_k = \sum_{j=1}^{k} a_j$ and $\Delta_k(z) = z - A_k + a(S)$ for $k \in S$. Then $\gamma_0(A_k) = \sum_{j=1}^{k} \rho_j(S \setminus j)$ and*

$$\gamma_0(z) = \begin{cases} g(\Delta_{k+1}(z)) + \gamma_0(A_{k+1}) - f(S) & \text{if } A_k \leq z \leq A_{k+1}, k = 0, \ldots, s-2, \\ g(z) + \gamma_0(A_s) - f(S) & \text{if } z \geq A_{s-1}. \end{cases}$$

Remark. Let $v_k = a(S) - a_{k+1}$, then $v_k + a_{k+1} = a(S)$ for $k = 0, 1, \ldots, s-1$. Given that $\gamma_0(A_k) = \sum_{j=1}^{k} \rho_j(S \setminus j)$ and $g(v_k) = g(a(S) - a_{k+1})$ for $k \in S$, it follows from Theorem 1 that the lifting function $\gamma_0(z)$ is subadditive on $\mathbb{R}_+$. Thus, the exact lifting is SI, and the resulting inequality from lifting is facet-defining for $\text{conv}(P_0)$, which is formally stated as follows.

Theorem 2. *For any $S \subseteq N$, the following inequality defines a facet of $\text{conv}(P_0)$*

$$w \leq f(S) - \sum_{j \in S} \rho_j(S \setminus j)(1 - x_j) + \sum_{j \in \bar{S}} \gamma_0(a_j) x_j. \tag{6}$$

Proof. It directly follows from the aforementioned Remark. $\square$

Note that Ahmed and Atamtürk [1] derive a subadditive approximation to the lifting function $\gamma_0(z)$ by applying the continuous relaxation and then using KKT conditions to solve the convex program. Nevertheless, as pointed out in Theorems 1 and 2 that the exact lifting function is naturally subadditive and we can directly obtain facet-defining inequalities without any approximations. Next, we provide an example to illustrate the difference between inequalities obtained by approximate lifting [1] and our exact lifting.

Example 2. Consider $P_0 = \{(w, x) \in \mathbb{R} \times \{0,1\}^n : w \leq -\exp(-a^T x)\}$, where $n = 7$, and $a = (0.8, 0.7, 0.7, 0.6, 0.5, 0.3, 0.2)^T$. Let $S = \{3, 4, 5, 6\}$. Figure 1(b) shows the lifting function $\gamma_0(z)$ and its approximation $\hat{\gamma}_0$ proposed in [1]. Specifically, the lifted inequality obtained by the approximate lifting in [1] is

$$w \leq -0.4695 + 0.1484x_1 + 0.1317x_2 + 0.1241x_3 + 0.1007x_4$$
$$+ 0.0794x_5 + 0.0428x_6 + 0.0447x_7.$$

The exact lifted inequality dominates the aforementioned one by carrying smaller coefficients for x_1 and x_2, which are **0.1454** and **0.1241**, respectively.

Denote by K_0 and F_0 as the mixed 0-1 knapsack set and single-node flow set when $\mathcal{X} = \{0,1\}^n$, respectively. As mentioned in Sect. 1, the well-known mixed 0-1 knapsack set K_0 is a special case of P_0. We then show that the lifted facet-defining inequalities of $\mathrm{conv}(K_0)$ can be obtained directly via (6). A class of valid inequalities for $\mathrm{conv}(F_0)$ then can be easily derived through the reduction to K_0, which has been well studied in [10]. Hence, our study on this submodular set unifies those classical results.

Corollary 1 ([10]). *Consider any $S \subseteq N$ such that $\lambda = \sum_{j \in S} a_j - b > 0$. Supposing $S = \{1, \ldots, s\}$ with $a_1 \geq \cdots \geq a_r > \lambda \geq \cdots \geq a_s$, the inequality $-\pi \leq -\lambda + \sum_{j=1}^r \lambda(1 - x_j) + \sum_{j=r+1}^s a_j(1 - x_j) + \sum_{j \in \bar{S}} \gamma_K(a_j)x_j$ is facet-defining for $\mathrm{conv}(K_0)$, where $\gamma_K(z)$ is the corresponding lifting function that can be computed from Proposition 1 with $g(z) = \min\{0, b - z\}$.*

3.2 Lifted Inequalities from Downlifting

For a set $S \subseteq N$, denote by $P_0(\emptyset, S) = \{(w, x) \in \mathbb{R} \times \{0,1\}^n : w \leq g(a^T x), x_j = 1 \ \forall j \in S\}$ the set P_0 with $x_j = 1$ for $j \in S$. It is shown in [1,11] that the submodular inequality $w \leq f(S) + \sum_{j \in \bar{S}} \rho_j(S)x_j$ defines a facet for the convex hull of $P(\emptyset, S)$. To lift that inequality for $\mathrm{conv}(P_0)$, the lifting function is given as

$$\eta_0(z) = \max w - \sum_{j \in \bar{S}} \rho_j(S)x_j - f(S)$$

$$\text{s.t. } w \leq g(\sum_{j \in \bar{S}} a_j x_j + a(S) - z), x_j \in \{0,1\} \ \forall j \in \bar{S},$$

where $z \in \mathbb{R}_+$. We next strengthen the result presented in [1] by showing that the lifting function is actually subadditive, and the resulting lifted inequality is facet-defining for $\mathrm{conv}(P_0)$. Again, it unifies some classical results for the mixed 0-1 knapsack set K_0 and the single-node flow set F_0.

Proposition 2. *Suppose $\bar{S} = \{1, 2, \ldots, \bar{s}\}$ and $a_1 \geq \cdots \geq a_{\bar{s}}$, let $A_0 = 0$, $A_k = \sum_{j=1}^k a_j$ and $\Delta_k(z) = a(S) + A_k - z$ for $k \in \bar{S}$. Then*

$$\eta_0(z) = \begin{cases} g(\Delta_{k+1}(z)) + \eta_0(A_{k+1}) - f(S) & \text{if } A_k \leq z \leq A_{k+1}, k = 0, \ldots, \bar{s} - 2, \\ g(a(N) - z) + \eta_0(A_{\bar{s}}) - f(S) & \text{if } z \geq A_{\bar{s}-1}, \end{cases}$$

where $\eta_0(A_k) = -\sum_{j=1}^k \rho_j(S)$. Furthermore, $\eta_0(z)$ is subadditive on $\mathbb{R}_+$.

Theorem 3. *For any $S \subseteq N$, the following inequality defines a facet of $conv(P_0)$*

$$w \le f(S) + \sum_{j \in \bar{S}} \rho_j(S)x_j + \sum_{j \in S} \eta_0(a_j)(1 - x_j). \qquad (7)$$

Corollary 2 ([10]). *Consider any $S \subseteq N$ such that $\mu = b - \sum_{j \in S} a_j > 0$. Supposing $\bar{S} = \{1, \ldots, \bar{s}\}$ with $a_1 \ge \cdots \ge a_{\bar{r}} > \mu \ge \cdots \ge a_{\bar{s}}$, the inequality $-\pi \le \sum_{j=1}^{\bar{r}}(\mu - a_j)x_j + \sum_{j \in S} \eta_K(a_j)(1 - x_j)$ is facet-defining for $conv(K_0)$, where $\eta_K(z)$ is the corresponding lifting function that can be computed from Proposition 2 with $g(z) = \min\{0, b - z\}$.*

4   Lifting for conv(P) with a Partition Matroid $\mathcal{X}$

Given a partition of N, $\{N_i\}_{i=1}^r$, and a sequence of positive integers $d_1, \ldots, d_r$, we study the convex hull of the submodular maximization set with a partition matroid constraints

$$P_{MC} = \{(w, x) \in \mathbb{R} \times \{0,1\}^n : w \le g(a^T x), \sum_{j \in N_i} x_j \le d_i \; \forall i \in [r]\}.$$

Define a mapping function $\sigma : N \rightarrow R$ such that $\sigma(j) = i$ if $j \in N_i$ for some $i \in [r]$. For any set $S \subseteq N$, let $S_i := S \cap N_i$ for any $i \in [r]$.

4.1   Multidimensional Lifting and Lifted Inequalities

Given a set $S \subseteq N$ such that $|S_i| \le d_i$, we start with the lifting procedure by restricting $x_j = 0$ for $j \in \bar{S}$. The lifting function of the submodular inequality $w \le f(S) - \sum_{j \in S} \rho_j(S \setminus j)(1 - x_j)$ is

$$\gamma\binom{z}{\mathbf{u}} = \max w + \sum_{j \in S} \rho_j(S \setminus j)(1 - x_j) - f(S)$$

$$s.t. \; w \le g(\sum_{j \in S} a_j x_j + z), x_j \in \{0,1\} \; \forall j \in S,$$

$$\sum_{j \in S_i} x_j + \mathbf{u}_i \le d_i \quad \forall i \in [r],$$

where $z \in \mathbb{R}_+$ and $\mathbf{u} \in \mathbb{Z}_+^r$. Note that if $r = 0$, then $\gamma\binom{z}{\mathbf{u}}$ reduces to $\gamma_0(z)$ studied in Sect. 3.1. Although $\mathbf{u}$ is multidimensional, the coefficient column of x_j is $\mathbf{e}_{\sigma(j)}$ for $j \in N$. As demonstrated in [2,15,16], we are not interested in all $\binom{z}{\mathbf{u}} \in \mathbb{R}_+ \times \mathbb{Z}_+^r$ but are only concerned with the value of $\mathbf{u} = \mathbf{e}_i$ for some $i \in [r]$. Furthermore, as our discussion in Sect. 4.2, the subadditivity does not hold for $\gamma\binom{z}{\mathbf{u}}$ on $\mathbb{R}_+ \times \mathbb{Z}_+^r$. On the other hand, our main result in Sect. 4 relies on the following sufficient and necessary conditions for SI lifting.

Proposition 3. *The lifting of $w \leq f(S) - \sum_{j \in S} \rho_j(S \setminus j)(1 - x_j)$ is SI for all possible values of $\binom{z}{u} \in \mathbb{R}_+ \times \{e_1, \ldots, e_r\}$ if and only if for any subset $\Gamma \subseteq \bar{S}$ such that $|\Gamma_i| \leq d_i$,*

$$\sum_{j \in \Gamma} \gamma \begin{pmatrix} z_j \\ e_{\sigma(j)} \end{pmatrix} \geq \gamma(\sum_{j \in \Gamma} \begin{pmatrix} z_j \\ e_{\sigma(j)} \end{pmatrix}) \quad \forall z_j \geq 0, j \in \Gamma. \tag{8}$$

Theorem 4. *Given any $S \subseteq N$ such that $|S_i| \leq d_i$ for all $i \in [r]$, the inequality*

$$w \leq f(S) - \sum_{j \in S} \rho_j(S \setminus j)(1 - x_j) + \sum_{j \in \bar{S}} \gamma \begin{pmatrix} a_j \\ e_{\sigma(j)} \end{pmatrix} x_j \tag{9}$$

is facet-defining for $conv(P_{MC})$.

Proof. Proof sketch is shown in Appendix B.   □

We next discuss a couple of immediate results of Theorem 4. If $N_1 = N$ and $d_1 = d$, then let $P_C = \{(w, x) \in \{0, 1\}^n : w \leq g(a^T x), \sum_{j \in N} x_j \leq d\}$. In [14], Yu and Ahmed provide approximation schemes by solving the continuous relaxation of the lifting problem for P_C. In fact, by Theorem 4, we show that the lifting is naturally SI and the approximations are no longer needed.

Proposition 4. *For any $S \subseteq N$ such that $|S| \leq d$, the inequality $w \leq f(S) - \sum_{j \in S} \rho_j(S \setminus j)(1 - x_j) + \sum_{j \in \bar{S}} \gamma_0(a_j) x_j$ defines a facet of $conv(P_C)$.*

Also, on the mixed 0-1 knapsack set with a partition matroid denoted by K_{MC}, which is a rather complex set that has not been investigated in the literature, we can easily derive a set of facet-defining inequalities by Theorem 4.

Proposition 5. *Consider any $S \subseteq N$ such that $|S_i| \leq d_i$ and $\lambda = \sum_{j \in S} a_j - b > 0$. Supposing $S = \{1, \ldots, s\}$ with $a_1 \geq \cdots \geq a_r > \lambda \geq \cdots \geq a_s$, the inequality $-\pi \leq -\lambda + \sum_{j=1}^r \lambda(1 - x_j) + \sum_{j=r+1}^s a_j(1 - x_j) + \sum_{j \in \bar{S}} \gamma_K \binom{a_j}{e_{\sigma(j)}} x_j$ defines a facet of $conv(K_{MC})$, where $\gamma_K \binom{z}{e_i}$ is the corresponding lifting function.*

4.2 Computing the Exact Lifting Function

To compute $\gamma \binom{z}{u}$, we introduce a simpler problem that replaces the cardinality constraints by fixing some variables in a given set. For some $T \subseteq S$, define

$$\gamma(z, T) = \max \left\{ \begin{array}{l} w + \sum_{j \in S} \rho_j(S \setminus j)(1 - x_j) - f(S) : x_j = 0 \, \forall j \in T, \\ \\ w \leq g(\sum_{j \in S} a_j x_j + z), x_j \in \{0, 1\} \, \forall j \in S \end{array} \right\}.$$

Note that if $T = \emptyset$, then $\gamma(z, T) = \gamma_0(z)$. For any $\Gamma \subseteq \bar{S}$, we can conclude that

$$\gamma \begin{pmatrix} z \\ \sum_{j \in \Gamma} e_{\sigma(j)} \end{pmatrix} = \max_{T \subseteq S} \{\gamma(z, T) : |T_i| = \max\{0, |S_i| + |\Gamma_i| - d_i\} \, \forall i \in [r]\}. \tag{10}$$

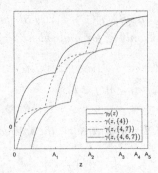

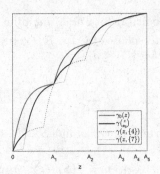

Fig. 2. The function $\gamma(z, T)$.   **Fig. 3.** The lifting function $\gamma\binom{z}{\mathbf{u}}$.

Then our question now is reduced to computing $\gamma(z, T)$. The explicit formula of $\gamma(z, T)$ is given in Lemma 3, see Appendix B. Note that $\gamma(z, T)$ (depicted in Fig. 2) is not subadditive on $z \in \mathbb{R}_+$ in general for a fixed $T \neq \emptyset$.

Proposition 6. *For any $i \in [r]$, if $|S_i| < d_i$, then $\gamma\binom{z}{\mathbf{e}_i} = \gamma_0(z)$. If $|S_i| = d_i$, and assume $S_i = \{i_1, \ldots, i_\ell\}$ such that $a_{i_1} \geq \cdots \geq a_{i_\ell}$, then $\gamma\binom{z}{\mathbf{e}_i} = \max\{\gamma(z, \{i_1\}), \gamma(z, \{i_\ell\})\}$.*

Example 3. Consider $P_{MC} = \{(w, x) \in \mathbb{R} \times \{0, 1\}^n : w \leq -\exp(-a^T x), x_3 + x_5 + x_6 \leq 2, x_1 + x_4 + x_7 \leq 2\}$, where $n = 7$ and $a = (1, 0.8, 0.7, 0.6, 0.5, 0.3, 0.2)^T$. Let $S = \{2, 3, 4, 6, 7\}$. If the cardinality constraints are ignored, then the lifted inequality (6) for $\mathrm{conv}(P_0)$ is calculated as:

$$w \leq -0.3441 + 0.1181x_1 + 0.091x_2 + 0.0753x_3 + 0.0611x_4$$
$$+ 0.065x_5 + 0.026x_6 + 0.0164x_7.$$

If multidimensional lifting is performed with respect to $\mathrm{conv}(P_{MC})$, then coefficients of x_1 and x_5 can be strengthened to **0.1156** and **0.0589**, respectively.

Remark. From Fig. 3, for a fixed $\mathbf{u}$, it can be seen that $\gamma\binom{z}{\mathbf{u}}$ is not a subadditive function on $z \in \mathbb{R}_+$. Furthermore, lifting function $\gamma\binom{z}{\mathbf{u}}$ is not subadditive on $\binom{z}{\mathbf{u}} \in \mathbb{R}_+ \times \mathbb{Z}_+^r$. As an example, $\gamma\binom{2}{\mathbf{e}_1} + \gamma\binom{0.2}{\mathbf{e}_1 + \mathbf{e}_2} = 0.2196 + 0.0164 < 0.2369 = \gamma\binom{2.2}{2\mathbf{e}_1 + \mathbf{e}_2}$. Therefore, the subadditivity in its standard form, which is a sufficient condition to ensure SI lifting, is not applicable to prove Theorem 4.

5   Computational Experiments

We perform a preliminary study for evaluating the lifted inequalities using the expected utility maximization problem considered in [1, 14]. Given a set of investment options N, let $v_i \in \mathbb{R}^n$ be the value of investments in the future at scenario i with probability $\pi_i, i = 1, \ldots, m$. We use the exponential function $1 - \exp(z/\lambda)$

as the utility function with risk tolerance λ. Then the expected utility maximization problem is formulated as

$$\max\{\sum_{i=1}^{m} \pi_i w_i : w_i \leq \exp(-\frac{v_i^T x}{\lambda}) \ \forall i \in [m], \sum_{j \in N} x_j \leq d, x \in \{0,1\}^n\},$$

where d is the cardinality budget. We generate the values of v_i following the settings in [1,14], with $\pi_i = \frac{1}{m}$ for $i \in [m]$ and $d = 15$. Then we use the exactly same algorithm to generate the initial submodular inequalities as in [14].

Table 1. The average performance of lifted inequalities.

n	m	λ	Lifted ineqs. in [1]			Ineqs. (6) & (7)			n	m	λ	Lifted ineqs. in [1]				Ineqs. (6) & (7)		
			Cuts	Nodes	Time	Cuts	Nodes	Time				Cuts	Nodes	Time	Egap	Cuts	Nodes	Time
50	50	0.8	400	0	0.96	400	0	0.54	150	50	0.8	2970	72	32		400	0	1.29
		1	340	0	0.82	340	0	0.45			1	2740	48	27		400	0	1.29
		2	200	0	0.51	200	0	0.27			2	745	7	9		200	0	0.66
	100	0.8	800	0	1.83	800	0	1.02		100	0.8	2920	27	28		800	0	2.57
		1	660	0	1.6	660	0	0.85			1	3340	21	32		800	0	2.6
		2	440	0	1.31	400	0	0.54			2	1400	6	17		400	0	1.31
100	50	0.8	400	0	1.82	400	0	0.9	200	50	0.8	12430	477	225		400	0	1.68
		1	620	3	4	400	0	0.91			1	6930	163	99		400	0	1.69
		2	410	3	3	200	0	0.45			2	1780	22	24		200	0	0.84
	100	0.8	1950	8	11	800	0	1.77		100	0.8	14310	314	379	0.08[1]	800	0	3.42
		1	2360	11	14	800	0	1.82			1	13150	250	284		800	0	3.47
		2	980	4	8	400	0	0.89			2	1840	11	28		400	0	1.72

Our experiments are conducted using Gurobi 8.1.1 on a Windows 10 PC with a 3.2 GHz CPU and 8 GB of RAM. We set the time limit to 1800 s. For each values of n, m and λ, the average performance over 10 randomly generated instances is reported, including the average solution time in seconds (time), the number of added cuts (cuts) and branch-and-cut explored nodes (nodes). For the instances that are not solved within the time limit, we show the percentage gap between the best known upper and the lower bounds at termination (egap); its superscript reports the number of unsolved instances.

The computational results that compare the approximate lifted inequalities in [1] and our exact lifted inequalities (6) and (7) are presented in Table 1. Note that the lifted inequalities developed in [14] are the same as those in [1] under the cardinality constraint $\mathcal{X}$. Observe that the parameter λ is an important factor to affect the overall performance, which is consistent with the computations in [1]. We highlight that all of the considered instances are solved to optimality at the root node with negligible computational time, when the exact lifted inequalities are applied. On the contrary, it often incurs many branch-and-bound operations or fails to generate optimal solutions due to time limit, when the approximate lifted inequalities are applied. Such a comparison clearly indicates that the exact lifted inequalities are fundamental and effective, and they can drastically improve the computational performance when solving submodular maximization problems.

A   Proof of Theorem 1

Lemma 1. *Let $z \in [A_k, A_{k+1}]$ for some $k \geq 1$. Then for any $\Delta \geq 0$ and $z + \Delta \leq A_{k+1}$, we have*

$$\phi(z + \Delta) - \phi(z) \leq \phi\left(z - \sum_{j=\ell}^{k+1} a_j + \Delta\right) - \phi\left(z - \sum_{j=\ell}^{k+1} a_j\right) \quad \forall \ell = 2, \ldots, k+1.$$

Proof. It suffices to show the statement holds for $\ell = k+1$. Since $z \in [A_k, A_{k+1}]$ and $z + \Delta \leq A_{k+1}$, then $z - a_{k+1} \in [A_{k-1}, A_k]$ and $z + \Delta - a_{k+1} \leq A_k$. Therefore,

$$\begin{aligned}
\phi(z + \Delta) - \phi(z) &= g(\Omega + v_k - a_k + \Delta) - g(\Omega + v_k - a_k) \\
&\leq g(\Omega + v_{k-1} - a_{k+1} + \Delta) - g(\Omega + v_{k-1} - a_{k+1}) \\
&= \phi(z - a_{k+1} + \Delta) - \phi(z - a_{k+1}),
\end{aligned}$$

where $\Omega = z - A_{k-1}$, the inequality follows from $v_{k-1} + a_k \leq v_k + a_{k+1}$ and the concavity of g (i.e., $g(z_0 + \Delta) - g(z_0) \geq g(z_1 + \Delta) - g(z_1)$ if $z_0 \leq z_1$ and $\Delta \geq 0$). □

Lemma 2. *Let $\Delta \in [0, a_{k+1}]$ for some $k \geq 0$. Then for any $z \geq A_k$, we have*

$$\phi(A_k + \Delta) - \phi(A_k) \geq \phi(z + \Delta) - \phi(z).$$

Proof. Suppose $z \in [A_{k_1}, A_{k_1+1}]$ and $z + \Delta \in [A_{k_2}, A_{k_2+1}]$, where $k \leq k_1 \leq k_2$. We establish the result for each $k_2 - k_1$ by induction.

- If $k_2 = k_1$, then based on Lemma 1, we have

$$\begin{aligned}
\phi(z + \Delta) - \phi(z) &\leq \phi\left(z - \sum_{j=k+2}^{k_1+1} a_j + \Delta\right) - \phi\left(z - \sum_{j=k+2}^{k_1+1} a_j\right) \\
&= g(\Omega + v_k + \Delta) - g(\Omega + v_k) \\
&\leq g(v_k + \Delta) - g(v_k) = \phi(A_k + \Delta) - \phi(A_k),
\end{aligned}$$

where $\Omega = z - A_k - \sum_{j=k+2}^{k_1+1} a_j$, the second inequality follows from $a_{k+1} \geq a_{k_1+1}$ and $\Omega = z - A_k - \sum_{j=k+2}^{k_1+1} a_j \geq z - A_k - \sum_{j=k+1}^{k_1} a_j \geq 0$.
- If the statement holds for $k_2 - k_1 = m$, then we show that the statement also holds for $k_2 - k_1 = m + 1$. Let $\Delta' = A_{k_1+1} - z$, then $\Delta \geq \Delta'$. Since $\Delta \in [0, a_{k+1}]$ and $A_{k_1+1} - \sum_{j=k+2}^{k_1+1} a_j = A_{k+1} \geq A_k + \Delta$, we have

$$\begin{aligned}
\phi(A_k + \Delta) - \phi(A_k + \Delta - \Delta') &\geq \phi(A_{k+1}) - \phi(A_{k+1} - \Delta') \\
&= \phi\left(A_{k_1+1} - \sum_{j=k+2}^{k_1+1} a_j\right) - \phi\left(z - \sum_{j=k+2}^{k_1+1} a_j\right) \\
&\geq \phi(A_{k_1+1}) - \phi(z),
\end{aligned}$$

where the first inequality follows from that ϕ is concave on $[A_k, A_{k+1}]$.

Let $z' = A_{k_1+1}$, then $z' \in [A_{k_1+1}, A_{k_1+2}]$. Since $k_2 - (k_1 + 1) = m$, by the induction hypothesis, we have

$$\phi(A_k + \Delta - \Delta') - \phi(A_k) \geq \phi(z' + \Delta - \Delta') - \phi(z') = \phi(z + \Delta) - \phi(A_{k_1+1}).$$

Summing the above two inequalities, we obtain the desired inequality.   □

Proof (Theorem 1). To show that ϕ is subadditive on $[0, +\infty)$, it is sufficient to prove that the inequality $\phi(z) - \phi(0) \geq \phi(z + \Delta) - \phi(\Delta)$ holds for any $z, \Delta \geq 0$.
    First, by Lemma 2, we have that for any $\Delta \geq 0$,

$$\phi(A_{j+1}) - \phi(A_j) \geq \phi(A_{j+1} + \Delta) - \phi(A_j + \Delta) \quad \forall j = 0, 1, \ldots. \qquad (11)$$

Suppose $z \in [A_k, A_{k+1}]$ for some k and let $\Delta' = z - A_k \in [0, a_{k+1}]$. By Lemma 2 we have

$$\phi(A_k + \Delta') - \phi(A_k) \geq \phi(z + \Delta) - \phi(z + \Delta - \Delta'). \qquad (12)$$

Note that $A_k + \Delta' = z$ and $z + \Delta - \Delta' = A_k + \Delta$. Summing equations (11) over all $j = 0, 1, \ldots, k - 1$ and (12), yields the desired result.   □

B   Proof Sketch of Theorem 4

We first give the explicit formula to compute $\gamma(z, T)$.

Lemma 3. *Suppose $T = \{\ell_1, \ldots, \ell_{|T|}\} \subseteq S$ such that $\ell_1 < \cdots < \ell_{|T|}$. Denote $A_k = \sum_{j=1}^{k} a_j$ for $k \in S$ and $A_0 = 0$. Denote $A_t^T = a(T) - \sum_{j=1}^{t} a_{\ell_j}$ for $t = 1, \ldots, |T|$, and $A_0^T = a(T)$, $\ell_0 = 0$. There have three cases to consider: (i) if $0 \leq z \leq a(T)$, then $\gamma(z, T) = g(a(S) - a(T) + z) + \sum_{j \in T} \rho_j(S \setminus j) - f(S)$. (ii) if $A_k + A_t^T \leq z \leq A_{k+1} + A_t^T$ for $k = \ell_t, \ldots, \ell_{t+1} - 2$ and $t = 0, 1, \ldots, |T| - 1$, then $\gamma(z, T) = g(a(S) - A_{k+1} - A_t^T + z) + \sum_{j \in [k] \cup \{\ell_j\}_{j=t+1}^{|T|}} \rho_j(S \setminus j) - g(a(S) - a_{k+1})$. (iii) if $z \geq A_{\ell_{|T|}}$, then $\gamma(z, T) = \gamma_0(z)$.*

To establish the proof of Theorem 4, our basic idea is to exploit (8) in Proposition 3. To show (8), we first establish that inequalities similar in spirit to (8) hold for $\gamma(z, T)$ in Lemmas 4 and 5. Then we use the fact that $\gamma\binom{z}{u}$ can be computed through (10) and $\gamma(z, T)$ to complete the proof of Theorem 4.

Lemma 4. *For any $T \subseteq S$, we have $\sum_{j \in T} \gamma(z_j, \{j\}) \geq \gamma(z(T), T), \forall z_j \geq 0$.*

Lemma 5. *For any $i \in [r]$, we have $\gamma_0(z_0) + \gamma\binom{z_1}{e_i} \geq \gamma\binom{z_0 + z_1}{e_i}, \forall z_0, z_1 \geq 0$.*

Proof (Theorem 4). Let $\Gamma \subseteq \bar{S}$ and $|\Gamma_i| \leq d_i$, recall equation (10), we have

$$\sum_{j \in \Gamma} \gamma\binom{z_j}{e_{\sigma(j)}} = \sum_{j \in \Gamma} \max_{T^j} \{\gamma(z_j, T^j) : |T^j| = \max\{0, |S_i| + 1 - d_i\}, T^j \subseteq S_{\sigma(j)}\}$$

$$= \max_{\{T^j\}_{j \in \Gamma}} \{\sum_{j \in \Gamma} \gamma(z_j, T^j) : |T^j| = \max\{0, |S_i| + 1 - d_i\}, T^j \subseteq S_{\sigma(j)}\}$$

$$\geq \max_{T \subseteq S} \{\gamma(z(\Gamma), T) : |T_i| = \max\{0, |S_i| + |\Gamma_i| - d_i\}\} = \gamma\binom{z(\Gamma)}{\sum_{j \in \Gamma} e_{\sigma(j)}},$$

where the inequality is based on Lemmas 4 and 5.   □

References

1. Ahmed, S., Atamtürk, A.: Maximizing a class of submodular utility functions. Math. Program. **128**(1–2), 149–169 (2011). https://doi.org/10.1007/s10107-009-0298-1
2. Angulo, A., Espinoza, D., Palma, R.: Sequence independent lifting for mixed Knapsack problems with GUB constraints. Math. Program. **154**(1–2), 55–80 (2015). https://doi.org/10.1007/s10107-015-0902-5
3. Atamtürk, A., Küçükyavuz, S., Tezel, B.: Path cover and path pack inequalities for the capacitated fixed-charge network flow problem. SIAM J. Optimiz. **27**(3), 1943–1976 (2017)
4. Chen, L., Xu, J., Lu, Z.: Contextual combinatorial multi-armed bandits with volatile arms and submodular reward. In: Advances in Neural Information Processing Systems, pp. 3247–3256 (2018)
5. Dolhansky, B.W., Bilmes, J.A.: Deep submodular functions: definitions and learning. In: Advances in Neural Information Processing Systems, pp. 3404–3412 (2016)
6. El-Arini, K., Veda, G., Shahaf, D., Guestrin, C.: Turning down the noise in the blogosphere. In: Proceedings of the 15th ACM SIGKDD International Conference on Knowledge Discovery and Data Mining, pp. 289–298. ACM (2009)
7. Gu, Z., Nemhauser, G.L., Savelsbergh, M.W.: Sequence independent lifting in mixed integer programming. J. Comb. Optimiz. **4**(1), 109–129 (2000). https://doi.org/10.1023/A:1009841107478
8. Li, J., Deshpande, A.: Maximizing expected utility for stochastic combinatorial optimization problems. In: 2011 IEEE 52nd Annual Symposium on Foundations of Computer Science, pp. 797–806. IEEE (2011)
9. Lin, H., Bilmes, J.: A class of submodular functions for document summarization. In: Proceedings of the 49th Annual Meeting of the Association for Computational Linguistics: Human Language Technologies-Volume 1, pp. 510–520. Association for Computational Linguistics (2011)
10. Marchand, H., Wolsey, L.A.: The 0-1 Knapsack problem with a single continuous variable. Math. Program. **85**(1), 15–33 (1999). https://doi.org/10.1007/s101070050044
11. Nemhauser, G.L., Wolsey, L.A.: Integer and Combinatorial Optimization. Wiley, New York (1988)
12. Stobbe, P., Krause, A.: Efficient minimization of decomposable submodular functions. In: Advances in Neural Information Processing Systems, pp. 2208–2216 (2010)
13. Wolsey, L.A.: Submodularity and valid inequalities in capacitated fixed charge networks. Oper. Res. Lett. **8**(3), 119–124 (1989)
14. Yu, J., Ahmed, S.: Maximizing a class of submodular utility functions with constraints. Math. Program. **162**(1–2), 145–164 (2017). https://doi.org/10.1007/s10107-016-1033-3
15. Zeng, B., Richard, J.-P.P.: A framework to derive multidimensional superadditive lifting functions and its applications. In: Fischetti, M., Williamson, D.P. (eds.) IPCO 2007. LNCS, vol. 4513, pp. 210–224. Springer, Heidelberg (2007). https://doi.org/10.1007/978-3-540-72792-7_17
16. Zeng, B., Richard, J.P.P.: A polyhedral study on 0–1 knapsack problems with disjoint cardinality constraints: strong valid inequalities by sequence-independent lifting. Discrete Optimiz. **8**(2), 259–276 (2011)

A Fast $(2 + 2/7)$-Approximation Algorithm for Capacitated Cycle Covering

Vera Traub[1]([✉]) and Thorben Tröbst[2]

[1] Research Institute for Discrete Mathematics and Hausdorff Center for Mathematics,
University of Bonn, Bonn, Germany
`traub@dm.uni-bonn.de`
[2] Department of Computer Science, University of California, Irvine, Irvine, CA, USA
`t.troebst@uci.edu`

Abstract. We consider the *capacitated cycle covering problem*: given an undirected, complete graph G with metric edge lengths and demands on the vertices, we want to cover the vertices with vertex-disjoint cycles, each serving a demand of at most one. The objective is to minimize a linear combination of the total length and the number of cycles. This problem is closely related to the capacitated vehicle routing problem (CVRP) and other cycle cover problems such as min-max cycle cover and bounded cycle cover. We show that a greedy algorithm followed by a post-processing step yields a $(2 + 2/7)$-approximation for this problem by comparing the solution to a polymatroid relaxation. We also show that the analysis of our algorithm is tight and provide a $2 + \epsilon$ lower bound for the relaxation.

Keywords: Cycle cover · Vehicle routing · Greedy algorithms ·
Approximation algorithms · Polymatroids

1 Introduction

Our work is motivated by the classical and well-studied capacitated vehicle routing problem (CVRP) which was introduced by Dantzig and Ramser [7]. In this problem we are given an undirected, complete graph $G = (V, E)$ with metric edge lengths $\ell : E \to \mathbb{R}_{\geq 0}$ and a distinguished vertex $s \in V$ which is called the *depot*. Moreover, every vertex is assigned a demand $b(v)$. The goal is to cover V with cycles $C_1, \ldots, C_k$ such that each cycle visits s, satisfies $b(C_i) \leq 1$ and the total length $\sum_{i=1}^{k} \ell(C_i)$ is minimum. Here $b(C_i) := \sum_{v \in V(C_i)} b(v)$ is the total demand of the vertices of C_i and $\ell(C_i) := \sum_{e \in E(C_i)} \ell(e)$ is the total length of the edges of C_i.

The CVRP has received a large amount of attention in the last 60 years. While there has been much progress regarding computational results (see e.g. [16, 17,19]), from the viewpoint of approximation algorithms little progress has been made. The simple optimal tour partitioning algorithm by Altinkemer and Gavish [1], which achieves an approximation ratio of 3.5, has not been improved in the

© Springer Nature Switzerland AG 2020
D. Bienstock and G. Zambelli (Eds.): IPCO 2020, LNCS 12125, pp. 391–404, 2020.
https://doi.org/10.1007/978-3-030-45771-6_30

past 30 years. (In fact, the approximation ratio is $2+\alpha$ where α is the best known approximation ratio for TSP.) For the case where all vertices have demand $1/Q$ for some $Q \in \mathbb{N}$, the tour partitioning algorithm by Haimovich and Kan [12] from 1985 has approximation ratio $1 + \alpha$, which is currently 2.5, and this result is also still the best known.

Significant improvements have only been achieved in special cases, such as when the metric is Euclidean [8,13] or arises from graphs with special structure [2–4,14]. The only result for the general case is by Bompadre et al. [5] who improved the approximation guarantee by $\Theta(1/Q^3)$ where Q is the least common denominator of the (rational) demands b.

In this paper we study a variant of the CVRP, where we do not have a depot vertex that must be visited by every tour, but instead have a fixed opening cost $\gamma > 0$ per tour. Formally, this problem, which we call the *capacitated cycle covering problem* (CCCP), is defined as follows. We are given an undirected, complete graph $G = (V, E)$ with metric edge lengths $\ell : E \to \mathbb{R}_{\geq 0}$, vertex demands $b : V \to [0, 1]$, and an opening cost $\gamma \in \mathbb{R}_{\geq 0}$. The goal is to compute a *capacitated cycle cover*, i.e. cycles $C_1, \ldots, C_k$ in G, such that every $v \in V$ is contained in exactly one cycle and $b(C_i) \leq 1$ for all i, minimizing the total cost $\sum_{i=1}^{k} \ell(C_k) + \gamma k$. Here it is allowed that a cycle contains only one or two vertices.

To the best of our knowledge, this precise problem formulation has not appeared in the literature. However, besides the capacitated vehicle routing problem, the CCCP is also closely related to other cycle covering problems. This includes min-max cycle cover and bounded cycle cover which were first studied by Even et al. [10]. In the former problem we are asked to compute a cycle cover $C_1, \ldots, C_k$ which minimizes $\max_{i=1}^{k} \ell(C_i)$ where k is part of the input. In the latter we wish to find a cycle cover $C_1, \ldots, C_k$ with $\ell(C_i) \leq 1$ for all i with minimum k. Recently, Yu et al. [20,21] provided new approximation algorithms for the these problems with approximation ratios of 5 and $4 + 4/7$ respectively. Their algorithms need $O(n^5)$ time, where here and in the following $n := |V|$.

Even more recently, Das et al. [9] studied the min-max variant of the capacitated cycle covering problem. In this problem we wish to find a capacitated cycle cover $C_1, \ldots, C_k$ where k is part of the input such that $\max_{i=1}^{k} \ell(C_i)$ is minimized. They provide a 196-approximation algorithm for min-max capacitated tree cover which implies a 392-approximation algorithm for the cycle cover variant.

1.1 Our Results and Techniques

Note that the capacitated cycle covering problem includes both the TSP (for $b \equiv 0$ and suitably large γ) and bin packing (for $\ell \equiv 0$) and is thus NP-hard to approximate within a factor of $3/2 - \epsilon$. Hence, we are primarily interested in approximation algorithms and relaxations for the problem. Our main result is the following theorem.

Theorem 1. *Given an instance of the capacitated cycle covering problem, we can compute a $(2 + 2/7)$-approximate solution in $O(n^2)$ time.*

We remark that if the pairwise distances between all vertices are given explicitly, the input has size n^2 and hence the runtime is linear.

The first step of our algorithm is to compute a carefully chosen spanning forest in our input graph. Having such a forest, we turn it into a capacitated cycle cover as follows. We first ensure that every connected component of the forest contains vertices of total demand at most 1. This is done by splitting large components into smaller ones if necessary. Then from every connected component of the forest we can compute a cycle of at most twice the length of the forest component. See Sect. 2.

The most important part of our algorithm is to choose the initial spanning forest. We do not solve a tree covering problem as a black box but anticipate that we will have to double edges and split up large components. To compute our spanning forest we use a linear programming relaxation, which we call the tree cover LP. This LP is closely related to a natural LP relaxation for the capacitated vehicle routing problem. Moreover, the tree cover LP has the important property that the set of feasible solutions is a polymatroid. This allows us to solve the LP very efficiently using the polymatroid greedy algorithm. See Sect. 3.

We then analyze a simple randomized rounding algorithm that rounds a fractional LP solution to a spanning forest. For this we exploit that the extreme point solutions of our LP relaxation are highly structured. As a result, we obtain a randomized $(2 + 2/7)$-approximation algorithm for the CCCP and also show that the ratio between our solution for CCCP and the value of the tree cover LP is at most $2 + 2/7$. See Sect. 4.

Then we show that we can derandomize our algorithm and obtain a simple and deterministic greedy algorithm for computing our spanning forest (Sect. 5). This will complete the proof of Theorem 1.

Finally, we provide two forms of lower bounds for our analysis: we prove that the analysis of our deterministic algorithm is tight and we show a $2 + \epsilon$ lower bound on the gap between the tree cover LP and the capacitated cycle covering problem (Sect. 6).

2   Tree Splitting

In the following we will call a set U of vertices *large* if $b(U) := \sum_{u \in U} b(u) > 1$ and *small* otherwise. A common and useful technique for dealing with capacities in facility location and vehicle routing problems is to cluster vertices into clusters with demands between $1/2$ and 1 (see e.g. [10,14,15,20]). By making sure that the demand in each cluster is at least $1/2$, we can guarantee that we have at most twice as many clusters as necessary. This idea can be used to prove the following lemma.

Lemma 1 (Tree Splitting). *Let $T = (V, E)$ be a tree and $b : V \rightarrow [0, 1]$ some vertex demands with $b(V) > 1$, i.e. V is large. Then we can partition V into $k \leq 2b(V)$ many small sets $R_1, \ldots, R_k$ and find edge-disjoint connected subgraphs $T_1, \ldots, T_k$ of T such that $R_i \subseteq V(T_i)$, i.e. T_i is a Steiner tree with terminal set R_i, for all i. Moreover, this can be done in linear time.*

We defer the proof to Appendix A. As a corollary, we get a simple construction which turns any forest F in G into a solution to the capacitated cycle covering problem. For an edge set F, we denote by $\mathcal{C}(F)$ the collection of vertex sets of the connected components of (V, F).

Lemma 2. *Let (V, F) be a forest. Then we can compute in linear time a feasible solution $C_1, \ldots, C_k$ to the CCCP with cost bounded by*

$$2\ell(F) + \gamma \cdot \sum_{A \in \mathcal{C}(F)} u(A) \tag{1}$$

where $\ell(F) := \sum_{e \in F} \ell(e)$ and $u : 2^V \to \mathbb{R}_{\geq 0}$ is given by

$$u(A) := \begin{cases} 1 & \text{if } A \text{ is small,} \\ 2b(A) & \text{if } A \text{ is large.} \end{cases} \tag{2}$$

Proof. We first apply Lemma 1 to all large connected components of F. Together with the remaining small connected components, this yields a partition of V into k small sets $R_1, \ldots, R_k$ and Steiner trees $T_1, \ldots, T_k$ with terminal sets $R_1, \ldots, R_k$ respectively, where $k \leq \sum_{A \in \mathcal{C}(F)} u(A)$. Then we turn each Steiner tree T_i with terminal set R_i into a cycle C_i with vertex set R_i and $\ell(C_i) \leq 2\ell(T_i)$. This is accomplished by the standard technique of ordering the elements of R_i as they appear in a depth-first search of T_i. Equivalently, one can double all edges of T_i, find an Eulerian walk, and shortcut this walk to a cycle on R_i. Shortcutting does not increase the length since ℓ is metric.   □

Thus in the following sections we will discuss how to find a forest F such that (1) is at most $(2 + 2/7)$ times the cost of an optimum capacitated cycle cover.

3   The Tree Cover LP

To obtain a lower bound on the cost of an optimum solution to the CCCP, we use the following linear program.

$$\begin{aligned} \min \quad & \ell(x) + \gamma(|V| - x(E)) \\ \text{s.t.} \quad & x(E[A]) \leq |A| - \max\{1, b(A)\} \quad \forall \emptyset \neq A \subseteq V \\ & x \geq 0, \end{aligned} \tag{3}$$

where $\ell(x) := \sum_{e \in E} x_e \ell(e)$, $x(E) := \sum_{e \in E} x_e$, and $E[A]$ denotes the set of edges in E that have both endpoints in A.

Note that LP (3) is rather a relaxation of a tree covering problem than of capacitated cycle covering: integral solutions are edge sets of forests in which every connected component contains vertices of total demand at most 1. Nonetheless, it provides a lower bound for the cost of an optimum CCCP solution because every feasible solution to the CCCP contains such a forest. Hence we get the following.

Lemma 3. *Let (G, ℓ, b, γ) be an instance of the CCCP. Then the optimum value of the LP (3) is a lower bound on the cost of an optimum solution of the CCCP.*

We would like to remark that the tree cover LP (3) is related to the following natural LP relaxation for the CVRP:

$$
\begin{aligned}
\min \quad & \ell(x) \\
\text{s.t.} \quad & x(\delta(A)) \geq 2 && \forall \emptyset \neq A \subseteq V \setminus \{s\} \\
& x(\delta(A)) \geq 2b(A) && \forall \emptyset \neq A \subseteq V \setminus \{s\} && (4) \\
& x(\delta(v)) = 2 && \forall v \in V \setminus \{s\} \\
& x \geq 0,
\end{aligned}
$$

where $\delta(A)$ denotes the set of edges with exactly one endpoint in A and $\delta(v) := \delta(\{v\})$. (To see that (4) is a relaxation of the CVRP, note that we need at least $b(A)$ cycles for covering the vertices in A and each cycle contains the depot s.)

The optimal tour partitioning algorithm for the CVRP [1] computes a solution of cost at most 3.5 times the value of (4) (see e.g. [18]). In particular, the integrality gap of (4) is at most 3.5. The following LP is equivalent to (4) in the sense that every feasible solution to one of the LPs is also a feasible solution for the other.

$$
\begin{aligned}
\min \quad & \ell(x) \\
\text{s.t.} \quad & x(E[A]) \leq |A| - \max\{1, b(A)\} && \forall \emptyset \neq A \subseteq V \setminus \{s\} \\
& x(\delta(v)) = 2 && \forall v \in V \setminus \{s\} \\
& x \geq 0.
\end{aligned}
$$

Therefore, for every feasible solution x to (4), the restriction of x to $G - s$ is a feasible solution to the tree cover LP (3).

In the remaining part of this section we explain how one can solve the tree cover LP (3) by a greedy algorithm. The key insight for proving this is that (3) is equivalent to optimizing over a polymatroid.

Lemma 4. *Let P be the set of feasible solutions to the LP (3). Then*

$$
P = \left\{ x \in \mathbb{R}^E \;\middle|\; \begin{array}{ll} x(F) \leq r(F) & \forall F \subseteq E, \\ x \geq 0 \end{array} \right\}
$$

where $r(F) := \sum_{A \in \mathcal{C}(F)} (|A| - \max\{1, b(A)\})$. Moreover, r is monotone, submodular, and satisfies $r(\emptyset) = 0$. Thus P is a polymatroid.

We sketch the proof of Lemma 4 in Appendix B. Algorithm 1 formally describes the polymatroid greedy algorithm for solving (3).

Note that $\mathcal{C}$ remains a partition of the vertex set. At the end of iteration i it contains the vertex sets of the connected components of $(V, \{e_1, \ldots, e_i\})$. Moreover, the support $\{e \in E : x_e > 0\}$ of the returned LP solution x is the edge set of a forest (by the condition in line 5). This structure will be useful in the next section, where we analyze an algorithm for rounding x to an integral vector.

Algorithm 1: Polymatroid greedy algorithm for the tree cover LP

Input: An instance (G, ℓ, b, γ) of the CCCP with $G = (V, E)$.
Output: An optimum solution x to the linear program (3).

1 Let $x_e := 0$ for all $e \in E$.
2 Let $e_1, \ldots, e_m \in E$ be the edges with $\ell(e_i) < \gamma$ sorted such that
  $\ell(e_1) \leq \cdots \leq \ell(e_m)$.
3 Let $\mathcal{C} := \{\{v\} \mid v \in V\}$.
4 **for** $i := 1, \ldots, m$ **do**
5   **if** e_i connects two distinct $C, C' \in \mathcal{C}$ **then**
6        $\mathcal{C} := \mathcal{C} \setminus \{C, C'\} \cup \{C \cup C'\}$

7        $x_{e_i} := \begin{cases} 1 & \text{if } C \cup C' \text{ is small} \\ 1 - b(C) & \text{if } C \text{ is small and } C' \text{ is large} \\ 1 - b(C') & \text{if } C' \text{ is small and } C \text{ is large} \\ 2 - b(C) - b(C') & \text{if } C, C' \text{ are small, but } C \cup C' \text{ is large} \\ 0 & \text{else, i.e. } C, C' \text{ large.} \end{cases}$

8 **return** x.

Lemma 5. *Algorithm 1 computes an optimum solution of the LP* (3).

Proof (Sketch). By Lemma 4 we know that LP (3) optimizes over a polymatroid. Thus the polymatroid greedy algorithm which sets $x_{e_i} := r(\{e_1, \ldots, e_i\}) - r(\{e_1, \ldots, e_{i-1}\})$ for every $i \leq m$ produces an optimal solution. One can verify that Algorithm 1 sets x to exactly those values. $\qquad\square$

4   Randomized Rounding

We will now show how we can round the fractional solution x generated by Algorithm 1 to a forest F while bounding the cost (1) of the resulting CCCP solution. More precisely, we will prove the following theorem.

Theorem 2 (Randomized rounding). *Let x be a solution of the tree-covering LP* (3) *computed by Algorithm 1. Define a random edge set $F \subseteq E$ by independently picking each edge e with probability $\min\{1, (1 + 1/7)x_e\}$. Then*

$$\mathbb{E}\left[\sum_{A \in \mathcal{C}(F)} u(A)\right] \leq \left(2 + \frac{2}{7}\right)(|V| - x(E)),$$

where u is defined by (2), *and $\mathbb{E}[2\ell(F)] \leq (2 + 2/7)\,\ell(x)$.*

Note that this implies that the total cost (1) is at most $2 + 2/7$ times the objective value $\ell(x) + \gamma(|V| - x(E))$ of our optimum LP solution x. The scaling factor $1 + 1/7$ on the probabilities x_e is chosen to decrease the expected number of components of (V, F) (while increasing the expected length) such that we lose the same factor in both cost terms wrt. the LP. By Lemmas 2 and 3, Theorem 2 yields a randomized $(2+2/7)$-approximation algorithm for the CCCP.

In the rest of this section we prove Theorem 2. We may assume wlog. that $(V, \{e_1, \ldots, e_m\})$ is connected; otherwise we prove the statement for each connected component. Let E' be the set of edges e_i for which the condition in line 5 of Algorithm 1 was fulfilled. Every such edge $e_i = \{v, w\} \in E'$ connected two sets $C, C' \in \mathcal{C}$ in iteration i of Algorithm 1. Let $C^v_{e_i} \in \{C, C'\}$ be the set containing v and let $C^w_{e_i} \in \{C, C'\}$ be the other set (containing w). By construction of $\mathcal{C}$ in Algorithm 1, (V, E') is a spanning tree. Thus, F is always a forest. Moreover, the subgraphs of $(V, \{e_1, \ldots, e_{i-1}\} \cap E')$ induced by $C^v_{e_i}$ and $C^w_{e_i}$ are connected.

Lemma 6. *For every set $F \subseteq E'$, we have*

$$\sum_{A \in \mathcal{C}(F)} u(A) \leq 2 \cdot (|V| - x(E)) + \sum_{e \in E' \setminus F} \sum_{u \in e} \max\{1 - 2b(C^u_e), 0\}.$$

Proof. We first consider the case $\mathcal{C}(F) = \{V\}$ and hence $F = E'$. Then we have $x(E) \leq |V| - \max\{1, b(V)\}$ since x is a feasible solution to (3) and hence $u(V) \leq \max\{1, 2b(V)\} \leq 2(|V| - x(E))$. Now assume $\mathcal{C}(F) \neq \{V\}$ and compute

$$
\begin{aligned}
\sum_{A \in \mathcal{C}(F)} u(A) &= \sum_{\substack{A \in \mathcal{C}(F) \\ A \text{ large}}} 2b(A) + \sum_{\substack{A \in \mathcal{C}(F) \\ A \text{ small}}} 1 \\
&= 2b(V) + \sum_{\substack{A \in \mathcal{C}(F) \\ A \text{ small}}} (1 - 2b(A)) \\
&\leq 2b(V) + \sum_{A \in \mathcal{C}(F)} \max\{1 - 2b(A), 0\} \\
&\leq 2 \cdot (|V| - x(E)) + \sum_{A \in \mathcal{C}(F)} \max\{1 - 2b(A), 0\},
\end{aligned}
\tag{5}
$$

where we used in the last inequality that x is a feasible solution to (3).

Recall that $\mathcal{C}(F) \neq \{V\}$ and (V, E') is a spanning tree. Consider some $A \in \mathcal{C}(F)$ and let i minimum such that $e_i = \{v, w\} \in \delta(A) \cap E'$, where wlog. $v \in A$. So $v \in A \cap C^v_{e_i} \neq \emptyset$. Since the subgraphs of $(V, \{e_1, \ldots, e_{i-1}\} \cap E')$ induced by $C^v_{e_i}$ and $C^w_{e_i}$ are connected and i was chosen minimal, we have $C^v_{e_i} \subseteq A$. Hence, $\max\{1 - 2b(C^v_{e_i}), 0\} \geq \max\{1 - 2b(A), 0\}$. Note that $e_i \in E' \setminus F$ because $e_i \in \delta(A)$ and $A \in \mathcal{C}(F)$. Hence,

$$
\begin{aligned}
\sum_{A \in \mathcal{C}(F)} \max\{1 - 2b(A), 0\} &\leq \sum_{A \in \mathcal{C}(F)} \sum_{e \in E' \setminus F} \sum_{\substack{u \in e \\ C^u_e \subseteq A}} \max\{1 - 2b(A), 0\} \\
&\leq \sum_{e \in E' \setminus F} \sum_{u \in e} \max\{1 - 2b(C^u_e), 0\},
\end{aligned}
$$

because $\mathcal{C}(F)$ is a partition of V. Together with (5) this completes the proof. $\square$

Lemma 7. *Let x be a solution of the tree-covering LP* (3) *computed by Algorithm 1. Define a random edge set $F \subseteq E$ by independently picking each edge e with probability $\min\{1, (1 + 1/7)x_e\}$. Then*

$$\mathbb{E}\left[\sum_{e \in E' \setminus F} \sum_{u \in e} \max\{1 - 2b(C_e^u), 0\}\right] \leq 2/7 \cdot (|V| - x(E)).$$

Proof. We consider an edge $e \in E'$ and a vertex $u \in e$. If $x_e < 1$, by the definition of x_e in Algorithm 1 we have $x_e \geq 1 - b(C_e^u)$ and therefore

$$\begin{aligned}
&\mathbb{P}[e \notin F] \cdot \max\{1 - 2b(C_e^u), 0\} \\
&= \max\{1 - (1 + 1/7) \cdot x_e, \, 0\} \cdot \max\{1 - 2b(C_e^u), 0\} \\
&\leq \max\{1 - (1 + 1/7) \cdot (1 - b(C_e^u)), \, 0\} \cdot \max\{1 - 2b(C_e^u), 0\} \\
&\leq 2/7 \cdot b(C_e^u).
\end{aligned}$$

Hence,

$$\begin{aligned}
&\mathbb{E}\left[\sum_{e \in E' \setminus F} \sum_{u \in e} \max\{1 - 2b(C_e^u), 0\}\right] \\
&= \sum_{e \in E' : x_e < 1} \mathbb{P}[e \notin F] \cdot \sum_{u \in e} \max\{1 - 2b(C_e^u), 0\} \quad (6) \\
&\leq \sum_{e \in E' : x_e < 1} \sum_{u \in e : C_e^u \text{ small}} 2/7 \cdot b(C_e^u).
\end{aligned}$$

Let $1 \leq i < j \leq m$ with $e_i = \{u, v\}, e_j = \{u', v'\} \in E'$ with $x_{e_i}, x_{e_j} < 1$. We claim that if the vertex sets $C_{e_i}^u$ and $C_{e_j}^{u'}$ are both small, then they are disjoint. In iteration i of Algorithm 1, we merge C_e^u and C_e^v into a single component $C_e^u \cup C_e^v$. This new component must be large because $x_{e_i} < 1$. During the course of the algorithm we only merge components of the partition $\mathcal{C}$ of V. Therefore either $C_{e_i}^u$ and $C_{e_j}^{u'}$ are disjoint, or $C_e^u \cup C_e^v \subseteq C_{e_j}^{u'}$ which implies that $C_{e_j}^{u'}$ is large. Hence,

$$\sum_{e \in E' : x_e < 1} \sum_{u \in e : C_e^u \text{ small}} b(C_e^u) \leq b(V) \leq |V| - x(E),$$

where $b(V) \leq |V| - x(E)$ holds because x is a feasible solution to (3). Together with (6) this completes the proof. □

The bound $\mathbb{E}[2\ell(F)] \leq (2 + 2/7)\ell(x)$ follows directly from the linearity of expectation. Hence, Lemmas 6 and 7 imply Theorem 2.

5   A Fast and Deterministic Algorithm

In this section we show how one can derandomize our $(2 + 2/7)$-approximation algorithm. Algorithm 2 formally describes the computation of the forest (V, F). The partition $\mathcal{C}$ is updated exactly as in Algorithm 1. However, now we do not compute the value x_{e_i} but instead directly round it in a deterministic way (lines 7–10).

Algorithm 2: Solving the relaxation and rounding deterministically

---

Input: An instance (G, ℓ, b, γ) of the CCCP with $G = (V, E)$.
Output: A forest (V, F).

1   Let $F := \emptyset$.
2   Let $e_1, \ldots, e_m \in E$ be the edges with $\ell(e_i) < \gamma$ sorted such that
    $\ell(e_1) \leq \cdots \leq \ell(e_m)$.
3   Let $\mathcal{C} := \{\{v\} \mid v \in V\}$.
4   **for** $i := 1, \ldots, m$ **do**
5   **if** e_i connects two distinct $C, C' \in \mathcal{C}$ **then**
6         $\mathcal{C} := \mathcal{C} \setminus \{C, C'\} \cup \{C \cup C'\}$
7   **if** $C \cup C'$ is small **then**
8            $F := F \cup \{e_i\}$
9   **if** $\gamma \cdot (\max\{1 - 2b(C), 0\} + \max\{1 - 2b(C'), 0\}) > 2\ell(e_i)$ **then**
10           $F := F \cup \{e_i\}$

11   **return** (V, F).

Lemma 8. *Algorithm 2 computes a forest (V, F) with*

$$2\ell(F) + \gamma \cdot \sum_{A \in \mathcal{C}(F)} u(A) \leq \left(2 + \frac{2}{7}\right) \cdot \mathrm{LP}, \tag{7}$$

where LP *denotes the value of* (3).

We defer the proof to Appendix C. Note that the runtime of Algorithm 2 is dominated by sorting the edges E in line 2. In a preprocessing step, one can compute a minimum spanning tree wrt. to ℓ and remove all edges not contained in this tree. This yields a total runtime of $O(\theta + n \log n)$ where $\theta = O(n^2)$ is the time needed to compute an MST. Lemmas 8, 3, and 2 thus directly imply our main result Theorem 1.

6   Lower Bounds

In this section we show that the approximation ratio of Algorithm 2 followed by the Algorithm from Lemma 2 is at least $(2 + 2/7)$, i.e. we show that the analysis in the preceding sections is tight. Moreover, we show that the cost of an optimum solution to the CCCP might be more than twice the value of the tree cover LP (3).

Theorem 3. *For any $\epsilon > 0$ there is a CCCP instance where Algorithm 2 computes an edge set $F \subseteq E$, such that there is no capacitated cycle cover $C_1, \ldots, C_k$ with cost at most $(2 + 2/7 - \epsilon)$LP and where $V(C_i)$ is connected in (V, F) for all $i \in \{1, \ldots, k\}$.*

We defer the proof of Theorem 3 to Appendix D. We remark that although Theorem 3 shows that our analysis of Algorithm 2 followed by the Algorithm from Lemma 2 is tight, it might be that the analysis of our randomized rounding algorithm is not.

We now show that the cost of an optimum solution to the CCCP might be more than twice the value of the tree cover LP (3). We define

$$\rho := \sup \left\{ \frac{\text{OPT}(\mathcal{I})}{\text{LP}(\mathcal{I})} \;\middle|\; \mathcal{I} \text{ is a CCCP instance} \right\}.$$

Here we use $\text{OPT}(\mathcal{I})$ to refer to the minimum cost of a CCCP solution on the instance $\mathcal{I} = (G, \ell, b, \gamma)$. Similarly, $\text{LP}(\mathcal{I})$ refers to the solution value of the tree cover LP (3) for the instance $\mathcal{I}$.

Theorem 4. $\rho \geq 2 + \frac{62}{11745} > 2.005$.

To prove Theorem 4 we use the following lemma that can be proven by an argument similar to Goemans [11], and Carr and Vempala [6].

Lemma 9. *Let* $G = (V, E)$ *a complete graph and* $b : V \to [0, 1]$ *some vertex demands. Moreover, let* $(x_e)_{e \in E}$ *be a feasible solution to the tree cover LP (3) such that the support of* x *is the edge set of a tree* T. *Then there are weights* $\lambda_1, \dots, \lambda_k > 0$, *small sets* $R_1, \dots, R_k \subseteq V$ *and trees* $T_1, \dots, T_k$ *in* T *such that* $R_i \subseteq V(T_i)$ *for all* i *and*

- $\sum_{i=1}^{k} \lambda_i \leq \rho(|V| - x(E))$,
- $\sum_{i : e \in T_i} \lambda_i \leq \frac{\rho}{2} x_e$ *for every* $e \in E$, *and*
- $\sum_{i : v \in R_i} \lambda_i \geq 1$ *for every* $v \in V$.

We consider the family of LP solutions depicted in Fig. 1. One can show that if $\rho < 2 + \frac{62}{11745}$ and k is sufficiently large, it is not possible to find weights λ_i, vertex sets R_i, and trees T_i as in Lemma 9. This implies Theorem 4.

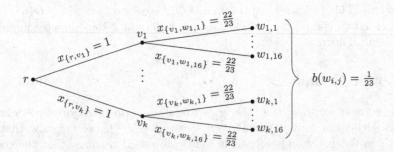

Fig. 1. A family of LP solutions x that together with Lemma 9 proves Theorem 4. Here, the demands for the vertex r and the vertices $v_1, \dots, v_k$ are 0 and the demand of the vertices $w_{i,j}$ for $i = 1, \dots, k$ and $j = 1, \dots, 16$ are 1/23. The constants are chosen to maximize the lower bound obtained from this family of instances. In the figure edges e with $x_e > 0$ are shown. One can verify that this is indeed a feasible solution to the tree cover LP (3).

A   Sketch of the Proof of Lemma 1

Pick an arbitrary root r for T. Then we perform the following splitting-off procedure (similar to Algorithm A in [15]).

As long as the vertex set $V(T)$ of the tree T remains large, we iterate the following. Let v be maximally far away from r with the property that $V(T_v)$ is large, where T_v is the subtree rooted at v. Let $w_1, \ldots, w_l$ be the children of v. Since $b(V(T_v)) = b(v) + \sum_{i=1}^{l} b(V(T_{w_i}))$, we must have that $b(v) \geq 1/2$ or there exists a set $N \subseteq \{1, \ldots, l\}$ with $\sum_{i \in N} b(V(T_{w_i})) \in [1/2, 1]$. In the first case we split off a singleton tree $(\{v\}, \emptyset)$ covering the vertex v and replace v in T by a Steiner vertex, i.e. we set its demand to zero. In the second case we split off a tree covering all vertices contained in the subtrees T_{w_i} for $i \in N$; the Steiner tree for this set of terminals contains v as a Steiner vertex and for $i \in N$ contains the edge $\{v, w_i\}$ and the subtree T_{w_i}. Thus we then remove these subtrees from T.

Let $T_1, \ldots, T_{k-1}$ be the Steiner trees split off during this algorithm and let T_k be the remaining tree. Moreover, let $R_1, \ldots, R_k$ be the respective terminal sets of these Steiner trees. Then we know that $b(R_i) \geq 1/2$ for all $i \leq k-2$ and $b(R_{k-1}) + b(R_k) \geq 1$. Thus $2b(V) = 2\sum_{i=1}^{k} b(R_i) \geq k$.

B   Sketch of the Proof of Lemma 4

The key part is showing that r is indeed submodular. For this let $F' \subseteq F \subseteq E$ be arbitrary and $e \in E \setminus F$. We need to show that

$$r(F' \cup \{e\}) - r(F') \geq r(F \cup \{e\}) - r(F).$$

Let $A_1, A_2 \in \mathcal{C}(F)$ be the two components of F joined by e. Moreover, let $A_1', A_2' \in \mathcal{C}(F')$ be the same for F'. We can assume that $A_1' \subseteq A_1$ and $A_2' \subseteq A_2$ since $F' \subseteq F$. Then one can show that

$$r(F \cup \{e\}) - r(F) = \max\{1, b(A_1)\} + \max\{1, b(A_2)\} - \max\{1, b(A_1 \cup A_2)\}$$

and

$$r(F' \cup \{e\}) - r(F') = \max\{1, b(A_1')\} + \max\{1, b(A_2')\} - \max\{1, b(A_1' \cup A_2')\}.$$

So the submodularity of r reduces to observation that the expression

$$\max\{1, x\} + \max\{1, y\} - \max\{1, x+y\}$$

is non-increasing in x and y for $x, y \geq 0$.

C   Sketch of the Proof of Lemma 8

Note that the partition $\mathcal{C}$ in iteration i of Algorithm 2 is the same as in iteration i of Algorithm 1; here we assume wlog. that the edges are sorted in the same order in both algorithms. Hence, we apply lines 7–10 of Algorithm 2 precisely for those edges e_i for which we set x_{e_i} in line 7 of Algorithm 1. We define E' as in Sect. 4 and also define $C_e^u \subseteq V$ for an edge $e \in E'$ and a vertex $u \in e$ as before.

Let x be the output of Algorithm 1. It is easy to verify that the choice of which edges we include in the forest F in lines 7–10 of Algorithm 2 is such that we minimize

$$\sum_{e \in F} 2 \cdot \ell(e) + \sum_{e \in E \setminus F} \sum_{u \in e} \gamma \cdot \max\{1 - 2b(C_e^u), 0\} \tag{8}$$

among all set F with $\{e \in E : x_e = 1\} \subseteq F \subseteq \{e \in E : x_e > 0\}$. By Lemma 7 there exists such an edge set F where (8) is at most $(2 + 2/7) \cdot \ell(x) + 2/7 \cdot \gamma \cdot (|V| - x(E))$. Hence, also the edge set F computed by Algorithm 2 fulfills this bound. By Lemma 6 this implies the claimed bound (7).

D   Proof of Theorem 3

For $n \in \mathbb{N}$ with $n \geq 4$, let $G = (V, E)$ be the complete graph on the vertices $v_1, \ldots, v_n$ with the metric ℓ on V given by $\ell(v_i, v_j) := \frac{1}{4}|i - j|$, i.e. (G, ℓ) is the metric closure of a path. Assign uniform demands of $b(v) := 1/4$ to every vertex v and let $\gamma := 1$. Then we observe that $\mathrm{LP}(G, \ell, b, \gamma) = \frac{7}{16}n$. See Fig. 2.

But now consider what Algorithm 2 does on this instance. Assume that the edges are sorted such that $e_i = \{v_i, v_{i+1}\}$ for all $i \in \{1, \ldots, n-1\}$. The algorithm will then buy the edges e_1 to e_3. But it will not buy any other edge as

$$\gamma \max\{1 - 2b(v_{i+1}), 0\} = \tfrac{1}{2} = 2\ell(\{v_i, v_{i+1}\})$$

for all $i \in \{1, \ldots, n - 1\}$. So the condition in line 9 is never satisfied except for the first three iterations of the loop. Hence, any CCCP solution which is "contained" in the connected components of F (i.e. it does not contain a cycle C_i where $V(C_i)$ is not connected in (V, F)), must contain at least $n - 4$ singleton cycles.

Finally, we conclude that any such CCCP solution has a cost of at least

$$n - 4 = \frac{n - 4}{\frac{7}{16}n}\mathrm{LP} \geq \left(\frac{16}{7} - \epsilon\right)\mathrm{LP} = \left(2 + \frac{2}{7} - \epsilon\right)\mathrm{LP}$$

for n large enough.

Fig. 2. An optimum solution to the tree cover LP (3) for instance from the proof of Theorem 3 for $n = 12$. For every solid edge e we have $x_e = 1$ and for every dotted edge e we have $x_e = 3/4$.

References

1. Altinkemer, K., Gavish, B.: Heuristics for unequal weight delivery problems with a fixed error guarantee. Oper. Res. Lett. **6**(4), 149–158 (1987)
2. Becker, A.: A tight 4/3 approximation for capacitated vehicle routing in trees. In: Blais, E., Jansen, K., Rolim, J.D.P., Steurer, D. (eds.) Approximation, Randomization, and Combinatorial Optimization, Algorithms and Techniques (APPROX/RANDOM 2018). Leibniz International Proceedings in Informatics (LIPIcs), vol. 116, pp. 3:1–3:15. Schloss Dagstuhl-Leibniz-Zentrum fuer Informatik, Dagstuhl (2018)
3. Becker, A., Klein, P.N., Saulpic, D.: Polynomial-time approximation schemes for k-center, k-median, and capacitated vehicle routing in bounded highway dimension. In: Azar, Y., Bast, H., Herman, G. (eds.) 26th Annual European Symposium on Algorithms (ESA 2018). Leibniz International Proceedings in Informatics (LIPIcs), vol. 112, pp. 8:1–8:15. Schloss Dagstuhl-Leibniz-Zentrum fuer Informatik, Dagstuhl (2018)
4. Becker, A., Klein, P.N., Schild, A.: A PTAS for bounded-capacity vehicle routing in planar graphs. In: Friggstad, Z., Sack, J.-R., Salavatipour, M.R. (eds.) WADS 2019. LNCS, vol. 11646, pp. 99–111. Springer, Cham (2019). https://doi.org/10.1007/978-3-030-24766-9_8
5. Bompadre, A., Dror, M., Orlin, J.B.: Improved bounds for vehicle routing solutions. Discrete Optim. **3**(4), 299–316 (2006)
6. Carr, R., Vempala, S.: On the Held-Karp relaxation for the asymmetric and symmetric traveling salesman problems. Math. Program. **100**(3), 569–587 (2004). https://doi.org/10.1007/s10107-004-0506-y
7. Dantzig, G.B., Ramser, J.H.: The truck dispatching problem. Manag. Sci. **6**(1), 80–91 (1959)
8. Das, A., Mathieu, C.: A quasipolynomial time approximation scheme for Euclidean capacitated vehicle routing. Algorithmica **73**(1), 115–142 (2015)
9. Das, S., Jain, L., Kumar, N.: A constant factor approximation for capacitated min-max tree cover. arXiv:1907.08304 (2019)
10. Even, G., Garg, N., Koenemann, J., Ravi, R., Sinha, A.: Min-max tree covers of graphs. Oper. Res. Lett. **32**(4), 309–315 (2004)
11. Goemans, M.X.: Worst-case comparison of valid inequalities for the TSP. Math. Program. **69**(1), 335–349 (1995). https://doi.org/10.1007/BF01585563
12. Haimovich, M., Kan, A.H.G.R.: Bounds and heuristics for capacitated routing problems. Math. Oper. Res. **10**(4), 527–542 (1985)
13. Khachay, M., Dubinin, R.: PTAS for the Euclidean capacitated vehicle routing problem in R^d. In: Kochetov, Y., Khachay, M., Beresnev, V., Nurminski, E., Pardalos, P. (eds.) DOOR 2016. LNCS, vol. 9869, pp. 193–205. Springer, Cham (2016). https://doi.org/10.1007/978-3-319-44914-2_16
14. Labbé, M., Laporte, G., Mercure, H.: Capacitated vehicle routing on trees. Oper. Res. **39**(4), 616–622 (1991)
15. Maßberg, J., Vygen, J.: Approximation algorithms for network design and facility location with service capacities. In: Chekuri, C., Jansen, K., Rolim, J.D.P., Trevisan, L. (eds.) APPROX/RANDOM-2005. LNCS, vol. 3624, pp. 158–169. Springer, Heidelberg (2005). https://doi.org/10.1007/11538462_14
16. Pecin, D., Pessoa, A., Poggi, M., Uchoa, E.: Improved branch-cut-and-price for capacitated vehicle routing. Math. Program. Comput. **9**(1), 61–100 (2016). https://doi.org/10.1007/s12532-016-0108-8

17. Pessoa, A., Sadykov, R., Uchoa, E., Vanderbeck, F.: A generic exact solver for vehicle routing and related problems. In: Lodi, A., Nagarajan, V. (eds.) IPCO 2019. LNCS, vol. 11480, pp. 354–369. Springer, Cham (2019). https://doi.org/10.1007/978-3-030-17953-3_27

18. Tröbst, T.: Capacitated vehicle routing and cycle covering problems. Master's thesis, Research Institute for Discrete Mathematics, University of Bonn (2019)

19. Vidal, T., Crainic, T.G., Gendreau, M., Prins, C.: A unified solution framework for multi-attribute vehicle routing problems. Eur. J. Oper. Res. **234**(3), 658–673 (2014)

20. Yu, W., Liu, Z.: Better approximability results for min-max tree/cycle/path cover problems. J. Comb. Optim. **37**(2), 563–578 (2019). https://doi.org/10.1007/s10878-018-0268-8

21. Yu, W., Liu, Z., Bao, X.: New approximation algorithms for the minimum cycle cover problem. Theor. Comput. Sci. **793**, 44–58 (2019)

Graph Coloring Lower Bounds
from Decision Diagrams

Willem-Jan van Hoeve[✉] (iD)

Carnegie Mellon University, Pittsburgh, PA 15213, USA
vanhoeve@andrew.cmu.edu

Abstract. We introduce an iterative framework for computing lower bounds to graph coloring problems. We utilize relaxed decision diagrams to compactly represent an exponential set of color classes, or independent sets, some of which may contain edge conflicts. Our procedure uses minimum network flow models to compute lower bounds on the coloring number and identify conflicts. Infeasible color classes associated with these conflicts are removed by refining the decision diagram. We prove that in the best case, our approach may use exponentially smaller diagrams than exact diagrams for proving optimality. We also provide an experimental evaluation on benchmark instances, and report an improved lower bound for one open instance.

1 Introduction

Graph coloring is a fundamental combinatorial optimization problem that asks to color the vertices of a given graph with a minimum number of colors, such that adjacent vertices are colored differently. Graph coloring is a core component of many applications, in particular those related to timetabling or scheduling [3,16,18,26]. The most efficient exact solution methods are the Randall-Brown algorithm using the Dsatur vertex ordering [8,22,23], integer linear programming [15], and column generation (branch-and-price) [13,14,19–21].

A major challenge for exact graph coloring methods is to find strong lower bounds to help accelerate the proof of optimality. A natural lower bound is the clique number of a graph—the size of the largest complete subgraph, which requires all its vertices to be colored differently. In this work, we explore an alternative approach that does not directly rely on maximal cliques, but instead makes use of relaxed decision diagrams [2]. Relaxed decision diagrams provide a graphical discrete relaxation of a solution set and can be used to derive optimization bounds [2,6,7].

For the graph coloring problem, we let the decision diagram compactly represent the collection of independent sets of the input graph, each of which corresponds to a color class (a subset of vertices with the same color). We obtain a graph coloring lower bound by solving a constrained minimum network flow

Partially supported by Office of Naval Research Grant No. N00014-18-1-2129 and National Science Foundation Award #1918102.

D. Bienstock and G. Zambelli (Eds.): IPCO 2020, LNCS 12125, pp. 405–418, 2020.
https://doi.org/10.1007/978-3-030-45771-6_31

over the decision diagram that ensures that each vertex only appears in one color class. However, in our relaxed decision diagram, not all color classes may be exact. We therefore identify conflicts (adjacent vertices) in each color class, which are subsequently removed from the diagram by a refinement step. We iteratively apply this process until no more conflicts are found. We show that an integral conflict-free solution to the constrained minimum network flow problem is guaranteed to be optimal.

Our approach is relatively generic, in that the iterative refinement process based on conflicts is not restricted to graph coloring problems. For example, we can define a very similar procedure for bin packing problems, in which case a subset of items is conflicting if its weight exceeds the capacity of the bin. In fact, it can be viewed as a 'dual' form of column generation: instead of iteratively generating new columns (color classes), our approach iteratively removes infeasible color classes from consideration. In both cases, however, a solution is defined as a subset of columns.

Contributions. The main contributions of this work include (1) the introduction of a new framework for obtaining graph coloring lower bounds based on relaxed decision diagrams, and proving its correctness, (2) a proof that relaxed decision diagrams can be exponentially smaller than their exact versions for finding optimal solutions, and (3) an experimental evaluation of our lower bounds on benchmark instances, with an improved lower bound for one open instance.

2   Graph Coloring by Independent Sets

We first present a formal definition of graph coloring [24]. Let $G = (V, E)$ be an undirected simple graph with vertex set V and edge set E. We define $n = |V|$ and $m = |E|$. We denote by N_i the set of neighbors of $i \in V$. For convenience, we label the vertices V as integers $\{1, \ldots, n\}$. A *vertex coloring* of G is a mapping of each vertex to a color such that adjacent vertices are assigned different colors. We refer to the subset of vertices with the same color as a *color class*. The graph coloring problem is to find a vertex coloring with the minimum number of colors. The minimum number of colors to color G is called the coloring number or the *chromatic number* of G, denoted by $\chi(G)$.

Observe that each color class is defined as a subset of variables that are pairwise non-adjacent. In other words, a color class corresponds to an *independent set* of G, and conversely each independent set of G can be used as a color class. This allows to formulate the graph coloring problem as follows. Let I be the collection of all independent sets of G. We introduce a binary variable y_i for each independent set $i \in I$, representing whether i is used as a color class in a solution. We let binary parameter a_{ij} represent whether vertex $j \in V$ belongs to independent set $i \in I$. The graph coloring problem can then be formulated as the following integer program:

$$
\begin{aligned}
\min \ & \textstyle\sum_{i \in I} y_i \\
\text{s.t.} \ & \textstyle\sum_{i \in I} a_{ij} y_i = 1 \ \forall j \in V, \\
& y_i \in \{0, 1\} \qquad \forall i \in I,
\end{aligned}
\tag{1}
$$

where the equality constraint ensures that each vertex belongs to one color class. This formulation forms the basis of the column generation approaches for graph coloring, as first proposed in [20]. Instead of enumerating all exponentially many independent sets I, column generation iteratively adds new independent sets (with negative reduced cost) to an initial collection. In our approach, we start with all subsets sets I and iteratively remove sets that contain adjacent vertices.

3   Decision Diagrams

Decision diagrams were originally developed to represent switching circuits and, more generally, Boolean functions [1,17,25]. They became particularly popular after the introduction of efficient compilation methods for Reduced Ordered Binary Decision Diagrams [9,10], and have been applied widely to verification and configuration problems. More recently, decision diagrams have been applied to solve optimization problems [6], which is the context we follow in this paper.

Definitions. For our purposes, a decision diagram will represent the set of solutions to an optimization problem P defined on an ordered set of decision variables $X = \{x_1, \ldots, x_n\}$. In this paper, we assume that each variable is binary. The feasible set of P is denoted by $\mathrm{Sol}(P)$.

A *decision diagram* for P is a layered directed acyclic graph $D = (N, A)$ with node set N and arc set A. D has $n + 1$ layers that represent state-dependent decisions for the variables. The first layer (layer 1) is a single root node r, while the last layer (layer $n + 1$) is a single terminal node t. Layer j is a collection of nodes associated with variable $x_j \in X$, for $j = 1, \ldots, n$. Arcs are directed from a node u in layer j to a node v in layer $j + 1$, and have an associated label $\ell(u, v)$ which can be either 0 or 1. We refer to the former as 0-arcs and to the latter as 1-arcs. The layer of node u is denoted by $L(u)$. Each arc, and each node, must belong to a path from r to t. Each arc-specified r-t path $p = (a_1, a_2, \ldots, a_n)$ defines a variable assignment by letting $x_j = \ell(a_j)$ for $j = 1, \ldots, n$. We slightly abuse notation and denote by $\mathrm{Sol}(D)$ the collection of variable assignments for all r-t paths in D.

Definition 1. *A decision diagram D is* exact *for problem P if* $\mathrm{Sol}(D) = \mathrm{Sol}(P)$. *A decision diagram D is* relaxed *for problem P if* $\mathrm{Sol}(D) \supseteq \mathrm{Sol}(P)$.

The benefit of using decision diagrams for representing solutions is that *equivalent* nodes, i.e., nodes with the same set of completions, can be merged. A decision diagram is called *reduced* if no two nodes in a layer are equivalent. A key property is that for a given fixed variable ordering, there exists a *unique* reduced ordered decision diagram [9]. Nonetheless, even reduced decision diagrams may be exponentially large to represent all solutions for a given problem.

Exact Compilation. In this work we apply top-down compilation methods that depend on state-dependent information (similar to state variables in

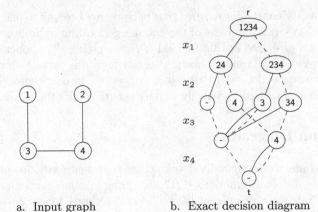

a. Input graph   b. Exact decision diagram

Fig. 1. Input graph for Example 1 (a) and the associated exact decision diagram representing all independent sets (b). The diagram uses the lexicographic ordering of the vertices. Dashed arcs represent 0-arcs, while solid arcs represent 1-arcs. For convenience, the set of eligible vertices (the state information) is given in each node.

dynamic programming models). We limit the exposition to the compilation of decision diagrams for independent set problems [4,5]. We define a binary variable x_i for each $i \in V$ representing whether i is selected. The state information we maintain is the set of 'eligible vertices', i.e., the set of graph vertices that can be added to the independent sets represented by paths into the node.

Formally, for each node u in the decision diagram we recursively define a set $S(u) \subseteq V$, and we initialize $S(r) = V$, $S(t) = \varnothing$. For node u in layer $L(u) = j$ we distinguish two cases. If $j \notin S(u)$, we define a 0-arc (or transition) from u to v, with $S(v) = S(u)$. Otherwise, if $j \in S(u)$, we define both a 1-arc and a 0-arc out of u, with

$$S(v) = \begin{cases} S(u) \setminus (\{j\} \cup N_j) & \text{if } (u,v) \text{ is a 1-arc} \\ S(u) \setminus \{j\} & \text{if } (u,v) \text{ is a 0-arc} \end{cases}$$

The top-down compilation procedure starts at the root node, creates all nodes in the next layer (following the 0-arcs and 1-arcs), and merges the nodes that are equivalent. In our case, two nodes u and v are equivalent if $S(u) = S(v)$. This top-down compilation procedure yields the unique reduced decision diagram for representing all independent sets (for a given ordering), as shown in [4,5].

Example 1. Consider the graph in Fig. 1a. We depict the exact decision diagram representing all independent sets for this graph in Fig. 1b.

Compilation by Separation. As an alternative to exact compilation, we can apply *constraint separation* to iteratively construct the decision diagram [6,11]. We will apply this method to compile *relaxed* decision diagrams, and again describe it in the context of independent sets.

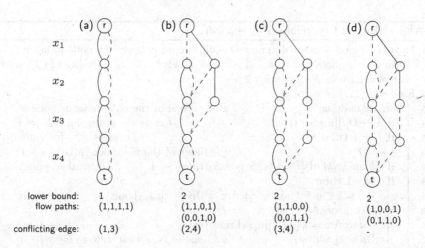

	(a)	(b)	(c)	(d)
lower bound:	1	2	2	2
flow paths:	(1,1,1,1)	(1,1,0,1)	(1,1,0,0)	(1,0,0,1)
		(0,0,1,0)	(0,0,1,1)	(0,1,1,0)
conflicting edge:	(1,3)	(2,4)	(3,4)	-

Fig. 2. Applying constraint separation to the input graph of Example 1. The initial relaxed diagram (a) is iteratively refined until the optimal solution (the flow paths) no longer contains infeasible color classes.

We initialize each layer j of D as a single node u_j, with state information $S(u_j) = \{j, \ldots, n\}$, for $j = 1, \ldots, n$, and $S(u_{n+1}) = \varnothing$. We define a 0-arc and a 1-arc between nodes u_j and u_{j+1}, for $j = 1, \ldots, n$. The input to our separation algorithm is a relaxed decision diagram D together with a path $p = u_j, u_{j+1}, \ldots, u_k$ with associated arc labels $l_j, l_{j+1}, \ldots, l_{k-1}$, and a conflict, i.e., the edge between vertices j and k, where $j < k$. The goal of the separation algorithm is to resolve the conflict along path p by splitting nodes, and arcs, appropriately. This is described in Algorithm 1.

We represent D as a two-dimensional vector D[][] of nodes, such that D is a vector of 'layers' and D[] is a vector of 'nodes', one for each layer. Both are indexed starting from 1. The size of vector D is fixed to $n + 1$, while we dynamically update the size of the layers D[]. The root is represented as D[1][1] and the terminal as D[$n+1$][1]. Each node $u =$ D[j][i] (u is the i-th node in layer j) has state information D[j][i].S $= S(u)$, a reference to the node in layer $j + 1$ that represents the endpoint of its 1-arc D[j][i].oneArc, and a similar reference for its 0-arc D[j][i].zeroArc. If the 1-arc does not exist, the reference holds value -1.

Algorithm 1 considers each node D[i][u_i] along the path in sequence (line 2) and splits off the next node in the path. This is done by first creating a temporary node w (lines 3–5). If an equivalent node already exists in layer $i + 1$, we direct the path to its index (lines 6–7). Otherwise we complete the definition of w by copying the outgoing arcs of node D[$i + 1$][u_{i+1}] (lines 9–11), and we add w to layer $i + 1$ (lines 12–13). Lastly, we redirect the path from u_i to the new node with index t (lines 15–17). Figure 2 gives an illustration of the algorithm. When we apply the algorithm to the decision diagram (a), with the all-ones path and conflicting edge $(1, 3)$ as input, we obtain the decision diagram (b).

Algorithm 1: Compilation by separation.

1 **Input:** relaxed decision diagram D (D[j][i] represents the ith node in layer j),
   path node indices $u_j, \ldots, u_{k-1}$, path arc labels $l_j, \ldots, l_{k-1}$, conflict (j,k), and
   list of neighbors N_i for each $i \in V$

2 **for** $i = j, \ldots, k-1$ **do**

3   create node w // goal is to split the path towards node w

4     $w.S \leftarrow D[i][u_i].S \setminus \{i\}$ // copy the state and remove i

5   **if** $l_i = 1$ **then** $S(w) \leftarrow S(w) \setminus N_i$ // remove N_i for 1-arc

6     $t \leftarrow -1$ // t is index of the new node in layer $i+1$

7   **if** $\exists k$ *such that* D[i+1][k].S = w.S **then** $t \leftarrow k$ // equivalent node

8   **if** $t = -1$ **then**

9   **if** $i+1 \in w.S$ **then** w.oneArc = D[i+1][u_{i+1}].oneArc   // copy 1-arc

10   **else** w.oneArc = -1

11   w.zeroArc = D[i+1][u_{i+1}].zeroArc   // copy 0-arc

12   D[i+1].add(w)   // append w as new node to layer $i+1$

13       $t \leftarrow |D[i+1]|$ // update t to last index of layer $i+1$

14   **end**

15   **if** $l_i = 1$ **then** D[i][u_i].oneArc= t // re-direct path in case of 1-arc

16   **else** D[i][u_i].zeroArc= t // re-direct path in case of 0-arc

17     $u_{i+1} = t$ // update path index

18 **end**

Compilation by constraint separation proceeds in iterations by gradually refining the relaxed decision diagram. For our model, it has some useful properties:

Lemma 1. *Each decision diagram that is compiled by separation is reduced, i.e., no two nodes on any layer are equivalent.*

Lemma 2. *Compilation by separation will terminate with an exact decision diagram if each possible conflict, along any path, is separated.*

These two lemmas follow from the fact that for independent set problems, the state information is sufficient to prove equivalence [4,5]. Lemma 1 will ensure that our approach produces the smallest decision diagram at each iteration, while Lemma 2 will guarantee termination and optimality.

4 Network Flow Model

We next formulate the graph coloring problem based on independent sets as a network flow problem on the decision diagram. We let $\delta^+(u)$ and $\delta^-(u)$ denote the set of arcs leaving, respectively entering node $u \in N$. For each arc $a \in A$ we introduce a variable y_a that represents the 'flow' through a. We then define:

$$(F) = \min \sum_{a \in \delta^+(r)} y_a \tag{2}$$

$$\text{s.t.} \sum_{a=(u,v)|L(u)=j, \ell(a)=1} y_a = 1 \qquad \forall j \in V \tag{3}$$

$$\sum_{a \in \delta^-(u)} y_a - \sum_{a \in \delta^+(u)} y_a = 0 \qquad\qquad \forall u \in N \setminus \{r, t\} \qquad (4)$$

$$y_a \in \{0, 1, \ldots, n\} \qquad\qquad\qquad \forall a \in A \qquad (5)$$

The objective function (2) minimizes the total amount of flow. Constraints (3) define that in each layer exactly one 1-arc is selected. Constraints (4) ensure flow conservation. Constraints (5) make sure the flow is integer.

Lemma 3. *A solution to model (F) corresponds to a partition of vertex set V.*

Proof. The partition can be found by decomposing the flow into paths. For vertex $i \in V$ there is exactly one arc $a = (u, v)$ for which $y_a = 1$, by constraints (3). By constraints (4) there exists an r-u path and a v-t path for which $y_a \geq 1$ for each arc a along the path, which together with arc (u, v) forms an r-t path. Let P be the collection of such r-t paths; the cardinality of P is given by the objective function (2). Each path $(a_1, a_2, \ldots, a_n) \in P$ corresponds to a subset of vertices $\{i \mid \ell(a_i) = 1\}$. By constraints (3), these subsets are disjoint. $\qquad\square$

Theorem 1. *If the decision diagram is exact, model (F) finds an optimal solution to the graph coloring problem.*

Proof. Since each path in the exact decision diagram corresponds to an independent set, the theorem follows from Lemma 3. $\qquad\square$

By the definition of relaxed decision diagrams, we have the following corollary:

Corollary 1. *If the decision diagram is relaxed, the objective value of model (F) is a lower bound on the graph coloring problem.*

One may wonder whether model (F) can be solved in polynomial time, since the NP-hardness of graph coloring may be accounted for by the worst-case exponential size of the decision diagram. The answer, however, is negative:

Theorem 2. *Solving model (F) for an arbitrary decision diagram is NP-hard.*

The proof follows from a reduction from minimum set partitioning and is given in the appendix. Note that paths in the solution are not explicitly given, but instead follow from the arc flow values. Moreover, paths can share nodes (and 0-arcs). We can therefore restrict the proof of Theorem 2 to be paths corresponding to a solution to the minimum set partitioning problem:

Corollary 2. *If the decision diagram is relaxed, deciding whether a given solution to model (F) corresponds to a feasible graph coloring solution is NP-hard.*

5   Iterative Refinement Procedure

While the worst-case complexity results in the previous section are negative, we can, however, efficiently identify a conflict in a given solution:

Algorithm 2: Iterative refinement by conflict detection and separation.

1 **Input:** decision diagram D and input graph data
2 foundSol ← false
3 **while** *foundSol = false* **do**
4   solve model (F) with decision diagram D
5   lowerBound ← obj(F)
6   **if** *flow decomposition algorithm finds no conflict* **then** foundSol ← true
7   **else** separate conflict in D using Algorithm 1
8 **end**

Theorem 3. *For a given solution to (F) we can in polynomial time (in the size of the decision diagram) either determine feasibility or identify a subset of vertices that creates an infeasibility.*

The proof, given in the appendix, is constructive and based on a path decomposition of the network flow. We apply this path decomposition to identify and separate conflicts inside an iterative refinement procedure, described at a high-level in Algorithm 2. The algorithm repeatedly solves the flow model, the objective of which provides a lower bound. If the solution contains a conflict, the decision diagram is refined, and we repeat the process. Otherwise, we found an optimal solution and terminate. Note that depending on the flow decomposition, the same solution to (F) might either return a feasible solution or a conflict.

Example 2. Figure 2 depicts the iterative refinement procedure when applied to the input graph given in Fig. 1a. We start with the trivial relaxation in (a), which encodes all possible subsets of V, and find the all-ones solution as a single path. Thus, the lower bound is 1. The first conflict we identify is edge (1,3) which is subsequently separated. An optimal flow for diagram (b) contains two paths which yields lower bound 2, and we can identify conflict (2,4) to be separated. This continues until we find the diagram (d) and identify flow paths without conflicts.

Example 2 shows that iterative refinement may yield relaxed decision diagram that are smaller than exact decision diagrams, to prove optimality. This is formalized in the following key result, the proof of which is given in the appendix:

Theorem 4. *The iterative refinement procedure can find a provably optimal solution with a relaxed decision diagram that is exponentially smaller than the exact diagram that is defined on the same variable ordering.*

6   Implementation and Experimental Results

We implemented our method in C++, and performed an experimental evaluation on the 137 graph coloring benchmark instances obtained from [12], which includes all DIMACS graph coloring instances [16]. We use CPLEX 12.9 as integer and linear programming solver. The decision diagrams have a given maximum size (either 100,000 or 1,000,000 nodes).

Table 1. Evaluating the impact of solving linear programming models before solving the integer programming models (a) and single conflict and multiple conflict resolution (b). In each case, we first list the number of instances for which either method finds a better bound. For those cases with equal bounds, we report the ratio of the average time to the best solution and the standard deviation.

IP > LP+IP	2		single > multiple	3
LP+IP > IP	31		multiple > single	43
LP+IP = IP	104		multiple = single	91
(LP+IP)/IP	0.53 (s.d.=0.38)		multiple/single	0.36 (s.d.=0.39)

a. Impact of adding LP to IP   b. Single vs. multiple conflicts

Implementation Details. The single most important parameter that influences the performance of the algorithm is the variable ordering of the decision diagram. We apply the following variable ordering heuristic, which dominated all other heuristics we tested: among all unselected vertices, select the one that is connected to the most vertices that have been selected so far. In case of ties, select a vertex with the highest degree.

Second, observe that solving the continuous linear programming relaxation of model (F) provides a valid lower bound, as well as a path decomposition that can be used to identify conflicts. While the LP bound may be weaker than the IP bound, it is faster to compute and may therefore speed up the overall process. We implemented this by starting with LP-based iterative refinement, which is continued until a conflict-free LP optimum solution is found. After that, we continue with the IP-based iterative refinement to prove integer optimality.

Third, instead of resolving a single conflict (separation) in each iteration, it is possible to resolve multiple conflicts; one for each path in the decomposition (if it exists). The validity of identifying and separating multiple conflicts per iteration relies on the specific state information for independent sets, as well as on the indexing of our data structure D[][].

We ran experiments to assess the impact of adding the LP-based iterative refinement and multiple conflicts. The results are shown in Table 1. For each feature, we ran our algorithm with and without that feature, on all 137 benchmark instances for a maximum of 300 s per instance. The impact is measured by the quality of the bound (number of instances with a better bound), and when the bounds are equal, by the computation time to the best bound. Both using LP and multiple conflicts have a substantial positive impact on the performance, reducing the time to the best bound by a factor 0.53 (LP), resp. 0.38 (multiple conflicts).

We added two other features to streamline the solving process. After reading in the data, we run the Dsatur heuristic to quickly find an upper bound, to help prove optimality in some cases. Furthermore, before running the LP and IP-based iterative refinement, we run a refinement procedure based on the longest path (with respect to 1-arcs) on the decision diagram, for at most 100 iterations.

Table 2. Performance of the relaxed decision diagram on a selection of open instances. For each instance we list the number of nodes (n) and edges (m), edge density (d), and the best known lower bound $(\underline{\chi})$ and upper bound $(\overline{\chi})$. For the relaxed decision diagram, we report the lower bound (LB) and time to best bound (TTB). The time limit was set to 3,600 s (the maximum size of 1,000,000 was never exceeded). Upper bounds marked with an asterisk were computed with the Dsatur heuristic.

						Relaxed DD	
Instance	n	m	d	$\underline{\chi}$	$\overline{\chi}$	LB	TTB
abb313GPIA	1,557	53,356	0.04	8	10	7	1.71
C2000.9	2,000	1,799,532	0.90	0	400	**98**	1,826.92
DSJC250.1	250	3,218	0.10	6	8	5	0.01
latin_square_10	900	307,350	0.76	90	97	90	0.73
wap01a	2,368	110,871	0.04	41	48*	40	1.20
wap02a	2,464	111,742	0.04	40	45*	40	1.61
wap03a	4,730	286,722	0.03	40	56*	40	0.88
wap04a	5,231	294,902	0.02	40	42	40	1.05
wap06a	947	43,571	0.10	40	54*	40	0.15
wap07a	1,809	103,368	0.06	40	41	39	705.66
wap08a	1,870	104,176	0.06	40	45*	39	107.23

Experimental Analysis. The aim of our first experiment is to compare the performance of the iterative refinement and the exact compilation. We consider all instances that are solved to optimality by either method—47 instances in total. The details can be found in Table 3 in the Appendix. The exact decision diagram can be remarkably small, which allows solving 37 instances directly. Perhaps even more remarkable is that the relaxed diagram can sometimes be orders of magnitude smaller than the exact diagram for proving optimality, demonstrating the value of Theorem 4 in practice.

The second set of experiments, presented in Table 2, investigates the quality of the bounds of the iterative refinement procedure. The table considers a selection of open instances.[1] We report the lower bound (LB) and the time to the best lower bound in seconds (TTB). We were able to improve the lower bound for instance C2000.9 (marked in bold).

7   Conclusion

We introduced a new approach for obtaining lower bounds to graph coloring problems, by solving a minimum network flow problem defined over a relaxed decision diagram. By separating conflicts in the network flow solution, the relaxed decision diagram is iteratively refined, which results in stronger bounds. We showed both theoretically and experimentally that relaxed decision diagrams can be orders of magnitude smaller than exact diagrams when proving optimality. This allowed to find an improved lower bound for one open benchmark instance.

---

[1] According to the website with benchmark results [12] – accessed November 29, 2019.

Graph Coloring Lower Bounds from Decision Diagrams: Appendix

Proof of Theorem 2. By reduction from minimum set partitioning. We are given a collection S of sets based on a universe of elements E, and need to find a subset of S of minimum cardinality such that each element in E belongs to exactly one subset. We define a polynomial-size decision diagram with $|E| + 1$ node layers, such that layer i represents the i-th element from E following an arbitrary but fixed ordering of E. We then define an r-t path for each set in S by introducing nodes and arcs between each layer i and $i+1$, with arc label 1 if the i-th element of E is in S and 0 otherwise. An optimal solution to model (F) directly corresponds to solving the minimum set partitioning problem.   □

Proof of Theorem 3. We apply a flow decomposition algorithm that iteratively finds r-t paths with flow value 1, starting from the root. The algorithm maintains the set of vertices i for which $y_a = 1$, $\ell(a) = 1$ and $L(a) = i$. Each time a vertex i is added to this set, we inspect whether there exists an edge $(i, j) \in E$ with j in the set. If so, we terminate and report that the current (partial) path violates the edge constraint for (i, j). Otherwise, when the r-t path is completed, we subtract the minimum flow value along the path from each of its arcs, and repeat the process until all arcs out of the root have flow value zero. If in the process none of the paths violates an edge constraint, we report that the solution is feasible. Finding one r-t path takes linear time (in the size of the decision diagram, $|D|$). The edge inspection takes $O(n)$ time per event, which makes the total time for identifying a single path $O(n \cdot |D|)$. Since there are at most n paths, the total time is $O(n^2|D|)$.   □

Proof of Theorem 4. Consider a graph $G = (V, E)$ with vertex labels $V = \{1, \ldots, n\}$ and edge set $\{(i, i+1) \mid i \in \{1, \ldots, n-1\}\}$, i.e., G is a path from vertex 1 to n, where n is an odd integer. We define the following fixed variable ordering to compile the decision diagrams. For layers $i = 1, \ldots, \lceil n/2 \rceil$, we associate vertex i if i is odd, and vertex $\lceil n/2 \rceil + i$ if i is even. The remaining layers are defined in the 'reverse order'; layer $i = \lceil n/2 \rceil + 1, \ldots, n$ is associated with vertex $n - i + 2$ if i is odd, and vertex $n - i + \lceil n/2 \rceil$ if i is even.

Observation 1: up to layer $\lceil n/2 \rceil - 1$, vertex i appears in each state of layer $i - 1$ since the vertices associated with these layers are not adjacent in G. Therefore, each of these states has two outgoing arcs, the 0-arc and the 1-arc.

Observation 2: up to layer $\lceil n/2 \rceil - 1$, each 1-arc eliminates one element from the set $\{\lceil n/2 \rceil + 1, \ldots, n\}$. Therefore, the states of each layer, up to layer $\lceil n/2 \rceil - 1$, are distinct. These two observations imply that the exact decision diagram requires at least $O(2^{\lceil n/2 \rceil})$ states (the size of layer $\lceil n/2 \rceil$).

Without loss of generality, we assume that the iterative refinement procedure applies a lexicographic search in each iteration to find an optimal solution to model (F), and refines the decision diagram whenever an edge conflict is detected. In iteration i, the algorithm will consider the conflict associated with edge $(i, i + \lceil n/2 \rceil)$. Each of these conflicts is refined by adding one more node to state $i + \lceil n/2 \rceil + 1$. After $\lceil n/2 \rceil$ iterations, no more conflicts are found. The iterative refinement procedure therefore terminates with a decision diagram of $O(n)$ size.   □

Table 3. A comparison of relaxed and exact decision diagrams on instances that were optimally solved by either method (indicated by *). For each instance we list the number of nodes (n) and edges (m), and edge density (d). We report the lower bound (LB), upper bound (UB), solving time (in seconds), and the size of the decision diagram. The last column (R/E) represents the ratio of the relaxed and exact diagram sizes. The time limit was set to 1,800 s, and the maximum size was set to 100,000 nodes.

instance	n	m	d	Relaxed DD LB	UB	time	size	Exact DD LB	UB	time	size	R/E
1-FullIns_3	30	100	0.23	4	4 *	0.03	249	4	4 *	0.04	747	0.33
2-FullIns_3	52	201	0.15	5	5 *	0.26	1,206	5	5 *	0.59	12,866	0.09
david	87	406	0.11	11	11 *	0.01	246	11	11 *	1.60	37,029	0.01
DSJC125.9	125	6,961	0.90	44	44 *	9.52	9,433	44	44 *	0.37	9,868	0.96
fpsol2.i.1	496	11,654	0.09	65	65 *	3.19	7,135	65	65 *	0.09	8,295	0.86
fpsol2.i.2	451	8,691	0.09	30	30 *	0.06	957	30	30 *	0.14	10,167	0.09
fpsol2.i.3	425	8,688	0.10	30	30 *	0.06	970	30	30 *	0.19	10,257	0.09
huck	74	301	0.11	11	11 *	0.09	786	11	11 *	0.02	1,077	0.73
inithx.i.1	864	18,707	0.05	54	54 *	0.83	3,993	54	54 *	0.21	15,804	0.25
inithx.i.2	645	13,979	0.07	31	31 *	34.57	21,787	31	31 *	1.05	24,588	0.89
inithx.i.3	621	13,969	0.07	31	31 *	51.19	22,777	31	31 *	0.79	24,550	0.93
jean	80	254	0.08	10	10 *	0.01	291	10	10 *	0.32	5,251	0.06
miles1000	128	3,216	0.40	42	42 *	0.58	4,104	42	42 *	0.11	8,031	0.51
miles1500	128	5,198	0.64	73	73 *	0.31	2,697	73	73 *	0.04	4,007	0.67
miles250	128	387	0.05	8	8 *	0.01	294	8	8 *	0.17	2,812	0.10
miles500	128	1,170	0.14	20	20 *	0.02	361	20	20 *	0.39	15,272	0.02
miles750	128	2,113	0.26	31	31 *	0.04	710	31	31 *	0.29	13,153	0.05
mulsol.i.1	197	3,925	0.20	49	49 *	0.12	1,107	49	49 *	0.02	2,487	0.45
mulsol.i.2	188	3,885	0.22	31	31 *	0.07	793	31	31 *	0.03	2,611	0.30
mulsol.i.3	184	3,916	0.23	31	31 *	0.06	789	31	31 *	0.03	2,621	0.30
mulsol.i.4	185	3,946	0.23	31	31 *	0.06	806	31	31 *	0.03	2,636	0.31
mulsol.i.5	186	3,973	0.23	31	31 *	0.06	807	31	31 *	0.03	2,649	0.30
myciel3	11	20	0.36	4	4 *	0.02	59	4	4 *	0.01	62	0.95
myciel4	23	71	0.28	5	5 *	2.31	454	5	5 *	0.35	459	0.99
queen5_5	25	160	0.53	5	5 *	0.01	194	5	5 *	0.01	560	0.35
queen6_6	36	290	0.46	7	7 *	1.03	1,927	7	7 *	0.11	2,686	0.72
queen7_7	49	476	0.40	7	7 *	0.89	3,267	7	7 *	0.23	13,838	0.24
queen8_8	64	728	0.36	9	9 *	164.56	30,606	9	9 *	38.82	81,574	0.38
r125.1	125	209	0.03	5	5 *	0.01	339	5	5 *	0.02	920	0.37
r125.1c	125	7,501	0.97	46	46 *	0.65	3,569	46	46 *	0.04	4,007	0.89
r125.5	125	3,838	0.50	36	36 *	57.31	18,924	36	36 *	0.49	23,242	0.81
r250.1c	250	30,227	0.97	64	64 *	11.54	18,059	64	64 *	0.15	20,322	0.89
zeroin.i.1	211	4,100	0.19	49	49 *	0.08	1,024	49	49 *	0.03	2,769	0.37
zeroin.i.2	211	3,541	0.16	30	30 *	0.05	774	30	30 *	0.05	3,470	0.22
zeroin.i.3	206	3,540	0.17	30	30 *	0.05	769	30	30 *	0.05	3,457	0.22
anna	138	493	0.05	11	11 *	0.01	356	0	11	2.80	$\geq$ 100k	$\leq$ 0.00
DSJR500.1	500	3,555	0.03	12	12 *	0.01	626	0	12	1.43	$\geq$ 100k	$\leq$ 0.01
games120	120	638	0.09	9	9 *	40.77	43,074	0	9	1.90	$\geq$ 100k	$\leq$ 0.39
le450_25a	450	8,260	0.08	25	25 *	0.05	943	0	25	1.90	$\geq$ 100k	$\leq$ 0.01
le450_25b	450	8,263	0.08	25	25 *	0.04	827	0	25	1.51	$\geq$ 100k	$\leq$ 0.01
le450_5d	450	9,757	0.10	5	5 *	0.02	700	0	5	2.45	$\geq$ 100k	$\leq$ 0.01
r1000.1	1,000	14,378	0.03	20	20 *	0.06	1,234	0	20	1.56	$\geq$ 100k	$\leq$ 0.01
r250.1	250	867	0.03	8	8 *	3.03	11,614	0	8	1.40	$\geq$ 100k	$\leq$ 0.11
school1	385	19,095	0.26	14	14 *	23.87	22,322	0	15	1.01	$\geq$ 100k	$\leq$ 0.21
school1_nsh	352	14,612	0.24	14	14 *	23.80	23,430	0	16	1.23	$\geq$ 100k	$\leq$ 0.21
2-Insertions_3	37	72	0.11	3	4	1,800	2,713	4	4 *	362.45	2,963	0.92
DSJC250.9	250	27,897	0.90	71	93	1,800	79,637	72	72 *	1,749.80	80,681	0.99

References

1. Akers, S.B.: Binary decision diagrams. IEEE Trans. Comput. **27**, 509–516 (1978)
2. Andersen, H.R., Hadzic, T., Hooker, J.N., Tiedemann, P.: A constraint store based on multivalued decision diagrams. In: Bessière, C. (ed.) CP 2007. LNCS, vol. 4741, pp. 118–132. Springer, Heidelberg (2007). https://doi.org/10.1007/978-3-540-74970-7_11
3. Barnier, N., Brisset, P.: Graph coloring for air traffic flow management. Ann. Oper. Res. **130**, 163–178 (2004)
4. Bergman, D., Cire, A.A., van Hoeve, W.-J., Hooker, J.N.: Variable ordering for the application of BDDs to the maximum independent set problem. In: Beldiceanu, N., Jussien, N., Pinson, É. (eds.) CPAIOR 2012. LNCS, vol. 7298, pp. 34–49. Springer, Heidelberg (2012). https://doi.org/10.1007/978-3-642-29828-8_3
5. Bergman, D., Cire, A.A., van Hoeve, W.-J., Hooker, J.N.: Optimization bounds from binary decision diagrams. INFORMS J. Comput. **26**(2), 253–268 (2014)
6. Bergman, D., Cire, A.A., van Hoeve, W.-J., Hooker, J.N.: Decision Diagrams for Optimization. Springer, Heidelberg (2016). https://doi.org/10.1007/978-3-319-42849-9
7. Bergman, D., van Hoeve, W.-J., Hooker, J.N.: Manipulating MDD relaxations for combinatorial optimization. In: Achterberg, T., Beck, J.C. (eds.) CPAIOR 2011. LNCS, vol. 6697, pp. 20–35. Springer, Heidelberg (2011). https://doi.org/10.1007/978-3-642-21311-3_5
8. Brélaz, D.: New methods to color the vertices of a graph. Commun. ACM **22**(4), 251–256 (1979)
9. Bryant, R.E.: Graph-based algorithms for Boolean function manipulation. IEEE Trans. Comput. **C–35**, 677–691 (1986)
10. Bryant, R.E.: Symbolic Boolean manipulation with ordered binary decision diagrams. ACM Comput. Surv. **24**, 293–318 (1992)
11. Cire, A.A., Hooker, J.N.: The separation problem for binary decision diagrams. In: Proceedings of ISAIM (2014)
12. Gualandi, S., Chiarandini, M.: Graph Coloring Benchmarks. https://sites.google.com/site/graphcoloring/home
13. Gualandi, S., Malucelli, F.: Exact solution of graph coloring problems via constraint programming and column generation. INFORMS J. Comput. **24**(1), 81–100 (2012)
14. Held, S., Cook, W., Sewell, E.C.: Maximum-weight stable sets and safe lower bounds for graph coloring. Math. Program. Comput. **4**(4), 363–381 (2012)
15. Jabrayilov, A., Mutzel, P.: New integer linear programming models for the vertex coloring problem. In: Bender, M.A., Farach-Colton, M., Mosteiro, M.A. (eds.) LATIN 2018. LNCS, vol. 10807, pp. 640–652. Springer, Cham (2018). https://doi.org/10.1007/978-3-319-77404-6_47
16. Johnson, D.S., Trick, M.A. (eds.): Cliques, Coloring, and Satisfiability: Second DIMACS Implementation Challenge. DIMACS Series in Discrete Mathematics and Theoretical Computer Science, 11–13 October 1993, vol. 26. American Mathematical Society (1996)
17. Lee, C.Y.: Representation of switching circuits by binary-decision programs. Bell Syst. Tech. J. **38**, 985–999 (1959)
18. Lewis, R., Thompson, J.: On the application of graph colouring techniques in round-robin sports scheduling. Comput. Oper. Res. **38**(1), 190–204 (2011)
19. Malaguti, E., Monaci, M., Toth, P.: An exact approach for the Vertex Coloring Problem. Discrete Optim. **8**, 174–190 (2011)

20. Mehrotra, A., Trick, M.A.: A column generation approach for graph coloring. INFORMS J. Comput. **8**(4), 344–354 (1996)
21. Morrison, D.R., Sewell, E.C., Jacobson, S.H.: Solving the pricing problem in a branch-and-price algorithm for graph coloring using zero-suppressed binary decision diagrams. INFORMS J. Comput. **28**(1), 67–82 (2016)
22. Peemöller, J.: A correction to Brelaz's modification of Brown's coloring algorithm. Commun. ACM **26**(8), 595–597 (1983)
23. Randall-Brown, J.: Chromatic scheduling and the chromatic number problem. Manag. Sci. **19**(4), 456–463 (1972)
24. Schrijver, A.: Combinatorial Optimization: Polyhedra and Efficiency. Springer, Heidelberg (2003)
25. Wegener, I.: Branching Programs and Binary Decision Diagrams: Theory and Applications. SIAM Monographs on Discrete Mathematics and Applications. Society for Industrial and Applied Mathematics (2000)
26. Wood, D.C.: A technique for coloring a graph applicable to large-scale timetabling problems. Comput. J. **12**(4), 317–322 (1969)

On Convex Hulls of Epigraphs of QCQPs

Alex L. Wang$^{(\boxtimes)}$ and Fatma Kılınç-Karzan

Carnegie Mellon University, Pittsburgh, PA 15213, USA
alw1@cs.cmu.edu

Abstract. Quadratically constrained quadratic programs (QCQPs) are a fundamental class of optimization problems well-known to be NP-hard in general. In this paper we study sufficient conditions for a *convex hull result* that immediately implies that the standard semidefinite program (SDP) relaxation of a QCQP is tight. We begin by outlining a general framework for proving such sufficient conditions. Then using this framework, we show that the convex hull result holds whenever the quadratic eigenvalue multiplicity, a parameter capturing the amount of symmetry present in a given problem, is large enough. Our results also imply new sufficient conditions for the tightness (as well as convex hull exactness) of a second order cone program relaxation of simultaneously diagonalizable QCQPs.

Keywords: Quadratically constrained quadratic programming · Semidefinite program · Convex hull · Relaxation · Lagrange function

1 Introduction

In this paper we study *quadratically constrained quadratic programs* (QCQPs) of the following form

$$\text{Opt} := \inf_{x \in \mathbb{R}^N} \left\{ q_0(x) : \begin{array}{l} q_i(x) \leq 0, \ \forall i \in [\![m_I]\!] \\ q_i(x) = 0, \ \forall i \in [\![m_I + 1, m_I + m_E]\!] \end{array} \right\}, \tag{1}$$

where for every $i \in [\![0, m_I + m_E]\!]$, the function $q_i : \mathbb{R}^N \to \mathbb{R}$ is a (possibly nonconvex) quadratic function. We will write $q_i(x) = x^\top A_i x + 2b_i^\top x + c_i$ where $A_i \in \mathbb{S}^N$, $b_i \in \mathbb{R}^N$, and $c_i \in \mathbb{R}$. We will assume that the number of constraints $m := m_I + m_E$ is at least 1.

QCQPs arise naturally in many areas. A non-exhaustive list of applications contains facility location, production planning, pooling, max-cut, max-clique, and certain robust optimization problems (see [2,7,21] and references therein).

Although QCQPs are NP-hard to solve in general, they admit tractable convex relaxations. One natural relaxation is the standard (Shor) semidefinite program (SDP) relaxation [34]. There is a vast literature on approximation guarantees associated with this relaxation [9,27,30,40], however, less is known about its exactness. Recently, a number of exciting results in phase retrieval [17] and clustering [1,28,31] have shown that under various assumptions on the data, the

© Springer Nature Switzerland AG 2020
D. Bienstock and G. Zambelli (Eds.): IPCO 2020, LNCS 12125, pp. 419–432, 2020.
https://doi.org/10.1007/978-3-030-45771-6_32

QCQP formulation of the corresponding problem has a tight SDP relaxation. In contrast to these results, which address QCQPs arising from particular problems, Burer and Ye [16] very recently gave appealing deterministic sufficient conditions under which the standard SDP relaxation of *general* QCQPs is tight. In our paper, we continue this vein of research for general QCQPs. More precisely, we will provide sufficient conditions under which the *convex hull* of the epigraph of the QCQP is given by the projection of the epigraph of its SDP relaxation. Note that such a result immediately implies that the optimal objective value of the QCQP is equal to the optimal objective value of its SDP relaxation. We will refer to these two types of results as "convex hull results" and "SDP tightness results." In this paper we will focus mainly on conditions that imply the convex hull result. See the full paper [38] for additional new conditions which imply the SDP tightness result directly.

Convex hull results will necessarily require stronger assumptions than SDP tightness results, however they are also more broadly applicable because they may be used to derive strong convex relaxations for complex problems. In fact, the convexification of commonly occurring substructures has been critical in advancing the state-of-the-art computational approaches for mixed integer linear programs and general nonlinear nonconvex programs [18,36]. For computational purposes, conditions guaranteeing simple convex hull descriptions are particularly favorable. As we will discuss later, a number of our sufficient conditions will guarantee that the desired convex hulls are given by a finite number of easily computable convex quadratic constraints in the original space of variables.

Related Work. Convex hull results are well-known for simple QCQPs such as the Trust Region Subproblem (TRS) and the Generalized Trust Region Subproblem (GTRS). Recall that the TRS is a QCQP with a single strictly convex inequality constraint and that the GTRS is a QCQP with a single (possibly nonconvex) inequality constraint. A celebrated result due to Fradkov and Yakubovich [19] implies that the SDP relaxation of the GTRS is tight. More recently, Ho-Nguyen and Kılınç-Karzan [22] and Wang and Kılınç-Karzan [37] showed that the (closed) convex hulls of the TRS and GTRS epigraphs are given exactly by the projection of the SDP epigraphs. In both cases, the projections of the SDP epigraphs can also be described in the original space of variables with at most two convex quadratic inequalities. As a result, the TRS and the GTRS can be solved without explicitly running costly SDP-based algorithms.

A different line of research has focused on providing explicit descriptions for the convex hull of the intersection of a single nonconvex quadratic region with convex sets such as convex quadratic regions, second-order cones (SOCs), or polytopes, or with one other nonconvex quadratic region [14,24,29,32,42,43]. For example, the convex hull of the intersection of a two-term disjunction, which is a nonconvex quadratic constraint under mild assumptions, with the second-order cone (SOC) or its cross sections has received much attention in mixed integer programming; see [14,24,43] and references therein. In contrast to these results, we will not limit the number of nonconvex quadratic constraints in our QCQPs.

On the other hand, the nonconvex sets that we study in this paper will arise as epigraphs of QCQPs. In particular, the epigraph variable will play a special role in our analysis. Therefore, we view our developments as complementary to these results.

The convex hull question has also received attention for certain strengthened relaxations of simple QCQPs [12,13,15,35]. In this line of work, the standard SDP relaxation is strengthened by additional inequalities derived using the Reformulation-Linearization Technique (RLT). For example, Sturm and Zhang [35] showed that the standard SDP relaxation strengthened with an additional SOC constraint derived from RLT gives the convex hull of the epigraph of the TRS with one additional linear inequality. See [12] for a survey of some results in this area. In this paper, we restrict our attention to the standard SDP relaxation of QCQPs. Nevertheless, exactness conditions for strengthened SDP relaxations of QCQPs are clearly of great interest and are a direction for future research.

A number of SDP tightness results are known for variants of the TRS [6, 22,23,41], for simultaneously diagonalizable QCQPs [26], quadratic matrix programs [4,5], and random general QCQPs [16]. See the full version of this paper for a more complete survey of the related SDP tightness results.

Overview and Outline of Paper. In contrast to the literature, which has mainly focused on simple QCQPs or QCQPs under certain structural assumptions, in this paper, we will consider general QCQPs and develop sufficient conditions for both the convex hull result and the SDP tightness result.

We first introduce the epigraph of the QCQP by writing

$$\text{Opt} = \inf_{(x,t)\in\mathbb{R}^{N+1}} \{2t : (x,t) \in \mathcal{D}\},$$

where $\mathcal{D}$ is the epigraph of the QCQP in (1), i.e.,

$$\mathcal{D} := \left\{ (x,t) \in \mathbb{R}^N \times \mathbb{R} : \begin{array}{l} q_0(x) \leq 2t \\ q_i(x) \leq 0, \, \forall i \in [\![m_I]\!] \\ q_i(x) = 0, \, \forall i \in [\![m_I + 1, m]\!] \end{array} \right\}. \tag{2}$$

As $(x,t) \mapsto 2t$ is linear, we may replace the (potentially nonconvex) epigraph $\mathcal{D}$ with its convex hull $\text{conv}(\mathcal{D})$. Then,

$$\text{Opt} = \inf_{(x,t)\in\mathbb{R}^{N+1}} \{2t : (x,t) \in \text{conv}(\mathcal{D})\}.$$

A summary of our contributions[1], along with an outline of the paper, is as follows. In Sect. 2, we introduce and study the standard SDP relaxation of QCQPs [34] along with its optimal value Opt_{SDP} and projected epigraph $\mathcal{D}_{\text{SDP}}$. We set up a framework for deriving sufficient conditions for the "convex hull

---

[1] Due to space constraints, we omit full proofs, more detailed comparisons of our results with the literature, and our SDP tightness results in this extended abstract. The full version of this paper can be found at [38].

result," $\mathrm{conv}(\mathcal{D}) = \mathcal{D}_{\mathrm{SDP}}$, and the "SDP tightness result," $\mathrm{Opt} = \mathrm{Opt}_{\mathrm{SDP}}$. This framework is based on the Lagrangian function $(\gamma, x) \mapsto q_0(x) + \sum_{i=1}^{m} \gamma_i q_i(x)$ and the eigenvalue structure of a dual object $\Gamma \subseteq \mathbb{R}^m$. This object Γ, which consists of the convex Lagrange multipliers, has been extensively studied in the literature (see [39, Chapter 13.4] and more recently [33]). In Sect. 3, we define an integer parameter k, the quadratic eigenvalue multiplicity, that captures the amount of symmetry in a given QCQP. We then give examples where the quadratic eigenvalue multiplicity is large. Specifically, vectorized reformulations of quadratic matrix programs [4] are such an example. In Sect. 4, we use our framework to derive sufficient conditions for the convex hull result: $\mathrm{conv}(\mathcal{D}) = \mathcal{D}_{\mathrm{SDP}}$. Theorem 2 states that if Γ is polyhedral and k is sufficiently large, then $\mathrm{conv}(\mathcal{D}) = \mathcal{D}_{\mathrm{SDP}}$. This theorem actually follows as a consequence of Theorem 1, which replaces the assumption on the quadratic eigenvalue multiplicity with a weaker assumption regarding the dimension of zero eigenspaces related to the A_i matrices. Furthermore, our results in this section establish that if Γ is polyhedral, then $\mathcal{D}_{\mathrm{SDP}}$ is SOC representable; see Remark 3. In particular, when the assumptions of Theorems 1 or 2 hold, we have that $\mathrm{conv}(\mathcal{D}) = \mathcal{D}_{\mathrm{SDP}}$ is SOC representable. We provide several classes of problems that satisfy the assumptions of these theorems. In particular, we recover a number of results regarding the TRS [22], the GTRS [37], and the solvability of systems of quadratic equations [3].

To the best of our knowledge, our results are the first to provide a unified explanation of many of the exactness guarantees in the literature. Moreover, we provide significant generalizations of known results in a number of settings.

Notation. For nonnegative integers $m \leq n$ let $[\![n]\!] := \{1, \ldots, n\}$ and $[\![m, n]\!] := \{m, m+1, \ldots, n-1, n\}$. Let $\mathbf{S}^{n-1} = \{x \in \mathbb{R}^n : \|x\| = 1\}$ denote the $n-1$ sphere. Let $\mathbb{S}^n$ denote the set of real symmetric $n \times n$ matrices. For a positive integer n, let $I = I_n$ denote the $n \times n$ identity matrix. When the dimension is clear, we will simply write I. Given two matrices A and B, let $A \otimes B$ denote their Kronecker product. For a set $\mathcal{D} \subseteq \mathbb{R}^n$, let $\mathrm{conv}(\mathcal{D})$, $\mathrm{cone}(\mathcal{D})$, $\mathrm{extr}(\mathcal{D})$, $\dim(\mathcal{D})$ and $\mathrm{aff}\,\dim(\mathcal{D})$ denote the convex hull, conic hull, extreme points, dimension, and affine dimension of $\mathcal{D}$, respectively.

2   A General Framework

In this section, we introduce a general framework for analyzing the standard Shor SDP relaxation of QCQPs. We will examine how both the objective value and the feasible domain change when moving from a QCQP to its SDP relaxation.

We make an assumption that can be thought of as a primal feasibility and dual strict feasibility assumption. This assumption (or a slightly stronger version of it) is standard and is routinely made in the literature on QCQPs [4, 10, 41].

Assumption 1. *Assume the feasible region of* (1) *is nonempty and there exists* $\gamma^* \in \mathbb{R}^m$ *such that* $\gamma_i^* \geq 0$ *for all* $i \in [\![m_I]\!]$ *and* $A_0 + \sum_{i=1}^{m} \gamma_i^* A_i \succ 0$. $\qquad\qquad \square$

The standard SDP relaxation to (1) is

$$
\text{Opt}_{\text{SDP}} := \inf_{x \in \mathbb{R}^N, X \in \mathbb{S}^N} \left\{ \langle Q_0, Y \rangle : \begin{array}{l} Y := \begin{pmatrix} 1 & x^\top \\ x & X \end{pmatrix} \\ \langle Q_i, Y \rangle \leq 0, \ \forall i \in [\![m_I]\!] \\ \langle Q_i, Y \rangle = 0, \ \forall i \in [\![m_I + 1, m]\!] \\ Y \succeq 0 \end{array} \right\}, \quad (3)
$$

where $Q_i \in \mathbb{S}^{N+1}$ is the matrix $Q_i := \begin{pmatrix} c_i & b_i^\top \\ b_i & A_i \end{pmatrix}$. Let $\mathcal{D}_{\text{SDP}}$ denote the epigraph of the relaxation (3) projected away from the X variables, i.e., define

$$
\mathcal{D}_{\text{SDP}} := \left\{ (x, t) \in \mathbb{R}^{N+1} : \begin{array}{l} \exists X \in \mathbb{S}^N : \\ Y := \begin{pmatrix} 1 & x^\top \\ x & X \end{pmatrix} \\ \langle Q_0, Y \rangle \leq 2t \\ \langle Q_i, Y \rangle \leq 0, \ \forall i \in [\![m_I]\!] \\ \langle Q_i, Y \rangle = 0, \ \forall i \in [\![m_I + 1, m]\!] \\ Y \succeq 0 \end{array} \right\}. \quad (4)
$$

By taking $X = xx^\top$ in both (3) and (4), we see that $\mathcal{D} \subseteq \mathcal{D}_{\text{SDP}}$ and $\text{Opt} \geq \text{Opt}_{\text{SDP}}$. Noting that $\mathcal{D}_{\text{SDP}}$ is convex (it is the projection of a convex set), we further have that $\text{conv}(\mathcal{D}) \subseteq \mathcal{D}_{\text{SDP}}$. The framework that we set up in the remainder of this section allows us to reason about when equality occurs in either relation.

2.1 Rewriting the SDP in Terms of a Dual Object

For $\gamma \in \mathbb{R}^m$, define

$$
A(\gamma) := A_0 + \sum_{i=1}^m \gamma_i A_i, \quad b(\gamma) := b_0 + \sum_{i=1}^m \gamma_i b_i, \quad c(\gamma) := c_0 + \sum_{i=1}^m \gamma_i c_i,
$$

$$
q(\gamma, x) := q_0(x) + \sum_{i=1}^m \gamma_i q_i(x).
$$

Our framework for analyzing (3) is based on the dual object

$$
\Gamma := \left\{ \gamma \in \mathbb{R}^m : \begin{array}{l} A(\gamma) \succeq 0 \\ \gamma_i \geq 0, \ \forall i \in [\![m_I]\!] \end{array} \right\}.
$$

This object will play a key role our analysis for the following fundamental reason.

Lemma 1. *Suppose Assumption 1 holds. Then*

$$
\mathcal{D}_{\text{SDP}} = \left\{ (x, t) : \sup_{\gamma \in \Gamma} q(\gamma, x) \leq 2t \right\} \quad and \quad \text{Opt}_{\text{SDP}} = \min_{x \in \mathbb{R}^N} \sup_{\gamma \in \Gamma} q(\gamma, x).
$$

The second identity is well-known; see e.g., Fujie and Kojima [20].

2.2   The Eigenvalue Structure of Γ

We now define a number of objects related to Γ. Noting that $\gamma \mapsto q(\gamma, \hat{x})$ is linear and that Γ is closed leads to the following observation.

Observation 1. *Let $\hat{x} \in \mathbb{R}^N$. If $\sup_{\gamma \in \Gamma} q(\gamma, \hat{x})$ is finite, then $q(\gamma, \hat{x})$ achieves its maximum value in Γ on some face $\mathcal{F}$ of Γ.*

In particular, the following definition is well-defined.

Definition 1. *For any $\hat{x} \in \mathbb{R}^N$ such that $\sup_{\gamma \in \Gamma} q(\gamma, \hat{x})$ is finite, define $\mathcal{F}(\hat{x})$ to be the face of Γ maximizing $q(\gamma, \hat{x})$.*

Definition 2. *Let $\mathcal{F}$ be a face of Γ. We say that $\mathcal{F}$ is a definite face if there exists $\gamma \in \mathcal{F}$ such that $A(\gamma) \succ 0$. Otherwise, we say that $\mathcal{F}$ is a semidefinite face and let $\mathcal{V}(\mathcal{F})$ denote the shared zero eigenspace of $\mathcal{F}$, i.e.,*

$$\mathcal{V}(\mathcal{F}) := \left\{ v \in \mathbb{R}^N : A(\gamma)v = 0, \ \forall \gamma \in \mathcal{F} \right\}.$$

It is possible to show that for $\mathcal{F}$ semidefinite, the set $\mathcal{V}(\mathcal{F})$ is nontrivial. As a sketch, suppose otherwise, then for every v on the unit sphere, we can associate a $\gamma_v \in \mathcal{F}$ such that $v^\top A(\gamma_v)v > 0$. Then we can produce a positive definite matrix $A(\bar{\gamma})$ where $\bar{\gamma}$ is an "average" over the γ_v, a contradiction. See Lemma 2 in the full version of this paper for a formal proof.

2.3   The Framework

Our framework consists of two parts: an "easy part" that only requires Assumption 1 to hold and a "hard part" that may require much stronger assumptions. The "easy part" consists of the following lemma and observation.

Lemma 2. *Suppose Assumption 1 holds and let $(\hat{x}, \hat{t}) \in \mathcal{D}_{\mathrm{SDP}}$. If $\mathcal{F}(\hat{x})$ is a definite face of Γ, then $(\hat{x}, \hat{t}) \in \mathcal{D}$.*

Observation 2. *Suppose Assumption 1 holds and let $\mathcal{F}$ be a face of Γ. If aff $\dim(\mathcal{F}) = m$, then $\mathcal{F}$ is definite.*

The "hard part" of the framework works as follows: In order to show the convex hull result $\mathcal{D}_{\mathrm{SDP}} = \mathrm{conv}(\mathcal{D})$, it suffices to guarantee that every $(\hat{x}, \hat{t}) \in \mathcal{D}_{\mathrm{SDP}}$ can be decomposed as a convex combination of pairs (x_α, t_α) for which $\mathcal{F}(x_\alpha)$ is definite. Then, by Lemma 2, we will have that $(x_\alpha, t_\alpha) \in \mathcal{D}$. We give examples of such sufficient conditions in Sect. 4. Our decomposition procedures will be recursive and we will use Observation 2 to show that they terminate.

Remark 1. Consider performing an invertible affine transformation on the space $\mathbb{R}^N$, i.e. let $y = U(x + z)$ where $U \in \mathbb{R}^{N \times N}$ is an invertible linear transformation and $z \in \mathbb{R}^N$. Define the quadratic functions $q'_0, \ldots, q'_m : \mathbb{R}^N \to \mathbb{R}$ such that $q'_i(y) = q'_i(U(x + z)) = q_i(x)$ for all $x \in \mathbb{R}^N$. We will use an apostrophe to denote all the quantities corresponding to the QCQP in the variable y.

Define the map $\ell : \mathbb{R}^{N+1} \to \mathbb{R}^{N+1}$ by $(x, t) \mapsto (U(x+z), t)$. Note that Opt$'$ = Opt and conv$(\mathcal{D}') = \ell(\text{conv}(\mathcal{D}))$. Furthermore a straightforward application of Lemma 1 gives Opt$'_{\text{SDP}}$ = Opt$_{\text{SDP}}$ and $\mathcal{D}'_{\text{SDP}} = \ell(\mathcal{D}_{\text{SDP}})$. We deduce that the questions conv$(\mathcal{D}) \overset{?}{=} \mathcal{D}_{\text{SDP}}$ and Opt $\overset{?}{=}$ Opt$_{\text{SDP}}$ are invariant under invertible affine transformation of the x-space. In particular, the sufficient conditions that we will present in Theorems 1 and 2 only need to hold after some invertible affine transformation. In this sense, the SDP relaxation will "find" structure in a given QCQP even if it is "hidden" by an affine transformation. $\square$

3   Symmetries in QCQPs

In this section, we examine a parameter k that captures the amount of symmetry present in a QCQP of the form (1).

Definition 3. *The quadratic eigenvalue multiplicity of a QCQP of the form (1) is the largest integer k such that for every $i \in [\![0, m]\!]$ there exists $\mathcal{A}_i \in \mathbb{S}^n$ for which $A_i = I_k \otimes \mathcal{A}_i$. Let $\mathcal{A}(\gamma) := \mathcal{A}_0 + \sum_{i=1}^m \gamma_i \mathcal{A}_i$.*

This value is well-defined: k is always at least 1 as we can write $A_i = I_1 \otimes A_i$. On the other hand, k must also be a divisor of N.

The next lemma states the crucial structure inherent in QCQPs with large quadratic eigenvalue multiplicities.

Lemma 3. *If $\mathcal{F}$ is a semidefinite face of Γ, then $\dim(\mathcal{V}(\mathcal{F})) \geq k$.*

Remark 2. In quadratic matrix programming [4,5], we are asked to optimize

$$\inf_{X \in \mathbb{R}^{n \times k}} \left\{ \begin{array}{l} \text{tr}(X^\top \mathcal{A}_0 X) + 2\text{tr}(B_0^\top X) + c_0 : \\ \quad \text{tr}(X^\top \mathcal{A}_i X) + 2\text{tr}(B_i^\top X) + c_i \leq 0, \, \forall i \in [\![m_I]\!] \\ \quad \text{tr}(X^\top \mathcal{A}_i X) + 2\text{tr}(B_i^\top X) + c_i = 0, \, \forall i \in [\![m_I + 1, m]\!] \end{array} \right\}, \quad (5)$$

where $\mathcal{A}_i \in \mathbb{S}^n$, $B_i \in \mathbb{R}^{n \times k}$ and $c_i \in \mathbb{R}$ for all $i \in [\![0, m]\!]$. We can transform this program to an equivalent QCQP in the vector variable $x \in \mathbb{R}^{nk}$. Then $\text{tr}(X^\top \mathcal{A}_i X) + 2\text{tr}(B_i^\top X) + c_i = x^\top (I_k \otimes \mathcal{A}_i) x + 2b_i^\top x + c_i$, where $b_i \in \mathbb{R}^{nk}$ has entries $(b_i)_{(t-1)n+s} = (B_i)_{s,t}$. In particular, the vectorized reformulation of (5) has quadratic eigenvalue value multiplicity k. $\square$

4   Convex Hull Results

We now present new sufficient conditions for the convex hull result $\mathcal{D}_{\text{SDP}}$ = conv$(\mathcal{D})$. We analyze the case where the geometry of Γ is particularly nice.

Assumption 2. *Assume that Γ is polyhedral.* $\square$

We remark that although Assumption 2 is rather restrictive, it is general enough to cover the case where the set of quadratic forms $\{A_i\}_{i \in [\![0,m]\!]}$ is diagonal or simultaneously diagonalizable—a class of QCQPs which have been studied extensively in the literature [8, 25, 26]. See the full version of this paper for convex hull and SDP tightness results without Assumption 2 as well as a discussion on the difficulties in removing it.

Our main result in this paper is the following theorem.

Theorem 1. *Suppose Assumptions 1 and 2 hold. If for every semidefinite face $\mathcal{F}$ of Γ we have $\dim(\mathcal{V}(\mathcal{F})) \geq \mathrm{aff}\dim(\{b(\gamma) : \gamma \in \mathcal{F}\})+1$, then $\mathrm{conv}(\mathcal{D}) = \mathcal{D}_{\mathrm{SDP}}$.*

Assumption 1 allows us to apply Lemma 2 to handle any $(\hat{x}, \hat{t}) \in \mathcal{D}_{\mathrm{SDP}}$ for which $\mathcal{F}(\hat{x})$ is definite. Therefore, in order to prove Theorem 1, it suffices to prove the following lemma.

Lemma 4. *Suppose Assumptions 1 and 2 hold. Let $(\hat{x}, \hat{t}) \in \mathcal{D}_{\mathrm{SDP}}$ and let $\mathcal{F} = \mathcal{F}(\hat{x})$. If $\mathcal{F}$ is a semidefinite face of Γ and $\dim(\mathcal{V}(\mathcal{F})) \geq \mathrm{aff}\dim(\{b(\gamma) : \gamma \in \mathcal{F}\}) + 1$, then $(\hat{x}, \hat{t})$ can be written as a convex combination of points (x_α, t_α) satisfying the following properties:*

1. *$(x_\alpha, t_\alpha) \in \mathcal{D}_{\mathrm{SDP}}$, and*
2. *$\mathrm{aff}\dim(\mathcal{F}(x_\alpha)) > \mathrm{aff}\dim(\mathcal{F}(\hat{x}))$.*

We give a proof sketch of Lemma 4 in Appendix A.

The proof of Theorem 1 follows at once from Lemmas 2 and 4 and Observation 2. Indeed, Lemma 4 guarantees that $\mathrm{aff}\dim(\mathcal{F}(x_\alpha)) > \mathrm{aff}\dim(\mathcal{F}(\hat{x}))$. Thus, by Observation 2, we will have successfully decomposed $(\hat{x}, \hat{t})$ as a convex combination of (x_α, t_α), where $(x_\alpha, t_\alpha) \in \mathcal{D}_{\mathrm{SDP}}$ and $\mathcal{F}(x_\alpha)$ is definite, after at most $m - 1$ rounds of applying Lemma 4. Finally, Lemma 2 guarantees that each pair (x_α, t_α) is an element of $\mathcal{D}$, the epigraph of the QCQP.

The next theorem follows as a corollary to Theorem 1.

Theorem 2. *Suppose Assumptions 1 and 2 hold. If for every semidefinite face $\mathcal{F}$ of Γ we have $k \geq \mathrm{aff}\dim(\{b(\gamma) : \gamma \in \mathcal{F}\}) + 1$, then $\mathrm{conv}(\mathcal{D}) = \mathcal{D}_{\mathrm{SDP}}$.*

Remark 3. We remark that when Γ is polyhedral (Assumption 2), the set $\mathcal{D}_{\mathrm{SDP}}$ is actually SOC representable: By the Minkowski-Weyl Theorem, we can decompose $\Gamma = \Gamma_e + \mathrm{cone}(\Gamma_r)$ where both Γ_e and Γ_r are polytopes. Let $\breve{q}(\gamma, x) = \sum_{i=1}^m \gamma_i q_i(x)$. Then, by Lemma 1 we can write

$$\mathcal{D}_{\mathrm{SDP}} = \left\{ (x,t) : \sup_{\gamma \in \Gamma} q(\gamma, x) \leq 2t \right\} = \left\{ (x,t) : \begin{array}{l} q(\gamma_e, x) \leq 2t, \, \forall \gamma_e \in \mathrm{extr}(\Gamma_e) \\ \breve{q}(\gamma_f, x) \leq 0, \, \forall \gamma_f \in \mathrm{extr}(\Gamma_r) \end{array} \right\}.$$

That is, $\mathcal{D}_{\mathrm{SDP}}$ is defined by finitely many convex quadratic inequalities Thus the assumptions of Theorems 1 and 2 imply that $\mathrm{conv}(\mathcal{D})$ is SOC representable. □

We now give examples of problems where our assumptions hold.

Corollary 1. *Suppose $m = 1$ and Assumption 1 holds. Then,* $\mathrm{conv}(\mathcal{D}) = \mathcal{D}_{\mathrm{SDP}}$.

Corollary 1 recovers results associated with the epigraph of the TRS[2] and the GTRS (see [22, Theorem 13] and [37, Theorems 1 and 2]).

Corollary 2. *Suppose Assumptions 1 and 2 hold. If $b_i = 0$ for all $i \in [\![m]\!]$, then* $\mathrm{conv}(\mathcal{D}) = \mathcal{D}_{\mathrm{SDP}}$.

Example 1. Consider the following optimization problem.

$$\inf_{x \in \mathbb{R}^2} \left\{ x_1^2 + x_2^2 + 10x_1 : \begin{array}{l} x_1^2 - x_2^2 - 5 \leq 0 \\ -x_1^2 + x_2^2 - 50 \leq 0 \end{array} \right\}$$

We check that the conditions of Corollary 2 hold. Assumption 1 holds as $A(0) = A_0 = I \succ 0$ and $x = 0$ is feasible. Next, Assumption 2 holds as

$$\Gamma = \left\{ \gamma \in \mathbb{R}^2 : \begin{array}{l} 1 + \gamma_1 - \gamma_2 \geq 0 \\ 1 - \gamma_1 + \gamma_2 \geq 0 \\ \gamma \geq 0 \end{array} \right\}.$$

One can verify that $\Gamma = \mathrm{conv}\left(\{(0,0),(1,0),(0,1)\}\right) + \mathrm{cone}(\{(1,1)\})$. Finally, we note that $b_1 = b_2 = 0$. Hence, Corollary 2 and Remark 3 imply that

$$\mathrm{conv}(\mathcal{D}) = \mathcal{D}_{\mathrm{SDP}} = \left\{ (x,t) : \begin{array}{l} x_1^2 + x_2^2 + 10x_1 \leq 2t \\ 2x_1^2 + 10x_1 - 5 \leq 2t \\ 2x_2^2 + 10x_1 - 50 \leq 2t \end{array} \right\}.$$

We plot $\mathcal{D}$ and $\mathrm{conv}(\mathcal{D}) = \mathcal{D}_{\mathrm{SDP}}$ in Fig. 1.   □

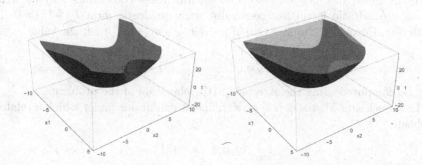

Fig. 1. The sets $\mathcal{D}$ (in orange) and $\mathrm{conv}(\mathcal{D})$ (in yellow) from Example 1 (Color figure online)

---

[2] Corollary 1 fails to recover the full extent of [22, Theorem 13]. Indeed, [22, Theorem 13] also gives a description of the convex hull of the epigraph of the TRS with an additional conic constraint under some assumptions.

Remark 4. Barvinok [3] shows that one can decide in polynomial time (in N) whether a constant number, m_E, of quadratic forms $\{A_i\}_{i \in [\![m_E]\!]}$ has a joint nontrivial zero. That is, whether the system $x^\top A_i x = 0$ for $i \in [\![m_E]\!]$ and $x^\top x = 1$ is feasible. We can recast this as asking whether the following optimization problem

$$\min_{x \in \mathbb{R}^N} \left\{ -x^\top x : \begin{array}{l} x^\top x \leq 1 \\ x^\top A_i x = 0, \forall i \in [\![m_E]\!] \end{array} \right\}$$

has objective value -1 or 0.

Thus, the feasibility problem studied in [3] reduces to a QCQP of the form we study in this paper. It is easy to verify that Assumption 1 holds. Then when Γ is polyhedral (Assumption 2), Corollary 2 implies that the feasibility problem (even in a variable number of quadratic forms) can be decided using a semidefinite programming approach. Nevertheless, Assumption 2 may not necessarily hold in general and so Corollary 2 does not recover the full result of [3].   □

Corollary 3. *Suppose Assumption 1 holds and for every $i \in [\![0, m]\!]$, there exists α_i such that $A_i = \alpha_i I_N$. If $m \leq N$, then conv($\mathcal{D}$) $= \mathcal{D}_{\text{SDP}}$.*

Remark 5. Consider the problem of finding the distance between the origin $0 \in \mathbb{R}^N$ and a piece of Swiss cheese $C \subseteq \mathbb{R}^N$. We will assume that C is nonempty and defined as

$$C = \left\{ x \in \mathbb{R}^N : \begin{array}{l} \|x - y_i\| \leq s_i, \forall i \in [\![m_1]\!] \\ \|x - z_i\| \geq t_i, \forall i \in [\![m_2]\!] \\ \langle x, b_i \rangle \geq c_i, \forall i \in [\![m_3]\!] \end{array} \right\},$$

where $y_i, z_i, b_i \in \mathbb{R}^N$ and $s_i, t_i, c_i \in \mathbb{R}$ are arbitrary. In other words, C is defined by m_1-many "inside-ball" constraints, m_2-many "outside-ball" constraints, and m_3-many linear inequalities. Note that each of these constraints may be written as a quadratic inequality constraint with quadratic form I, $-I$, or 0. In particular, Corollary 3 implies that if $m_1 + m_2 + m_3 \leq N$, then the value

$$\inf_{x \in \mathbb{R}^N} \left\{ \|x\|^2 : x \in C \right\}$$

may be computed using the standard SDP relaxation of the problem.

Bienstock and Michalka [11] give sufficient conditions under which a related problem

$$\inf_{x \in \mathbb{R}^N} \left\{ q_0(x) : x \in C \right\},$$

is polynomial-time solvable. Here, $q_0 : \mathbb{R}^N \to \mathbb{R}$ may be an arbitrary quadratic function however m_1 and m_2 must be constant. They devise an enumerative algorithm for this problem and prove its correctness under different assumptions. In contrast, our work deals only with the standard SDP relaxation and does not assume that the number of quadratic forms is constant.   □

Acknowledgments. This research is supported in part by NSF grant CMMI 1454548.

A   Proof Sketch of Lemma 4

For simplicity, we will assume that Γ is a polytope in this proof sketch. Let $(\hat{x}, \hat{t})$ satisfy the assumptions of Lemma 4. Without loss of generality, we may assume that $\sup_{\gamma \in \Gamma} q(\gamma, \hat{x}) = 2\hat{t}$.

We claim that the following system in variables v and s

$$\begin{cases} \langle b(\gamma), v \rangle = s, \, \forall \gamma \in \mathcal{F} \\ v \in \mathcal{V}(\mathcal{F}), \, s \in \mathbb{R} \end{cases}$$

has a nonzero solution. Indeed, we may replace the first constraint with at most

$$\text{aff dim}(\{b(\gamma) : \gamma \in \mathcal{F}\}) + 1 \leq \dim(\mathcal{V}(\mathcal{F}))$$

homogeneous linear equalities in the variables v and s. The claim then follows by noting that the equivalent system is an under-constrained homogeneous system of linear equalities and thus has a nonzero solution (v, s). It is easy to verify that $v \neq 0$ and hence, by scaling, we may take $v \in \mathbf{S}^{N-1}$.

We will modify $(\hat{x}, \hat{t})$ in the (v, s) direction. For $\alpha \in \mathbb{R}$, define

$$(x_\alpha, t_\alpha) := (\hat{x} + \alpha v, \, \hat{t} + \alpha s).$$

We will sketch the existence of an $\alpha > 0$ such that (x_α, t_α) satisfies the conclusions of Lemma 4. A similar line of reasoning will produce an analogous $\alpha < 0$. This will complete the proof sketch.

Suppose $\gamma \in \mathcal{F}$. Then, by our choice of v and s, the function $\alpha \mapsto q(\gamma, x_\alpha) - 2t_\alpha = q(\gamma, \hat{x}) - 2t = 0$ is identically zero. Now suppose $\gamma \in \Gamma \setminus \mathcal{F}$. Then, the function $\alpha \mapsto q(\gamma, x_\alpha) - 2t_\alpha$ is a convex quadratic function which is negative at $\alpha = 0$.

We conclude that the following set

$$\mathcal{Q} := \{\alpha \mapsto q(\gamma, x_\alpha) - 2t_\alpha : \gamma \in \text{extr}(\Gamma)\} \setminus \{0\},$$

consists of convex quadratic functions which are negative at $\alpha = 0$. The finiteness of this set follows from the assumption that Γ is polyhedral.

Assumption 1 implies that at least one of the functions in $\mathcal{Q}$ is strictly convex. Then as $\mathcal{Q}$ is a finite set, there exists an $\alpha_+ > 0$ such that $q(\alpha_+) \leq 0$ for all $q \in \mathcal{Q}$ with at least one equality. We emphasize that this is the step where Assumption 2 cannot be dropped.

Finally, it is easy to check that $(x_{\alpha_+}, t_{\alpha_+})$ satisfies the conclusions of Lemma 4.

References

1. Abbe, E., Bandeira, A.S., Hall, G.: Exact recovery in the stochastic block model. IEEE Trans. Inf. Theory **62**(1), 471–487 (2015)
2. Bao, X., Sahinidis, N.V., Tawarmalani, M.: Semidefinite relaxations for quadratically constrained quadratic programming: a review and comparisons. Math. Program. **129**(1), 129 (2011). https://doi.org/10.1007/s10107-011-0462-2
3. Barvinok, A.I.: Feasibility testing for systems of real quadratic equations. Discrete Comput. Geom. **10**(1), 1–13 (1993). https://doi.org/10.1007/BF02573959
4. Beck, A.: Quadratic matrix programming. SIAM J. Optim. **17**(4), 1224–1238 (2007)
5. Beck, A., Drori, Y., Teboulle, M.: A new semidefinite programming relaxation scheme for a class of quadratic matrix problems. Oper. Res. Lett. **40**(4), 298–302 (2012)
6. Beck, A., Eldar, Y.C.: Strong duality in nonconvex quadratic optimization with two quadratic constraints. SIAM J. Optim. **17**(3), 844–860 (2006)
7. Ben-Tal, A., El Ghaoui, L., Nemirovski, A.: Robust Optimization. Princeton Series in Applied Mathematics. Princeton University Press, Philadehia (2009)
8. Ben-Tal, A., den Hertog, D.: Hidden conic quadratic representation of some nonconvex quadratic optimization problems. Math. Program. **143**(1), 1–29 (2014). https://doi.org/10.1007/s10107-013-0710-8
9. Ben-Tal, A., Nemirovski, A.: Lectures on Modern Convex Optimization. MPS-SIAM Series on Optimization. SIAM, Philadehia (2001)
10. Ben-Tal, A., Teboulle, M.: Hidden convexity in some nonconvex quadratically constrained quadratic programming. Math. Program. **72**(1), 51–63 (1996). https://doi.org/10.1007/BF02592331
11. Bienstock, D., Michalka, A.: Polynomial solvability of variants of the trust-region subproblem. In: Proceedings of the Twenty-Fifth Annual ACM-SIAM Symposium on Discrete Algorithms, pp. 380–390 (2014)
12. Burer, S.: A gentle, geometric introduction to copositive optimization. Math. Program. **151**(1), 89–116 (2015). https://doi.org/10.1007/s10107-015-0888-z
13. Burer, S., Anstreicher, K.M.: Second-order-cone constraints for extended trust-region subproblems. SIAM J. Optim. **23**(1), 432–451 (2013)
14. Burer, S., Kılınç-Karzan, F.: How to convexify the intersection of a second order cone and a nonconvex quadratic. Math. Program. **162**(1), 393–429 (2017). https://doi.org/10.1007/s10107-016-1045-z
15. Burer, S., Yang, B.: The trust region subproblem with non-intersecting linear constraints. Math. Program. **149**(1), 253–264 (2015). https://doi.org/10.1007/s10107-014-0749-1
16. Burer, S., Ye, Y.: Exact semidefinite formulations for a class of (random and nonrandom) nonconvex quadratic programs. Math. Program., 1–17 (2018). https://doi.org/10.1007/s10107-019-01367-2
17. Candes, E.J., Eldar, Y.C., Strohmer, T., Voroninski, V.: Phase retrieval via matrix completion. SIAM Rev. **57**(2), 225–251 (2015)
18. Conforti, M., Cornuéjols, G., Zambelli, G.: Integer Programming, vol. 271. Springer, Cham (2014). https://doi.org/10.1007/978-3-319-11008-0
19. Fradkov, A.L., Yakubovich, V.A.: The S-procedure and duality relations in nonconvex problems of quadratic programming. Vestn. LGU Ser. Mat. Mekh. Astron **6**(1), 101–109 (1979)

20. Fujie, T., Kojima, M.: Semidefinite programming relaxation for nonconvex quadratic programs. J. Glob. Optim. **10**(4), 367–380 (1997). https://doi.org/10.1023/A:1008282830093. ISSN 1573-2916

21. Phan-huy Hao, E.: Quadratically constrained quadratic programming: some applications and a method for solution. Zeitschrift für Oper. Res. **26**(1), 105–119 (1982)

22. Ho-Nguyen, N., Kılınç-Karzan, F.: A second-order cone based approach for solving the trust region subproblem and its variants. SIAM J. Optim. **27**(3), 1485–1512 (2017)

23. Jeyakumar, V., Li, G.Y.: Trust-region problems with linear inequality constraints: exact SDP relaxation, global optimality and robust optimization. Math. Program. **147**(1), 171–206 (2014). https://doi.org/10.1007/s10107-013-0716-2

24. Kılınç-Karzan, F., Yıldız, S.: Two-term disjunctions on the second-order cone. Math. Program. **154**(1), 463–491 (2015). https://doi.org/10.1007/s10107-015-0903-4

25. Locatelli, M.: Some results for quadratic problems with one or two quadratic constraints. Oper. Res. Lett. **43**(2), 126–131 (2015)

26. Locatelli, M.: Exactness conditions for an SDP relaxation of the extended trust region problem. Optim. Lett. **10**(6), 1141–1151 (2016). https://doi.org/10.1007/s11590-016-1001-0

27. Megretski, A.: Relaxations of quadratic programs in operator theory and system analysis. In: Borichev, A.A., Nikolski, N.K. (eds.) Systems, Approximation, Singular Integral Operators, and Related Topics, vol. 129, pp. 365–392. Birkhäuser Basel, Basel (2001). https://doi.org/10.1007/978-3-0348-8362-7_15. ISBN 978-3-0348-8362-7

28. Mixon, D.G., Villar, S., Ward, R.: Clustering subgaussian mixtures by semidefinite programming. arXiv preprint arXiv:1602.06612 (2016)

29. Modaresi, S., Vielma, J.P.: Convex hull of two quadratic or a conic quadratic and a quadratic inequality. Math. Program. **164**(1–2), 383–409 (2017). https://doi.org/10.1007/s10107-016-1084-5

30. Nesterov, Y.: Quality of semidefinite relaxation for nonconvex quadratic optimization. Technical report, Université catholique de Louvain, Center for Operations Research and Econometrics (CORE) (1997)

31. Rujeerapaiboon, N., Schindler, K., Kuhn, D., Wiesemann, W.: Size matters: cardinality-constrained clustering and outlier detection via conic optimization. SIAM J. Optim. **29**(2), 1211–1239 (2019)

32. Santana, A., Dey, S.S.: The convex hull of a quadratic constraint over a polytope. arXiv preprint arXiv:1812.10160 (2018)

33. Sheriff, J.L.: The convexity of quadratic maps and the controllability of coupled systems. Ph.D. thesis (2013)

34. Shor, N.Z.: Dual quadratic estimates in polynomial and boolean programming. Ann. Oper. Res. **25**(1), 163–168 (1990). https://doi.org/10.1007/BF02283692

35. Sturm, J.F., Zhang, S.: On cones of nonnegative quadratic functions. Math. Oper. Res. **28**(2), 246–267 (2003)

36. Tawarmalani, M., Sahinidis, N.V., Sahinidis, N.: Convexification and Global Optimization in Continuous and Mixed-Integer Nonlinear Programming: Theory, Algorithms, Software, and Applications, vol. 65. Springer, Dordrecht (2002). https://doi.org/10.1007/978-1-4757-3532-1

37. Wang, A.L., Kılınç-Karzan, F.: The generalized trust region subproblem: solution complexity and convex hull results. Technical report (2019). https://arxiv.org/abs/1907.08843

38. Wang, A.L., Kılınç-Karzan, F.: On the tightness of SDP relaxations of QCQPs. Technical report (2019). https://arxiv.org/abs/1911.09195
39. Wolkowicz, H., Saigal, R., Vandenberghe, L.: Handbook of Semidefinite Programming: Theory, Algorithms, and Applications, vol. 27. Springer, New York (2012). https://doi.org/10.1007/978-1-4615-4381-7
40. Ye, Y.: Approximating quadratic programming with bound and quadratic constraints. Math. Program. **84**(2), 219–226 (1999)
41. Ye, Y., Zhang, S.: New results on quadratic minimization. SIAM J. Optim. **14**(1), 245–267 (2003)
42. Yıldıran, U.: Convex hull of two quadratic constraints is an LMI set. IMA J. Math. Control Inf. **26**(4), 417–450 (2009)
43. Yıldız, S., Cornuéjols, G.: Disjunctive cuts for cross-sections of the second-order cone. Oper. Res. Lett. **43**(4), 432–437 (2015)

On the Convexification of Constrained Quadratic Optimization Problems with Indicator Variables

Linchuan Wei[1] , Andrés Gómez[2]([⊠]) , and Simge Küçükyavuz[1]

[1] Department of Industrial Engineering and Management Sciences, Northwestern University, Evanston, IL, USA
LinchuanWei2022@u.northwestern.edu, simge@northwestern.edu
[2] Daniel J. Epstein Department of Industrial and Systems Engineering, University of Southern California, Los Angeles, CA, USA
gomezand@usc.edu

Abstract. Motivated by modern regression applications, in this paper, we study the convexification of quadratic optimization problems with indicator variables and combinatorial constraints on the indicators. Unlike most of the previous work on convexification of sparse regression problems, we simultaneously consider the nonlinear objective, indicator variables, and combinatorial constraints. We prove that for a separable quadratic objective function, the perspective reformulation is ideal independent from the constraints of the problem. In contrast, while rank-one relaxations cannot be strengthened by exploiting information from k-sparsity constraint for $k \geq 2$, they can be improved for other constraints arising in inference problems with hierarchical structure or multicollinearity.

Keywords: Convexification · Perspective formulation · Indicator variables · Quadratic optimization · Combinatorial constraints

1 Introduction

Given a data matrix $X = [x_1, \ldots, x_p] \in \mathbb{R}^{n \times p}$ of features and a response vector $y \in \mathbb{R}^n$, we study constrained regression problems of the form

$$\min_{z,\beta} \|y - X\beta\|_2^2 + \lambda f(\beta) \tag{1a}$$

$$\text{subject to } \beta_i(1 - z_i) = 0, \qquad i \in [p] \tag{1b}$$

$$\beta \in \mathbb{R}^p, \ z \in Q \subseteq \{0,1\}^p, \tag{1c}$$

where β is a vector of regression coefficients, z is a vector of indicator variables with $z_i = 1$ if $\beta_i \neq 0$ (through indicator constraint (1b)), the set Q in constraints (1c) encodes combinatorial constraints on the indicator variables and

Andrés Gómez is supported, in part, by grant 1930582 of the National Science Foundation. Simge Küçükyavuz is supported, in part, by ONR grant N00014-19-1-2321.

D. Bienstock and G. Zambelli (Eds.): IPCO 2020, LNCS 12125, pp. 433–447, 2020.
https://doi.org/10.1007/978-3-030-45771-6_33

$[p] = \{1, 2 \ldots, p\}$. The objective (1a) is to minimize the squared loss function plus a regularization term $\lambda f(\beta)$. Typical choices of f include L0, L1 or L2 regularizations.

If Q is defined via a k-sparsity constraint, $Q = \{z \in \{0, 1\}^p \mid \sum_{i=1}^{p} z_i \leq k\}$, then problem (1) reduces to the best subset selection problem [37], a fundamental problem in statistics. Nonetheless, constraints other than the cardinality constraint arise in several statistical problems. Bertsimas and King [9] suggest imposing constraints of the form $\sum_{i \in S} z_i \leq 1$ for some $S \subseteq [p]$ to prevent multicollinearity. Constraints of the form $z_i \leq z_j$ can be used to impose hierarchy constraints [11]. In group variable selection, indicator variables of regression coefficients of variables in the same group are linked, see [33]. Manzour et al. [36] impose that the indicator variables, which correspond to edges in an underlying graph, do not define cycles – a necessary constraint for inference problems with causal graphs. Cozad et al. [17] suggest imposing a variety of constraints in both the continuous and discrete variables to enforce priors from human experts.

Problem (1) is $\mathcal{NP}$-hard [39], and is often approximated with a convex surrogate such as lasso [30,41]. Solutions with better statistical properties than lasso can be obtained from non-convex continuous approximations [23,47]. Alternatively, it is possible to solve (1) to optimality via branch-and-bound methods [10,16]. In all cases, most of the approaches for (1) have focused on the k-sparsity constraint (or its Lagrangian relaxation). For example, a standard technique to improve the relaxations of (1) revolves around the use of the *perspective reformulation* [1,15,20,21,24–27,29,32,45,48], an ideal formulation of a separable quadratic function with indicators (but no additional constraints). Recent work on obtaining ideal formulations for non-separable quadratic functions [3–5,21,28,34] also ignores additional constraints in Q.

There is a recent research thrust on studying constrained versions of (1). Dong et al. [19] study problem (1) from a continuous optimization perspective (after projecting out the discrete variables), see also [18]. Hazimeh and Mazumder [31] give specialized algorithms for the natural convex relaxation of (1) where Q is defined via hierarchy constraints. Several results exist concerning the convexification of nonlinear optimization problems with constraints [2,7,12–14,35,38,40,42–44], but such methods in general do not deliver ideal, compact or closed-form formulations for the specific case of problem (1) with structured feasible regions. In a recent work closely related to the setting considered here, Xie and Deng [46] proves that the perspective formulation is *ideal* if the objective is separable and Q is defined with a k-sparsity constraint. In a similar vein, Bacci et al. [6] show that the perspective reformulations of the form $zg(\beta/z)$ for convex differentiable functions are tight for 1-sum compositions, and they use this result to show that they are ideal under unit commitment constraints. However, similar results for more general (non-separable) objective functions or constraints are currently not known.

Our Contributions and Outline. In this paper, we provide a first study (from a convexification perspective) of the interplay between convex quadratic objectives and combinatorial constraints on the indicator variables. Specifically, we generalize the result in Xie and Deng [46] to arbitrary constraints on z. We also show that the rank-one strengthening given in [4] is ideal for k-sparsity with $k \geq 2$. However, we show that the rank-one strengthening can be improved if $k = 1$, or for hierarchy constraints [11,31]. We conclude our work with a preliminary numerical study on problems with hierarchy constraints showing that the resulting formulations achieve strong relaxations with only a modest increase in the computational effort required to solve the resulting convex formulations.

Notation. Throughout the paper, we adopt the convention that for $a \in \mathbb{R}$, $\frac{a^2}{0} = +\infty$ if $a \neq 0$ and $\frac{a^2}{0} = 0$ when $a = 0$. We let $\mathbf{1}$ be the vector of all ones, and let e_i denote the ith unit vector of appropriate dimension with 1 in ith component and zeros elsewhere. For a set Q, we denote by $\mathrm{conv}(Q)$ its convex hull and by $\mathrm{cl}\,\mathrm{conv}(Q)$ the closure of its convex hull.

2   Convex Hull Results

We present our convex hull results first for separable quadratic functions, followed by the non-separable case.

2.1   Separable Quadratic Function

Consider the mixed-integer epigraph of a separable quadratic function with arbitrary constraints, $z \in Q \subseteq \{0,1\}^p$, on the indicator variables:

$$W = \left\{ (z, \beta, t) \in Q \times \mathbb{R}^p \times \mathbb{R} \mid \sum_{i \in [p]} \beta_i^2 \leq t,\ \beta_i(1 - z_i) = 0,\ \forall i \in [p] \right\}.$$

As Theorem 1 below shows, ideal formulations of W can be obtained by applying the perspective reformulation on the separable quadratic term and, *independently*, strengthening the continuous relaxation of Q. This generalizes the result of Xie and Deng [46] for $Q = \{z \in \{0,1\}^p \mid \sum_{i=1}^p z_i \leq k\}$. Let

$$Y = \left\{ (z, \beta, t) \in \mathbb{R}^{2p+1} \mid \sum_{i \in [p]} \frac{\beta_i^2}{z_i} \leq t,\quad z \in \mathrm{conv}(Q) \right\}.$$

Theorem 1. Y *is the closure of the convex hull of* W: $\mathrm{cl}\,\mathrm{conv}(W) = Y$.

Proof. Note that inequality $\frac{\beta_i^2}{z_i} \leq t_i$ is precisely the perspective reformulation [25] of a single quadratic term $t_i = \beta_i^2$, thus the validity of the corresponding inequality in Y follows immediately. For any $(a, b, c) \in \mathbb{R}^{2p+1}$ consider the following two problems

$$\min\quad a^\top z + b^\top \beta + ct \qquad \text{subject to}\quad (z, \beta, t) \in W, \tag{2}$$

and

$$\min \quad a^\top z + b^\top \beta + ct \qquad \text{subject to} \quad (z, \beta, t) \in Y. \tag{3}$$

It suffices to show that (2) and (3) are equivalent, i.e., there exists an optimal solution of (3) that is optimal for (2) with the same objective value. If $c = 0, b = \mathbf{0}$, then both (2) and (3) are equivalent to $\min_{z \in Q} a^\top z$. If either $c = 0$ and $b \neq \mathbf{0}$, or $c < 0$, then (2) and (3) are unbounded. When $c > 0$, without loss of generality, we may assume that $c = 1$ by scaling. For any $(z_i, \beta_i) \in [0, 1] \times \mathbb{R}$, $i \in [p]$

$$\max_{\alpha_i \in \mathbb{R}} -\alpha_i \beta_i - \frac{\alpha_i^2}{4} z_i = \begin{cases} \frac{\beta_i^2}{z_i} & \text{if } z_i \neq 0, \\ 0 & \text{if } z_i = \beta_i = 0, \\ +\infty & \text{otherwise.} \end{cases} \tag{4}$$

Identity (4) can be proven by taking derivatives with respect to α_i and setting to 0, see also [8]. Hence, for any $\alpha \in \mathbb{R}^p$,

$$-\alpha^\top \beta - \sum_{i \in [p]} \frac{\alpha_i^2}{4} z_i \leq \sum_{i \in [p]} \frac{\beta_i^2}{z_i}. \tag{5}$$

In particular, consider the relaxation of (3) obtained by replacing the constraint that $\sum_{i \in [p]} \frac{\beta_i^2}{z_i} \leq t$ with $-b^\top \beta - \sum_{i \in [p]} \frac{b_i^2}{4} z_i \leq t$ (where we let $\alpha = b$ in (5)), i.e.,

$$\min \quad a^\top z + b^\top \beta + t \tag{6a}$$

$$\text{subject to} \quad -b^\top \beta - \sum_{i \in [p]} \frac{b_i^2}{4} z_i \leq t \tag{6b}$$

$$z \in \text{conv}(Q). \tag{6c}$$

Due to constraint (6b), problem (6) is equivalent to

$$\min \quad a^\top z - \sum_{i \in [p]} \frac{b_i^2}{4} z_i \qquad \text{subject to} \quad z \in \text{conv}(Q).$$

Since (6) is equivalent to a linear program (LP) over an integral polyhedron, it must have an integral optimal solution $z^* \in Q$. Let β^* be such that

$$\beta_i^* = \begin{cases} 0 & \text{if } z_i^* = 0, \\ -\frac{b_i}{2} & \text{if } z_i^* = 1. \end{cases}$$

Now if we let $t^* = \sum_{i \in [p]} (\beta_i^*)^2$, then $(z^*, \beta^*, t^*) \in W$ and $b^\top \beta^* + t^* = -\sum_{i \in [p]} \frac{b_i^2}{4} z_i^*$. Thus the optimal values of (2) and (6) coincide. And since (6) is also a relaxation of (3), the optimal values of (2) and (3) coincide. $\qquad \square$

2.2   Rank-One Quadratic Function

In this section, we study the epigraph of a (non-separable) rank-one quadratic function with constraints

$$Z_Q = \{(z, \beta, t) \in Q \times \mathbb{R}^p \times \mathbb{R} \mid (\mathbf{1}^\top \beta)^2 \leq t, \beta_i(1 - z_i) = 0, \forall i \in [p]\}$$

for some $Q \subseteq \{0, 1\}^p$. We note that ideal formulations for the unconstrained case $Z_{\{0,1\}^p}$ were provided in [4]:

Proposition 1 (Atamtürk and Gómez [4]). *The closure of the convex hull of $Z_{\{0,1\}^p}$ is*

$$\text{cl conv}(Z_{\{0,1\}^p}) = \left\{ (z, \beta, t) \in [0, 1]^p \times \mathbb{R}^{p+1} \Big| (\mathbf{1}^\top \beta)^2 \leq t, \frac{(\mathbf{1}^\top \beta)^2}{\sum_{i \in [p]} z_i} \leq t \right\}.$$

k-sparsity constraint We first study sets defined by the k-sparsity constraint,

$$Q_1 = \left\{ z \in \{0, 1\}^p : \sum_{i \in [p]} z_i \leq k \right\},$$

and prove that, under mild conditions, a generalization of the result of Xie and Deng [46] also holds in this case, that is, ideal formulations are achieved by focusing only on the nonlinear objective and indicator constraints.

Theorem 2. *If $k \geq 2$ and integer, then*

$$\text{cl conv}(Z_{Q_1}) = \left\{ (z, \beta, t) \in [0, 1]^p \times \mathbb{R}^{p+1} \mid (\mathbf{1}^\top \beta)^2 \leq t, \frac{(\mathbf{1}^\top \beta)^2}{\sum_{i \in [p]} z_i} \leq t, \sum_{i \in [p]} z_i \leq k \right\}.$$

The proof of Theorem 2 is given in the appendix. The assumption that $k \geq 2$ in Theorem 2 is necessary. As we show next, if $k = 1$, then it is possible to strengthen the formulation with a valid inequality that uses the information from the cardinality constraint, which was not possible for $k > 1$. Note that the case $k = 1$ is also of practical interest, as set Q_1 with $k = 1$ arises for example when preventing multi-collinearity, see [9].

Proposition 2. *If $k = 1$, then the following inequality is valid for Z_{Q_1}*

$$\sum_{i \in [p]} \frac{\beta_i^2}{z_i} \leq t. \tag{7}$$

Proof. If $k = 1$, then for any $(z, \beta, t) \in Z_{Q_1}$, if $\sum_{i \in [p]} z_i = 0$, then $z_i = \beta_i = 0, \forall i \in [p]$. Hence, by our convention $0 = \sum_{i \in [p]} \frac{\beta_i^2}{z_i} = (\mathbf{1}^\top \beta)^2 \leq t$. Otherwise, $z_j = 1$ for some $j \in [p]$, and $z_i = \beta_i = 0, \forall i \in [p], i \neq j$. Hence $\sum_{i \in [p]} \frac{\beta_i^2}{z_i} = \beta_j^2 = (\mathbf{1}^\top \beta)^2 \leq t$. □

Observe that inequality (7) is not valid if $k \geq 2$, as for example $(\beta_i + \beta_j)^2 < \beta_i^2 + \beta_j^2 \leq \frac{\beta_i^2}{z_i} + \frac{\beta_j^2}{z_j}$ whenever $\beta_i \beta_j < 0$. As we now show, the addition of (7) leads to an ideal formulation of Z_{Q_1} if $k = 1$.

Theorem 3. *If $k = 1$, then*

$$\mathrm{cl}\,\mathrm{conv}(Z_{Q_1}) = \left\{ (z, \beta, t) \in [0,1]^p \times \mathbb{R}^{p+1} \mid \sum_{i \in [p]} \frac{\beta_i^2}{z_i} \leq t, \sum_{i \in [p]} z_i \leq 1 \right\}.$$

Proof. First, consider another mixed integer epigraph:

$$W_{Q_1} = \left\{ (z, \beta, t) \in \{0,1\}^p \times \mathbb{R}^{p+1} \mid \sum_{i \in [p]} \beta_i^2 \leq t, \beta_i(1 - z_i) = 0, \forall i \in [p], \sum_{i \in [p]} z_i \leq 1 \right\}.$$

For $\forall (z, \beta, t) \in Z_{Q_1}$, there exists at most one β_i, $i \in [p]$ such that $\beta_i \neq 0$. Hence, $(\mathbf{1}^\top \beta)^2 = \sum_{i \in [p]} \beta_i^2$ and the result follows from Theorem 1. $\square$

Hierarchy Constraints. We now consider the hierarchy constraints. Hierarchy constraints arise from regression problems under the model

$$y = X\beta + \sum_{S \in \mathcal{P}} \left(\prod_{j \in S} X_j \right) \theta_S + \epsilon, \tag{8}$$

where $\mathcal{P}$ is a collection of subsets of $[p]$ – usually consisting of all pairs of elements of $[p]$ –, X_j is the j-th column of X, $\prod_{j \in S} X_j$ denotes the entry-wise multiplication of vectors X_j, $\theta_S \in \mathbb{R}$ is a regression variable and ϵ is a noise vector. Under this setting, the strong hierarchy constraints

$$\theta_S \neq 0 \implies \beta_i \neq 0, \forall i \in S$$

have been shown to improve statistical performance [11,31]. Strong hierarchy constraints can be enforced via the constraints $z_S \leq z_i$ for all $i \in S$, where $z_S \in \{0,1\}$ is an appropriate indicator variable such that $\theta_S(1 - z_S) = 0$. Thus, in order to devise strong convex relaxations of problems with hierarchy constraints, we study the set

$$Q_2 = \{z \in \{0,1\}^p \mid z_p \leq z_i, \forall i \in [p-1]\}.$$

Note that in Q_2 we identify S with $[p-1]$, z_S with z_p and θ_S with β_p; since p is arbitrary, this identification is without loss of generality. First, we give a valid inequality for the set Z_{Q_2}, and then show that it is sufficient to describe $\mathrm{cl}\,\mathrm{conv}(Z_{Q_2})$, when added to the continuous relaxation of the original formulation.

Proposition 3. *The following inequality is valid for* Z_{Q_2}

$$\frac{(\mathbf{1}^\top \beta)^2}{\sum_{i \in [p-1]} z_i - (p-2)z_p} \leq t.$$

Proof. For any $(z, \beta, t) \in Z_{Q_2}$, if $z_p = 1$, then $z_i = 1, \forall i \in [p]$. In this case, $\frac{(\mathbf{1}^\top \beta)^2}{\sum_{i \in [p-1]} z_i - (p-2)z_p} = (\mathbf{1}^\top \beta)^2 \leq t$. If $z_p = 0$, then $\frac{(\mathbf{1}^\top \beta)^2}{\sum_{i \in [p-1]} z_i - (p-2)z_p} = \frac{(\mathbf{1}^\top \beta)^2}{\sum_{i \in [p]} z_i} \leq$ t. If $z_i = 0, \forall i$, then $\beta = \mathbf{0}$, and by our convention $0 = \frac{(\mathbf{1}^\top \beta)^2}{\sum_{i \in [p]} z_i} = (\mathbf{1}^\top \beta)^2 \leq t$. If $z_i = 1$ for some $i \in [p]$, then $\frac{(\mathbf{1}^\top \beta)^2}{\sum_{i \in [p]} z_i} \leq (\mathbf{1}^\top \beta)^2 \leq t$. $\quad\square$

To establish the convex hull of Z_{Q_2}, we first give a lemma whose proof is in the Appendix.

Lemma 1. *The extreme points of the polyhedron*

$$Q_g = \left\{ z \in [0,1]^p \mid \sum_{i \in [p-1]} z_i - (p-2)z_p \geq 1, \ z_p \leq z_i, \ \forall i \in [p-1] \right\}$$

are integral.

Now we are ready to give an ideal formulation for Z_{Q_2}.

Theorem 4. *The closure of the convex hull of* Z_{Q_2} *is given by*

$$\mathrm{cl} \, \mathrm{conv}(Z_{Q_2}) = \Big\{ (z, \beta, t) \in [0,1]^p \times \mathbb{R}^{p+1} \mid (\mathbf{1}^\top \beta)^2 \leq t, \ z_p \leq z_i, \forall i \in [p-1],$$

$$\frac{(\mathbf{1}^\top \beta)^2}{\sum_{i \in [p-1]} z_i - (p-2)z_p} \leq t \Big\}.$$

Proof. For $a, b \in \mathbb{R}^p$ and $c \in \mathbb{R}$, consider the optimization problems:

$$\min \quad a^\top z + b^\top \beta + ct \qquad \text{subject to} \quad (z, \beta, t) \in Z_{Q_2}. \tag{9}$$

and

$$\min \quad a^\top z + b^\top \beta + ct \tag{10a}$$
$$\text{subject to} \quad (\mathbf{1}^\top \beta)^2 \leq t \tag{10b}$$
$$\frac{(\mathbf{1}^\top \beta)^2}{\sum_{i \in [p-1]} z_i - (p-2)z_p} \leq t \tag{10c}$$
$$z_p \leq z_i, \quad \forall i \in [p-1] \tag{10d}$$
$$z \in [0,1]^p. \tag{10e}$$

Following similar arguments to those in the beginning of the proof of Theorem 2 (with the exception of letting $\bar{z} = \mathbf{1}$ in the corresponding case), we can assume that $c = 1$ and $b = \kappa \mathbf{1}^\top$ for some $\kappa \in \mathbb{R}$; in this case, (9) and (10) have finite optimal value. Suppose (z^*, β^*, t^*) is an optimal solution of (10), then it suffices to show that (z^*, β^*, t^*) is integral in z^*. If $0 = \sum_{i \in [p-1]} z_i^* - (p-2)z_p^* = z_1^* + \sum_{i=2}^{p-1}(z_i^* - z_p^*)$, then by the constraint $z_i \geq z_p$ and non-negativity of z_i we must have $z_1^* = 0$ and $z_i^* = z_p^*$ for $i = 2, \ldots, p-1$. Furthermore, since $z_1^* \geq z_p^*$, we find that $z_p^* = 0$ and $z_i^* = 0, \forall i \in [p-1]$.

If $0 < \sum_{i \in [p-1]} z_i^* - (p-2)z_p^* < 1$ and the corresponding optimal objective value is 0 (or positive), then by letting $z^* = \mathbf{0}$, $\beta^* = \mathbf{0}$ and $t^* = 0$, we obtain a feasible solution with the same (or better) objective value and integral in z^*. Now suppose (z^*, β^*, t^*) attains a negative objective value in (10); let $\gamma = \frac{1}{\sum_{i \in [p-1]} z_i^* - (p-2)z_p^*} > 1$, then $(\gamma z^*, \gamma \beta^*, \gamma t^*)$ is also a feasible solution of (10) because $\gamma^2 (\mathbf{1}^\top \beta^*)^2 = \gamma \frac{(\mathbf{1}^\top \beta^*)^2}{\sum_{i \in [p-1]} z_i^* - (p-2)z_p^*} \leq \gamma t^*$, for each $i \in [p-1]$ we have $\gamma z_i^* = \frac{z_i^*}{z_i^* + \sum_{j \neq i, j \in [p-1]}(z_j^* - z_p^*)} \leq 1$, and $\gamma z_p^* \leq \gamma z_i^* \leq 1$. Furthermore, the solution $(\gamma z^*, \gamma \beta^*, \gamma t^*)$ has a strictly smaller objective value than the solution (z^*, β^*, t^*), which is a contradiction.

Finally, consider the case where $\sum_{i \in [p-1]} z_i^* - (p-2)z_p^* \geq 1$. In this case, because the constraint $(\mathbf{1}^\top \beta^*)^2 \leq t$ is active, the optimal value is attained when $\mathbf{1}^\top \beta^* = -\frac{\kappa}{2}$ and $t^* = (\mathbf{1}^\top \beta^*)^2$ and (10) has the same optimal value as

$$
\begin{aligned}
\min \quad & a^\top z - \frac{\kappa^2}{4} \\
\text{subject to} \quad & \sum_{i \in [p-1]} z_i - (p-2)z_p \geq 1 \\
& z_p \leq z_i, \quad \forall i \in [p-1] \\
& z_i \in [0, 1]^p.
\end{aligned}
$$

From Lemma 1, the extreme points of this problem are integral, which completes the proof.   □

3   Computations

We provide preliminary computations of the proposed strengthening derived in Sect. 2.2 with second-order *hierarchy* constraints, that is, $\mathcal{P} = \{\{i, j\} : i < j\}$ in (8). All computations were performed on a laptop with Intel Core i7-8550U CPU and 16 GB memory using Mosek 8.1 solver.

Specifically, there is a set of 3-tuples H, such that if $(i, j, k) \in H$ then $\beta_k \neq 0$ implies $\beta_i \neq 0$ and $\beta_j \neq 0$, resulting in the optimization problem [31]

$$\min \frac{1}{2}\|y - X\beta\|_2^2 + \lambda \sum_{i=1}^{p} z_i \tag{11a}$$

$$\text{s.t. } \beta_i(1 - z_i) = 0, \qquad\qquad\qquad \forall i \in [p] \tag{11b}$$

$$z_k \leq z_i, \ z_k \leq z_j, \qquad\qquad\qquad \forall (i,j,k) \in H \tag{11c}$$

$$z \in \{0,1\}^p, \tag{11d}$$

with L0 regularization with parameter $\lambda > 0$. We consider the following strong (and big-M free) semi-definite relaxations of (11):

- **Dynamic perspective relaxation (persp)** The dynamic perspective reformulation was proposed in [20] and involves the introduction of additional variables $B = \beta\beta^\top$:

$$\min \frac{1}{2}\|y\|_2^2 - y^\top X\beta + \frac{1}{2}\langle X^\top X, B\rangle + \lambda \sum_{i=1}^{p} z_i \tag{12a}$$

$$\text{s.t. } z_k \leq z_i, \ z_k \leq z_j, \qquad\qquad\qquad \forall (i,j,k) \in H \tag{12b}$$

$$\beta_i^2 \leq z_i B_{ii}, \qquad\qquad\qquad\qquad \forall i \in [p] \tag{12c}$$

$$\begin{pmatrix} 1 & \beta^\top \\ \beta & B \end{pmatrix} \succeq 0 \tag{12d}$$

$$z \in [0,1]^p. \tag{12e}$$

It corresponds to the best perspective reformulation that can be attained by decomposing the matrix $X^\top X = D + R$ with $D, R \succeq 0$ and D diagonal, and using the perspective reformulation to strengthen the term $\beta^\top D\beta$. This relaxation, depending on the diagonal dominance of matrix $X^\top X$, can be substantially stronger than the natural convex relaxation of (11). In light of Theorem 1 – and since the constraints (11c) are totally unimodular –, this formulation cannot be strengthened unless non-separable quadratic terms are accounted for.

- **Two-dimensional rank-one relaxation (R1)** The rank-one relaxation, proposed in [4], is a strengthening of the perspective reformulation by optimally decomposing $X^\top X = T + R$ where T is a sum of low-dimensional rank-one matrices, and using perspective and rank-one strengthening to strengthen the term $\beta^\top T\beta$. If all rank-one matrices are two-dimensional, then the resulting formulation involves the addition of constraints

$$\begin{pmatrix} z_i + z_j & \beta_i & \beta_j \\ \beta_i & B_{ii} & B_{ij} \\ \beta_j & B_{ij} & B_{jj} \end{pmatrix} \succeq 0$$

for all $i < j$. Observe that this formulation requires adding $O(p^2)$ constraints.

- **Hierarchical strengthening (Hier)** Corresponds to strengthening the rank-one formulation by exploiting the constraints of Theorem 4. Suppose that $(i, \ell, j) \in H$; then using the same techniques used in [4], we obtain three valid inequalities

$$\begin{pmatrix} z_i & \beta_i & \beta_j \\ \beta_i & B_{ii} & B_{ij} \\ \beta_j & B_{ij} & B_{jj} \end{pmatrix} \succeq 0, \quad \begin{pmatrix} z_\ell & \beta_\ell & \beta_j \\ \beta_\ell & B_{\ell\ell} & B_{\ell j} \\ \beta_j & B_{\ell j} & B_{jj} \end{pmatrix} \succeq 0 \text{ and } \begin{pmatrix} z_i + z_\ell - z_j & \beta_i & \beta_\ell & \beta_j \\ \beta_i & B_{ii} & B_{i\ell} & B_{ij} \\ \beta_\ell & B_{i\ell} & B_{\ell\ell} & B_{\ell j} \\ \beta_j & B_{ij} & B_{\ell j} & B_{jj} \end{pmatrix} \succeq 0.$$

(13)

The hierarchical strengthening corresponds to adding constraints (13) for every element in H to formulation (12); it requires adding only $O(p)$ constraints.

- **Full strengthening (Hier+R1)** Corresponds to adding all constraints from the rank-one relaxation and the hierarchical strengthening.

Results We compare the strength of the formulations as well as the time required to solve the SDP relaxations on the Diabetes dataset [4,10,22] which involves second order interactions between variables ($p = 64$). Figure 1 depicts the result. Figure 1(a) shows the optimal objective values of the different convex relaxations of (11) as a function of λ, thus larger values indicate stronger (and better) relaxations. Figure 1(b) depicts the time required to solve the relaxations; we did not observe any correlation between the value of λ and the time required, so we report aggregated times across all values of λ tested.

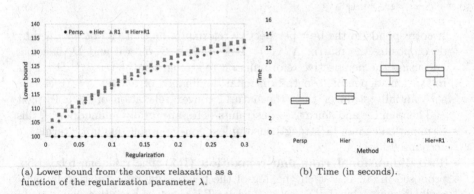

(a) Lower bound from the convex relaxation as a function of the regularization parameter λ.

(b) Time (in seconds).

Fig. 1. Lower bounds obtained from the convex relaxation and time required to solve the relaxation. In (a) the values were scaled so that the objective value obtained from the perspective relaxation [20] is 100.

We observe that just using the hierarchical strengthening (Hier) achieves almost the same improvement in terms of the lower bound as that using the rank-one strengthening, despite only requiring $O(p)$ additional constraints instead of $O(p^2)$. Indeed we observe from Fig. 1(b) that the Hierarchical strengthening results in only a modest increase in the computational time with respect to the perspective relaxation, while the rank-one strengthening requires almost double that time. In addition, if the Hierarchical strengthening is used on top of the rank-one strengthening (`Hier+R1`), then we notice a small but noticeable improvement in the quality of the lower bound across all values of λ, with no apparent increase in the computational time. These preliminary results suggest that by exploiting the constraints of the optimization problems, it is possible to achieve stronger relaxations without substantially increasing the difficulty of solving the convex problems.

Appendix

Proof (Theorem 2). First, note that the validity of the new inequality defining $\operatorname{cl\,conv}(Z_{Q_1})$ follows from Proposition 1. For $a, b \in \mathbb{R}^p$ and $c \in \mathbb{R}$, consider the following two optimization problems:

$$\min \quad a^\top z + b^\top \beta + ct \qquad \text{subject to} \quad (z, \beta, t) \in Z_{Q_1}. \tag{14}$$

and

$$\min \quad a^\top z + b^\top \beta + ct \tag{15a}$$

$$\text{subject to} \quad (\mathbf{1}^\top \beta)^2 \le t \tag{15b}$$

$$\frac{(\mathbf{1}^\top \beta)^2}{\sum_{i \in [p]} z_i} \le t \tag{15c}$$

$$\sum_{i \in [p]} z_i \le k \tag{15d}$$

$$z \in [0, 1]^p. \tag{15e}$$

The analysis for cases where $c = 0$ and $c < 0$ is similar to the proof of Theorem 1, and we can proceed with assuming $c = 1$ and $b \in \mathbb{R}^p$. First suppose that b is not a multiple of all-ones vector, then $\exists b_i < b_j$ for some $i, j \in [p], i \ne j$. Let $\bar{z} = e_i + e_j$, $\bar{\beta} = \tau(e_i - e_j)$ for some scalar τ, and $\bar{t} = 0$. Note that $(\bar{z}, \bar{\beta}, \bar{t})$ is feasible for both (14) and (15), and if we let τ go to infinity the objective value goes to minus infinity. So (14) and (15) are unbounded.

Now suppose that $b = \kappa \mathbf{1}^\top$ for some $\kappa \in \mathbb{R}$ and $c = 1$; in this case both (14) and (15) have finite optimal value. It suffices to show that there exists an optimal solution (z^*, β^*, t^*) of (15) that is integral in z^*. If $\sum_{i \in [p]} z_i^* = 0$, then we know $z_i^* = \beta_i^* = 0, \forall i \in [p]$ for both (14) and (15), and we are done. If $0 < \sum_{i \in [p]} z_i^* < 1$ and the corresponding optimal objective value is 0 (or positive), then by letting $z^* = \mathbf{0}$, $\beta^* = \mathbf{0}$ and $t^* = 0$, we get a feasible solution with the same objective value (or better). If $0 < \sum_{i \in [p]} z_i^* < 1$ and (z^*, β^*, t^*) attains a negative objective value, then let $\gamma = \frac{1}{\sum_{i \in [p]} z_i^*}$: $(\gamma z^*, \gamma \beta^*, \gamma t^*)$ is also a feasible solution of (15) with a strictly smaller objective value, which is a contradiction.

Finally, consider the case where $\sum_{i \in [p]} z_i^* \geq 1$. In this case, the constraint $(\mathbf{1}^\top \beta)^2 \leq t$ is active and the optimal value is attained when $\mathbf{1}^\top \beta^* = -\frac{\kappa}{2}$ and $t^* = (\mathbf{1}^\top \beta^*)^2$, and (15) has the same optimal value as the LP:

$$\min \quad a^\top z - \frac{\kappa^2}{4} \quad \text{subject to} \quad 1 \leq \sum_{i \in [p]} z_i \leq k, z \in [0,1]^p.$$

The constraint set of this LP is an interval matrix, so the LP has an integral optimal solution, z^*, hence, so does (15). □

Proof (Lemma 1). Suppose z^* is an extreme point of Q_g and z^* has a fractional entry. If $\sum_{i \in [p-1]} z_i^* - (p-2)z_p^* > 1$, let us consider the two cases where $z_p^* = 0$ and $z_p^* > 0$. When $z_p^* = 0$ and there exists a fractional coordinate z_i^* where $i \in [p-1]$, we can perturb z_i^* by a sufficient small quantity ϵ such that $z^* + \epsilon e_i$ and $z^* - \epsilon e_i$ are in Q_g. Then, $z^* = \frac{1}{2}(z^* + \epsilon e_i) + \frac{1}{2}(z^* - \epsilon e_i)$ which contradicts the fact that z^* is an extreme point of Q_g. When $1 > z_p^* > 0$ we can perturb z_p^* and all other z_i^* with $z_i^* = z_p^*$ by a sufficiently small quantity ϵ and stay in Q_g. Similarly, we will reach a contradiction.

Now suppose $\sum_{i \in [p-1]} z_i^* - (p-2)z_p^* = 1$, and let us consider again the two cases where $z_p^* = 0$ and $z_p^* > 0$. When $z_p^* = 0$, $z^* = z_1^* e_1 + \cdots + z_{(p-1)}^* e_{(p-1)}$, which is a contradiction since we can write z^* as a convex combination of points $e_i \in Q_g, i \in [p-1]$ and there exists at least two indices $i, j \in [p-1], i \neq j$ such that $1 > z_i^*, z_j^* > 0$ by the fact that z^* has a fractional entry and $\sum_{i \in [p-1]} z_i^* = 1, 0 \leq z_i^* \leq 1, \forall i$. When $1 > z_p^* > 0$, we first show that there exists at most one 1 in $z_1^*, z_2^*, \ldots, z_{(p-1)}^*$. Suppose we have $z_i^* = 1$ and $z_j^* = 1$ for $i, j \in [p-1]$ with $i \neq j$, then $\sum_{i \in [p-1]} z_i^* - (p-2)z_p^* = z_i^* + \sum_{l \in [p-1], l \neq i}(z_l^* - z_p^*) \geq z_i^* + (z_j^* - z_p^*) > z_i^* = 1$, which is a contradiction. We now show that we can perturb z_p^* and the $p-2$ smallest elements in $z_i^*, i \in [p-1]$ by a small quantity ϵ and remain in Q_g. The equality $\sum_{i \in [p-1]} z_i - (p-2)z_p = 1$ clearly holds after the perturbation. And, adding a small quantity ϵ to z_p^* and the $p-2$ smallest elements in $z_i^*, i \in [p-1]$ will not violate the hierarchy constraint since the largest element in $z_i^*, i \in [p-1]$ has to be strictly greater than z_p^*. (Note that if $z_i^* = z_p^*, \forall i \in [p]$, $\sum_{i \in [p-1]} z_i^* - (p-2)z_p^* = z_p^* < 1$.) Since $z_i^* \geq z_p^* > 0, \forall i \in [p-1]$ subtracting a small quantity ϵ will not violate the non-negativity constraint. Thus, we can write z^* as a convex combination of two points in Q_g, which is a contradiction. □

References

1. Aktürk, M.S., Atamtürk, A., Gürel, S.: A strong conic quadratic reformulation for machine-job assignment with controllable processing times. Oper. Res. Lett. **37**(3), 187–191 (2009)
2. Anstreicher, K.M.: On convex relaxations for quadratically constrained quadratic programming. Math. Program. **136**(2), 233–251 (2012). https://doi.org/10.1007/s10107-012-0602-3
3. Atamtürk, A., Gómez, A.: Strong formulations for quadratic optimization with M-matrices and indicator variables. Math. Program. **170**(1), 141–176 (2018). https://doi.org/10.1007/s10107-018-1301-5

4. Atamtürk, A., Gómez, A.: Rank-one convexification for sparse regression (2019). http://www.optimization-online.org/DB_HTML/2019/01/7050.html
5. Atamtürk, A., Gómez, A., Han, S.: Sparse and smooth signal estimation: convexification of L0 formulations (2018). http://www.optimization-online.org/DB_HTML/2018/11/6948.html
6. Bacci, T., Frangioni, A., Gentile, C., Tavlaridis-Gyparakis, K.: New MINLP formulations for the unit commitment problems with ramping constraints. Optimization (2019). http://www.optimization-online.org/DB_FILE/2019/10/7426.pdf
7. Belotti, P., Góez, J.C., Pólik, I., Ralphs, T.K., Terlaky, T.: A conic representation of the convex hull of disjunctive sets and conic cuts for integer second order cone optimization. In: Al-Baali, M., Grandinetti, L., Purnama, A. (eds.) Numerical Analysis and Optimization. SPMS, vol. 134, pp. 1–35. Springer, Cham (2015). https://doi.org/10.1007/978-3-319-17689-5_1
8. Bertsimas, D., Cory-Wright, R., Pauphilet, J.: A unified approach to mixed-integer optimization: nonlinear formulations and scalable algorithms. arXiv preprint arXiv:1907.02109 (2019)
9. Bertsimas, D., King, A.: OR forum - an algorithmic approach to linear regression. Oper. Res. $64(1)$, 2–16 (2016)
10. Bertsimas, D., King, A., Mazumder, R.: Best subset selection via a modern optimization lens. Ann. Stat. $44(2)$, 813–852 (2016)
11. Bien, J., Taylor, J., Tibshirani, R.: A lasso for hierarchical interactions. Ann. Stat. $41(3)$, 1111 (2013)
12. Bienstock, D., Michalka, A.: Cutting-planes for optimization of convex functions over nonconvex sets. SIAM J. Optim. $24(2)$, 643–677 (2014)
13. Burer, S.: On the copositive representation of binary and continuous nonconvex quadratic programs. Math. Program. $120(2)$, 479–495 (2009). https://doi.org/10.1007/s10107-008-0223-z
14. Burer, S., Kılınç-Karzan, F.: How to convexify the intersection of a second order cone and a nonconvex quadratic. Math. Program. $162(1-2)$, 393–429 (2016). https://doi.org/10.1007/s10107-016-1045-z
15. Ceria, S., Soares, J.: Convex programming for disjunctive convex optimization. Math. Program. 86, 595–614 (1999). https://doi.org/10.1007/s101070050106
16. Cozad, A., Sahinidis, N.V., Miller, D.C.: Learning surrogate models for simulation-based optimization. AIChE J. $60(6)$, 2211–2227 (2014)
17. Cozad, A., Sahinidis, N.V., Miller, D.C.: A combined first-principles and data-driven approach to model building. Comput. Chem. Eng. 73, 116–127 (2015)
18. Dong, H.: On integer and MPCC representability of affine sparsity. Oper. Res. Lett. $47(3)$, 208–212 (2019)
19. Dong, H., Ahn, M., Pang, J.-S.: Structural properties of affine sparsity constraints. Math. Program. $176(1-2)$, 95–135 (2019). https://doi.org/10.1007/s10107-018-1283-3
20. Dong, H., Chen, K., Linderoth, J.: Regularization vs. relaxation: a conic optimization perspective of statistical variable selection. arXiv preprint arXiv:1510.06083 (2015)
21. Dong, H., Linderoth, J.: On valid inequalities for quadratic programming with continuous variables and binary indicators. In: Goemans, M., Correa, J. (eds.) IPCO 2013. LNCS, vol. 7801, pp. 169–180. Springer, Heidelberg (2013). https://doi.org/10.1007/978-3-642-36694-9_15
22. Efron, B., Hastie, T., Johnstone, I., Tibshirani, R.: Least angle regression. Ann. Stat. $32(2)$, 407–499 (2004)

23. Fan, J., Li, R.: Variable selection via nonconcave penalized likelihood and its oracle properties. J. Am. Stat. Assoc. **96**(456), 1348–1360 (2001)

24. Frangioni, A., Furini, F., Gentile, C.: Approximated perspective relaxations: a project and lift approach. Comput. Optim. Appl. **63**(3), 705–735 (2015). https://doi.org/10.1007/s10589-015-9787-8

25. Frangioni, A., Gentile, C.: Perspective cuts for a class of convex 0–1 mixed integer programs. Math. Program. **106**, 225–236 (2006). https://doi.org/10.1007/s10107-005-0594-3

26. Frangioni, A., Gentile, C.: SDP diagonalizations and perspective cuts for a class of nonseparable MIQP. Oper. Res. Lett. **35**(2), 181–185 (2007)

27. Frangioni, A., Gentile, C., Grande, E., Pacifici, A.: Projected perspective reformulations with applications in design problems. Oper. Res. **59**(5), 1225–1232 (2011)

28. Frangioni, A., Gentile, C., Hungerford, J.: Decompositions of semidefinite matrices and the perspective reformulation of nonseparable quadratic programs. Math. Oper. Res. (2019). https://doi.org/10.1287/moor.2018.0969. Article in Advance (October)

29. Günlük, O., Linderoth, J.: Perspective reformulations of mixed integer nonlinear programs with indicator variables. Math. Program. **124**, 183–205 (2010). https://doi.org/10.1007/s10107-010-0360-z

30. Hastie, T., Tibshirani, R., Wainwright, M.: Statistical Learning with Sparsity: The Lasso and Generalizations. Monographs on Statistics and Applied Probability, vol. 143. Chapman and Hall/CRC, Boca Raton (2015)

31. Hazimeh, H., Mazumder, R.: Learning hierarchical interactions at scale: a convex optimization approach. arXiv preprint arXiv:1902.01542 (2019)

32. Hijazi, H., Bonami, P., Cornuéjols, G., Ouorou, A.: Mixed-integer nonlinear programs featuring "on/off" constraints. Comput. Optim. Appl. **52**(2), 537–558 (2012). https://doi.org/10.1007/s10589-011-9424-0

33. Huang, J., Breheny, P., Ma, S.: A selective review of group selection in high-dimensional models. Stat. Sci.: Rev. J. Inst. Math. Stat. **27**(4), 481–499 (2012)

34. Jeon, H., Linderoth, J., Miller, A.: Quadratic cone cutting surfaces for quadratic programs with on-off constraints. Discrete Optim. **24**, 32–50 (2017)

35. Kılınç-Karzan, F., Yıldız, S.: Two-term disjunctions on the second-order cone. In: Lee, J., Vygen, J. (eds.) IPCO 2014. LNCS, vol. 8494, pp. 345–356. Springer, Cham (2014). https://doi.org/10.1007/978-3-319-07557-0_29

36. Manzour, H., Küçükyavuz, S., Shojaie, A.: Integer programming for learning directed acyclic graphs from continuous data. arXiv preprint arXiv:1904.10574 (2019)

37. Miller, A.: Subset Selection in Regression. Chapman and Hall/CRC, Boca Raton (2002). https://doi.org/10.1201/9781420035933

38. Modaresi, S., Kılınç, M.R., Vielma, J.P.: Intersection cuts for nonlinear integer programming: convexification techniques for structured sets. Math. Program. **155**(1), 575–611 (2015). https://doi.org/10.1007/s10107-015-0866-5

39. Natarajan, B.K.: Sparse approximate solutions to linear systems. SIAM J. Comput. **24**(2), 227–234 (1995)

40. Richard, J.-P.P., Tawarmalani, M.: Lifting inequalities: a framework for generating strong cuts for nonlinear programs. Math. Program. **121**(1), 61–104 (2010). https://doi.org/10.1007/s10107-008-0226-9

41. Tibshirani, R.: Regression shrinkage and selection via the lasso. J. Roy. Stat. Soc.: Ser. B (Methodol.) **58**, 267–288 (1996)

42. Vielma, J.P.: Small and strong formulations for unions of convex sets from the Cayley embedding. Math. Program. **177**(1–2), 21–53 (2019). https://doi.org/10.1007/s10107-018-1258-4

43. Wang, A.L., Kılınç-Karzan, F.: The generalized trust region subproblem: solution complexity and convex hull results. arXiv preprint arXiv:1907.08843 (2019a)

44. Wang, A.L., Kılınç-Karzan, F.: On the tightness of SDP relaxations of QCQPs. Optimization Online preprint (2019b). http://www.optimization-online.org/DB_FILE/2019/11/7487.pdf

45. Wu, B., Sun, X., Li, D., Zheng, X.: Quadratic convex reformulations for semicontinuous quadratic programming. SIAM J. Optim. **27**(3), 1531–1553 (2017)

46. Xie, W., Deng, X.: The CCP selector: scalable algorithms for sparse ridge regression from chance-constrained programming. arXiv preprint arXiv:1806.03756 (2018)

47. Zhang, C.-H.: Nearly unbiased variable selection under minimax concave penalty. Ann. Stat. **38**, 894–942 (2010)

48. Zheng, X., Sun, X., Li, D.: Improving the performance of MIQP solvers for quadratic programs with cardinality and minimum threshold constraints: a semidefinite program approach. INFORMS J. Comput. **26**(4), 690–703 (2014)

Author Index

Printed in the United States
By Bookmasters